AF356205

Path Planning, Trajectory Tracking and Guidance for UAVs

Path Planning, Trajectory Tracking and Guidance for UAVs

Guest Editors

Heng Shi
Jihong Zhu
Zheng Chen
Minchi Kuang

Basel • Beijing • Wuhan • Barcelona • Belgrade • Novi Sad • Cluj • Manchester

Guest Editors

Heng Shi	Jihong Zhu	Zheng Chen
Tsinghua University	Tsinghua University	Zhejiang University
Beijing	Beijing	Hangzhou
China	China	China

Minchi Kuang
Tsinghua University
Beijing
China

Editorial Office
MDPI AG
Grosspeteranlage 5
4052 Basel, Switzerland

This is a reprint of the Special Issue, published open access by the journal *Drones* (ISSN 2504-446X), freely accessible at: https://www.mdpi.com/journal/drones/special_issues/CW6S65ZS54.

For citation purposes, cite each article independently as indicated on the article page online and as indicated below:

Lastname, A.A.; Lastname, B.B. Article Title. *Journal Name* **Year**, *Volume Number*, Page Range.

ISBN 978-3-7258-2947-7 (Hbk)
ISBN 978-3-7258-2948-4 (PDF)
https://doi.org/10.3390/books978-3-7258-2948-4

© 2025 by the authors. Articles in this book are Open Access and distributed under the Creative Commons Attribution (CC BY) license. The book as a whole is distributed by MDPI under the terms and conditions of the Creative Commons Attribution-NonCommercial-NoDerivs (CC BY-NC-ND) license (https://creativecommons.org/licenses/by-nc-nd/4.0/).

Contents

About the Editors . . . . **vii**

Preface . . . . **ix**

Xinghui Yan, Yuzhong Tang, Yulei Xu, Heng Shi and Jihong Zhu
Multi-Constrained Geometric Guidance Law with a Data-Driven Method
Reprinted from: *Drones* **2023**, *7*, 639, https://doi.org/10.3390/drones7100639 . . . . **1**

Ziyu Cao, Zhihui Du and Jianhua Yang
Topological Map-Based Autonomous Exploration in Large-Scale Scenes for Unmanned Vehicles
Reprinted from: *Drones* **2024**, *8*, 124, https://doi.org/10.3390/drones8040124 . . . . **19**

Ruiping Zheng, Qi Zhu, Shan Huang, Zhihui Du, Jingping Shi and Yongxi Lyu
Extended State Observer-Based Sliding-Mode Control for Aircraft in Tight Formation
Considering Wake Vortices and Uncertainty
Reprinted from: *Drones* **2024**, *8*, 165, https://doi.org/10.3390/drones8040165 . . . . **37**

Yanbo Fu, Wenjie Zhao and Liu Liu
Safe Reinforcement Learning for Transition Control of Ducted-Fan UAVs
Reprinted from: *Drones* **2023**, *7*, 332, https://doi.org/10.3390/drones7050332 . . . . **54**

Xiao Zhang, Wenjie Zhao, Changxuan Liu and Jun Li
Distributed Multi-Target Search and Surveillance Mission Planning for Unmanned Aerial
Vehicles in Uncertain Environments
Reprinted from: *Drones* **2023**, *7*, 355, https://doi.org/10.3390/drones7060355 . . . . **75**

Zhixiong Xu, Li Fan, Wei Qiu, Guangwei Wen and Yunhan He
A Robust Disturbance-Rejection Controller Using Model Predictive Control for Quadrotor UAV
in Tracking Aggressive Trajectory
Reprinted from: *Drones* **2023**, *7*, 557, https://doi.org/10.3390/drones7090557 . . . . **98**

Lexu Du, Yankai Fan, Mingzhen Gui and Dangjun Zhao
A Multi-Regional Path-Planning Method for Rescue UAVs with Priority Constraints
Reprinted from: *Drones* **2023**, *7*, 692, https://doi.org/10.3390/drones7120692 . . . . **119**

Zhilong Xi, Haoran Han, Jian Cheng and Maolong Lv
Reducing Oscillations for Obstacle Avoidance in a Dense Environment Using Deep
Reinforcement Learning and Time-Derivative of an Artificial Potential Field
Reprinted from: *Drones* **2024**, *8*, 85, https://doi.org/10.3390/drones8030085 . . . . **151**

Zhilan Zhang, Yufeng Wang, Yizhe Luo, Hang Zhang, Xiaorong Zhang and Wenrui Ding
Iterative Trajectory Planning and Resource Allocation for UAV-Assisted Emergency
Communication with User Dynamics
Reprinted from: *Drones* **2024**, *8*, 149, https://doi.org/10.3390/drones8040149 . . . . **174**

Jiabin Lou, Rong Ding and Wenjun Wu
HHPSO: A Heuristic Hybrid Particle Swarm Optimization Path Planner for Quadcopters
Reprinted from: *Drones* **2024**, *8*, 221, https://doi.org/10.3390/drones8060221 . . . . **196**

Xiaoxiong Liu, Wanhan Xue, Xinlong Xu, Minkun Zhao and Bin Qin
Research on Unmanned Aerial Vehicle (UAV) Visual Landing Guidance and Positioning
Algorithms
Reprinted from: *Drones* **2024**, *8*, 257, https://doi.org/10.3390/drones8060257 . . . . **214**

Long Chen, Guangrui Liu, Xia Zhu and Xin Li
A Heuristic Routing Algorithm for Heterogeneous UAVs in Time-Constrained MEC Systems
Reprinted from: *Drones* **2024**, *8*, 379, https://doi.org/10.3390/drones8080379 . . . . **239**

Yong Liao, Yuxin Wu, Shichang Zhao and Dan Zhang
Unmanned Aerial Vehicle Obstacle Avoidance Based Custom Elliptic Domain
Reprinted from: *Drones* **2024**, *8*, 397, https://doi.org/10.3390/drones8080397 . . . . **260**

Delong Shi, Jinrong Shen, Mingsheng Gao and Xiaodong Yang
A Multi-Waypoint Motion Planning Framework for Quadrotor Drones in Cluttered
Environments
Reprinted from: *Drones* **2024**, *8*, 414, https://doi.org/10.3390/drones8080414 . . . . **292**

Jia Guo, Minggang Gan and Kang Hu
Cooperative Path Planning for Multi-UAVs with Time-Varying Communication and Energy
Consumption Constraints
Reprinted from: *Drones* **2024**, *8*, 654, https://doi.org/10.3390/drones8110654 . . . . **316**

About the Editors

Heng Shi

Heng Shi is currently an Assistant Research Fellow in the Department of Precision Instrument at Tsinghua University. He earned his Ph.D. in Computer Science and Technology from Tsinghua University in 2020, following M.S. and B.S. degrees in Astronautics from Beihang University in 2015 and 2012, respectively. Dr. Shi's research primarily focuses on the guidance and control of vehicles, particularly in the context of unmanned aerial vehicles (UAVs). His current projects involve advancing cooperative guidance techniques and enhancing intelligent control systems through innovative applications of sensor fusion and reinforcement learning. Throughout his career, Dr. Shi has led several national scientific projects, published over 30 papers in esteemed journals and conferences, and holds 14 authorized national invention patents. He also contributes to the academic community as a Guest Editor and Topical Advisory Panel member for the journal *Drones*, and he has served as a reviewer for numerous top-tier journals. His work significantly impacts the field of autonomous systems, driving advancements in UAV technology and applications.

Jihong Zhu

Jihong Zhu is a Professor in the Department of Precision Instrument at Tsinghua University, where he leads research in advanced control systems and precision measurement technologies. His academic journey includes a rich background in automation and control, focusing on applications that bridge the gap between theoretical development and practical implementation. Dr. Zhu is currently involved in several innovative projects that explore the intersection of control engineering and intelligent systems, aiming to enhance the reliability and performance of autonomous vehicles and robotic systems. His scientific interests encompass adaptive control, estimation theory, and sensor fusion, with a particular emphasis on their applications in aerospace and robotics. Dr. Zhu has published extensively, contributing to over 200 research papers in leading academic journals and conference proceedings. His work has garnered several honors, reflecting his influence in the field. As an active member of the academic community, he serves on editorial boards and is a reviewer for numerous prestigious journals, promoting advancements in technology and fostering collaboration within the research community.

Zheng Chen

Zheng Chen received a Ph.D. degree in Applied Mathematics from the University Paris-Saclay in 2016, and received M.Sc. and B.Sc. degrees in Aerospace Engineering from Northwestern Polytechnical University in 2013 and 2010, respectively. Zheng is currently a researcher with the School of Aeronautics and Astronautics at Zhejiang University and is the director of the Laboratory of AI for Flight at Zhejiang University. Zheng has participated in more than 10 scientific projects and was the coordinator of 7 projects. Zheng's general interests include the development and application of optimal control theory and methods to aerospace engineering, particularly in dynamic and time-critical environments. To date, Zheng has contributed to more than 50 scientific publications, 36 of which are in international journals with referees.

Minchi Kuang

Minchi Kuang is an Associate Research Fellow in the Department of Precision Instrument at Tsinghua University, specializing in intelligent control systems, with a particular focus on high-performance guidance and control for advanced aerospace applications. After earning his B.S. and Ph.D. in Computer Science from Tsinghua University, Dr. Kuang has dedicated his career to developing robust control solutions for national scientific projects, including intelligent control and digital twin systems. His achievements include leading several national research projects and receiving a prestigious national science and technology award for his innovative work. Recognized as a Young Talent by the China Association for Science and Technology, Dr. Kuang holds numerous patents, with several applied in critical national projects.

Preface

This Special Issue on *Path Planning, Trajectory Tracking, and Guidance for UAVs* brings together recent advancements in the algorithms and control systems enabling Unmanned Aerial Vehicles to operate autonomously in complex environments. The collected works address critical aspects of autonomous UAV functionality, from path optimization and real-time trajectory adjustments to precise guidance and obstacle avoidance. Given the increasing demand for robust UAV applications across various fields, this collection provides essential insights for researchers, engineers, and practitioners focused on enhancing UAV safety, stability, and adaptability. The authors contribute a wealth of innovative approaches and practical solutions to the challenges of UAV autonomy, reflecting collective efforts to refine and expand UAV technology. We hope it will serve as a valuable resource for anyone interested in the development and application of autonomous UAV systems.

Heng Shi, Jihong Zhu, Zheng Chen, and Minchi Kuang
Guest Editors

drones

Article

Multi-Constrained Geometric Guidance Law with a Data-Driven Method

Xinghui Yan [1], Yuzhong Tang [1], Yulei Xu [1], Heng Shi [2,*] and Jihong Zhu [2]

[1] School of Power and Energy, Northwestern Polytechnical University, Xi'an 710129, China; yanxh@nwpu.edu.cn (X.Y.); tangyz@mail.nwpu.edu.cn (Y.T.)

[2] Department of Precision Instrument, Tsinghua University, Beijing 100084, China

[*] Correspondence: shiheng@tsinghua.edu.cn

Abstract: A data-driven geometric guidance method is proposed for the multi-constrained guidance problem of variable-velocity unmanned aerial vehicles (UAVs). Firstly, a two-phase flight trajectory based on a log-aesthetic space curve (LASC) is designed. The impact angle is satisfied by a specified straight-line segment. The impact time is controlled by adjusting the phase switching point. Secondly, a deep neural network is trained offline to establish the mapping relationship between the initial conditions and desired trajectory parameters. Based on this mapping network, the desired flight trajectory can be generated rapidly and precisely. Finally, the pure pursuit and line-of-sight (PLOS) algorithm is employed to generate guidance commands. The numerical simulation results validate the effectiveness and superiority of the proposed method in terms of impact time and angle control under time-varying velocity.

Keywords: data-driven; multi-constrained guidance; impact time control; impact angle control; time-varying velocity

Citation: Yan, X.; Tang, Y.; Xu, Y.; Shi, H.; Zhu, J. Multi-Constrained Geometric Guidance Law with a Data-Driven Method. *Drones* **2023**, *7*, 639. https://doi.org/10.3390/drones7100639

Academic Editor: Sanjay Sharma

Received: 14 September 2023
Revised: 8 October 2023
Accepted: 13 October 2023
Published: 18 October 2023

Copyright: © 2023 by the authors. Licensee MDPI, Basel, Switzerland. This article is an open access article distributed under the terms and conditions of the Creative Commons Attribution (CC BY) license (https:// creativecommons.org/licenses/by/ 4.0/).

1. Introduction

Classical guidance laws, such as proportional navigation, do not give enough consideration to impact time and angle, which results in poor cooperative performance. For example, an impact angle constraint should be imposed for the UAVs to approach the target from a specified direction in some cases [1–4]. When multiple UAVs need to be gathered in a formation, the impact time should be coordinated [5–8]. Three performance indicators, namely miss distance, impact time, and impact angle, should be satisfied simultaneously during cluster flight. Therefore, impact time and angle control guidance (ITACG) laws have attracted extensive attention in both the academic and industrial worlds [9–18].

Scholars have conducted a series of studies based on different theories and methods, including optimal control theory, sliding mode control theory, computational geometry, and those methods based on proportional navigation. Lee et al. [9] proposed the first ITACG law based on optimal control theory, which consists of a feedback loop and an additional control command. In [10], a two-phase optimal ITACG law based on the virtual target method is proposed, which is suitable for all-around control in planar formations under admissible initial conditions and predetermined terminal constraints. The guidance problem is described as an optimal control model with discontinuities in [11], and an iterative guidance method is used to realize multi-constrained control. In [12], a new robust second-order sliding mode control law was developed by incorporating backstepping control, the line-of-sight rate shaping method, and second-order sliding mode control. Hou et al. [13] designed two different terminal sliding surfaces based on nonsingular terminal sliding mode control theory. This sliding mode control-based guidance law enables the UAV to satisfy both time and angle constraints while reaching the target. Reference [14], which combined the impact angle control guidance law based on backstepping control and the impact time control guidance law based on proportional navigation, achieved simultaneous control of both impact angle and time. A new ITACG law with adjustable coefficients was proposed in [15]. The bias proportional guidance with impact angle constraint is extended with feedback control of impact time error. However, the UAVs' flight trajectories under the aforementioned guidance laws are implicit, making it difficult to accurately estimate the time-to-go in advance. Zhang et al. [16] proposed a trajectory planning method based on three circular arcs to control impact time and angle. An explicit two-phase flight trajectory based on second-order Bezier curves was proposed in [17], allowing for precise impact time and angle requirements. In [18], the geometric ITACG law was extended to three-dimensional space using log-aesthetic space curves (LASCs). However, the offline trajectory design methods in [16–18] cannot be computed quickly enough to achieve an optimal solution and do not adequately consider time-varying velocity.

Besides the classical theories and methods mentioned above, data-driven methods [19–21] have gained significant attention in the field of guidance law design in recent years. Data-driven methods focus on the relationships between input and output data in a system and approximate the behavior of the system without analytical modeling. After collecting large amounts of input and output data, a data-driven model can be established using machine learning techniques. In [22], a deep neural network (DNN) was trained to learn the mapping relationship between flight states and distances. Based on this DNN, a multi-constrained prediction–correction guidance algorithm was proposed. Guo et al. [23] proposed a guidance law that combines data-driven methods and proportional navigation. The data-driven methods were used to compensate for the impact time error of proportional navigation guidance. Based on this, a biased proportional navigation guidance law was designed to control the impact time. In [24], another impact time control guidance law was designed based on proportional navigation and data-driven methods, which do not rely on time-to-go information. Reference [25] designed a two-phase guidance law based on data-driven methods to control both impact time and angle.

However, the aforementioned guidance laws are developed based on the assumption of constant velocity and two-dimensional kinematics. In practical applications, the UAV's velocity varies with engine thrust and air drag in three dimensions. As a result, it is challenging to control impact time and angle precisely under time-varying velocity. This manuscript addresses the guidance problem of impact time and angle control in three dimensions. A guidance law based on data-driven methods and computational geometry is proposed. The main contributions of this manuscript are summarized as follows:

1. The proposed three-dimensional, multi-constrained guidance method develops a geometric trajectory planning framework. It uses data-driven methods to solve the trajectory parameters, taking impact time, impact angle, maximum overload, and time-varying velocity into account.
2. Different from references [9–15], the proposed guidance method overcomes the reliance on accurate time-to-go estimation, which is quite difficult under time-varying velocities. The proposed method achieves better performance in terms of impact time and angle under time-varying velocity.
3. Compared to references [16–18], the proposed guidance law adopts data-driven methods to solve trajectory parameters, which greatly speeds up the trajectory generation process. This improvement benefits the update of trajectories due to different mission requirements.

The remainder of this manuscript is organized as follows: Section 2 introduces foundational concepts, including the LASC and guidance model. In Section 3, a novel ITACG law is proposed based on geometric planning and data-driven methods. Numeric simulations and results are given in Section 4. Section 5 provides the conclusions.

2. Preliminaries

2.1. Log-Aesthetic Space Curves

LASCs are widely applied in many fields, such as industrial design [26,27] and artistic aesthetics [28,29]. Each LASC has a linear, logarithmic curvature graph (LCG) and a linear, logarithmic torsion graph (LTG), as represented by Equations (1) and (2) [30].

$$\log(\rho \frac{ds}{d\rho}) = \alpha \log \rho + c_1 \tag{1}$$

$$\log(\mu \frac{ds}{d\mu}) = \beta \log \mu + c_2 \tag{2}$$

where s, ρ, and μ are the curve length, radius of curvature, and radius of torsion, respectively; α and β are the slopes of the segments in LCG and LTG, respectively; c_1 and c_2 are constants. Setting $\Lambda = e^{-c_1}$ and $\Omega = e^{-c_2}$, we obtain

$$\frac{ds}{d\rho} = \frac{\rho^{\alpha-1}}{\Lambda} \tag{3}$$

$$\frac{ds}{d\mu} = \frac{\mu^{\beta-1}}{\Omega} \tag{4}$$

Setting $\rho = 1$ and $\mu = v$ at $s = 0$, and integrating Equations (3) and (4), results in

$$\rho = \begin{cases} e^{\Lambda s} & , \text{if } \alpha = 0 \\ (\Lambda \alpha s + 1)^{\frac{1}{\alpha}}, & \text{otherwise} \end{cases} \tag{5}$$

$$\mu = \begin{cases} e^{(\Omega s + \log v)} & , \text{if } \beta = 0 \\ (\Omega \beta s + v^\beta)^{\frac{1}{\beta}}, & \text{otherwise} \end{cases} \tag{6}$$

Hence, the curvature and torsion of a curve can be represented as

$$\kappa(s) = \begin{cases} e^{-\Lambda s} & , \text{if } \alpha = 0 \\ (\Lambda \alpha s + 1)^{-\frac{1}{\alpha}}, & \text{otherwise} \end{cases} \tag{7}$$

$$\tau(s) = \begin{cases} e^{-(\Omega s + \log v)} & , \text{if } \beta = 0 \\ (\Omega \beta s + v^\beta)^{-\frac{1}{\beta}}, & \text{otherwise} \end{cases} \tag{8}$$

Then, the LASC can be established by integrating the Frenet–Serret formula as follows [30,31]:

$$\begin{bmatrix} \frac{dT(s)}{ds} \\ \frac{dN(s)}{ds} \\ \frac{dB(s)}{ds} \end{bmatrix} = \begin{bmatrix} 0 & \kappa(s) & 0 \\ -\kappa(s) & 0 & \tau(s) \\ 0 & -\tau(s) & 0 \end{bmatrix} \begin{bmatrix} T(s) \\ N(s) \\ B(s) \end{bmatrix} \tag{9}$$

where $T(s)$, $N(s)$, and $B(s)$ are unit tangent vector, unit normal vector, and unit binormal vector, respectively.

2.2. Guidance Problem Statement

To facilitate analytical analysis, the following assumptions are made [10,14,18]: (1) The UAV and target are assumed to be mass points. (2) Compared with the UAV, the target's velocity is low and can be considered static. (3) Since the autopilot dynamics are much faster than the UAV dynamics, the autopilot lag is neglected in guidance law design.

The guidance scenario in three-dimensional space is shown in Figure 1. The position coordinates in the inertial frame of the UAV and target are $I(x_m, y_m, z_m)$ and $T(x_t, y_t, z_t)$, respectively. The inertial frame is denoted as $O - X - Y - Z$. The subscripts 0 and f indicate the initial and final moments, respectively. The relative kinematic equations are:

$$\begin{aligned}
\dot{x}_m &= V \cos\theta_m \cos\psi_m \\
\dot{y}_m &= V \cos\theta_m \sin\psi_m \\
\dot{z}_m &= V \sin\theta_m \\
\dot{\theta}_m &= \frac{a_{zm}}{V} \\
\dot{\psi}_m &= \frac{a_{ym}}{V \cos\theta_m}
\end{aligned} \tag{10}$$

where V is the UAV velocity. a_{zm} and a_{ym} are the corresponding pitch and yaw acceleration. θ_m and ψ_m represent the pitch angle and the yaw angle, respectively. R is the range between I and T.

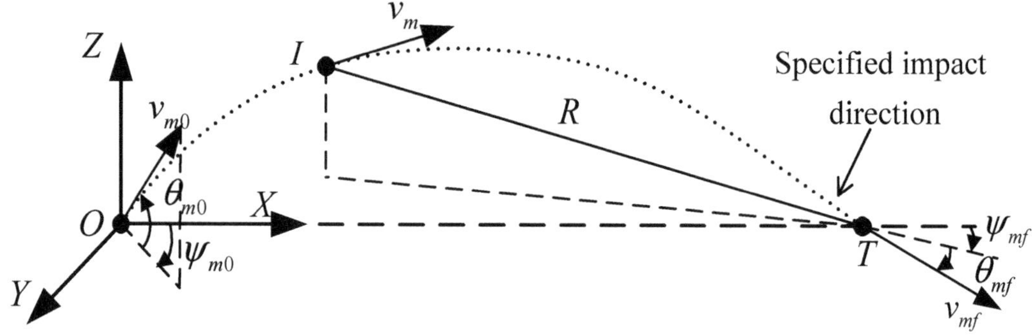

Figure 1. Three-dimensional guidance geometry.

Since air drag has a great influence on the UAV velocity, the drag model in [32,33] is adopted. The velocity change in the UAV can be expressed as

$$\dot{V} = \frac{1}{m}\left(T - \frac{1}{2}\rho V^2 S C_D - k_D a^2\right) \tag{11}$$

where m, T, and S represent the mass, thrust, and reference area of the UAV, respectively. C_D is the parasite drag coefficient, and its relationship with the UAV velocity V is presented in Figure 2 [17]. k_D is the induced drag coefficient. m and T are functions of time t, and can be represented as

$$T(t) = \begin{cases} T_{en}, & t \leq t_{en} \\ 0, & t > t_{en} \end{cases} \tag{12}$$

$$m(t) = \begin{cases} m_0 - \int_0^t \mu dt, & t \leq t_{en} \\ m_0 - m_{fu}, & t > t_{en} \end{cases} \tag{13}$$

where T_{en}, μ, m_{fu}, and t_{en} are the thrust, fuel consumption rate, fuel mass, and maximum working time of the UAV engine, respectively.

Considering the constraints of miss distance, acceleration limitation, impact time, and angle, the design objective of the guidance law can be mathematically defined as

$$\begin{cases} \lim\limits_{t \to t_f} x_m \to x_t, \ \lim\limits_{t \to t_f} y_m \to y_t, \ \lim\limits_{t \to t_f} z_m \to z_t \\ \lim\limits_{t \to t_f} \theta_m \to \theta_{mf}, \ \lim\limits_{t \to t_f} \psi_m \to \psi_{mf} \\ \sqrt{a_{zm}^2(t) + a_{ym}^2(t)} \leq a_{\max} \end{cases} \tag{14}$$

For the sake of simplicity, the launch position of the UAV is set to the origin of the inertial coordinate system $O(0,0,0)$. The target position is placed at $T(x_t,0,0)$ on the X-axis.

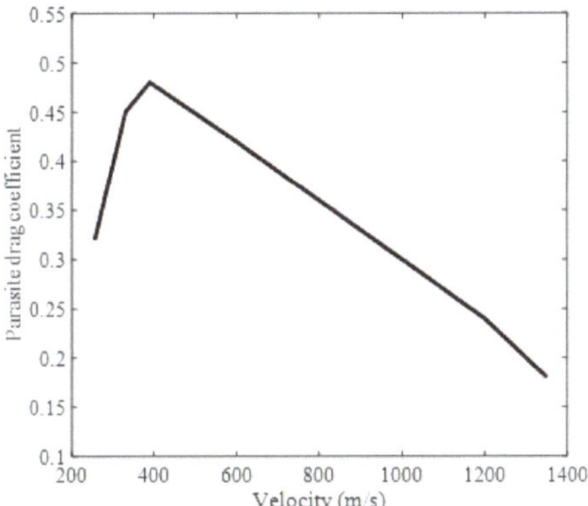

Figure 2. Profile of the parasitic drag coefficient of the UAV with respect to velocity.

3. Guidance Law Design

In this section, the design concept of a two-phase trajectory is initially introduced (Section 3.1). Then, the impact angle, impact time, and acceleration limitation are analyzed (Section 3.2). Afterward, a trajectory generation data-driven method is proposed to satisfy multiple constraints (Section 3.3). Lastly, a pure pursuit and line-of-sight (PLOS) tracking algorithm is employed to generate the guidance commands (Section 3.4). Figure 3 shows the block diagram of the guidance method.

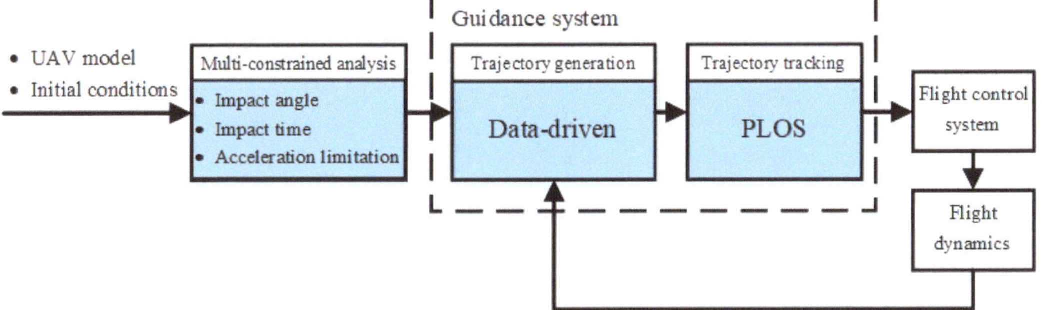

Figure 3. Guidance method design block diagram.

3.1. Trajectory Design Concept

The impact time and angle control are realized by adjusting the LASC trajectory's length, endpoint position, and tangent vector. Denote the initial launch angle of the UAV as (θ_{m0}, ψ_{m0}), and the desired angle at impact as (θ_{mf}, ψ_{mf}). Near the moment that the UAV approaches the target, the terminal-phase flight trajectory can be approximated as a straight line. If the angle formed by this straight line with respect to the target is equal to the desired impact angle, the impact angle constraint is satisfied. To ensure a smooth transition between the initial and terminal trajectories, these two trajectories must be tangent at the phase switching point. Moreover, by adjusting the position of this phase switching point along the straight-line trajectory, the length of the two-phase trajectory can be adjusted along with the impact time.

The parameters of the multi-constrained trajectory remain to be determined, including LASC parameters and phase switching point position. However, the complexity of the trajectory generation process makes it very difficult and even impractical to find an analytical solution. Consequently, a data-driven method is used to design the two-phase trajectory. The design framework is illustrated in Figure 4. The initial phase trajectory is a LASC that coincides with the launch angle at the origin, while the terminal phase trajectory is a straight-line segment that fulfills the impact angle constraint. The phase switching point is denoted as S. $O - X_L - Y_L - Z_L$ is the local coordinate system established for solving the LASC parameters under given conditions.

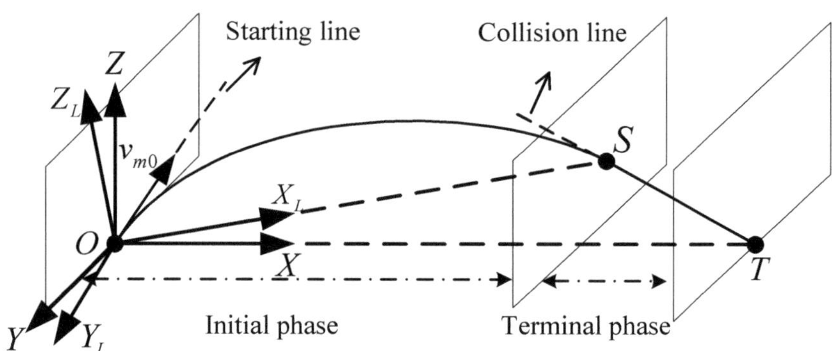

Figure 4. Two-phase guidance trajectory and geometric framework.

3.2. Multi-Constrained Analysis

In order to satisfy the impact angle constraint, it is necessary to make the initial phase trajectory tangent to the starting line at the origin point O and tangent to the collision line at the phase switching point S. Equations (5) and (6) provide insight that the LASC involves five trajectory parameters, namely α, β, Λ, Ω, and v. With the values of α, β, and Ω fixed, adjusting the values of Λ and v is adequate to satisfy the terminal positional and tangential constraints [30]. Thus, we set $\alpha = \beta = \Omega = 0$ in this manuscript.

For simplicity, the LASC can be normalized into a standard form by scaling and rotating transformations [18,30]. The coordinates of the two endpoints are transformed to $O(0,0,0)$ and $S'(2,0,0)$, with the scaling factor of $|OS|/2$. The initial condition is determined by the starting position $O(0,0,0)$ and the launch angle (θ_{m0}, ψ_{m0}). The terminal condition is determined by the endpoint position $S'(2,0,0)$ and the impact angle (θ_{mf}, ψ_{mf}). Thus, the LASC in the standard form can be obtained by integrating Equation (9). The generated LASC conforms to the following mapping relation f_1:

$$(L_{S_0}, \theta_{mf}, \psi_{mf}) = f_1(\theta_{m0}, \psi_{m0}, \Lambda, v, |OS|) \tag{15}$$

where L_{S_0} is the length of the LASC in the standard form.

To apply the standard-form trajectory to the inertial frame, the size of the LASC needs to be scaled as

$$L_S = L_{S_0} \cdot |OS|/2 \tag{16}$$

The length L of the entire two-phase trajectory is

$$L = L_S + |ST| \tag{17}$$

To satisfy the impact time constraint, the trajectory length needs to be adjusted by tuning the position of the phase switching point. The position of the phase switching point is determined by the length of the line segment $|ST|$. Therefore, the desired two-phase trajectory satisfies the following mapping relation f_2:

$$(L, \theta_{mf}, \psi_{mf}) = f_2(\theta_{m0}, \psi_{m0}, \Lambda, v, |OS|, |ST|) \tag{18}$$

The following geometric relationship can be obtained from Figure 4.

$$|OS|^2 = |ST|^2 + |OT|^2 - 2|ST| \cdot |OT|^2 \cos \angle OTS \tag{19}$$

Then, Equation (18) can be rewritten as

$$(L, \theta_{mf}, \psi_{mf}) = f_3(\theta_{m0}, \psi_{m0}, \Lambda, v, |ST|, |OT|) \tag{20}$$

Although the trajectory length and the trajectory parameters have been obtained, the impact time still needs to be calculated due to time-varying velocity. The following relationship between the trajectory length and the impact time needs to be satisfied.

$$\int_0^{t_f} V(t)dt - L = 0 \tag{21}$$

After obtaining the two-phase trajectory that satisfies the constraints of impact time and angle, it is necessary to evaluate the maximum acceleration demand during trajectory tracking. Because the acceleration demand imposes a constraint on the curvature of the trajectory as follows:

$$V^2 \cdot |\kappa(s)|_{\max} \leq a_{\max} \tag{22}$$

According to Equation (7), the curvature of the LASC is monotonically decreasing. Hence, we have

$$|\kappa(s)|_{\max} = |\kappa(0)| = |\kappa_0(0)| \cdot 2/|OS| \tag{23}$$

where $\kappa(s)$ and $\kappa_0(s)$ are the curvature in the inertial frame and the standard form, respectively.

Equation (20) can be supplemented as

$$(L, \theta_{mf}, \psi_{mf}, |\kappa(s)|_{\max}) = f_3(\theta_{m0}, \psi_{m0}, \Lambda, v, |ST|, |OT|) \tag{24}$$

While the UAV velocity varies with time, it is important to determine the exact maximum acceleration during trajectory tracking. To ensure the feasibility of flying along the trajectory, the maximum acceleration is evaluated using the maximum velocity and maximum curvature as

$$V_{\max} \cdot |\kappa(s)|_{\max} \leq a_{\max} \tag{25}$$

3.3. Trajectory Generation

3.3.1. Analysis of Mapping Relationship

The LASC plays a crucial role in ensuring a smooth transition between the two phases. To satisfy the impact angle constraint while maintaining the adjustability of the trajectory length, a predefined collision line is set. The trajectory length is adjusted by tuning the phase switching point, which further realizes control over the impact time. Consequently, the trajectory parameters that need to be determined include LASC parameters Λ and v, initial relative distance $|OT|$, and straight-line trajectory length $|ST|$.

If the trajectory scaling factor is σ, then the initial relative distance $|OT|$, straight-line trajectory length $|ST|$, and trajectory length L are all scaled in the same proportion. Moreover, the maximum curvature becomes $|\kappa(s)|_{\max}/\sigma$, while the LASC parameters Λ and v remain unchanged. For the sake of simplicity, we assume the initial distance between the objects is a definite value R_0 for further analysis. Thus, the Equation (24) can be rewritten as

$$(L, \theta_{mf}, \psi_{mf}, |\kappa(s)|_{\max}) = f_0(\theta_{m0}, \psi_{m0}, \Lambda, v, |ST|) \tag{26}$$

It can be inferred from Equation (26) that the trajectory parameters are determined by the initial angle, impact angle, and trajectory length. There is also a maximum curvature associated with these trajectory parameters. The corresponding mapping relation F_0 can be expressed as

$$(\Lambda, v, |ST|, |\kappa(s)|_{\max}) = F_0(\theta_{m0}, \psi_{m0}, \theta_{mf}, \psi_{mf}, L) \tag{27}$$

It is worth pointing out that when a curve rotates around the X-axis, its geometric features, such as relative distances and angles, remain unchanged. As shown in Figures 5 and 6, v'_{m0} and v'_{mf} are the projection vectors of the initial unit vector $v_{m0} = (x_0, y_0, z_0)$ and terminal unit vector $v_{mf} = \left(x_f, y_f, z_f\right)$ in the YOZ plane, respectively. θ_0 is the angle between v_{m0} and the YOZ plane, while θ_f is the angle between v_{mf} and the YOZ plane. ψ_0 is the angle between v'_{m0} and the Z-axis, while ψ_f is the angle between v'_{mf} and the Z-axis. The related relation expressions are as follows:

$$\begin{aligned}
\theta_0 &= \arcsin x_0 \\
\theta_f &= \arcsin x_f \\
\psi_0 &= \arctan(y_0, z_0) \\
\psi_f &= \arctan(y_f, z_f)
\end{aligned} \tag{28}$$

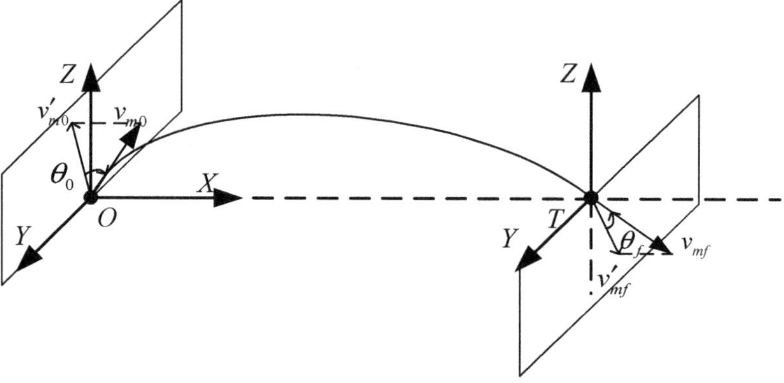

Figure 5. Illustration of the initial and terminal vectors of the curve.

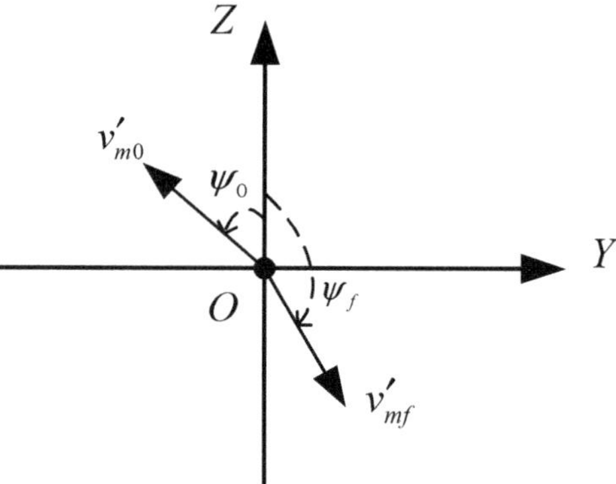

Figure 6. Illustration of the projection of the initial and terminal vectors onto the YOZ plane.

θ_0, θ_f, and $\Delta\psi$ represent the relative geometry of the initial vector and terminal vector in space. The corresponding values remain unchanged in spite of rotating around the X-axis. Therefore, the Equation (27) can be further simplified as follows:

$$(\Lambda, v, |ST|, |\kappa(s)|_{\max}) = F(\theta_0, \theta_f, \Delta\psi, L) \tag{29}$$

where $\Delta\psi = \left|\psi_0 - \psi_f\right|$.

Compared with Equation (27), Equation (29) reduces the dimensions of the input variables from 5 to 4, which significantly reduces the computational burden of constructing the database. Additionally, the proposed trajectory design method gains better flexibility by using fewer parameters in trajectory generation.

3.3.2. Acquisition of Trajectory Parameters

The LASC that satisfies the terminal constraints can be obtained by solving the LASC parameters Λ and v with the improved simplex method [30,34]. However, the existing trajectory generation problem involves time-varying velocity, which makes it difficult to solve analytically and rapidly. Local optimization algorithms like the improved simplex method would struggle to obtain the global optimal solution, while global optimization algorithms would suffer from time-consuming computation. Hence, a data-driven method is introduced here to solve the trajectory parameters efficiently. Instead of relying on complex analytical equations, the data-driven method focuses on establishing the relationship between input and output variables, which benefits solving the trajectory parameters accurately and efficiently.

Due to the strong capability of nonlinear fitting, DNN-based modeling is a widely adopted data-driven method and has gained more and more popularity in various fields [22,35,36]. DNN is composed of multiple layers and massive neurons, which allows it to model complex relationships and patterns through training data. Through learning from large datasets, DNN can perform nonlinear mappings between input and output variables, which makes it highly effective in handling complex and high-dimensional problems.

The general training process has three main steps [37]: forward propagation, loss function calculation, and backpropagation. In the step of forward propagation, input data are passed through the network's input layer to the output layer:

$$Y^m = f^m (W^m Y^{m-1} + b^m) \tag{30}$$

where W^m and b^m are the weight coefficients and biases of the mth layer, respectively. Y^m represents the output value of the mth layer. To achieve better performance in nonlinear mapping, the activation function f^m is applied.

The predicted output values are obtained through forward propagation. Then, these predicted values are compared with the actual output values to further assess the DNN's mapping performance. The prediction error is obtained by the loss function, while the weights and biases of each neuron are tuned according to the gradient of the loss function. The commonly used loss function is the mean squared error (MSE), which is defined as follows:

$$L = \frac{1}{N} \sum_{i=1}^{N} \left(Y_i^{pred} - Y_i^{actual} \right)^2 \tag{31}$$

where N is the number of samples. Y_i^{pred} and Y_i^{actual} represent the predicted output and actual output values of the ith sample, respectively.

In the step of backpropagation, the updates for weights and biases are performed as follows:

$$W_t^m = W_{t-1}^m - \eta \frac{\partial L}{\partial W_{t-1}^m} \tag{32}$$

$$b_t^m = b_{t-1}^m - \eta \frac{\partial L}{\partial b_{t-1}^m} \tag{33}$$

where W_t^m represents the weights of the mth layer at time t. b_t^m denotes the biases of the mth layer at time t. η is the learning rate that determines the step size for searching weight values. ∂ represents the gradient sign. The "$-$" sign indicates that the weight updates are made to reduce the loss function.

Therefore, a deep neural network, denoted as NET_{DNN} and shown in Figure 7, can be trained to establish the mapping in Equation (29). NET_{DNN} is used to solve the trajectory parameters Λ, v, $|ST|$ and the maximum curvature $|\kappa(s)|_{\max}$. Then, the multiple constraints can be checked according to Equations (9) and (25).

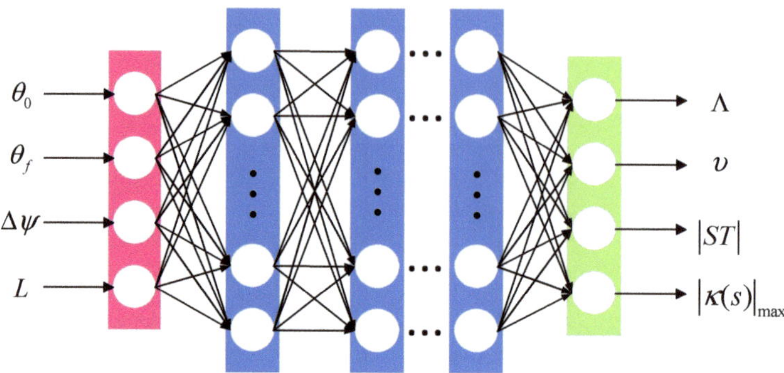

Figure 7. Structure diagram of a neural network.

3.4. Trajectory Tracking

It is crucial to guide the UAV along the designed trajectory in the presence of disturbances. Otherwise, the actual flight trajectory may deviate from the designed trajectory.
Therefore, a closed-loop trajectory tracking method is needed. Four commonly used
three-dimensional trajectory tracking algorithms are compared in [38], namely lookahead,
nonlinear guidance law (NLGL), pure pursuit and line-of-sight (PLOS), and vector field.
Among these algorithms, the PLOS algorithm performs the best in both accuracy and
computational efficiency. Therefore, the PLOS trajectory tracking method depicted as
Algorithm 1 is used to track the designed multi-constrained trajectory.

Algorithm 1 Pure Pursuit and Line-of-Sight algorithm.

1:    **for** each time step **do**
2:        Obtain the current waypoint $W_i = (x_1, y_1, z_1)$, next waypoint $W_{i+1} = (x_2, y_2, z_2)$, current UAV position $I = (x, y, z)$ and corresponding pitch and yaw angles θ, ψ in the inertial frame.
3:        Calculate $\alpha_y \leftarrow \arctan\left(\frac{W_{i+1}(2)-W_i(2)}{W_{i+1}(1)-W_i(1)}\right)$.
4:        Calculate $\psi_d \leftarrow \arctan(W_{i+1}(2) - I(2), W_{i+1}(1) - I(1))$.
5:        Calculate $W_i^R \leftarrow R_Z(-\alpha_y)W_i$, $W_{i+1}^R \leftarrow R_Z(-\alpha_y)W_{i+1}$, $I^R \leftarrow R_Z(-\alpha_y)I$.
6:        Calculate $\alpha_z \leftarrow \arctan\left(\frac{W_{i+1}^R(3)-W_i^R(3)}{W_{i+1}^R(1)-W_i^R(1)}\right)$.
7:        Calculate $\theta_d \leftarrow \arctan\left(W_{i+1}^R(3) - I^R(3), W_{i+1}^R(2) - I^R(2)\right)$.
8:        Calculate $e = R_Y(-\alpha_z)R_Z(-\alpha_y)(I - W_i)$.
9:        Calculate $\dot\psi_m \leftarrow k_{1,\psi}(\psi_d - \psi) + k_{2,\psi}(-e_y)$
       $\dot\theta_m \leftarrow k_{1,\theta}(\theta_d - \theta) + k_{2,\theta}(e_z)$
10:      Calculate a_{zm} and a_{ym} commands from Equation (10)
       $a_{zm} \leftarrow \left(k_{1,\theta}(\theta_d - \theta) + k_{2,\theta}(e_z)\right)V$ $a_{ym} \leftarrow \left(k_{1,\psi}(\psi_d - \psi) + k_{2,\psi}(-e_y)\right)V\cos\theta$
11:   **end for**

Where a_{zm} and a_{ym} are the pitch and yaw acceleration commands, respectively. α_z
and α_y are the rotation angles used to transform the inertial frame into the local frame
with respect to UAV velocity. These rotations can be represented by rotation matrices R_Z
and R_Y. e_z and e_y are the vertical and cross-track errors, respectively. $k_{1,\theta}$ and $k_{1,\psi}$ are the
proportional gains of the pure pursuit strategy. $k_{2,\theta}$ and $k_{2,\psi}$ are the proportional gains of
the LOS strategy.

4. Numerical Simulation

4.1. Simulation Setup

To assess the performance of the proposed guidance method, it is compared with the
guidance method in [14]. The simulation experiment involves four UAVs: I_1, I_2, I_3, and I_4.

Their simulation settings are depicted in Table 1. The UAV engine thrust T_{en} = 10,000 N, maximum working time of the engine t_{en} = 10 s, UAV's initial mass m_0 = 200 kg, fuel mass m_{fu} = 50 kg, fuel consumption rate μ = 5 kg/s, induced drag coefficient k_D = 0.05, and maximum acceleration limit $a_{\max}$ = 100 m/s^2. The four proportional gains of the PLOS algorithm are set as $k_{1,\theta}$ = $k_{1,\psi}$ = 20 and $k_{2,\theta}$ = $k_{2,\psi}$ = 0. The position of the target is (15,000, 0, 0) m. The autopilot is assumed to have a first-order lag with a time constant of 0.2 s. Thus, the flight velocity, distance, and time can be obtained through simulation.

Table 1. Comparison of simulation settings for UAVs.

UAV	Guidance	Launch Time	Initial Heading Direction	Desired Impact Direction
I_1	Proposed	$t = 0$ s	$(0.4082, 0.4082, 0.8165)$	$(0.6247, -0.6247, -0.4685)$
I_2	Proposed	$t = 2$ s	$(0.6667, -0.3333, 0.6667)$	$(0.4682, 0.8109, -0.3511)$
I_3	Ref. [14]	$t = 0$ s	$(0.4082, 0.4082, 0.8165)$	$(0.6247, -0.6247, -0.4685)$
I_4	Ref. [14]	$t = 2$ s	$(0.6667, -0.3333, 0.6667)$	$(0.4682, 0.8109, -0.3511)$

The simulation was performed on a hardware platform with a 2.5 GHz six-core processor and 16 GB RAM. It takes approximately 50 ms to generate a flight trajectory that satisfies the constraints of impact time, impact angle, and maximum acceleration limit.

4.2. Establishment of Mapping Network

Before training network NET_{DNN}, trajectory dataset samples need to be obtained first. Here, the optimal Latin hypercube method is employed to sample the trajectory data within the design space [39]. This sampling technique ensures that the samples are evenly distributed across the entire design space. The value ranges of the trajectory parameters are set to $\Lambda \in [0,1]$, $v \in (1.5,15]$ and $|ST| \in [0,6000 \text{ m}]$. Other trajectory parameters are set as $\alpha = \beta = \Omega = 0$ and R_0 = 15,000 m. Based on the trajectory parameters, the flight trajectories are obtained by integrating the Frenet–Serret Equation (9). Then the corresponding parameters θ_0, θ_f, $\Delta\psi$, L and $|\kappa(s)|_{\max}$ can be obtained from the generated trajectories. During the generation process of LASC, an undesired condition may occur in which the cumulative angular variation in tangent or binormal vectors is greater than 2π. Such conditions should be excluded because of the excessive loss of UAV velocity in practical applications. Then, a dataset of 1000 samples is obtained, as shown in Table 2. The input data S_1 are composed of parameters θ_0, θ_f, $\Delta\psi$, and L. The output data S_2 are composed of parameters Λ, v, $|ST|$, and $|\kappa(s)|_{\max}$. The mapping network NET_{DNN} is then trained using S_1 and S_2. All the data are normalized before training and split into two parts. A total of 70% of the data is used for training, while the remaining 30% is used for testing.

Table 2. Sample data for DNN mapping network training and testing.

NO.	Input Parameter				Output Parameter			
	θ_0	θ_f	$\Delta\psi$	L	Λ	v	$|ST|$	$|\kappa(s)|_{\max}$
1	24.7554	42.2519	149.8320	18,234.11	0.1860	2.1069	1284	1.5675×10^{-4}
2	47.3558	35.4730	167.8687	18,980.46	0.5441	5.7322	4577.62	3.7876×10^{-4}
3	36.0618	30.7972	174.3580	19,442.37	0.4036	13.3226	3539.06	2.5077×10^{-4}
$\vdots$	$\vdots$	$\vdots$	$\vdots$	$\vdots$	$\vdots$	$\vdots$	$\vdots$	$\vdots$
1000	35.5963	23.7348	170.6936	19,876.64	0.3774	9.4083	2949	2.5318×10^{-4}

Then, a feed-forward fully connected DNN is developed to learn the nonlinear relationship between S_1 and S_2. The appropriate selection of the hyperparameters is critical to achieving optimal performance. Here, the optimized hyperparameters are given in Table 3. The corresponding optimization process is similar to that in [36]. The training loss curve in Figure 8 shows that the DNN model keeps approaching the real model during the training process. The coefficient of determination R^2 is adopted to assess the approximation error [40]. Figure 9 illustrates the learning effect, which visualizes the prediction performance of DNN. The results demonstrate that the coefficients of determination of all four output parameters are above 0.99. Therefore, the prediction accuracy of the mapping network NET_{DNN} is high.

4.3. Case 1: UAV Guidance under Constant Velocity

In Case 1, the velocity is set to be constant and equal to $V = 300$ m/s. The desired impact time is set to $t_f = 65$ s. According to Equation (28), we have $\theta_{0,1} = 24.0948°$, $\theta_{f,1} = 38.6598°$, $\Delta\psi_1 = 153.4349°$, $L_1 = 19{,}500$ m, and $\theta_{0,2} = 41.8103°$, $\theta_{f,2} = 27.9152°$, $\Delta\psi_2 = 139.9783°$, $L_2 = 18{,}900$ m. Based on mapping network NET_{DNN}, we obtain $\Lambda_1 = 0.1553$, $v_1 = 2.2468$, $|ST|_1 = 3857.56$ m, $|\kappa(s)|_{max,1} = 1.6785 \times 10^{-4}$, and $\Lambda_2 = 0.3257$, $v_2 = 2.1061$, $|ST|_2 = 1931.00$ m, $|\kappa(s)|_{max,2} = 2.2483 \times 10^{-4}$. Both trajectories of the UAVs I_1 and I_2 satisfy the maximum acceleration constraint according to Equation (25).

The simulation results are shown in Figure 10 and Table 4. It can be observed that all the UAVs have successfully reached the target position. The impact time errors for UAVs I_1 and I_2 are both within 0.08 s, while the impact angle errors are both within 0.02°. Obviously, compared with UAVs I_3 and I_4, UAVs I_1 and I_2 achieve better accuracy in both impact time and impact angle. This is because the guidance method in [14] adopts a linear estimation method to calculate the time-to-go, which would degenerate under nonlinear kinematics.

4.4. Case 2: UAV Guidance under Time-Varying Velocity

In this case, the UAV velocities are time-varying according to Equation (11). The desired impact time is set to $t_f = 40$ s. All other simulation conditions are the same as in Case 1. According to Equation (28), we have $\theta_{0,1} = 24.0948°$, $\theta_{f,1} = 38.6598°$, $\Delta\psi_1 = 153.4349°$, $L_1 = 19{,}433.92$ m, and $\theta_{0,2} = 41.8103°$, $\theta_{f,2} = 27.9152°$, $\Delta\psi_2 = 139.9783°$, $L_2 = 18{,}786.81$ m. Based on mapping network NET_{DNN}, we obtain $\Lambda_1 = 0.1413$, $v_1 = 2.2106$, $|ST|_1 = 3495.75$ m, $|\kappa(s)|_{max,1} = 1.6338 \times 10^{-4}$, and $\Lambda_2 = 0.3193$, $v_2 = 2.0755$, $|ST|_2 = 1746.41$ m, $|\kappa(s)|_{max,2} = 2.1801 \times 10^{-4}$. Both trajectories of the UAVs I_1 and I_2 satisfy the maximum acceleration constraint according to Equation (25).

The simulation results are shown in Figure 11 and Table 5. It can be revealed that although all the UAVs have reached the target point, the guidance performances of UAVs I_3 and I_4 have significantly decreased. Their impact time errors are close to 2 s and the impact pitch angle errors exceed 10°. This is mainly because the guidance law in [14] adopts a linear estimation method for time-to-go prediction, which significantly degenerates under time-varying velocity. By contrast, the impact time errors of UAVs I_1 and I_2 are both less than 0.1 s, and the impact angle errors are both less than 0.05°. It can be seen that the time-varying velocity has little influence on the performance of the proposed guidance method. This is because the proposed method has taken the influence of variable velocity into account during neural network training.

Table 3. Hyperparameters for the DNN model.

Hyperparameters	Preferences
Optimizer	Adam
Number of hidden layers	4
Number of neurons per layer	4; 40; 40; 40; 40; 4
Activation functions	Hidden: Relu
Regularization	L2
Learning rate	0.01
Mini-batch size	150
Number of epochs	100
Weight initialization	He
Loss function	MSE

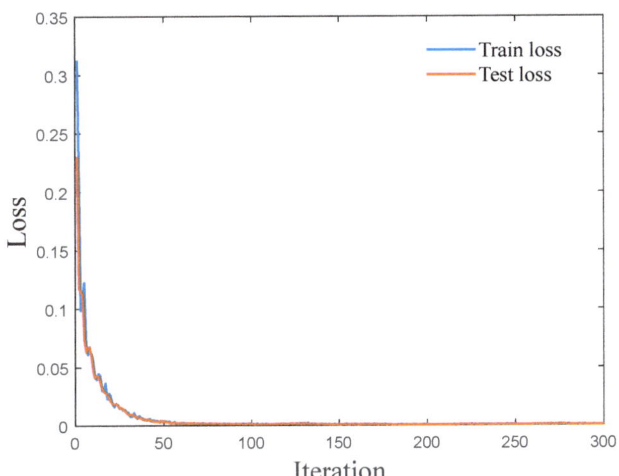

Figure 8. DNN training loss curve.

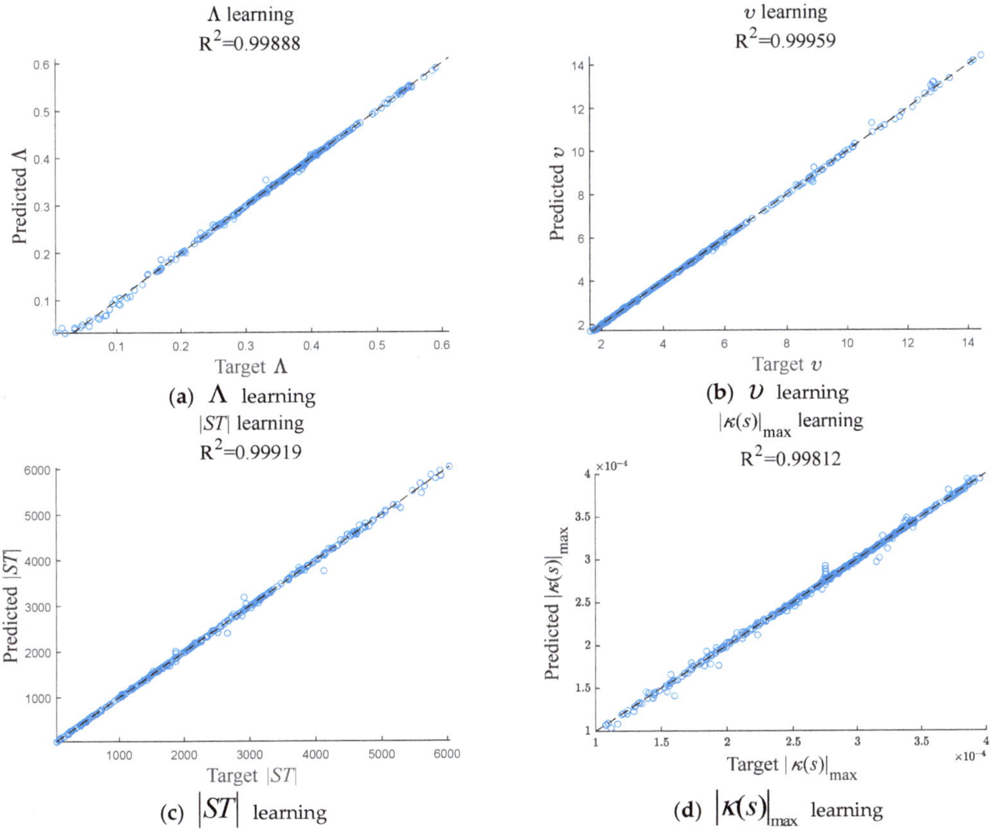

(**a**) Λ learning

(**b**) υ learning

(**c**) $\left|ST\right|$ learning

(**d**) $\left|\kappa(s)\right|_{max}$ learning

Figure 9. Learning effect of the network on parameter prediction.

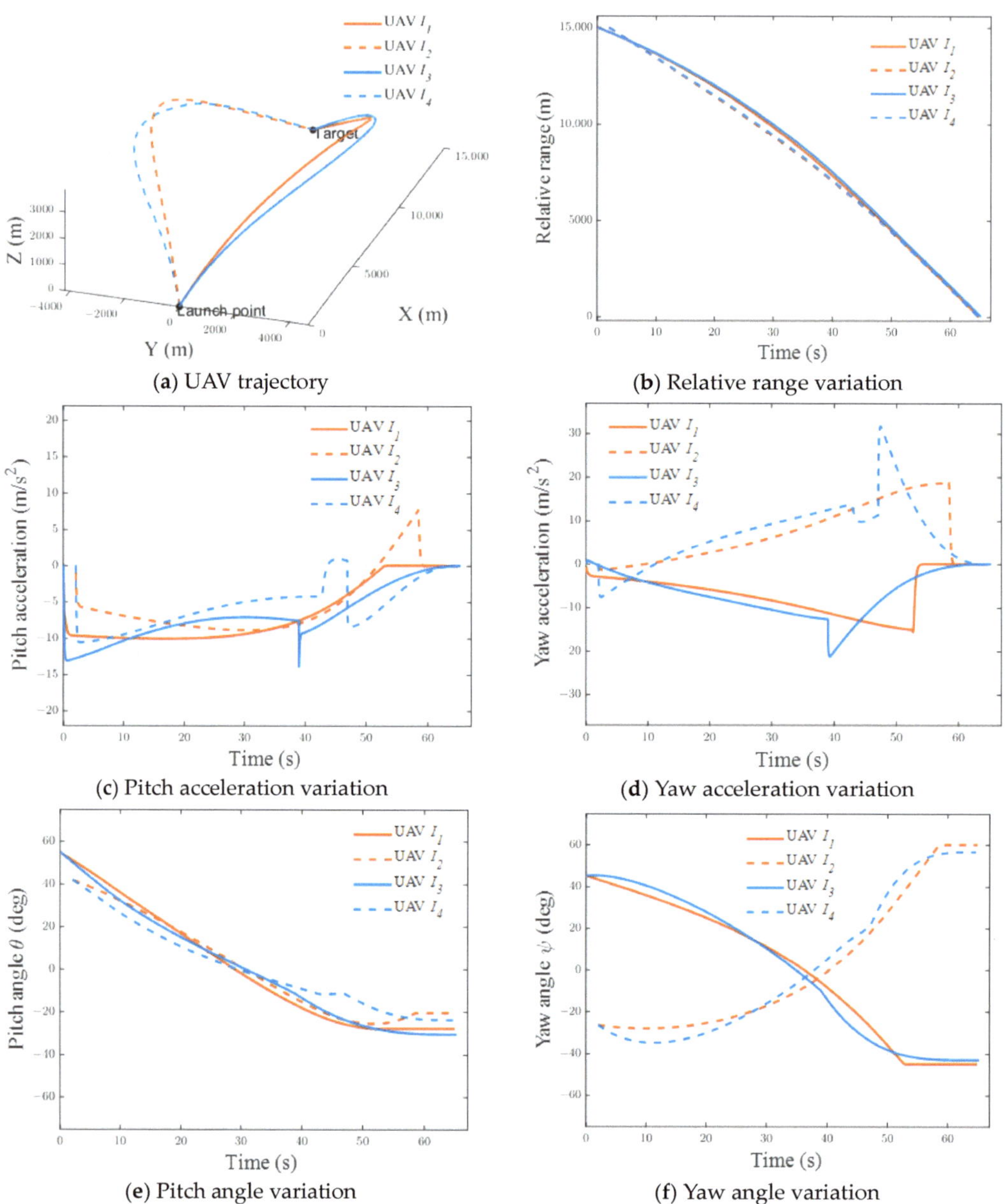

Figure 10. Simulation results of Case 1 (UAVs I_1 and I_2 adopt the proposed guidance method; UAVs I_3 and I_4 adopt the guidance method in Reference [14]).

Table 4. Simulation results of Case 1.

UAV	I_1	I_3	I_2	I_4
Desired impact time		65 s		
Impact time	64.92 s	65.29 s	64.94 s	64.73 s
Desired impact pitch angle	$-27.94°$		$-20.57°$	
Impact pitch angle	$-27.95°$	$-30.50°$	$-20.55°$	$-23.75°$
Desired impact yaw angle	$-45°$		$60°$	
Impact yaw angle	$-45.00°$	$-43.05°$	$60.02°$	$-56.55°$

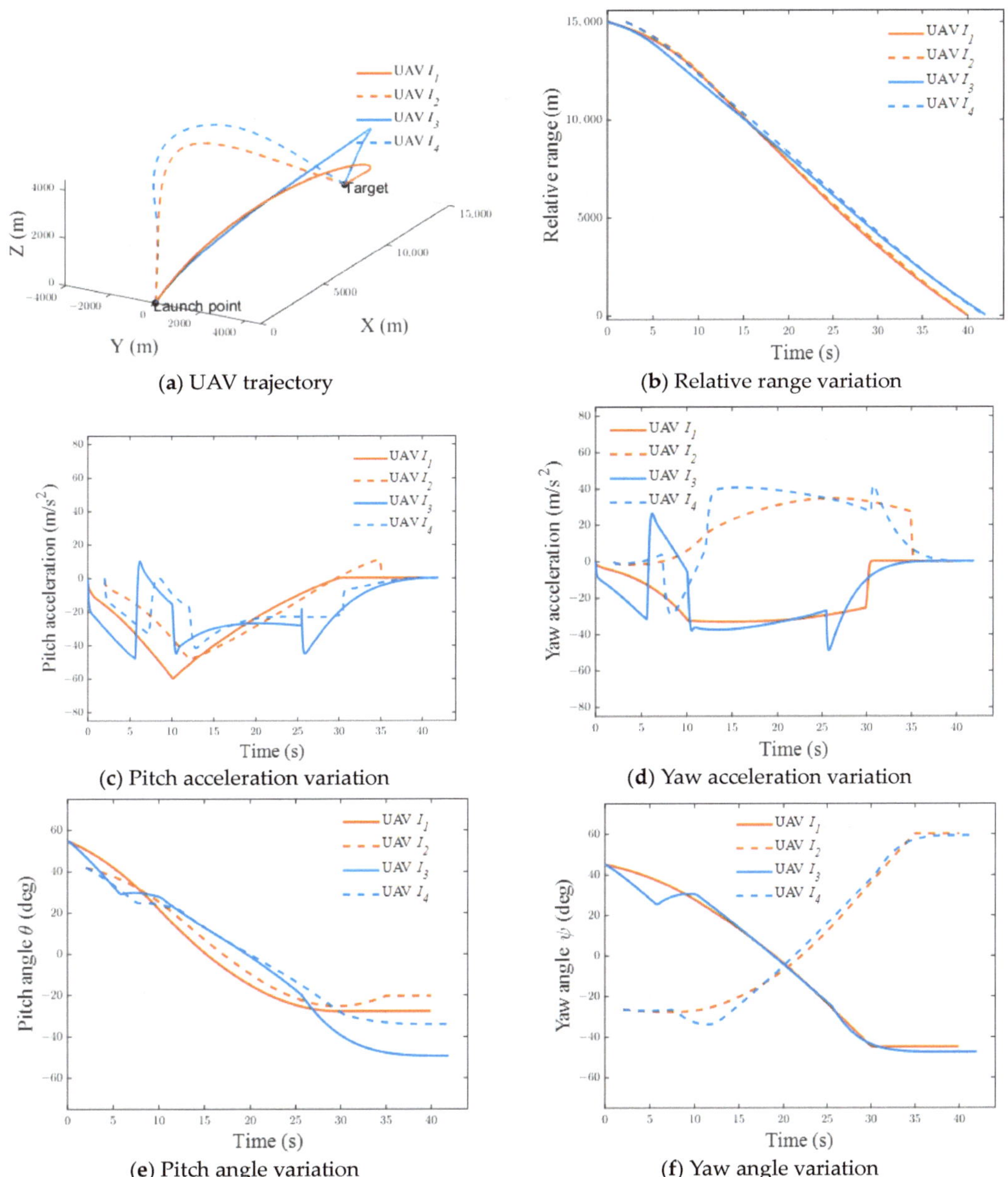

(**a**) UAV trajectory

(**b**) Relative range variation

(**c**) Pitch acceleration variation

(**d**) Yaw acceleration variation

(**e**) Pitch angle variation

(**f**) Yaw angle variation

Figure 11. Simulation results of Case 2 (UAVs I_1 and I_2 adopt the proposed guidance method; UAVs I_3 and I_4 adopt the guidance method in Reference [14]).

Table 5. Simulation results of Case 2.

UAV	I_1	I_3	I_2	I_4
Desired impact time		40 s		
Impact time	39.92 s	41.93 s	39.91 s	41.71 s
Desired impact pitch angle	$-27.94°$		$-20.57°$	
Impact pitch angle	$-27.94°$	$-49.37°$	$-20.56°$	$-34.19°$
Desired impact yaw angle	$-45°$		$60°$	
Impact yaw angle	$-44.99°$	$-47.53°$	$59.97°$	$-59.18°$

5. Conclusions

This manuscript proposes a data-driven method based on a multi-constrained geometric guidance law under time-varying velocity in three dimensions. The two-phase trajectory consists of an LASC and a collision line segment. By adjusting the phase switching points, the UAV's flight time and impact angle can be precisely controlled. The characteristic parameters of the desired trajectory are calculated through DNN using a data-driven method. Finally, the guidance commands are generated using a trajectory-tracking algorithm.

The proposed guidance can generate an explicit flight trajectory, which does not rely on accurate time-to-go information. Moreover, the trajectory generation processes are greatly shortened by data-driven-based mapping. The simulation results demonstrate the high performance of the proposed guidance method in terms of impact time and angle control under both constant and time-varying velocities.

Author Contributions: Conceptualization, X.Y. and Y.T.; methodology, X.Y. and Y.T.; software, X.Y. and Y.T.; formal analysis, X.Y. and H.S.; writing—original draft preparation, X.Y. and Y.T.; writing—review and editing, X.Y., Y.X., H.S. and J.Z. All authors have read and agreed to the published version of the manuscript.

Funding: This research was funded by the National Natural Science Foundation of China (Grant No. 62203362).

Data Availability Statement: The authors do not have permission to share data.

Acknowledgments: The authors are thankful to the anonymous reviewers for their useful suggestions.

Conflicts of Interest: The authors declare no conflict of interest.

References

1. Kim, M.; Grider, K.V. Terminal Guidance for Impact Attitude Angle Constrained Flight Trajectories. *IEEE Trans. Aerosp. Electron. Syst.* **1973**, *6*, 852–859. [CrossRef]
2. Ryoo, C.-K.; Cho, H.; Tahk, M.-J. Optimal Guidance Laws with Terminal Impact Angle Constraint. *J. Guid. Control Dyn.* **2005**, *28*, 724–732. [CrossRef]
3. Ratnoo, A.; Ghose, D. Impact Angle Constrained Guidance Against Nonstationary Nonmaneuvering Targets. *J. Guid. Control Dyn.* **2010**, *33*, 269–275. [CrossRef]
4. Song, K.-R.; Jeon, I.-S. Impact-Angle-Control Guidance Law with Terminal Constraints on Curvature of Trajectory. *Mathematics* **2023**, *11*, 974. [CrossRef]
5. Jeon, I.-S.; Lee, J.-I.; Tahk, M.-J. Impact-Time-Control Guidance Law for Anti-Ship Missiles. *IEEE Trans. Control Syst. Technol.* **2006**, *14*, 260–266. [CrossRef]
6. Cho, D.; Kim, H.J.; Tahk, M.-J. Nonsingular Sliding Mode Guidance for Impact Time Control. *J. Guid. Control Dyn.* **2016**, *39*, 61–68. [CrossRef]
7. Tekin, R.; Erer, K.S.; Holzapfel, F. Polynomial Shaping of the Look Angle for Impact-Time Control. *J. Guid. Control Dyn.* **2017**, *40*, 2668–2673. [CrossRef]
8. Shi, H.; Chen, Z.; Zhu, J.; Kuang, M. Model Predictive Guidance for Active Aircraft Protection from a Homing Missile. *IET Control Theory Appl.* **2022**, *16*, 208–218. [CrossRef]
9. Lee, J.-I.; Jeon, I.-S.; Tahk, M.-J. Guidance Law to Control Impact Time and Angle. *IEEE Trans. Aerosp. Electron. Syst.* **2007**, *43*, 301–310. [CrossRef]
10. Zhang, Z.; Ma, K.; Zhang, G.; Yan, L. Virtual Target Approach-Based Optimal Guidance Law with Both Impact Time and Terminal Angle Constraints. *Nonlinear Dyn.* **2022**, *107*, 3521–3541. [CrossRef]

11. Deng, Y.; Ren, J.; Wang, X.; Cai, Y. Midcourse Iterative Guidance Method for the Impact Time and Angle Control of Two-Pulse Interceptors. *Aerospace* **2022**, *9*, 323. [CrossRef]
12. Harl, N.; Balakrishnan, S.N. Impact Time and Angle Guidance With Sliding Mode Control. *IEEE Trans. Control Syst. Technol.* **2012**, *20*, 1436–1449. [CrossRef]
13. Hou, Z.; Yang, Y.; Liu, L.; Wang, Y. Terminal Sliding Mode Control Based Impact Time and Angle Constrained Guidance. *Aerosp. Sci. Technol.* **2019**, *93*, 105142. [CrossRef]
14. Jung, B.; Kim, Y. Guidance Laws for Anti-Ship Missiles Using Impact Angle and Impact Time. In Proceedings of the AIAA Guidance, Navigation, and Control Conference and Exhibit, Keystone, CO, USA, 21–24 August 2006; Guidance, Navigation, and Control and Co-located Conferences. American Institute of Aeronautics and Astronautics: Palo Alto, CA, USA, 2006; p. 6432.
15. Zhang, Y.; Ma, G.; Liu, A. Guidance Law with Impact Time and Impact Angle Constraints. *Chin. J. Aeronaut.* **2013**, *26*, 960–966. [CrossRef]
16. Zhang, Y.; Zhang, Y.; Li, H. Impact Time and Impact Angle Control Based on CCC Path Planning. In Proceedings of the 31st Chinese Control Conference, Hefei, China, 25–27 July 2012; pp. 4300–4305.
17. Yan, X.; Kuang, M.; Zhu, J. A Geometry-Based Guidance Law to Control Impact Time and Angle under Variable Speeds. *Mathematics* **2020**, *8*, 1029. [CrossRef]
18. Yan, X.; Zhu, J.; Kuang, M.; Yuan, X. A Computational-Geometry-Based 3-Dimensional Guidance Law to Control Impact Time and Angle. *Aerosp. Sci. Technol.* **2020**, *98*, 105672. [CrossRef]
19. Kirchdoerfer, T.; Ortiz, M. Data-Driven Computational Mechanics. *Comput. Methods Appl. Mech. Eng.* **2016**, *304*, 81–101. [CrossRef]
20. Hao, G.; Ni, W.; Tian, H.; Cao, L. Mobility-Aware Trajectory Design for Aerial Base Station Using Deep Reinforcement Learning. In Proceedings of the 2020 International Conference on Wireless Communications and Signal Processing (WCSP), Nanjing, China, 21–23 October 2020; pp. 1131–1136.
21. Arani, A.H.; Hu, P.; Zhu, Y. HAPS-UAV-Enabled Heterogeneous Networks: A Deep Reinforcement Learning Approach. *IEEE Open J. Commun. Soc.* **2023**, *4*, 1745–1760. [CrossRef]
22. Cheng, L.; Jiang, F.; Wang, Z.; Li, J. Multiconstrained Real-Time Entry Guidance Using Deep Neural Networks. *IEEE Trans. Aerosp. Electron. Syst.* **2021**, *57*, 325–340. [CrossRef]
23. Guo, Y.; Li, X.; Zhang, H.; Cai, M.; He, F. Data-Driven Method for Impact Time Control Based on Proportional Navigation Guidance. *J. Guid. Control Dyn.* **2020**, *43*, 955–966. [CrossRef]
24. Huang, J.; Chang, S.; Chen, Q.; Zhang, H. Data-Driven-Based Impact Time Control Guidance Law Independent of Time-to-Go. *Acta Armamentarii* **2022**, *44*, 2299. [CrossRef]
25. Huang, J.; Chang, S. Data-Driven Method Based Impact Time and Impact Angle Control Guidance Law. *Syst. Eng. Electron.* **2022**, *44*, 3213–3220. [CrossRef]
26. Ziatdinov, R.; Yoshida, N.; Kim, T. Analytic Parametric Equations of Log-Aesthetic Curves in Terms of Incomplete Gamma Functions. *Comput. Aided Geom. Des.* **2012**, *29*, 129–140. [CrossRef]
27. Gobithaasan, R.U.; Teh, Y.M.; Miura, K.T.; Ong, W.E. Lines of Curvature for Log Aesthetic Surfaces Characteristics Investigation. *Mathematics* **2021**, *9*, 2699. [CrossRef]
28. Miura, K.T.; Gobithaasan, R.U. Aesthetic Design with Log-Aesthetic CurvesLog-Aesthetic Curveand SurfacesLog-Aesthetic Curve and Surface. In *Mathematical Progress in Expressive Image Synthesis III*; Dobashi, Y., Ochiai, H., Eds.; Springer: Singapore, 2016; pp. 107–119.
29. Luca, L.; Popescu, I.; Cherciu, M.; Ghimisi, S.; Cirtina, M.L.; Pasare, M.M. Synthesis of Two New Mechanisms Which Generate a Highly Aesthetic Design Image—The Flower of Life. *Appl. Sci.* **2020**, *10*, 1670. [CrossRef]
30. Yoshida, N.; Fukuda, R.; Saito, T. Log-Aesthetic Space Curve Segments. In Proceedings of the 2009 SIAM/ACM Joint Conference on Geometric and Physical Modeling, San Francisco, CA, USA, 5–8 October 2009; Association for Computing Machinery: New York, NY, USA, 2009; pp. 35–46.
31. Struik, D.J. *Lectures on Classical Differential Geometry*; Courier Corporation: North Chelmsford, MA, USA, 1961; ISBN 978-0-486-65609-0.
32. Davidovitz, A.; Shinar, J. Two-Target Game Model of an Air Combat with Fire-and-Forget All-Aspect Missiles. *J. Optim. Theory Appl.* **1989**, *63*, 133–165. [CrossRef]
33. Alkaher, D.; Moshaiov, A. Dynamic-Escape-Zone to Avoid Energy-Bleeding Coasting Missile. *J. Guid. Control Dyn.* **2015**, *38*, 1908–1921. [CrossRef]
34. Nelder, J.A.; Mead, R. A Simplex Method for Function Minimization. *Comput. J.* **1965**, *7*, 308–313. [CrossRef]
35. LeCun, Y.; Bengio, Y.; Hinton, G. Deep Learning. *Nature* **2015**, *521*, 436–444. [CrossRef]
36. Cheng, L.; Wang, Z.; Jiang, F.; Li, J. Fast Generation of Optimal Asteroid Landing Trajectories Using Deep Neural Networks. *IEEE Trans. Aerosp. Electron. Syst.* **2020**, *56*, 2642–2655. [CrossRef]
37. Yang, M.; Liang, S.; Yi, M.; Tian, Y.; Guo, M.; Le, J.; Zhang, H. Identification of Turbulence Eddy Viscosity Coefficient in Supersonic Isolation Section Based on Deep Neural Network. *J. Aerosp. Power* **2023**, *38*, 312–324. [CrossRef]

38. Pelizer, G.V.; da Silva, N.B.F.; Branco, K.R.L.J. Comparison of 3D Path-Following Algorithms for Unmanned Aerial Vehicles. In Proceedings of the 2017 International Conference on Unmanned Aircraft Systems (ICUAS), Miami, FL, USA, 13–16 June 2017; pp. 498–505.
39. Jin, R.; Chen, W.; Sudjianto, A. An Efficient Algorithm for Constructing Optimal Design of Computer Experiments. *J. Stat. Plan. Inference* **2005**, *134*, 268–287. [CrossRef]
40. Nagelkerke, N.J.D. A Note on a General Definition of the Coefficient of Determination. *Biometrika* **1991**, *78*, 691–692. [CrossRef]

Disclaimer/Publisher's Note: The statements, opinions and data contained in all publications are solely those of the individual author(s) and contributor(s) and not of MDPI and/or the editor(s). MDPI and/or the editor(s) disclaim responsibility for any injury to people or property resulting from any ideas, methods, instructions or products referred to in the content.

 **drones**

Article

Topological Map-Based Autonomous Exploration in Large-Scale Scenes for Unmanned Vehicles

Ziyu Cao [1],*, Zhihui Du [2] and Jianhua Yang [1]

[1] School of Automation, Northwestern Polytechnical University, Xi'an 710129, China; yangjianhua@nwpu.edu.cn

[2] Department of Precision Instrument, Tsinghua University, Beijing 100084, China; zhihuidu@mail.tsinghua.edu.cn

* Correspondence: caoziyu@mail.nwpu.edu.cn

Abstract: Robot autonomous exploration is a challenging and valuable research field that has attracted widespread research interest in recent years. However, existing methods often encounter problems such as incomplete exploration, repeated exploration paths, and low exploration efficiency when facing large-scale scenes. Considering that many indoor and outdoor scenes usually have a prior topological map, such as road navigation maps, satellite road network maps, indoor computer-aided design (CAD) maps, etc., this paper incorporated this information into the autonomous exploration framework and proposed an innovative topological map-based autonomous exploration method for large-scale scenes. The key idea of the proposed method is to plan exploration paths with long-term benefits by tightly merging the information between robot-collected and prior topological maps. The exploration path follows a global exploration strategy but prioritizes exploring scenes outside the prior information, thereby preventing the robot from revisiting explored areas and avoiding the duplication of any effort. Furthermore, to improve the stability of exploration efficiency, the exploration path is further refined by assessing the cost and reward of each candidate viewpoint through a fast method. Simulation experimental results demonstrated that the proposed method outperforms state-of-the-art autonomous exploration methods in efficiency and stability and is more suitable for exploration in large-scale scenes. Real-world experimentation has also proven the effectiveness of our proposed method.

Keywords: autonomous exploration; path planning; PCATSP; prior information; topological map

Citation: Cao, Z.; Du, Z.; Yang, J. Topological Map-Based Autonomous Exploration in Large-Scale Scenes for Unmanned Vehicles. *Drones* **2024**, *8*, 124. https://doi.org/10.3390/drones8040124

Academic Editor: Diego González-Aguilera

Received: 29 February 2024
Revised: 23 March 2024
Accepted: 24 March 2024
Published: 27 March 2024

Copyright: © 2024 by the authors. Licensee MDPI, Basel, Switzerland. This article is an open access article distributed under the terms and conditions of the Creative Commons Attribution (CC BY) license (https://creativecommons.org/licenses/by/4.0/).

1. Introduction

Autonomous robot exploration technology requires robots to collect data within a given region and construct corresponding environmental maps. As a critical technology that reveals robotic autonomy, relevant research in robotics has garnered significant attention, driving widespread applications in geological exploration, 3D reconstruction, post-disaster rescue, and other fields.

Numerous autonomous exploration methods have been proposed in recent years and are divided into sampling-based and frontier-based categories. Sampling-based methods originated from the NBV (Next Best View) algorithm in the field of 3D reconstruction. RH-NBV (recurrent hybrid neural-based visual) first introduced the NBV algorithm into the autonomous exploration field [1], which consisted of the robot randomly sampling viewpoints in explored free space, constructing a rapidly exploring random tree (RRT), and evaluating the utility of each branch on the RRT. Finally, the robot focuses on the branch with the highest information reward and selects the first node of this branch as the local target. After that, numerous researchers have extended and improved the RH-NBV to meet the requirements of various application scenarios [2–6]. However, sampling-based autonomous exploration methods have lower exploration efficiency and lead to the robot being trapped. Ref. [7] first introduced the frontier-based exploration method, which groups free voxels adjacent to unknown voxels as frontier clusters and then drives the robot towards these frontier clusters to move to explore unknown areas. Since then, many frontier-based exploration methods have been proposed to meet various application requirements [8–10]. Ref. [11] proposed to select a viewpoint with minimal speed changes as the next goal within the Field Of View (FOV) of sensors, aiming to maintain the high movement speed of unmanned aerial vehicles (UAVs) and achieve efficient exploration. Fast UAV expLoration (FUEL) proposes the incremental frontier information structure (FIS) to address the problem of high computation of frontier extraction and low decision frequency of the path planner [12]. Based on FIS, UAVs can quickly and incrementally extract environmental information that the planner needs and promptly plan the exploration path.

However, most autonomous exploration methods tend to greedily guide the robot to exploration scenes with immediate rewards and neglect some targets with long-term rewards, resulting in lower efficiency in global exploration [12]. Although some methods plan paths from the global exploration standpoint, the robot inevitably overlooks some scenarios during exploration because of the limited perception range of sensors and the unpredictability of unknown environments [12,13]. To thoroughly explore a given region, the robot must revisit areas containing those missed scenarios, resulting in a waste of resources. Furthermore, when exploring large-scale scenes, the more information the robot collects with exploration, the more the path planner computes, which poses a significant challenge to onboard computers.

In order to solve the above problems, the work [14] proposed that supplying robots with prior information about a given region can aid them in making decisions that align with long-term benefits. Ref. [15] proposed a probabilistic information gain map as the prior knowledge to guide exploration. Ref. [16] introduced a general information theory framework to control multiple robots to search and rescue, wherein the prior knowledge of people is modeled to capture target positions and dynamics. Ref. [17] employs hand-drawn sketches as prior information, enabling the robot to explore even when the metric description of the environment is incomplete.

As the concision of a topological map, many researchers use them as prior information to guide robots in autonomous exploration. Ref. [18] proposed a novel autonomous exploration method based on a prior topometric graph, which verifies that prior information could aid the robot in swiftly completing the exploration of unknown environments. Ref. [19] proposed a path planning method based on topology information for 3D reconstruction, in which the multi-view stereo path planning is decomposed into a collection of overlapped viewing optimization problems that can be processed in parallel. In [14], the prior topometric map is employed to improve exploration efficiency and guide the robot to trigger a loop close, improving the localization accuracy of the Simultaneous Localization And Mapping (SLAM) system. Finally, the environmental information collected will be used to refine prior information.

Furthermore, some researchers focus on the generation of topological maps. Ref. [20] proposed a framework called "topomap" to provide robots with customized maps to simplify robot navigation tasks, transforming the sparse feature-based map from visual SLAM into a three-dimensional topological map. Ref. [21] proposed an efficient and flexible algorithm that generates a trajectory-independent 3D sparse topological skeleton graph captured from the spatial structure of free space.

Inspired by the abovementioned research, we select the topological map as the prior information to guide in robot exploration and employ the frontier-based exploration method suitable for exploring large-scale scenes to plan exploration paths. As a form of map representation, the topological map briefly provides relative position and connectivity between critical places in complex scenes, which could guide the robot in planning paths that follow long-term benefits [20]. In practical cases, many methods can easily acquire the skeleton structure of the environment as the topological map [20–23].

To fully take advantage of the guiding function of prior topological maps, we propose an autonomous exploration method based on topological maps. The proposed method employs a hierarchical path planning framework, integrates frontier information with prior topological maps, and plans the exploration path with long-term benefits. It first plans a global exploration path by solving the constructed Priority Constrained Asymmetric Traveling Salesman Problem (PCATSP). The global exploration path would follow optimal or customized global exploration strategies to guide the robot to cover frontiers but prioritizes exploring scenes outside the prior information, thus preventing the robot from revisiting previously visited areas. Then, the exploration path is refined from the global exploration path by quickly evaluating the rewards and costs of the candidate viewpoint for each frontier.

Because of the one-pass exploration process, our method will maintain the frontier at a small number, preventing excessive computational burden on the solver during exploration. The above properties make our method more suitable for autonomous exploration in large-scale scenes, and the contributions of this paper are as follows:

(1) An autonomous exploration method based on prior topological maps. The robot follows an optimal or customized strategy to explore a given region autonomously but prioritizes exploring scenes outside prior information, preventing the robot from revisiting the explored areas.

(2) A path planning method integrates information between frontiers and prior topological maps, which makes the topological map deeply involved in the path planning of robot exploration.

(3) A local path planning method, which quickly evaluates the rewards and costs of each candidate viewpoint to optimize the global exploration path, enhances the stability of exploration efficiency.

2. Design Objectives

Give the robot a region to autonomously explore, and provide it with a topological map of the region to be explored. The topological map should reflect the fundamental layout of the region but may not represent all of its spaces. The objectives we address are as follows:

Objective 1: The robot completes a comprehensive exploration of the given region. When there are no frontier clusters extractable within the given region, it indicates the completion of an information gathering task.

Objective 2: The robot utilizes real-time collected scene information and prior topological maps to plan exploration paths. When the robot encounters scenes that are not included in the priori information, the exploration path will guide the robot to prioritize exploring scenes beyond a priori information.

Objective 3: The exploration path enables the robot to complete an exploration of the visited area in a one-pass manner, preventing the robot from repeatedly visiting the areas that have been explored.

3. Methods

We define the topological map as follows. $G(S, E)$ consist of the global targets set $S = \{s_1, s_2, \ldots, s_n\}$ and the undirected edges set $E = \{e_1, e_2, \ldots, e_m\}$. s denotes a global target corresponding to a corner or intersection in the environment. $e_k = (s_i, s_j)$ is an undirected edge, connecting s_i and s_j, representing a straight-line scene, such as a road or corridor. With the support of a prior topological map, we can obtain a global exploration strategy $O = \{\overline{o_0}, \overline{o_1}, \overline{o_2}, \ldots, \overline{o_h}\}$ for the given region by customizing or solving the Chinese Postman Problem (CPP) [24], which is a priority queue. $\overline{o_k} = (s_i, s_j)$ denotes a directed line segment from s_i to s_j, corresponding to the exploration guidance. Define f as the frontier cluster, and $F = \{f^1, f^2, \ldots, f^n\}$ as a set of remaining frontier clusters in a scene. We adopt the incrementally frontier information structure (FIS) proposed by FUEL to update frontier clusters efficiently [12]. Viewpoint sequential queue $VP^k = \{vp_1^k, vp_2^k, \ldots, vp_m^k\}$ of frontier cluster f^k is extracted by random sampling, where vp_1^k is a viewpoint with the largest reward of f^k and will replace f^k in constructing PCATSP.

Figure 1 shows an overview of the proposed method, which operates upon a voxel grid map. We employ a hierarchical architecture to plan the exploration path, which consists of global path planning (Section 3.1) and local path planning (Section 3.4). The global path planning module takes a prior topological map, global exploration strategy, and frontier clusters as input to plan the global exploration path based on the Priority Constrained Asymmetric Traveling Salesman Problem (PCATSP). Nodes with priority in PCATSP will be extracted (Section 3.2), and the movement cost of some frontier clusters will also be updated (Section 3.3). Then, the global exploration path is given to the local path planning module, which refines the input path based on rewards and costs of each viewpoint candidate to improve the stability of exploration efficiency. Finally, the exploration path will output to the trajectory generation module. The exploration task will be completed when no frontier clusters can be extracted from environment.

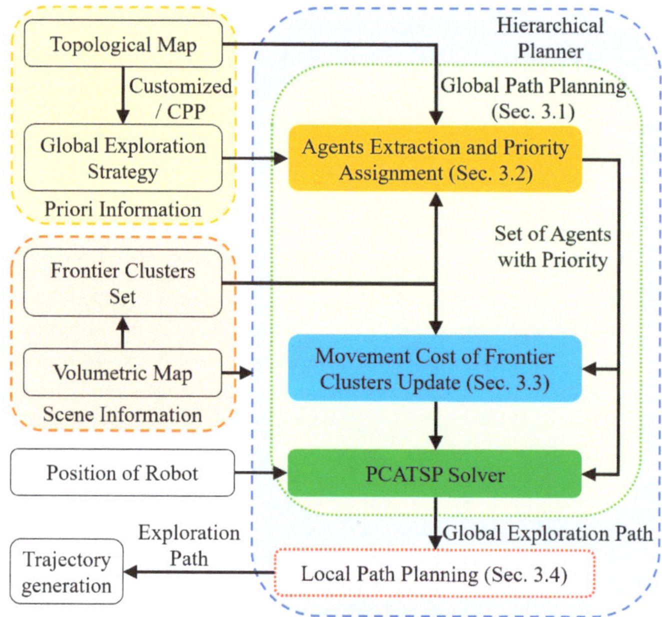

Figure 1. The overview of the proposed exploration method.

3.1. PCATSP-Based Global Path Planning

PCATSP is a variation of the classic Traveling Salesman Problem (TSP), which aims to find a minimum-cost Hamiltonian circuit, with the constraint that some nodes must be visited before others. If splitting the start and end points of PCATSP into two nodes, PCATSP is equivalent to finding a path between the start and end points that satisfy priority constraints, which is also considered a Sequential Ordering Problem (SOP) [25]. To address Objective 3, our basic idea of global path planning is to solve PCATSP with frontiers and priority-constrained global targets. It is equivalent to inserting frontiers into a sequential queue of global targets, utilizing the global targets to influence the covered sequence of some frontier clusters.

However, the construction of PCATSP faces the following challenges: First, constrained by the perception range of sensor and obstacle obstruction, the robot cannot accurately calculate the movement distance between global targets in unknown space and viewpoints inside the free space. Second, for TSP, the farther the metric distance between frontier cluster and global target, the less influence the global target can exert. So, we need to enable global targets to influence specific frontier clusters.

To address the above challenges, we define the following: if a global target s_i is inside the free space, and the other global target s_j connected to s_i is in unknown space, then the shortest path P_{p_k, s_j} between any node p_k in free space and global target s_j in unknown space is given by

$$P_{p_k, s_j} = P_{p_k, a} + P_{a, s_j},\qquad(1)$$

where a is an intersection point of frontier and undirected edge e_k that connects s_i and s_j, as shown in Figure 2a. P_{a, s_j} is a portion of e_k in unknown space, and $P_{p_k, a}$ is a search path from p_k to a.

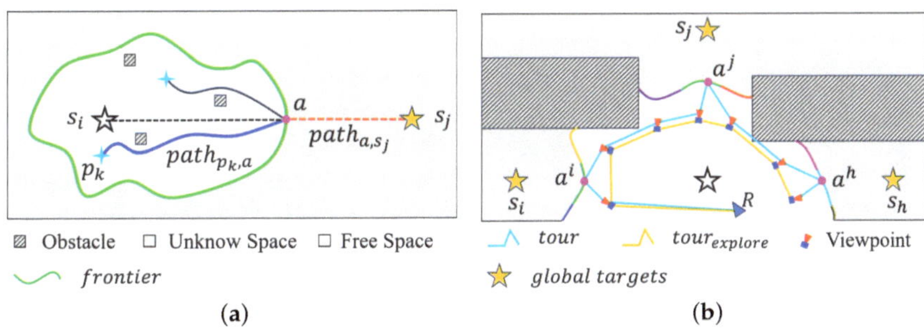

Figure 2. The basic scheme for autonomous exploration based on a prior topological map. (**a**) Extracting an agent for global target s_j that is in unknown space. (**b**) Obtain the exploration path $tour_{explore}$ from $tour$ provided by the PCATSP solver.

Based on the above definition and by incorporating the solving property of TSP, if all elements in a certain row or column of cost matrix D of TSP are subtracted by the same value, a new cost matrix D^* for the TSP will be obtained. However, D and D^* correspond to the same TSP solution [26]. Thus, we can subtract P_{a, s_j} from all P_{p, s_j}, and leave the TSP solution unchanged. It is equivalent to setting an agent point for the global target on the frontier. For constructing PCATSP, the intersection point a^j could be extracted on a frontier cluster as an agent for global target s_j and the priority can be set to a^j based on global exploration strategy O.

Hence, based on the above theory, we define agent a^j as the intersection point of a frontier cluster f^k and an undirected edge $e_k = (s_i, s_j)$, which is associated with a global target s_j in unknown space and possesses access priority. Then, the path between any point inside free space and an agent on frontier could be found by a path-searching algorithm, and then path length could also be accurately calculated. The method of agent extraction and priority assignment will be elaborated in Section 3.2.

Finally, we can naturally combine the prior topological map with real-time updated scene information in path planning based on PCATSP. We can then solve the PCATSP with the extracted agent set $A = \{a^i, a^j, \ldots, a^l\}$ and frontier cluster set F, where agents with priority will influence the visited order of nearby frontier clusters.

In this section, we suppose that the movement cost $c(p_i, p_j)$ between any two nodes p_i and p_j is calculated as follows:

$$c(p_i, p_j) = \frac{L\left(P_{p_i, p_j}\right)}{v_{\max}},\qquad(2)$$

where P_{p_i, p_j} is the shortest search path between nodes p_i and p_j; $L()$ denotes the length of search path; and v_{max} is the maximum velocity of the robot.

PCATSP solver provides a path $tour$ composed of input nodes. By removing A from $tour$, we obtain a global exploration path $tour_{explore} = \{vp_1^i, vp_1^j, \ldots, vp_1^l\}$ that satisfies global exploration strategy O, as shown in Figure 2b [27]. However, $tour_{explore}$ cannot guarantee that the robot with the priority will explore areas outside prior information, i.e., Objective 2. To address this objective, frontier clusters that guide the robot towards global targets will be recognized, and movement cost between these frontier clusters and the robot will be increased. Details are discussed in Section 3.3.

Figure 3 is a schematic diagram of the robot exploration process. The blue arrows represent the global exploration strategy, and green arrows are basic exert programs for global exploration paths. Based on our method, the robot explores the given region according to the global exploration strategy but prioritizes exploring scenarios outside prior information. Finally, following the global exploration strategy, the robot actively loops close and completes the exploration.

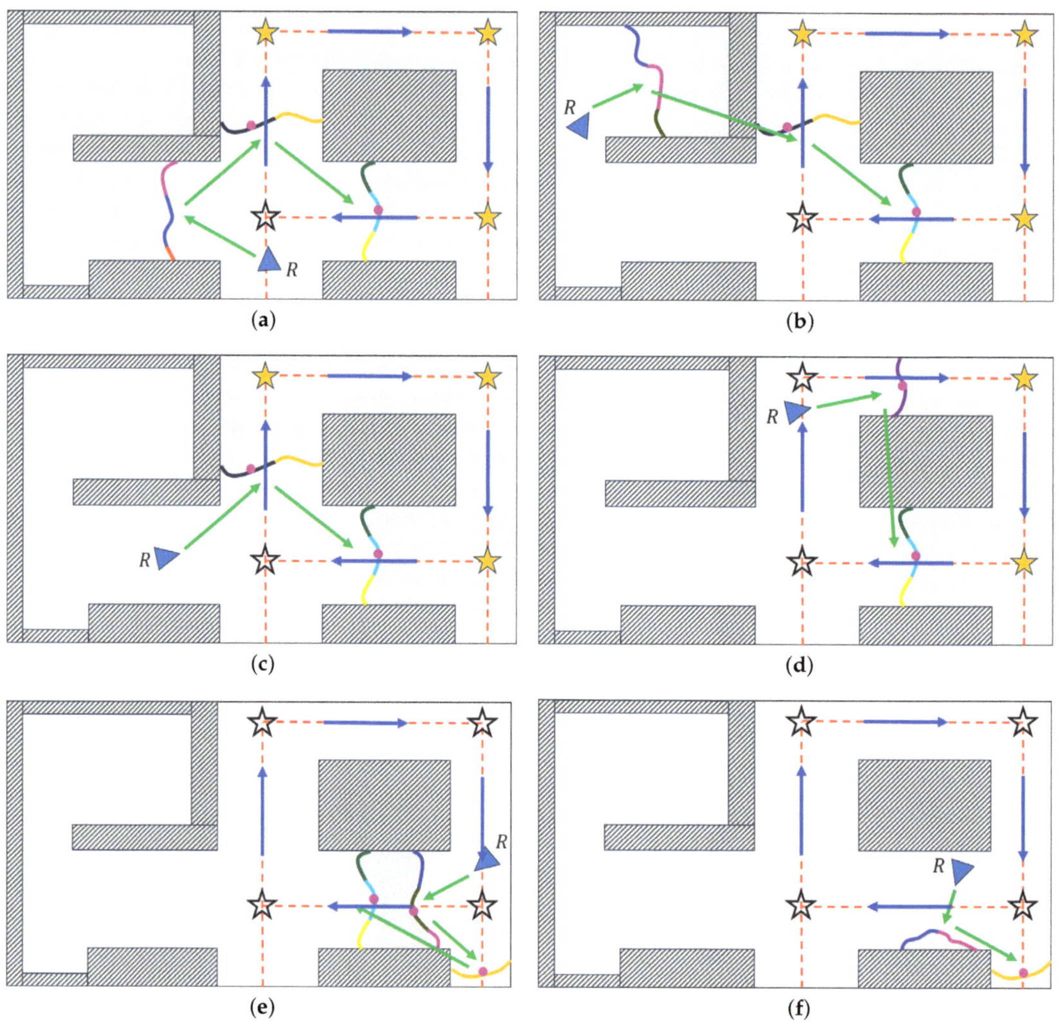

Figure 3. The schematic diagram of the exploration process. (**a**,**b**) Robot prioritizes exploring scenarios outside prior information. (**c**,**d**) After exploring scenarios outside prior information, the robot explores the unknown environment according to a global exploration strategy. (**e**,**f**) Robot actively conducts loop exploration according to the global exploration strategy and finally covers the frontier clusters near the lowest-priority agent.

3.2. Agent Extraction and Priority Assignment

When the position of a global target s_i is explored by the robot, other global targets s_j connected to s_i will be searched on a topological map. Then, the directed line segment $\overline{e_k} = (s_i, s_j)$ is constructed based on $e_k = (s_i, s_j)$, indicating from s_i to s_j. During implementation, we set up some satellite points for each global target to prevent the robot from missing them. The satellite points are evenly distributed at a set distance around the global target point.

The Oriented Bounding Box (OBB) of each new frontier cluster is extracted by principal component analysis (PCA), which is used to extract agents. As shown in Figure 4a, if f^k is crossed by $\overline{e_k}$, OBB of f^k must be crossed, and the following condition is met:

$$\left((s_j - s_i) \times (obb_p - s_i)\right)_z \left((s_j - s_i) \times (obb_q - s_i)\right)_z < 0,$$
$$p, q \in \{0, 1, 2, 3\}, \; p \neq q, \tag{3}$$

where *obb* are vertices of OBB. For the new frontier clusters that meet the condition, proceed with the following secondary evaluation to identify which global target the forthcoming agent will belong to:

$$\pi > \cos^{-1}\left((R - f_{ave}^k) \cdot \left(s_j - f_{ave}^k\right)\right) \geq \chi,$$
$$0 < \cos^{-1}\left((R - f_{ave}^k) \cdot \left(s_i - f_{ave}^k\right)\right) \leq \psi, \tag{4}$$

where R is the position of the robot, and f_{ave}^k is the average position of f^k, as shown in Figure 4b.

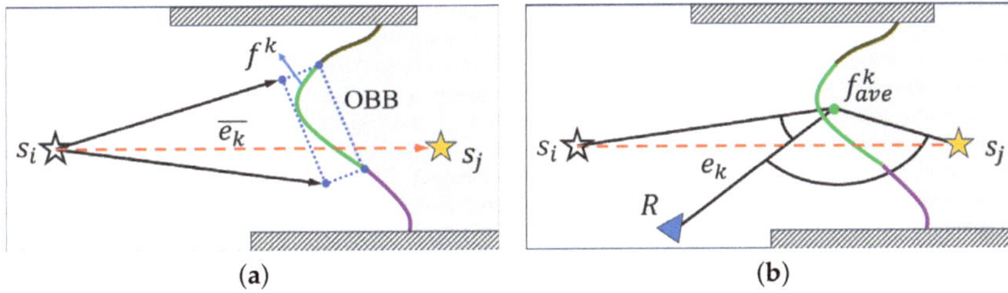

Figure 4. Method of agent extraction. (**a**) Crossing determination. (**b**) Angle determination.

If the above conditions are met with the newly extracted frontier cluster f^k, the cell closest to f_{ave}^k in f^k will be defined as an agent a^j for global target s_j. An agent of the global target is independent of the frontier cluster but is associated with it; if a frontier cluster is deleted or updated, the corresponding agent will be deleted.

The movement cost between an agent a^j and other nodes of PCATSP are calculated as follows:

$$c\left(R, a^j\right) = C,$$
$$c\left(a^j, R\right) = 0,$$
$$c\left(a^j, x\right) = c\left(x, a^j\right) = \frac{L\left(P_{a^j, x}\right)}{v_{max}}, \; x = vp_1^k, a^i, \tag{5}$$

where C is a large value, ensuring that the robot prioritizes exploring scenes outside prior information.

The priority pry_{a^j} of agent a^j is assigned as follows:

$$pry_{a^j} = \begin{cases} u, & \overline{e_k} = \overline{o_u} \\ size(O) + 1, & others \end{cases}, \tag{6}$$

which is determined by the sequential position of its $\overline{e_k}$ in O. $\overline{o_u}$ is a directed line segment in O, where u represents its ordinal position in O. The higher the sequential position of $\overline{e_k}$, the lower the priority of the agent extracted from $\overline{e_k}$. If $\overline{e_k}$ is not found in O, the priority of the agent equals $size(O) + 1$, and these agents with the lowest priority will not be actively explored, as shown in Figure 3. Additionally, during exploration, not only will one agent be generated by a directed line segment, but all agents will point towards the same global target. In such cases, we define the agent that is further away from the global target to have a higher priority, ensuring the robot does not miss the scene during exploration.

3.3. Update of Movement Cost of Frontier Clusters

As mentioned earlier, to ensure that the robot prioritizes exploring scenes outside prior information, the system needs to recognize frontier clusters that would guide the robot towards a global target and set a higher cost between them and the robot. To achieve this, we classify frontier clusters into three categories:

The first class includes frontier clusters with agents and other frontier clusters that are adjacent to these frontier clusters. They will guide the robot towards global targets.

The second class includes frontier clusters adjacent to the first class frontier clusters. These frontier clusters may guide the robot towards global targets. We utilize density-based spatial clustering of applications with noise (DBSCAN) algorithm to recognize these frontier clusters and rely on the following methods to identify whether they could guide the robot to global targets [28]:

$$e^{-r_{s_j,f^k}/r_{s_j,s_i}} e^{\cos \alpha^k - 1} \cos \beta^k \geq \varepsilon,$$
$$r_{s_j,f^k} < r_{s_j,R} \leq r_{s_j,s_i} + \gamma, \tag{7}$$

as shown in Figure 5a, α^k is the angle between vector $\overrightarrow{f_{ave}^k s_j}$ and $\overrightarrow{s_i s_j}$; β^k is the angle between vector $\overrightarrow{f_{ave}^k s_j}$ and $\overrightarrow{R f_{ave}^k}$; and r is the Euclidean distance between two points.

However, as the robot explores, the relative position between frontier clusters and robot changes, and frontier clusters cannot always satisfy Equation (7), as shown in Figure 5b,c. Thus, to keep the consistency of determination for these frontier clusters during exploration, we employ the Dynamic Time Warping (DTW) algorithm to evaluate similarities between path P_{R,a^j} and all paths of $P_{R,F} = \{P_{R,f^1}, P_{R,f^2}, \ldots, P_{R,f^n}\}$, as shown in Figure 5d [29]. If the similarity ranking of f^k satisfies the following condition, f^k is still believed to guide the robot towards global target s_j:

$$rank\left(dtw\left(P_{R,a^j}, P_{R,f^k}\right)\right) \leq \varphi. \tag{8}$$

The third class consists of the remaining frontier clusters. They will guide the robot to explore scenes outside prior information.

For the frontier clusters f^{k*} that guide the robot towards global targets, the movement cost from the robot to them is set as follows:

$$c\left(R, vp_1^{k*}\right) = C. \tag{9}$$

The movement cost from robot to other frontier clusters f^k is computed as follows:

$$c\left(R, vp_1^k\right) = \max\left\{\frac{L\left(P_{R,vp_1^k}\right)}{v_{max}}, \frac{\min(|\xi_1^k - \xi_R|, 2\pi - |\xi_1^k - \xi_R|)}{\omega_{max}}\right\} + \lambda \cos^{-1}\frac{(vp_1^k - R) \cdot v_R}{\|vp_1^k - R\| \|v_R\|}, \tag{10}$$

which considers the path length, yaw change, and motion consistency, where p_1^k and ξ_1^k are coordinates, and the yaw angle of viewpoint vp_1^k, ω_{max} is the maximum angular change rate; ξ_R is the yaw angle of the robot; and v_R is the current velocity.

As $c(vp_1^{k*}, R)$ and $c(vp_1^k, R)$ do not impact the solution results of open-loop path planning, we set them to 0. The movement cost between all frontier clusters are calculated as follows:

$$c\left(vp_1^i, vp_1^j\right) = c\left(vp_1^j, vp_1^i\right) = \max\left\{\frac{L\left(P_{vp_1^i, vp_1^j}\right)}{v_{\max}}, \frac{\min\left(\left|\xi_1^i - \xi_1^j\right|, 2\pi - \left|\xi_1^i - \xi_1^j\right|\right)}{\omega_{\max}}\right\}. \quad (11)$$

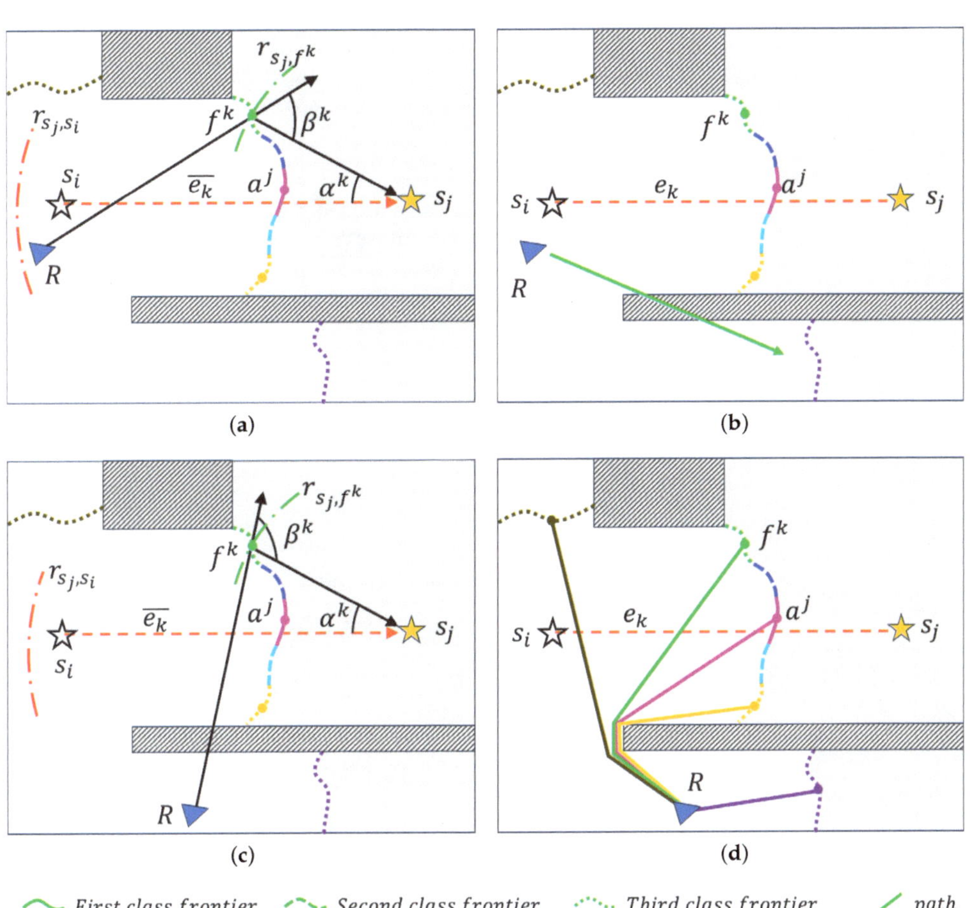

Figure 5. Determination of the second category of frontier clusters. (**a**) Determine whether the second-category frontier f^k guides the robot towards the global target based on angle and distance conditions. (**b**) Robot moves towards other scenes following the generated exploration path. (**c**) As the robot moves, f^k no longer meets the initial judgment criteria. (**d**) To maintain the consistency of the previous judgment to f^k, perform a secondary judgment of f^k by DTW.

3.4. Local Path Planning

The global path planning module aims to assist the robot in making decisions at a global standpoint for efficient exploration. The local path planning module aims to find the best viewpoints to make the robot to follow. Many previous works refine a path by evaluating the cost and reward for efficient exploration, but they consume significant computational resources in information evaluation [1,30]. Thus, we define the potential reward of a candidate viewpoint as a volume of unknown space within its a Field of View

(FoV) and propose a simple and fast reward evaluation method based on incremental frontier information structure (FIS). Finally, the local path is refined by synthesizing the reward and movement cost of each viewpoint candidate.

Cells of a frontier cluster are stored in FIS [12]. We use them to evaluate the volume of unknown space in FOV. As shown in Figure 6, each truncated pyramid is constructed based on cells that the candidate viewpoint could cover, and its volume is calculated as follows:

$$V = \left(h_{rw} \left(\frac{h_{rw} e_{cell}}{h_{cell}} \right)^2 - h_{cell}(e_{cell})^2 \right) / 3, \tag{12}$$

where e_{cell} is width of cell, h_{cell} is distance between cell and candidate viewpoint. h_{rw} is effective distance to calculate reward and computed by:

$$h_{rw} = \min(h_{\max}, h_{cell} + \delta), \tag{13}$$

h_{max} is maximum range of FOV, and δ is used to control the depth of truncated pyramid to balance movement cost and expected rewards.

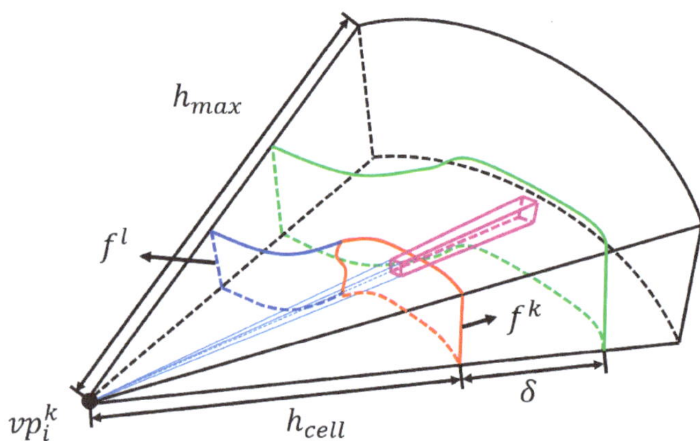

Figure 6. The schematic diagram of calculating the reward of a viewpoint.

The expected reward rw_i^k of a candidate viewpoint vp_i^k is evaluated by accumulating V:

$$rw_i^k = \sum_{x=1}^{n} V_x^k + \eta \sum_{y=1}^{m} V_y^l, \tag{14}$$

where V_x^k is taken from current frontier cluster f^k, and V_y^l is taken from next adjacent frontier cluster f^l to be visited. m and n, respectively, represent the number of cells that vp_i^k could cover. η is a weight coefficient.

We formulate local path planning as a graph search problem and refine an optimal path from the global exploration path by balancing expected reward and movement cost, where viewpoints of each frontier cluster serve as candidate points. Suppose that the optimal exploration path $path = \{vp_i^1, vp_k^2, \ldots, vp_j^{N_{rf}}, vp_1^{N_{rf}+1}\}$ provided by the Dijkstra algorithm will minimize the cost/reward ratio:

$$c\left(P_{R,vp_1^{N_{rf}+1}} \right) = \frac{c(R,vp_i^1)}{W_i^1} + \frac{c\left(vp_j^{N_{rf}}, vp_1^{N_{rf}+1} \right)}{W_1^{N_{rf}+1}} + \sum_{n=1}^{N_{rf}-1} \frac{c\left(vp_k^n, vp_k^{n+1} \right)}{W_k^{n+1}}, \tag{15}$$

where $N_{rf} + 1$ is the size of the frontier clusters to be optimized, and

$$W_k^n = \frac{rw_k^n}{V_{FOV}}, \tag{16}$$

V_{FOV} is the volume of FOV.

Finally, the exploration path *path* is output to the trajectory generation module.

4. Experiments

4.1. Implementation Details

We set $\psi = \pi/3$, $\chi = 2\pi/3$, and $C = 500$ in Section 3.2, $\varepsilon = 0.4$, $\gamma = 3$ m, and $\varphi = 6$ in Section 3.3, and $\delta = 1.5$ m and $\eta = 1.25$ in Section 3.4. Additionally, we employ the SOP solver from LKH-3.0 to solve PCATSP and implement the Chinese Postman Problem (CPP) solver ourselves for the global exploration strategy [27,31]. All simulation experiments are conducted in ROS-Kinetic Gazebo platform on Ubuntu 18.04 computer system, running on an CPU. For real-world experiments, the unmanned ground vehicle shown in Figure 7 was utilized to explore a given region. We equipped it with a depth camera, an inertial measurement unit, and an onboard computer with Ubuntu 18.04 computer system.

Figure 7. Real-world experiment vehicle platform.

4.2. Benchmark Comparisons

In this section, we conduct benchmark comparisons using simulation experiments to verify the effectiveness and exploration efficiency of the proposed method. Robot exploration in maze scenes is the most effective method to verify the efficiency of autonomous exploration [13]. Thus, we manually constructed two large-scale mazes, Maze-1 ($48 \times 63 \times 2$ m^3) and Maze-2 ($66 \times 62 \times 2$ m^3), in Gazebo simulation platform. The cross-sectional length of the road in Maze-1 is 4~6 m, and in Maze-2, 8~10 m. The topological maps of these mazes are generated by manually placing global targets on corners and intersections, and then connecting them according to the topology of mazes to simulate maze information [18]. But, we leave some space for the robot to explore autonomously without prior information. The mazes and their prior topological maps are shown in Figure 8.

We employ the optimal exploration strategy provided by the CPP solver to guide robot exploration and compare it with FUEL and FAEP [12,13]. They are state-of-the-art frontier-based methods which have been proposed in recent years, and which exhibit high exploration efficiency and have open-sourced their code to serve community. In all experiments, we utilize UAV as the exploration robot, with $v_{max} = 0.6$ m/s, $\omega_{max} = 0.9$ rad/s, and the maximum acceleration is 0.6 m/s^2. UAV equips depth camera to collect environmental information. FOV of depth camera is configured as $[80 \times 60]$ deg, h_{max} is 4.5 m. The grid map of the local update range is $5 \times 5 \times 2$ m^3. In all scenarios, three methods are run more than six times with the same configuration.

We evaluate the performance of the above methods based on exploration time, flight distance, and coverage efficiency. FAEP also utilizes FIS to update frontier cluster information and based TSP to plan global exploration paths. Therefore, we set the frontier cluster length limit of all methods to 2 m and counted the remaining number of frontier clusters during exploration. The number could directly reflect the computational burden of the path planner: the more frontier clusters remain in scene, the more enormous the solver computation.

Figure 8 displays exploration trajectories of different methods in two mazes. Our method prevents the robot from unnecessarily revisiting explored areas but ensures that the robot does not miss scenes outside prior information, even in scenes with large cross-sectional road lengths like Maze-2. In contrast, other methods re-explore already visited areas during exploration, leading to lower efficiency and inevitable resource wastage.

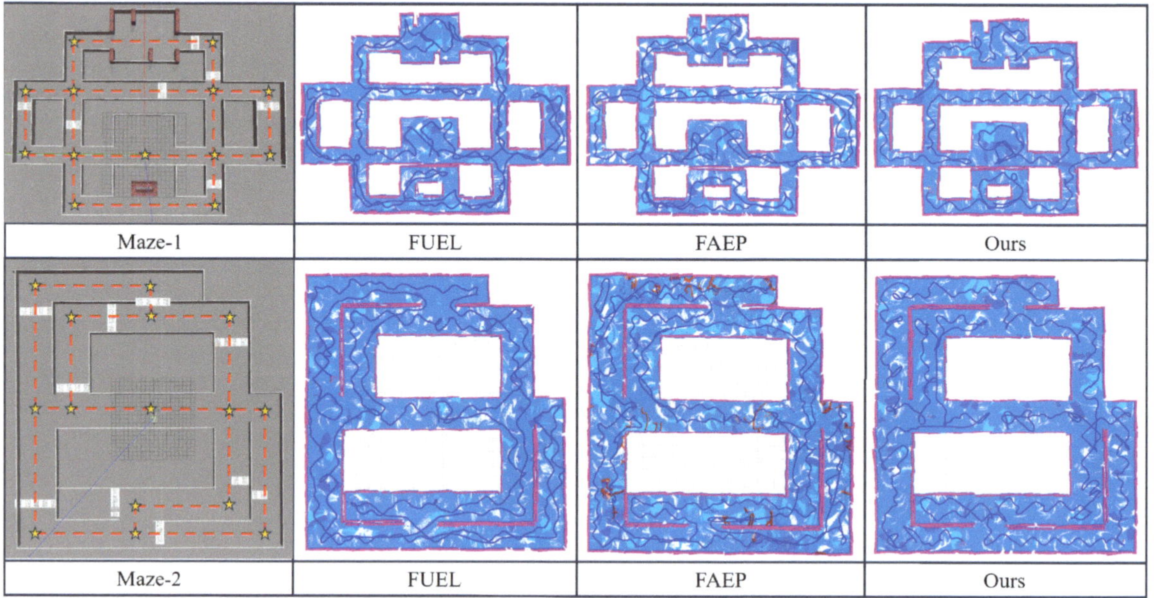

Figure 8. Benchmark comparison of exploration trajectories of the proposed method, FUEL, and FAEP, in two mazes.

The detailed exploration process of our method in Maze-1 is illustrated in Figure 9. The robot prioritizes exploring areas outside prior information in the given region. Subsequently, it follows the global exploration strategy to explore other areas, disregarding the direction of other frontier clusters during exploration. Finally, the robot actively loops close to cover the frontier clusters near the agent with the lowest priority. The exploration process aligns with the concept depicted in Figure 3.

Figure 9. Partial exploration trajectories generated by our method in Maze-1.

Figure 10 displays the process of volume coverage for all methods in two mazes. Our method demonstrates higher efficiency and nearly linear performance in conducting exploration. In contrast, other methods exhibit coverage stagnation or slower growth during exploration, meaning the robot moved towards a previously missed or visited space.

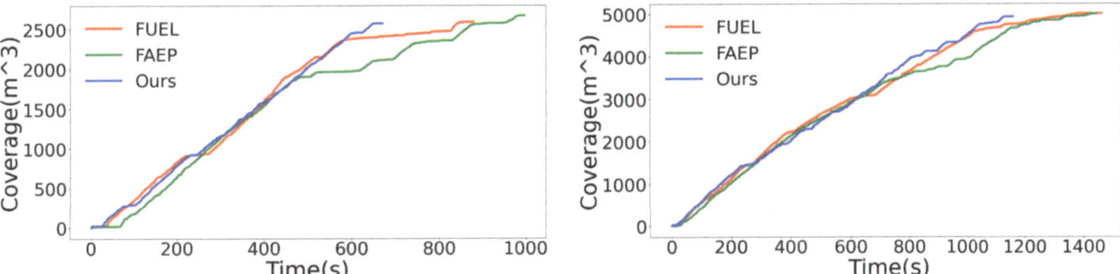

Figure 10. The exploration progress of three methods in Maze-1 (**left**) and Maze-2 (**right**).

The number of the remaining frontier clusters during exploration for all methods is shown in Figure 11. Compared with other methods, our method ensures a lower count of remaining frontier clusters. It proves that our method can avoid much computational burden for an onboard computer and is suitable for exploring large-scale scenes.

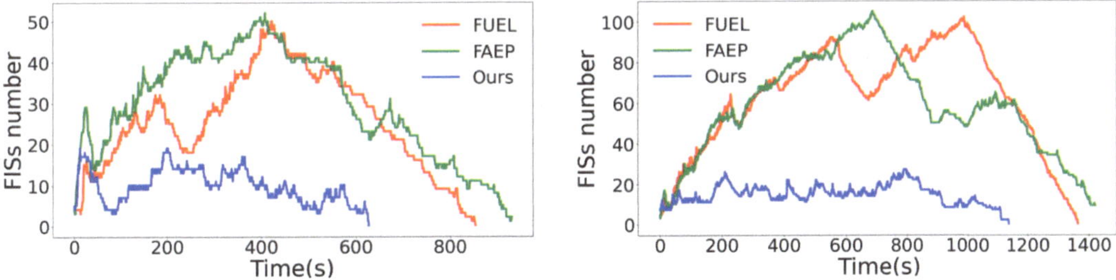

Figure 11. The number of remaining frontier clusters during exploration of the three methods in two mazes.

Table 1 presents a quantitative performance of three methods in two mazes. Compared with FUEL and FAEP, the average exploration time of our method is reduced by 18.76% and 20.87% in Maze-1, and 18.05% and 21.77% in Maze-2. Moreover, the average flight distance of our method is reduced by 18.93% and 17.19% in Maze-1, and 15.91% and 6.79% in Maze-2. In contrast, our proposed method outperforms other methods in exploration time, flight distance, and stability.

The average computation time of each module is shown in Table 2. The proposed method conducts a one-path planning with approximately 130 ms, meeting the frequency requirements for most robots. Furthermore, only 12 ms is consumed by local path planning, additionally proving the efficiency of the proposed reward evaluation method.

We deconstruct the proposed method to analyze the global and local path planning module performances. Ours^{Local} and Ours^{Global} correspond to implementing local and global path planning, both built upon FUEL framework. The exploration data of Ours^{Local}, Ours^{Global} in two mazes are list in Table 1.

Table 1. Exploration statistics in Maze-1 and Maze-2.

Scene	Method	Exploration Time (s)				Flight Distance (m)			
		Avg	Std	Max	Min	Avg	Std	Max	Min
Maze-1	FUEL	826.599	25.275	849.384	777.630	618.773	16.821	635.879	589.081
	FAEP	848.614	32.779	906.516	813.997	605.419	28.813	655.931	561.852
	Ours	671.101	13.880	694.328	663.532	501.523	12.490	520.579	486.109
	Ours^{Local}	826.881	10.485	855.695	814.620	616.217	8.390	635.330	605.435
	Ours^{Global}	670.742	24.009	693.757	626.820	497.563	16.467	517.316	469.384
Maze-2	FUEL	1407.048	46.729	1503.542	1356.909	1012.588	18.818	1036.984	984.151
	FAEP	1474.686	67.125	1569.142	1418.180	913.552	33.502	960.631	885.397
	Ours	1153.677	29.733	1200.470	1104.911	851.646	22.824	885.435	819.141
	Ours^{Local}	1396.996	27.841	1439.908	1339.974	1002.997	18.952	1037.262	980.147
	Ours^{Global}	1210.028	52.735	1277.481	1138.502	886.096	41.552	940.927	831.196

Table 2. Average computation time of each module.

Scene	Average Computation Time (ms)		
	Global Planning	Local Planning	Total Planning
Maze-1	141.79	17.39	159.19
Maze-2	127.45	12.35	130.85

Comparing the simulation results of Ours^{Global} and FUEL in two mazes, the standard deviations of Ours^{Global} in exploration time and flight distance are identical to FUEL. However, the average exploration time of Ours^{Global} is reduced by 18.88% and 14.01%, respectively, in Maze-1 and Maze-2. The average flight distance of Ours^{Global} is reduced by 19.57% and 12.45%, respectively, in Maze-1 and Maze-2. The average performance of Ours^{Global} outperforms FUEL, indicating that the primary contribution of global path planning lies in improving exploration efficiency. The same inference can also be drawn from comparing Ours and Ours^{Local}.

The average exploration time and flight distance of Ours^{Local} and FUEL in both mazes are approximate. But, Ours^{Local} maintains a low standard deviation, indicating more stability. Similar performance is observed between Ours^{Global} and Ours. It is demonstrated that the proposed local path planning method can effectively enhance the stability of exploration efficiency.

Finally, we can conclude that the exploration efficiency is primarily attributed to global path planning, while local path planning enhances the stability of exploration efficiency.

4.3. Real-World Experiment

In order to further validate the effectiveness of our proposed method, we conduct a real-world experiment with a ground vehicle. Based on the prior topological map, the ground vehicle will explore an indoor corridor of size $55 \times 15 \times 2$ m^3, and a cross-sectional length of road that is 2 m, as shown in Figure 12. The prior information we provided is a rectangular topological map that outlines the basic structure of the indoor corridor. In the indoor corridor, the open spaces and nooks served as regions beyond prior information, testing the effectiveness of our method.

In the experiment, we run VINS-Fusion on GPU to provide the positional state, while the proposed method runs on CPU to plan an exploration path [32]. We set $v_{max} = 0.5$ m/s, $\omega_{max} = 0.8$ rad/s, the maximum acceleration as 0.5 m/s^2, and the grid map of local update range as $4 \times 4 \times 2$ m^3. The FOV of the depth camera is set as $[80 \times 60]$ deg, and the maximum range h_{max} is 3.5 m.

Figure 13 shows the exploration trajectory of our method with the indoor environment, where the small labeled images represent the key nodes during exploration: (a) presents the robot exploring in the direction following a global exploration strategy; (b) and (c) present the instances where the vehicle prioritizes exploration directions beyond the prior information. The exploration time of the whole process is 738 s, and the movement distance is 167 m. It can be seen from the trajectory that the vehicle did not revisit the explored areas during exploration, which proves the effectiveness of our proposed method.

Figure 14, respectively, shows the process of volume coverage and the number of remaining frontier clusters during exploration. It reveals a stable and linear exploration process while the number of frontier clusters is maintained at a small level, demonstrating the efficiency of our proposed method.

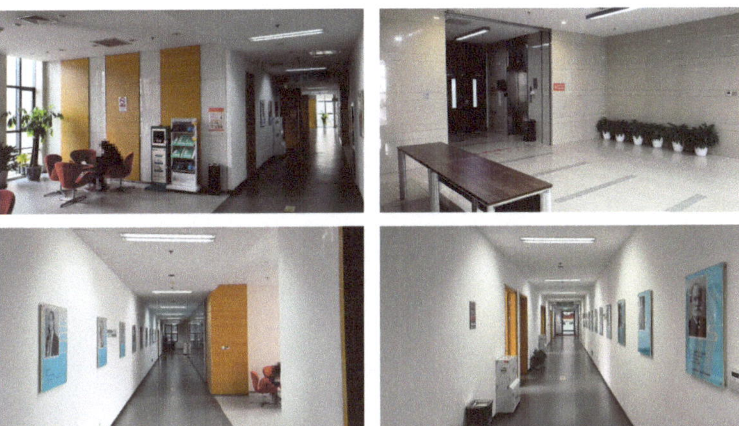

Figure 12. The experimental scene for real-world experiment.

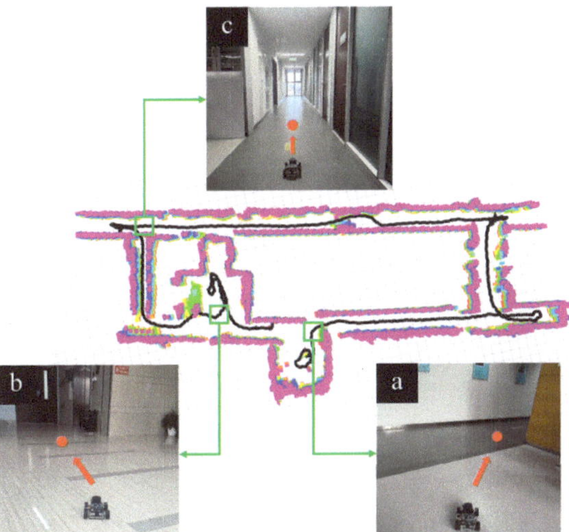

Figure 13. The exploration trajectory of the proposed method in real-world experiment. (**a**) shows the robot exploring in the direction following a global exploration strategy; (**b**,**c**) show the instances where the vehicle prioritizes exploration directions beyond the prior information.

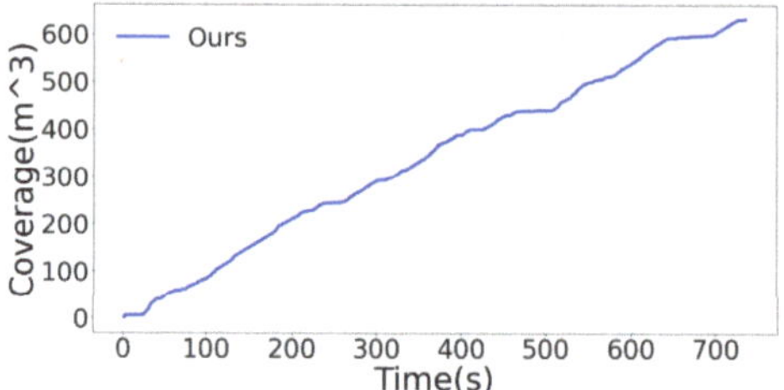
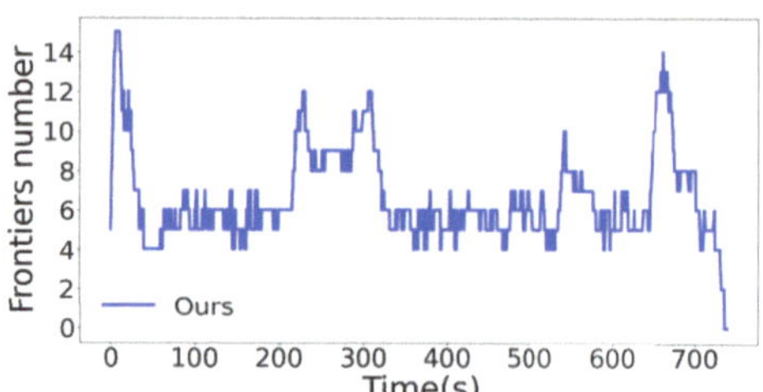

Figure 14. The process of volume coverage (**left**) and the number of remaining frontier clusters (**right**) during exploration in real-world scenario.

5. Conclusions

This paper introduces a novel autonomous exploration method based on a prior topological map, in which a robot explores the given large-scale region, following a global exploration strategy, but prioritizes exploring scenes outside prior information. The proposed method employs a hierarchical framework to plan exploration paths. Based on PCATSP, the global path planning module merges the prior topological map with real-time scene information, planning efficient global exploration paths. Subsequently, the local path planning module rapidly evaluates the rewards and movement costs for each candidate viewpoint to refine the input global exploration paths. Finally, the output exploration path is used to generate local trajectories. Simulation results prove that the proposed method enables the robot to efficiently and rapidly explore a given region and is suitable for operation in large-scale scenes. The ablation study also demonstrates that our proposed local path planning method could enhance the stability of exploration efficiency. The experiment conducted in the real world further validates the effectiveness of our method.

Author Contributions: Conceptualization, Z.C.; methodology, Z.C.; software, Z.C.; validation, Z.C.; formal analysis, Z.C.; investigation, Z.D.; resources, Z.D.; data curation, Z.C.; writing—original draft preparation, Z.C.; writing—review and editing, Z.C.; visualization, Z.C.; supervision, J.Y.; project administration, Z.C. All authors have read and agreed to the published version of the manuscript.

Funding: This research received no external funding.

Data Availability Statement: Data are contained within the article.

Conflicts of Interest: The authors declare no conflicts of interest.

References

1. Bircher, A.; Kamel, M.; Alexis, K.; Oleynikova, H.; Siegwart, R. Receding horizon "next-best-view" planner for 3d exploration. In Proceedings of the 2016 IEEE International Conference on Robotics and Automation (ICRA), Stockholm, Sweden, 16–21 May 2016; pp. 1462–1468.
2. Duberg, D.; Jensfelt, P. Ufoexplorer: Fast and scalable sampling-based exploration with a graph-based planning structure. *IEEE Robot. Autom. Lett.* **2022**, *7*, 2487–2494. [CrossRef]
3. Selin, M.; Tiger, M.; Duberg, D.; Heintz, F.; Jensfelt, P. Efficient autonomous exploration planning of large-scale 3-d environments. *IEEE Robot. Autom. Lett.* **2019**, *4*, 1699–1706. [CrossRef]
4. Xu, Z.; Deng, D.; Shimada, K. Autonomous UAV exploration of dynamic environments via incremental sampling and probabilistic roadmap. *IEEE Robot. Autom. Lett.* **2021**, *6*, 2729–2736. [CrossRef]
5. Respall, V.M.; Devitt, D.; Fedorenko, R.; Klimchik, A. Fast sampling-based next-best-view exploration algorithm for a MAV. In Proceedings of the 2021 IEEE International Conference on Robotics and Automation (ICRA), Xi'an, China, 30 May–5 June 2021; pp. 89–95.
6. Zhu, H.; Cao, C.; Xia, Y.; Scherer, S.; Zhang, J.; Wang, W. DSVP: Dual-stage viewpoint planner for rapid exploration by dynamic expansion. In Proceedings of the 2021 IEEE/RSJ International Conference on Intelligent Robots and Systems (IROS), Prague, Czech Republic, 27 September– 1 October 2021; pp. 7623–7630.
7. Yamauchi, B. A frontier-based approach for autonomous exploration. In Proceedings of the 1997 IEEE International Symposium on Computational Intelligence in Robotics and Automation CIRA'97. 'Towards New Computational Principles for Robotics and Automation', Monterey, CA, USA, 10–11 July 1997; pp. 146–151.

8. Dornhege, C.; Kleiner, A. A frontier-void-based approach for autonomous exploration in 3d. *Adv. Robot.* **2013**, *27*, 459–468. [CrossRef]
9. Zhong, P.; Chen, B.; Lu, S.; Meng, X.; Liang, Y. Information-driven fast marching autonomous exploration with aerial robots. *IEEE Robot. Autom. Lett.* **2021**, *7*, 810–817. [CrossRef]
10. Deng, D.; Duan, R.; Liu, J.; Sheng, K.; Shimada, K. Robotic exploration of unknown 2d environment using a frontier-based automatic-differentiable information gain measure. In Proceedings of the 2020 IEEE/ASME International Conference on Advanced Intelligent Mechatronics (AIM), Virtual, 6–10 July 2020; pp. 1497–1503.
11. Cieslewski, T.; Kaufmann, E.; Scaramuzza, D. Rapid exploration with multi-rotors: A frontier selection method for high speed flight. In Proceedings of the RSJ International Conference on Intelligent Robots and Systems (IROS), Vancouver, BC, Canada, 24–28 September 2017; pp. 2135–2142.
12. Zhou, B.; Zhang, Y.; Chen, X.; Shen, S. Fuel: Fast uav exploration using incremental frontier structure and hierarchical planning. *IEEE Robot. Autom. Lett.* **2021**, *6*, 779–786. [CrossRef]
13. Zhao, Y.; Yan, L.; Xie, H.; Dai, J.; Wei, P. Autonomous Exploration Method for Fast Unknown Environment Mapping by Using UAV Equipped with Limited FOV Sensor. *IEEE Trans. Ind. Electron.* **2023**, *5*, 4933–4943. [CrossRef]
14. Soragna, A.; Baldini, M.; Joho, D.; Kümmerle, R.; Grisetti, G. Active SLAM using connectivity graphs as priors. In Proceedings of the 2019 IEEE/RSJ International Conference on Intelligent Robots and Systems (IROS), Macau, China, 3–8 November 2019; pp. 340–346.
15. Zhang, Y.; McCalmon, J.; Peake, A.; Alqahtani, S.; Pauca, P. A Symbolic-AI Approach for UAV Exploration Tasks. In Proceedings of the 2021 7th International Conference on Automation, Robotics and Applications (ICARA), Prague, Czech Republic, 4–6 February 2021; pp. 101–105.
16. Krzysiak, R.; Butail, S. Information-based control of robots in search-and-rescue missions with human prior knowledge. *IEEE Trans.-Hum.-Mach. Syst.* **2021**, *52*, 52–63. [CrossRef]
17. Boniardi, F.; Valada, A.; Burgard, W.; Tipaldi, G.D. Autonomous indoor robot navigation using a sketch interface for drawing maps and routes. In Proceedings of the 2016 IEEE International Conference on Robotics and Automation (ICRA), Stockholm, Sweden, 16–21 May 2016; pp. 2896–2901.
18. Oßwald, S.; Bennewitz, M.; Burgard, W.; Stachniss, C. Speeding-up robot exploration by exploiting background information. *IEEE Robot. Autom. Lett.* **2016**, *1*, 716–723.
19. Shang, Z.; Shen, Z. Topology-based UAV path planning for multi-view stereo 3D reconstruction of complex structures. *Complex Intell. Syst.* **2023**, *9*, 909–926. [CrossRef]
20. Blochliger, F.; Fehr, M.; Dymczyk, M.; Schneider, T.; Siegwart, R. Topomap: Topological mapping and navigation based on visual slam maps. In Proceedings of the 2018 IEEE International Conference on Robotics and Automation (ICRA), Brisbane, Australia, 21–25 May 2018; pp. 3818–3825.
21. Chen, X.; Zhou, B.; Lin, J.; Zhang, Y.; Zhang, F.; Shen, S. Fast 3D Sparse Topological Skeleton Graph Generation for Mobile Robot Global Planning. In Proceedings of the 2022 IEEE/RSJ International Conference on Intelligent Robots and Systems (IROS), Kyoto, Japan, 23–27 October 2022; pp. 10283–10289.
22. Mei, J.; Li, R.J.; Gao, W.; Cheng, M.M. CoANet: Connectivity attention network for road extraction from satellite imagery. *IEEE Trans. Image Process.* **2021**, *30*, 8540–8552. [CrossRef]
23. Tan, Y.Q.; Gao, S.H.; Li, X.Y.; Cheng, M.M.; Ren, B. Vecroad: Point-based iterative graph exploration for road graphs extraction. In Proceedings of the IEEE/CVF Conference on Computer Vision and Pattern Recognition, Seattle, WA, USA, 13–19 June 2020; pp. 8910–8918.
24. Sokmen, O.C.; Emec, S.; Yilmaz, M.; Akkaya, G. An overview of Chinese postman problem. In Proceedings of the 3rd International Conference on Advanced Engineering Technologies, Bayburt, Turkey, 19–21 September 2019; Volume 10, pp. 1175–1184.
25. Gouveia, L.; Pesneau, P. On extended formulations for the precedence constrained asymmetric traveling salesman problem. *Netw. Int. J.* **2006**, *48*, 77–89. [CrossRef]
26. Morrison, D.R.; Jacobson, S.H.; Sauppe, J.J.; Sewell, E.C. Branch-and-bound algorithms: A survey of recent advances in searching, branching, and pruning. *Discret. Optim.* **2016**, *19*, 79–102. [CrossRef]
27. Helsgaun, K. An extension of the Lin-Kernighan-Helsgaun TSP solver for constrained traveling salesman and vehicle routing problems. *Roskilde Rosk. Univ.* **2017**, *12*, 966–980.
28. Ester, M.; Kriegel, H.P.; Sander, J.; Xu, X. A density-based algorithm for discovering clusters in large spatial databases with noise. In Proceedings of the Kdd, Portland, OR, USA, 2–4 August 1996; Volume 96, pp. 226–231.
29. Salvador, S.; Chan, P. Toward accurate dynamic time warping in linear time and space. *Intell. Data Anal.* **2007**, *11*, 561–580. [CrossRef]
30. Tao, Y.; Wu, Y.; Li, B.; Cladera, F.; Zhou, A.; Thakur, D.; Kumar, V. SEER: Safe efficient exploration for aerial robots using learning to predict information gain. In Proceedings of the 2023 IEEE International Conference on Robotics and Automation (ICRA), London, UK, 29 May–2 June 2023; pp. 1235–1241.

31. Helsgaun, K. An effective implementation of the Lin–Kernighan traveling salesman heuristic. *Eur. J. Oper. Res.* **2000**, *126*, 106–130. [CrossRef]
32. Qin, T.; Li, P.; Shen, S. Vins-mono: A robust and versatile monocular visual-inertial state estimator. *IEEE Trans. Robot.* **2018**, *34*, 1004–1020. [CrossRef]

Disclaimer/Publisher's Note: The statements, opinions and data contained in all publications are solely those of the individual author(s) and contributor(s) and not of MDPI and/or the editor(s). MDPI and/or the editor(s) disclaim responsibility for any injury to people or property resulting from any ideas, methods, instructions or products referred to in the content.

 *drones*

Article

Extended State Observer-Based Sliding-Mode Control for Aircraft in Tight Formation Considering Wake Vortices and Uncertainty

Ruiping Zheng [1], Qi Zhu [1], Shan Huang [1], Zhihui Du [2], Jingping Shi [1] and Yongxi Lyu [1,*]

[1] Department of Automatic Control, Northwestern Polytechnical University, Xi'an 710129, China; zhengruiping@mail.nwpu.edu.cn (R.Z.); shijingping@nwpu.edu.cn (J.S.)
[2] Department of Precision Instrument, Tsinghua University, Beijing 100084, China
* Correspondence: yongxilyu@nwpu.edu.cn

Abstract: The tight formation of unmanned aerial vehicles (UAVs) provides numerous advantages in practical applications, increasing not only their range but also their efficiency during missions. However, the wingman aerodynamics are affected by the tail vortices generated by the leading aircraft in a tight formation, resulting in unpredictable interference. In this study, a mathematical model of wake vortex was developed, and the aerodynamic characteristics of a tight formation were simulated using Xflow software. A robust control method for tight formations was constructed, in which the disturbance is first estimated with an extended state observer, and then a sliding mode controller (SMC) was designed, enabling the wingman to accurately track the position under conditions of wake vortex from the leading aircraft. The stability of the designed controller was confirmed. Finally, the controller was simulated and verified in mathematical simulation and semi-physical simulation platforms, and the experimental results showed that the controller has high tight formation accuracy and is robust.

Keywords: tight formation; unmanned aerial vehicles; xflow; extended state observer; sliding mode controller

Citation: Zheng, R.; Zhu, Q.; Huang, S.; Du, Z.; Shi, J.; Lyu, Y. Extended State Observer-Based Sliding-Mode Control for Aircraft in Tight Formation Considering Wake Vortices and Uncertainty. *Drones* **2024**, *8*, 165. https://doi.org/10.3390/drones8040165

Academic Editor: Andrey V. Savkin

Received: 29 February 2024
Revised: 6 April 2024
Accepted: 17 April 2024
Published: 21 April 2024

Copyright: © 2024 by the authors. Licensee MDPI, Basel, Switzerland. This article is an open access article distributed under the terms and conditions of the Creative Commons Attribution (CC BY) license (https://creativecommons.org/licenses/by/4.0/).

1. Introduction

In recent years, UAVs have been widely used in forest fire prevention, geological exploration, military applications, and other fields due to their low-cost, casualty-free, and flexible characteristics [1–4]. Multiple UAVs flying in formation are capable of dealing with more complex tasks than a single UAV and can increase the mission success rate [5]. When multiple UAVs fly in close formation, fuel consumption is reduced and therefore range is increased [6,7].

The notion of close-formation flight originated from migratory birds [8]. When migratory birds depart from their roosts, adopting a V formation or other forms of coordinated flight can substantially lengthen the flock's overall travel distance. Through extensive observation and study, researchers have determined that a flock consisting of 25 birds flying in formation cover 71% more distance than an individual bird flying solo [9]. The flight of the lead bird can generate upwash wake, and other birds flying in the correct position can minimize their energy consumption, thereby facilitating the conservation of physical strength and expanding the flock's activity range [10,11]. Tight-formation flight is defined as when the lateral distance between two aircraft is less than twice the span, and the aerodynamic coupling between the aircraft affects the wingmen's dynamics system. The aerodynamic forces and moments of the wingman are widely different from those of a single airplane, as demonstrated by Cho et al.'s experimentation with two small jet aircraft in a subsonic wind tunnel [12]. Researchers at NASA's Dryden Flight Research Center conducted tight-formation flight tests with two F/A-18s [13], with the wingman flying within the wingtip vortex of the leading aircraft; they found that the wingman's drag was reduced by more than 20 percent, and the maximum fuel reduction was more than 18 percent. Through aerodynamic calculations, Blake et al. found that the range of a formation of five aircraft could be increased by 60 percent, relative to that of one aircraft [14].

Researchers, including Thomas E. Kent, conducted a comprehensive case study on a representative sample of 210 transatlantic routes. The findings revealed that two-aircraft formations consumed less fuel by approximately 8.7% on average, whereas three-aircraft formations used even less fuel, with savings of 13.1%, compared with that of single-aircraft operations [15]. The C-17 [16] transport formation, studied by Pahle et al., involves two aircraft flying at a speed of 275 knots and an altitude of 25,000 ft. The wingmen were positioned 1000 and 3000 ft behind the lead aircraft. The replacement of the drag reduction with the consideration of fuel consumption and thrust in level flight achieved maximum average decreases in fuel consumption and thrust of approximately 6.8–7.8% and 9.2%, respectively, on both the left and right sides. The Air Force Research Laboratory and the U.S. Department of Defense's Advanced Research Projects Agency conducted a close-formation study based on the surfing aircraft vortex energy (SAVE) concept, resulting in fuel consumption savings exceeding 10 percent over a duration surpassing 90 min [17]. The National Aeronautics and Space Administration (NASA) Armstrong Flight Research Center (Edwards, CA, USA) completed a series of studies, in which a NASA Gulfstream C-20A airplane (Gulfstream Aerospace, Savannah, GA, USA) was flown as the trail airplane within the wake of a NASA Gulfstream III (G-III) airplane. The results showed fuel reductions in formation ranging from 3.5 to 8 percent, compared with that of a single aircraft [18]. The optimal positioning of a flying wing aircraft behind a refueling plane was extensively investigated by Okolo et al. [19], who revealed that any deviation from the static sweet spot, whether in the vertical or lateral direction, results in reductions in the lift-to-drag ratio benefit. The wingtip vortex field generated during leading-aircraft flight can strongly impact the wingman's aerodynamic performance. When the wingman is in the upwash area, the upwash velocity increases the wingman's angle of attack, which reduces drag and increases lift [20].

The desired formation flight involves placing the wingman in the optimal position when the wingman is under the maximum induced lift-to-drag ratio [21,22]. Keeping the UAV formation stable and using the aerodynamic benefits of the formation have become research challenges. Many researchers have studied the effects of wake vortices in tight formations and the control of UAVs [23]. Using suitable controllers on an airplane can reduce the operator's burden of operation. Zheng et al. [24] used a model predictive controller to control UAVs in tight formation. Zhang et al. studied a two-aircraft formation during level and straight flight and designed an adaptive controller, which was robust to external interference to some extent [25].

Pachter et al. [26] used a proportional–integral (PI) controller to allow the following UAV to maintain an optimal position during tight-formation flight; however, the robustness of the designed proportional–integral outer-loop controller was weak, due to the inaccuracy of the modeling of the wake vortex. Researchers [27,28] have used the extreme value search algorithm to study a linear formation controller to design an outer-loop navigation controller to guide the following UAV to follow at the optimal position during tight-formation flight. However, the use of the extreme value search algorithm is limited in practice: to ensure the convergence of the extreme value search algorithm, researchers added high-frequency oscillating signals to the extreme value control, which is unsuitable for controlling UAVs. These control algorithms share the assumption that the aerodynamic characteristics of tight formations are known or bounded, which are actually uncertain in practice, limiting the use of these control algorithms. The results of theoretical analyses [29,30] suggest that tracking accuracy can be increased and the tracking error reduced by using uncertainty and disturbance estimators. This paper presents the design of a new and robust tight formation controller that uses an expanded state observer to estimate uncertain disturbances in the system.

Sliding-mode variable structure control algorithms are widely used in control systems in various industries due to their simplicity, robustness, and reliability. Ren et al. applied sliding mode control for the trajectory tracking of a robot and designed a controller to control the trajectory of the robot [31]. Ding et al. controlled the speed of a permanent magnet synchronous motor in which a sliding mode control method was used. The results showed that the anti-interference performance of sliding mode control was strong [32]. However, sliding mode controllers are not always robust, not performing well during increased disturbances. In this study, an extended state observer was used to estimate the induced velocity to which the wingman is subjected. Then, a sliding mode controller was designed to control the tight formation, which includes interference compensation with a larger robust stability margin. The designed controller accurately estimated the value of the induced velocity when unknown to achieve high-accuracy and robust control performance. The main contributions of this study can be summarized as follows:

1. A mathematical model of the wake vortex was established, and the flight characteristics of two UAVs were calculated using Xflow software(The version number of the software is 2020x), which confirmed that the established mathematical model was relatively accurate.
2. A sliding mode controller based on an extended state observer was designed, through which tight-formation flights were accurately controlled.
3. Numerical simulations with the designed controller were conducted in MATLAB, and an experiment was conducted on a semi-physical platform, to verify the feasibility and reliability of the designed controller.

The remainder of this paper is organized as follows: In Section 2, the modeling of the induced wake vortices for tight-formation flight is described. In Section 3, the design of the controller is explained, and the stability and accuracy of the controller are demonstrated. In Section 4, the experimental results are provided with their analysis. In Section 5, the paper is summarized, and areas of future research are outlined.

2. Aerodynamic Modeling of Close-Formation UAVs

2.1. Vortex Mathematical Modeling

The studied UAV was XQ7B; three views of this UAV are shown in Figure 1.

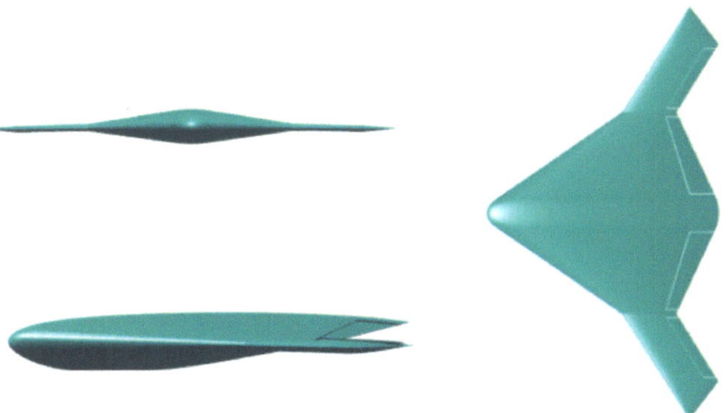

Figure 1. Three views of the aircraft.

As shown in Figure 2, the geometry of the UAV formation is determined by the wingman's position relative to the leading aircraft: longitudinal distance l_x, lateral distance l_y, and vertical distance l_z. In tight-formation flight, the effect of the longitudinal distance l_x on the induced forces and moments is much weaker than that of the lateral distance l_y and vertical distance l_z. Therefore, we do not discuss the effect of the longitudinal distance l_x on the wake vortex here.

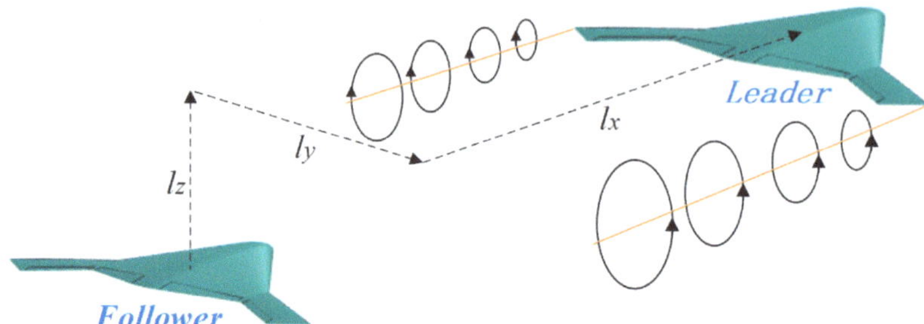

Figure 2. Three views of the aircraft.

First, the induced velocity $v(r)$ of the wake vortex was investigated. The induced velocity model used in this study is given in Equation (1). The vortex model [33] is based on a detailed analysis of LiDAR vortex tangential velocity observations, which draws conclusions that are valuable as a reference for the study of wake vortices in leading aircraft.

$$v(r) = \begin{cases} 1.0939\frac{\Gamma_0}{2\pi r}\left[1 - \exp\left(-10\left(\frac{1.4r_c}{b}\right)^{0.75}\right)\right] \times \left[1 - \exp\left(-1.2527\left(\frac{r}{r_c}\right)^2\right)\right] & \text{if } r \le 1.4r_c \\ \frac{\Gamma_0}{2\pi r}\left[1 - \exp\left(-10(r/b)^{0.75}\right)\right] & \text{if } r > 1.4r_c \end{cases} \tag{1}$$

where Γ_0 is the vortex strength of the wake vortex, b is the wing span, r_c is the radius of the vortex nucleus (generally $r_c = 5.82\%b$) [34], and r is the vertical distance from the point of induced velocity to the vortex line.

The vortex strength is calculated using Equation (2):

$$\Gamma_0 = \frac{L_{Leader}}{\rho V b'} = \frac{L_{Leader}}{\rho V(\pi/4)b} = \frac{\frac{1}{2}\rho V^2 S C_{L_{Leader}}}{\rho V(\pi/4)b} = \frac{2}{\pi}\frac{S}{b}C_{L_{Leader}}V \tag{2}$$

where $C_{L_{Leader}}$ is the lift coefficient of the leader, b is the wing span and S is the wing area.

Suppose one point located on the wing of the following aircraft is at distance s from the right wingtip. At this point, the induced upwash velocity produced by the left tail vortex of the leading aircraft is $w_{Left}(s) = v(r)\sin\beta$, where $r = \sqrt{\left(l_y - \frac{b'}{2} - s\right)^2 + l_z^2}$, $\sin\beta = \left(l_y - \frac{b'}{2} - s\right)/\sqrt{\left(l_y - \frac{b'}{2} - s\right)^2 + l_z^2}$.

Similarly, the induced upwash velocity produced by the right tail vortex of the leading aircraft at this point is $w_{Right}(s) = -v(r)\sin\beta$, where $r = \sqrt{\left(l_y + \frac{b'}{2} - s\right)^2 + l_z^2}$, $\sin\beta = \left(l_y + \frac{b'}{2} - s\right)/\sqrt{\left(l_y - \frac{b'}{2} - s\right)^2 + l_z^2}$.

The total velocity is

$$w(s) = w_{Left}(s) + w_{Right}(s) \tag{3}$$

As such, the average induced upwash velocity on the wing of the wingman is calculated by integrating Equation (3), as shown in Equation (4):

$$w_{UpWavg} = \frac{1}{b}\int_{-\frac{b}{2}}^{\frac{b}{2}} w(s)ds \tag{4}$$

As a result of the induced velocity, the lift force exerted on the wing of the wingman rotates, as shown in Figure 3.

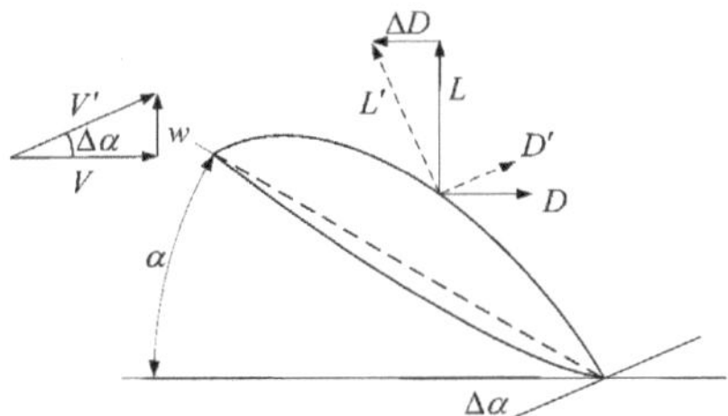

Figure 3. Side view of wingman's wing lift rotation.

V is the speed of the following aircraft, w is the induced upwash, and V' is the velocity of the wingman's wing surface air. The initial lift and drag are denoted by L and D, respectively; the rotated lift and drag vectors are denoted by L' and D', respectively. Figure 3 shows that the change in the angle of attack of the aircraft is $\Delta\alpha$.

$$\Delta\alpha = \arctan\left(\frac{\left|\vec{w}_{UpWavg}\right|}{V}\right) \tag{5}$$

Because $\Delta\alpha$ is small, Equation (5) can be approximated as $\Delta\alpha \approx \left|\vec{w}_{UpWavg}\right|/V$. Figure 3 shows that the rotation of the lift force leads to a change in the value of the drag force to be

$$\Delta D = -L\tan\Delta\alpha \approx -L\Delta\alpha = -L\left|\vec{w}_{UpWavg}\right|/V \tag{6}$$

The increase in the drag coefficient is

$$\Delta C_D = \frac{\Delta D}{qS} = \frac{-L\Delta\alpha}{qS} = -C_L\left|\vec{w}_{UpWavg}\right|/V \tag{7}$$

The following aircraft's increase in lift coefficient is

$$\Delta C_L = \Delta\alpha k_\alpha = k_\alpha\left|\vec{w}_{UpWavg}\right|/V \tag{8}$$

where k_α is the slope of the following aircraft's lift coefficient.

A schematic of the induced lift coefficients and the induced drag coefficients are shown in Figures 4 and 5, respectively. The figures show two maximum values of the induced lift coefficient and two minimum values of the induced drag coefficient, which both occur close to the wingtip of the leading aircraft and are symmetrical between the left and right. Therefore, when studying the flight of aircraft in close formation, only the characteristics of one of the sides of the leading aircraft need to be studied. The induced lift coefficient reaches its maximum near $l_y/b = 0.851$ and $l_z/b = 0$; the induced drag coefficient reaches its minimum near $l_y/b = 0.88$ and $l_z/b = 0$. To obtain both the maximum induced lift and maximum drag reduction, the lateral distance l_y and vertical distance l_z from the optimum point should be kept within 10% and 5% of the wingspan, respectively.

2.2. XFlow Software Calculation

Two UAVs flying in close formation were simulated using XFlow software, which allows the simulation of real atmospheric conditions through the setting of particle density. In the software, the wingman was placed to the right behind the leading aircraft, and the states of the leader and wingman at different positions were calculated by changing the relative positions of the wingman and leading aircraft. The main parameter settings of the XFlow software are shown in Table 1.

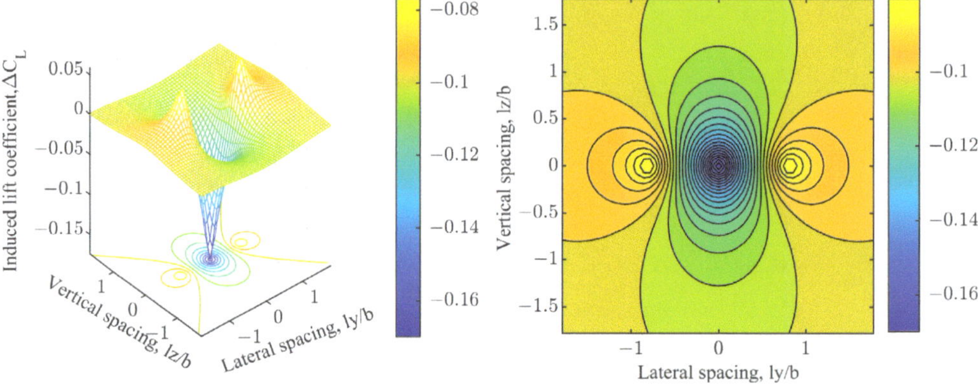

Figure 4. Variations in induced lift coefficient with lateral and vertical spacing.

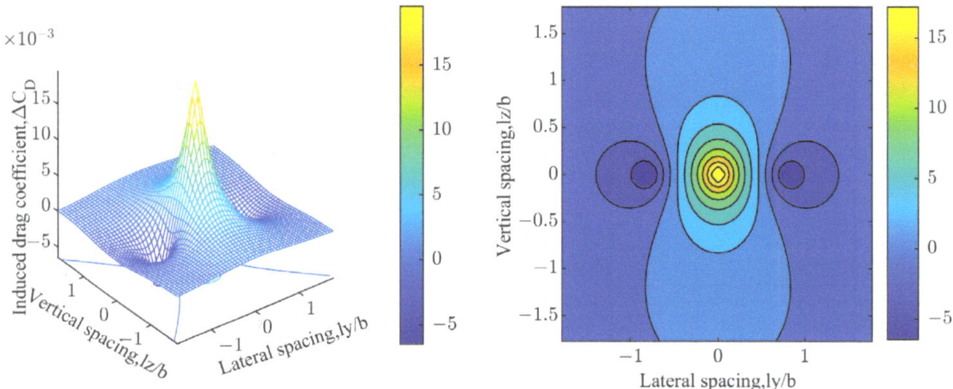

Figure 5. Variations in induced drag coefficient with lateral and vertical spacing.

Table 1. Parameters of the XFlow software.

Parameter	Value	Unit
Velocity	27.8	m/s
Mach number	0.082	
Reynolds number	827,600	
Particle resolution (Far field)	1.28	m
Particle resolution (Near field)	0.0025	m
Reference area	0.11175	m^2
Simulation time	0.06	s

The lift and drag coefficients of the leader and wingman are shown in Figures 6 and 7, respectively. When $l_x = 2b$, $l_z = 0$, by varying the relative positions of the wingman and leading aircraft, the maximum value of the lift coefficient occurs at $l_y = 0.875b$ (i.e., the wingtips of the leader and follower coincide by approximately $0.125b$). Similarly, the minimum drag coefficient occurs at t $l_y = 0.875b$. The results match those obtained with the developed mathematical model, with the aircraft obtaining the maximum tail vortex benefit when the wings of the two aircraft coincide at approximately $0.125b$.

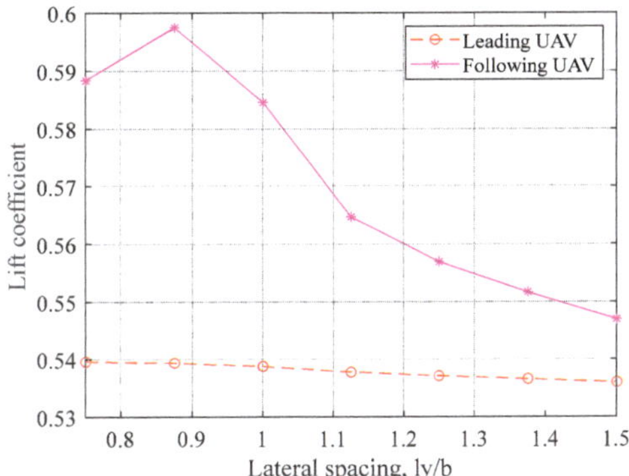

Figure 6. Lift coefficients for leader and follower ($l_x = 2b$, $l_z = 0$).

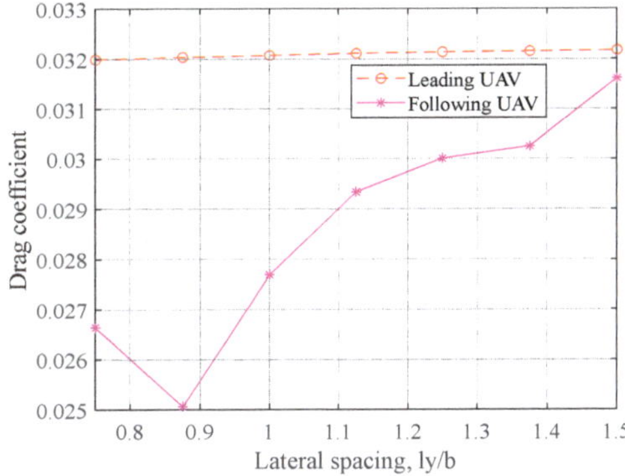

Figure 7. Drag coefficients for leader and follower ($l_x = 2b$, $l_z = 0$).

The following section describes the design of the robust controller that ensures that the follower maintains stable flight even in the presence of wake turbulence, thereby maximizing the advantages of flying in formation.

3. Design of Tight Formation Controller

The nonlinear kinematic equations for wingmen in close formation are shown in Equation (9).

$$\begin{cases} \dot{x}_f = V_f \cos \gamma_f \cos \chi_f + W_x \\ \dot{y}_f = V_f \cos \gamma_f \sin \chi_f + W_y \\ \dot{z}_f = -V_f \sin \gamma_f + W_z \end{cases} \tag{9}$$

where (x_f, y_f, z_f) is the wingman's position coordinates in the inertial coordinate system; V_f, γ_f, χ_f are the airspeed, flight path, and heading angle, respectively; and W_x, W_y, W_z is the induced wake velocity without considering external wind.

The formation flight system is described in an inertial coordinate system, as shown in Figure 8, where (x_l, y_l, z_l) is the position of the leading aircraft in the inertial coordinate system; (x_d, y_d, z_d) is the desired position for the wingman to follow, i.e., the optimal position for the follower in the formation flight system; and (x_f, y_f, z_f) is the position of the follower in the inertial coordinate system. (x_d, y_d, z_d) can be obtained from Equation (10).

$$\begin{bmatrix} x_d \\ y_d \\ z_d \end{bmatrix} = \begin{bmatrix} x_l \\ y_l \\ z_l \end{bmatrix} - L_{WI} \begin{bmatrix} l_x \\ l_y \\ l_z \end{bmatrix} \tag{10}$$

where L_{WI} is the rotation matrix

$$L_{WI} = \begin{bmatrix} \cos(\psi_l) & -\sin(\psi_l) & 0 \\ \sin(\psi_l) & \cos(\psi_l) & 0 \\ 0 & 0 & 1 \end{bmatrix}$$

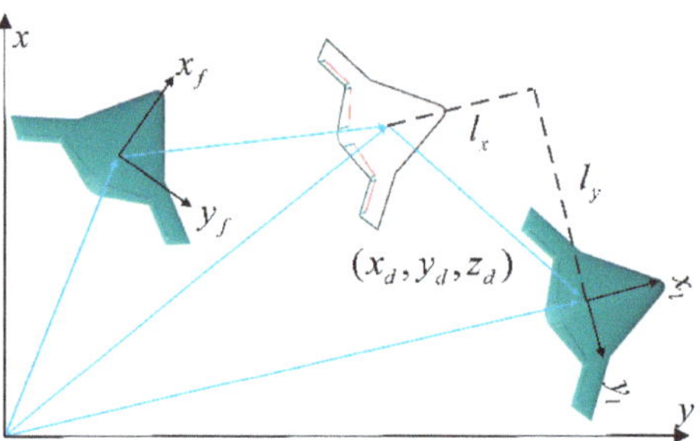

Figure 8. Schematic of the formation flight.

The follower-to-desired-position error is projected on the aircraft's body axis as

$$\begin{cases} e_x = \cos\chi_d(x_f - x_d) + \sin\chi_d(y_f - y_d) \\ e_y = -\sin\chi_d(x_f - x_d) + \cos\chi_d(y_f - y_d) \\ e_z = z_f - z_d \end{cases} \tag{11}$$

where χ_d is the desired heading angle of the following aircraft

After differentiating Equation (11), we obtain

$$\begin{cases} \dot{e}_x = \dot{\chi}_d e_y + v_f \cos\gamma_f \cos(\chi_d - \chi_f) - v_d \cos\gamma_d + W_x \cos\chi_d + W_y \sin\chi_d \\ \dot{e}_y = -\dot{\chi}_d e_x - v_f \cos\gamma_f \sin(\chi_d - \chi_f) - W_x \sin\chi_d + W_y \cos\chi_d \\ \dot{e}_z = -v_f \sin\gamma_f + v_r \sin\gamma_r + W_z \end{cases} \tag{12}$$

3.1. Design of Extended State Observer

Equation (12) can be written as follows

$$\dot{X}_1 = F + U \tag{13}$$

where $X_1 = \begin{bmatrix} e_x \\ e_y \\ e_z \end{bmatrix}$, $F = \begin{bmatrix} W_x \cos\chi_d + W_y \sin\chi_d \\ -W_x \sin\chi_d + W_y \cos\chi_d \\ W_z \end{bmatrix}$, $U = \begin{bmatrix} \dot{\chi}_d e_y + v_f \cos\gamma_f \cos(\chi_d - \chi_f) - v_d \cos\gamma_d \\ -\dot{\chi}_d e_x - v_f \cos\gamma_f \sin(\chi_d - \chi_f) \\ -v_f \sin\gamma_f + v_r \sin\gamma_r \end{bmatrix}$.

A nonlinear perturbation is used to estimate the disturbance of the wake vortex to which the follower is subjected, whereby the perturbation acting on the wingman is expanded into new state variable X_2.

The system shown in Equation (13) is expanded into the new control system shown in Equation (14).

$$\begin{cases} \dot{X}_1 = X_2 + U \\ \dot{X}_2 = W \\ Y = X_1 \end{cases} \tag{14}$$

The uncertainty term is involved as a state variable, and the following expansion state observer was built for the system:

$$\begin{cases} E_1 = Z_1 - Y \\ \dot{Z}_1 = Z_2 - \beta_{01}\mathrm{fal}(E_1, \tfrac{1}{2}, \delta) + U \\ \dot{Z}_2 = -\beta_{02}\mathrm{fal}(E_1, \tfrac{1}{4}, \delta) \end{cases} \tag{15}$$

where $\mathrm{fal}(E, \alpha, \delta) = \frac{E}{\delta^{1-\alpha}}s + |E|^{\alpha}\mathrm{sign}(E)(1 - s)$, $s = \frac{1}{2}\left(\mathrm{sign}(E + \delta \begin{bmatrix} 1 \\ 1 \\ 1 \end{bmatrix}) - \mathrm{sign}(E - \delta \begin{bmatrix} 1 \\ 1 \\ 1 \end{bmatrix}) \right)$.

The error of the state observer is discussed next. From Equations (14) and (15), the error system can be obtained as

$$\begin{cases} E_1 = Z_1 - X_1, E_2 = Z_2 - X_2 \\ \dot{E}_1 = E_2 - \beta_{01}E_1 \\ \dot{E}_2 = W - \beta_{02}fal(E_1, \tfrac{1}{4}, \delta) \end{cases} \tag{16}$$

This error system ultimately reaches a steady state. When the positions of the two aircraft are established, $W \leq W_0 = const$. As such, we have

$$W - \beta_{02}fal(E_1, \frac{1}{2}, \delta) = E_2 - \beta_{01}E_1 = 0$$

Furthermore, we can conclude that

$$E_1 = \left(\frac{W_0}{\beta_{02}}\right)^2, E_2 = \beta_{01}E_1 = \beta_{01}\left(\frac{W_0}{\beta_{02}}\right)^2$$

Thus, as long as β_{02} is sufficiently larger than W_0, these steady-state errors are of the same order of magnitude as $(W_0/\beta_{02})^2$.

3.2. Design of Sliding Mode Controller

The designed extended state observer accurately estimated the perturbation experienced by the wingman, i.e., $Z_2(t) \to F(t)$. By subtracting the estimated value of the disturbance from the control system, the controlled system is as follows:

$$\left\{ \begin{bmatrix} \dot{e}_x \\ \dot{e}_y \\ \dot{e}_z \end{bmatrix} = \begin{bmatrix} \dot{\chi}_d e_y + v_f \cos\gamma_f \cos(\chi_d - \chi_f) - v_d \cos\gamma_d \\ -\dot{\chi}_d e_x - v_f \cos\gamma_f \sin(\chi_d - \chi_f) \\ -v_f \sin\gamma_f + v_r \sin\gamma_r \end{bmatrix} \right. \tag{17}$$

Lemma 1 ([35]). $\forall a \in R, \exists \varepsilon > 0$,*The following inequality holds true*

$$0 \leq |a| - a\tanh\left(\frac{a}{\varepsilon}\right) \leq 0.2785\varepsilon \tag{18}$$

Lemma 1 is substantiated in [35] with a comprehensive proof.

In the subsequent analysis, the controller was designed to ensure that e_x, e_y, and e_z in Equation (17) converge to zero. The detailed design procedure and stability proof of the controller for lateral error e_y are presented below, with similar approaches applicable to the longitudinal and vertical errors.

The design sliding mode function $s(t) = e(t)$.

where $e(t)$ is the tracking error, $e(t) = e_y - 0 = e_y$.

We designed the control instruction as

$$\chi_c = \chi_d - \arcsin\left(\frac{-\dot{\chi}_d e_x + \eta s + D\tanh(\frac{s}{\varepsilon})}{v_f \cos\gamma_f}\right) \tag{19}$$

where η is a constant.

The stability was analyzed as follows:

Define the Lyapunov function as

$$W = \frac{1}{2}s^2 \tag{20}$$

Therefore,

$$\dot{s}(t) = \dot{e}_y(t) = -\dot{\chi}_d e_x - v_f \cos\gamma_f \sin(\chi_d - \chi_f) \tag{21}$$

$$\dot{W} = s\dot{s} = s(-\dot{\chi}_d e_x - v_f \cos\gamma_f \sin(\chi_d - \chi_f)) \tag{22}$$

From Lemma 1, we obtain

$$\left|s\right| - s\tanh\left(\frac{s}{\varepsilon}\right) \le 0.2785\varepsilon$$

So,

$$-s\tanh\left(\frac{s}{\varepsilon}\right) \le -\left|s\right| + 0.2785\varepsilon$$

Substituting the control law Equation (19) into Equation (22) produces

$$\begin{aligned}
\dot{W} &= s\dot{s} \\
&= s(d(t) - \eta s - D\tanh(\tfrac{s}{\varepsilon})) \\
&= s(-\eta s - D\tanh(\tfrac{s}{\varepsilon}) + d(t)) \\
&= -\eta s^2 - Ds\tanh(\tfrac{s}{\varepsilon}) + sd(t)) \\
&\le -\eta s^2 - D\left|s\right| + 0.2785D\varepsilon + sd(t)) \\
&\le -\eta s^2 + 0.2785D\varepsilon = -2\eta W + b
\end{aligned}$$

where $b = 0.2785D\varepsilon$.

The solution to inequality $\dot{W} \le -2\eta W + b$ is

$$\begin{aligned}
W(t) &\le e^{-2\eta(t-t_0)}W(t_0) + be^{-2\eta t}\int_{t_0}^{t} e^{2\eta\tau}d\tau \\
&= e^{-2\eta(t-t_0)}W(t_0) + \frac{be^{-2\eta t}}{2\eta}(e^{2\eta t} - e^{2\eta t_0}) \\
&= e^{-2\eta(t-t_0)}W(t_0) + \frac{b}{2\eta}(1 - e^{-2\eta(t-t_0)}) \\
&= e^{-2\eta(t-t_0)}W(t_0) + \frac{0.2785D\varepsilon}{2\eta\tau_y}(1 - e^{-2\eta(t-t_0)})
\end{aligned}$$

As such, $\lim\limits_{t\to\infty} W(t) \le \left(0.2785D\varepsilon / 2\eta\tau_y\right)$. The asymptotic convergence of $W(t)$ is proved.

After adding the disturbance compensation, the total control command is

$$\chi_c = \chi_d - \arcsin\left(\frac{-\dot{\chi}_d e_x - \left(-\eta e_y - D\tanh(\frac{e_y}{\varepsilon}) - Z_y\right)}{v_f \cos\gamma_f}\right)$$

Similarly, the control commands for the flight path angle and speed are respectively

$$\gamma_f = \arcsin\left(\frac{v_r \sin\gamma_r - \left(-\eta e_z - D\tanh(\frac{e_z}{\varepsilon}) - Z_z\right)}{v_f}\right)$$

$$v_f = \frac{v_d \cos \gamma_d - \eta e_x - D \tanh(\frac{e_x}{\varepsilon}) - Z_x - \dot{\chi}_d e_y}{\cos \gamma_f \cos(\chi_d - \chi_f)}$$

4. Simulation and Experimental Verification

We built upon the results of previous studies, where the nonlinear model of the aircraft was previously developed, and each aircraft had a separate inner-loop flight control system. The tight-formation flight control system is shown in Figure 9. In this study, the mathematical model of the studied wake vortex was added as a disturbance to the aerodynamic model of the following aircraft. Then, the designed tight formation controller was verified via mathematical and semi-physical simulations. The leading aircraft was flying on a predetermined trajectory, and the initial position of the following aircraft was far from the leading aircraft. By comparing the designed tight-formation controller with the previous controller, the reliability and practicability of the sliding mode control method based on the expanded state observer were verified. The parameters of the aircraft are presented in Table 2.

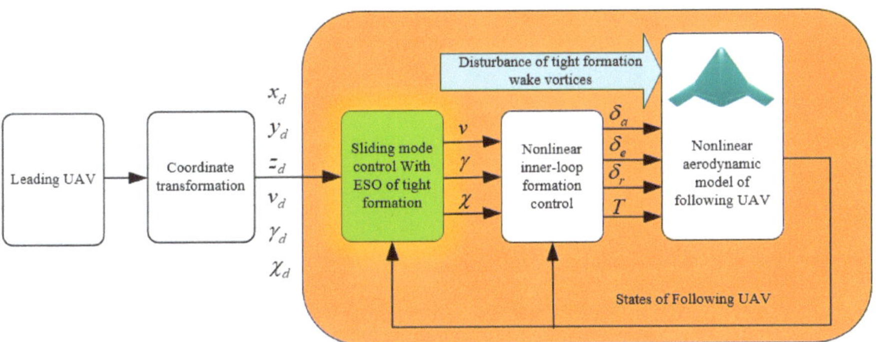

Figure 9. Schematic of tight formation flight control system.

Table 2. Parameters of the UAV.

Parameter	Symbol	Value	Unit
Wing area	S	1.546	m^2
Wing span	b	2.808	m
Mean aerodynamic chord	$\bar{c}$	0.78	m
Gross mass	m	15	kg
Roll moment of inertia	I_x	2.369	$Kg{\cdot}m^2$
Pitch moment of inertia	I_y	1.211	$Kg{\cdot}m^2$
Yaw moment of inertia	I_z	3.522	$Kg{\cdot}m^2$
Product moment of inertia	I_{xz}	0.022	$Kg{\cdot}m^2$

4.1. Numerical Simulation

The numerical simulations were performed on the Simulink platform of MATLAB.

The initial states of the leading aircraft are $x_l = 0$ m, $y_l = 1000$ m, $h_l = 1000$ m, $v_l = 27.8$ m/s, $\beta_l = 0$ deg, $p_l = q_l = r_l = 0$ rad/s, $\phi_l = \varphi_l = 0$ deg, $\mu_l = 0$ deg, $\gamma_l = \chi_l = 0$ deg, $\theta_l = \alpha_l = 5.73$ deg; Whereas the initial states of the following aircraft are $x_f = 0$ m, $y_f = 0$ m, $h_f = 1000$ m, $v_f = 27.8$ m/s, $\beta_f = 0$ deg, $p_f = q_f = r_f = 0$ rad/s, $\phi_f = 0$ deg, $\varphi_f = 0$ deg, $\mu_f = 0$ deg, $\gamma_f = \chi_f = 0$ deg, $\theta_f = \alpha_f = 5.73$ deg.

The relative positions of the aircraft at different moments of the formation flight are shown in Figure 10, and the aircraft trajectories are shown in Figure 11. From 0 to 250 s, the leading aircraft was flying straight and level, the following aircraft's initial lateral distance from the leading aircraft was 1000 m, the wingman's heading angle was negative (to reduce the lateral distance error), and, at 70 s, the lateral error was 0. At approximately

100 s, the longitudinal error reduced to zero. At this point, the following aircraft completed the approach to the leading aircraft from a far distance, and the two aircraft executed a tight formation. From 250 to 750 s, the aircraft turned, and the relative positions of the two aircraft slightly wavered but within acceptable ranges.

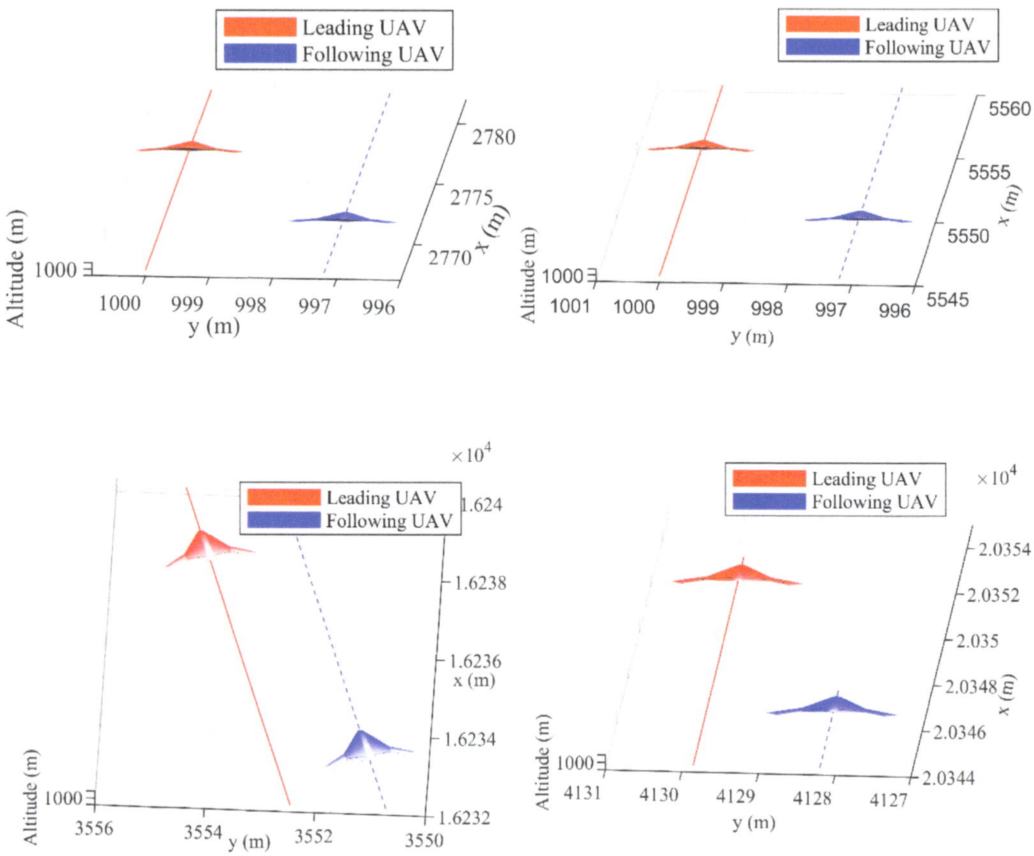

Figure 10. Relative positions of aircraft at different times during formation flight.

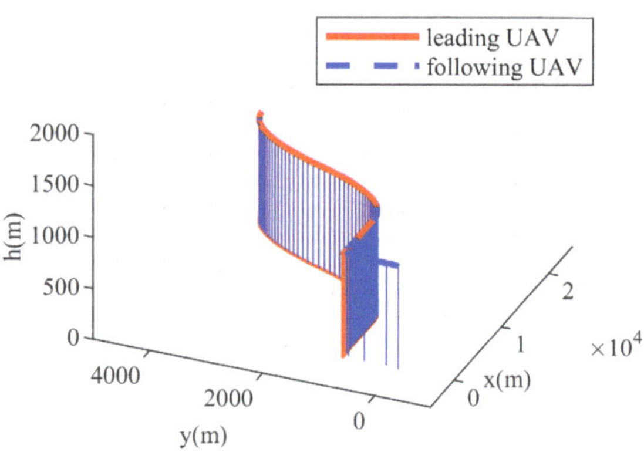

Figure 11. Formation trajectory.

The comparison of the results of the sliding mode control effect, with and without ESO, are shown in Figures 12–14. The results all indicate that the sliding mode control with ESO is more accurate, more strongly suppresses the influence of wake vortices, and shows more robustness. In comparing the model predictive control (MPC) algorithm with the proposed control algorithm in this paper, we see the robust controller designed in this paper provides advantages, including faster convergence and smaller static errors.

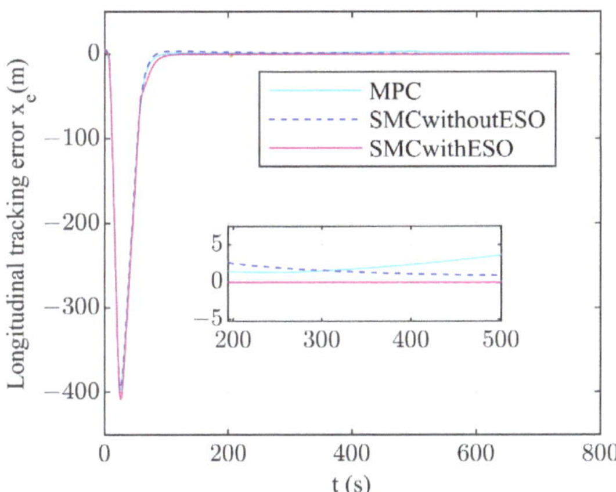

Figure 12. Longitudinal tracking error.

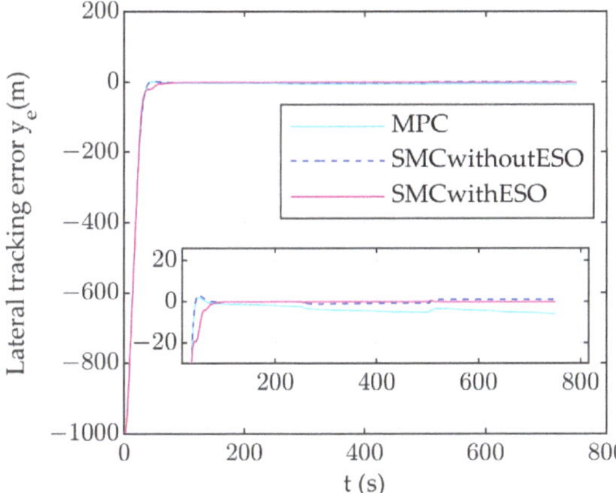

Figure 13. Lateral tracking error.

4.2. Experiments with Semi-physical Simulation Platform

The platform comprised the flight control system, radio communication equipment, UAV computer processing unit, and visual display unit, the structure of which is depicted in Figure 15. The flight control system generated precise control commands to govern the aircraft's movements. The radio facilitated seamless communication between the leading and following UAVs. The computer was responsible for solving complex UAV data computations (airspeed, angle of attack, sideslip angle, rolling angular rates, pitching angular rates, yawing angular rates, roll angle, pitch angle, yaw angle, flight-path angle, and positional coordinates of the aircraft). The visual display unit provided real-time information on the formation's position and status.

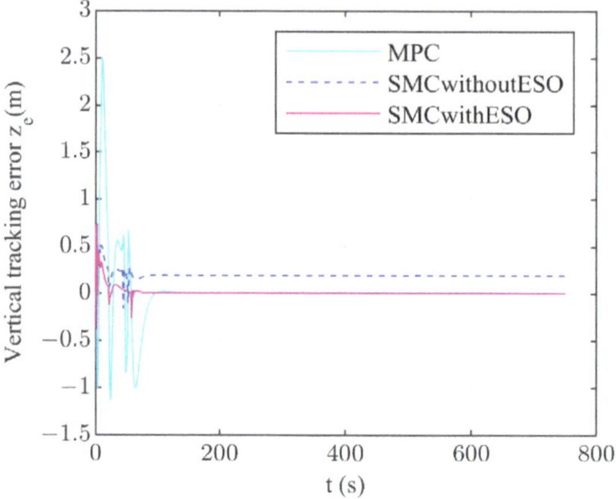

Figure 14. Vertical tracking error.

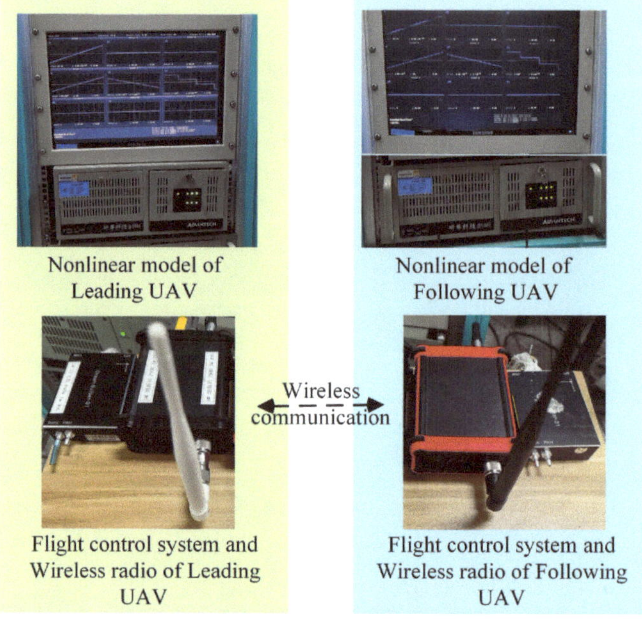

Figure 15. Architecture diagram of distributed semi-physical simulation platform.

The simulation results of the tight formation were displayed with Tacview software (The version number of the software is 1.9.3), as shown in Figure 16. Tacview is a versatile tool for analyzing flight data. Within this experimental platform, Tacview receives data from the UAV computer and presents detailed information on the formation's position and attitude.

The following aircraft consistently maintained the optimal relative position to the leading aircraft. According to the aforementioned experimental results, the proposed controller effectively mitigated interference and increased system robustness. The real-time nature of the system and the practicality of the control method were validated through experiments conducted on a distributed semi-physical simulation platform.

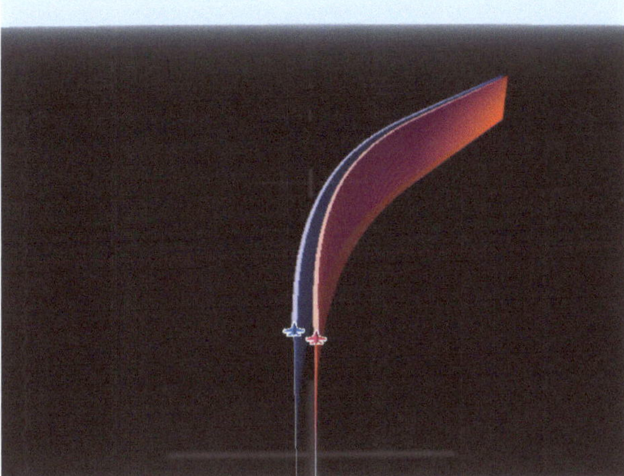

Figure 16. Display of close formation flights in Tacview software.

The semi-physical simulation verification demonstrated that the designed controller can successfully operate on this hardware platform and that the computational speed of the flight control board is sufficient for the control system. Communication delay is a critical factor for a formation system, as excessive communication delay can cause the instability in the tight formation flight system. The radios used in the semi-physical simulation experiments were capable of fulfilling the requirements of the control system.

5. Conclusions

This study investigated the scenario in which two aircraft fly in close formation. First, a mathematical model of the wake vortex was established. Second, Xflow software was employed to simulate the formation characteristics of the two UAVs. From the simulation results, we found that when the distance between the two UAVs is ($l_x = 2b$, $l_y = 0.875b$, $l_z = 0$), the wingman experiences the maximum formation aerodynamic benefit. This finding aligns with the conclusion derived from studying wake vortex. Third, the disturbance of the wake vortex experienced by the wingman was then added to the wingman's system, and a formation controller was developed that combines the extended state observer and sliding mode control methods. The controller considerably mitigated the effects on the following aircraft and enabled the following aircraft to maintain its optimal relative position to the leading aircraft. The designed formation control was validated, thus achieving the objectives through numerical and semi-physical simulations. In future studies, we will investigate the complex mechanisms of close-formation flight and the collision avoidance problems between the leading and following aircraft. Additionally, we will prepare for real close-formation flight experiments.

Author Contributions: Conceptualization, R.Z. and Q.Z.; Data curation, R.Z. and Y.L.; Methodology, R.Z. and S.H.; Software, R.Z. and Z.D.; Validation, R.Z. and J.S.; Writing—original draft, R.Z. All authors have read and agreed to the published version of the manuscript.

Funding: This research was funded by the National Natural Science Foundation of China, grant number 62373301 and 62173277; the Natural Science Foundation of Shaanxi Province, grant number 2023-JC-YB-526; the Aeronautical Science Foundation of China, grant number 20220058053002; and the Shaanxi Province Key Laboratory of Flight Control and Simulation Technology.

Data Availability Statement: The data presented in this study are available on request from the corresponding author.

Conflicts of Interest: The authors declare no conflicts of interest.

Nomenclature

Γ_0	Vortex circulation
$C_{L_{Leader}}$	Lift coefficient of the leader
(x_l, y_l, z_l)	The position coordinates of the leading aircraft
(x_f, y_f, z_f)	The position coordinates of the following aircraft
μ, γ, χ	The bank, flight path, and heading angles
W_x, W_y, W_z	The induced wake velocity
S	Wing area
b	Wing span
$\bar{c}$	Mean aerodynamic chord
m	Gross mass
I_x	Roll moment of inertia
I_y	Pitch moment of inertia
I_z	Yaw moment of inertia
I_{xz}	Product moment of inertia

References

1. Li, W.-H.; Shi, J.-P.; Wu, Y.-Y.; Wang, Y.-P.; Lyu, Y.-X. A Multi-UCAV cooperative occupation method based on weapon engagement zones for beyond-visual-range air combat. *Def. Technol.* **2022**, *18*, 1006–1022. [CrossRef]
2. Zhang, Q.; Liu, H.H.T. Aerodynamic model-based robust adaptive control for close formation flight. *Aerosp. Sci. Technol.* **2018**, *79*, 5–16. [CrossRef]
3. Zhang, Q.; Liu, H.H. Robust Design of Close Formation Flight Control via Uncertainty and Disturbance Estimator. In Proceedings of the AIAA Guidance, Navigation, and Control Conference, San Diego, CA, USA, 4–8 January 2016.
4. Frew, E.W.; Lawrence, D.A.; Morris, S. Coordinated standoff tracking of moving targets using Lyapunov guidance vector fields. *J. Guid. Control Dyn.* **2008**, *31*, 290–306. [CrossRef]
5. Hansen, J.L.; Cobleigh, B.R. Induced Moment Effects of Formation Flight Using Two F/A-18 Aircraft. In Proceedings of the AIAA Atmospheric Flight Mechanics Conference and Exhibit, Monterey, CA, USA, 5–8 August 2002.
6. Zheng, R.; Shi, J.; Qu, X. Modeling, Simulation and Control of Close Formation Flight. In Proceedings of the 2021 China Automation Congress (CAC), Beijing, China, 22–24 October 2021; pp. 3902–3907.
7. Saban, D.; Whidborne, J.F.; Cooke, A.K. Simulation of wake vortex effects for UAVs in close formation flight. *Aeronaut. J.-New Ser.* **2009**, *113*, 727–738. [CrossRef]
8. Lissaman, P.B.; Shollenberger, C.A. Formation flight of birds. *Science* **1970**, *168*, 1003–1005. [CrossRef] [PubMed]
9. Cutts, C.J.; Speakman, J.R. Energy Savings in Formation Flight of Pink-Footed Geese. *J. Exp. Biol.* **1994**, *189*, 251–261. [CrossRef] [PubMed]
10. Rayner, J. Estimating power curves of flying vertebrates. *J. Exp. Biol.* **1999**, *202*, 3449–3461. [CrossRef] [PubMed]
11. Weimerskirch, H.; Martin, J.; Clerquin, Y.; Alexandre, P.; Jiraskova, S. Energy saving in flight formation. *Nature* **2001**, *413*, 697–698. [CrossRef] [PubMed]
12. Cho, H.; Han, C. Effect of sideslip angle on the aerodynamic characteristics of a following aircraft in close formation flight. *J. Mech. Sci. Technol.* **2015**, *29*, 3691–3698. [CrossRef]
13. Vachon, M.J.; Ray, R.J.; Walsh, K.R.; Ennix, K. *F/A-18 Performance Benefits Measured during the Autonomous Formation Flight Project*; Armstrong Flight Research Center: Edwards Air Force Base, CA, USA, 2013.
14. Vicroy, D.; Vijgen, P.; Reimer, H.; Gallegos, J.; Spalart, P. Recent NASA wake-vortex flight tests, flow-physics database and wake-development analysis. In Proceedings of the AIAA and SAE, 1998 World Aviation Conference, Anaheim, CA, USA, 28–30 September 1998.
15. Kent, T.E.; Richards, A.G. Analytic approach to optimal routing for commercial formation flight. *J. Guid. Control Dyn.* **2015**, *38*, 1872–1884. [CrossRef]
16. Pahle, J.; Berger, D.; Venti, M.; Duggan, C.; Faber, J.; Cardinal, K. An initial flight investigation of formation flight for drag reduction on the C-17 aircraft. In Proceedings of the AIAA Atmospheric Flight Mechanics Conference, Minneapolis, MN, USA, 13–16 August 2012; p. 4802.
17. Bieniawski, S.R.; Rosenzweig, S.; Blake, W.B. Summary of flight testing and results for the formation flight for aerodynamic benefit program. In Proceedings of the 52nd Aerospace Sciences Meeting, National Harbor, MD, USA, 13–17 January 2014; p. 1457.
18. Hanson, C.E.; Pahle, J.; Reynolds, J.R.; Andrade, S.; Nelson, B. Experimental measurements of fuel savings during aircraft wake surfing. In Proceedings of the 2018 Atmospheric Flight Mechanics Conference, Atlanta, GA, USA, 25–29 June 2018; p. 3560.
19. Okolo, W.; Dogan, A.; Blake, W. Effect of trail aircraft trim on optimum location in formation flight. *J. Aircr.* **2015**, *52*, 1201–1213. [CrossRef]
20. Zhang, Q.; Pan, W.; Reppa, V. Model-reference reinforcement learning for collision-free tracking control of autonomous surface vehicles. *IEEE Trans. Intell. Transp. Syst.* **2021**, *23*, 8770–8781. [CrossRef]

21. Blake, W.B.; Gingras, D.R. Comparison of Predicted and Measured Formation Flight Interference Effects. *J. Aircr.* **2004**, *41*, 201–207. [CrossRef]
22. Ray, R.; Cobleigh, B.; Vachon, M.; St. John, C. Flight Test Techniques used to Evaluate Performance Benefits During Formation Flight. In Proceedings of the AIAA Atmospheric Flight Mechanics Conference and Exhibit, Monterey, CA, USA, 5–8 August 2002.
23. Wilson, D.B.; Goktogan, A.H.; Sukkarieh, S. A Vision Based Relative Navigation Framework for Formation Flight. In Proceedings of the 2014 IEEE International Conference on Robotics & Automation (ICRA), Hong Kong, China, 31 May–7 June 2014.
24. Zheng, R.; Lyu, Y. Nonlinear tight formation control of multiple UAVs based on model predictive control. *Def. Technol.* **2023**, *25*, 69–75. [CrossRef]
25. Zhang, Q.; Liu, H.H.T. Robust Nonlinear Close Formation Control of Multiple Fixed-Wing Aircraft. *J. Guid. Control Dyn.* **2021**, *44*, 572–586. [CrossRef]
26. Pachter, M.; D'Azzo, J.J.; Proud, A.W. Tight formation flight control. *J. Guid. Control Dyn.* **2001**, *24*, 246–254. [CrossRef]
27. Binetti, P.; Ariyur, K.B.; Krstic, M.; Bernelli, F. Formation flight optimization using extremum seeking feedback. *J. Guid. Control Dyn.* **2003**, *26*, 132–142. [CrossRef]
28. Chichka, D.F.; Speyer, J.L.; Fanti, C.; Park, C.G. Peak-seeking control for drag reduction in formation flight. *J. Guid. Control Dyn.* **2006**, *29*, 1221–1230. [CrossRef]
29. Zhang, Q.; Liu, H.H.T. UDE-Based Robust Command Filtered Backstepping Control for Close Formation Flight. *IEEE Trans. Ind. Electron.* **2018**, *65*, 8818–8827. [CrossRef]
30. Zhang, Q.; Liu, H.H. Robust cooperative close formation flight control of multiple unmanned aerial vehicles. In Proceedings of the Advances in Motion Sensing and Control for Robotic Applications: Selected Papers from the Symposium on Mechatronics, Robotics, and Control (SMRC'18)-CSME International Congress 2018, Toronto, ON, Canada, 27–30 May 2018; Springer: Cham, Switzerland, 2019; pp. 61–74.
31. Ren, C.; Li, X.; Yang, X.; Ma, S. Extended state observer-based sliding mode control of an omnidirectional mobile robot with friction compensation. *IEEE Trans. Ind. Electron.* **2019**, *66*, 9480–9489. [CrossRef]
32. Ding, S.; Hou, Q.; Wang, H. Disturbance-observer-based second-order sliding mode controller for speed control of PMSM drives. *IEEE Trans. Energy Convers.* **2022**, *38*, 100–110. [CrossRef]
33. Proctor, F.; Hamilton, D.; Han, J. Wake vortex transport and decay in ground effect-Vortex linking with the ground. In Proceedings of the 38th Aerospace Sciences Meeting and Exhibit, Reno, NV, USA, 10–13 January 2000; p. 757.
34. Proctor, F. Numerical simulation of wake vortices measured during the Idaho Falls and Memphis field programs. In Proceedings of the 14th Applied Aerodynamics Conference, New Orleans, LA, USA, 17–20 June 1996; p. 2496.
35. Polycarpou, M.M.; Ioannou, P.A. A robust adaptive nonlinear control design. In Proceedings of the 1993 American Control Conference, San Francisco, CA, USA, 2–4 June 1993; pp. 1365–1369.

Disclaimer/Publisher's Note: The statements, opinions and data contained in all publications are solely those of the individual author(s) and contributor(s) and not of MDPI and/or the editor(s). MDPI and/or the editor(s) disclaim responsibility for any injury to people or property resulting from any ideas, methods, instructions or products referred to in the content.

![MDPI]

Article

Safe Reinforcement Learning for Transition Control of Ducted-Fan UAVs

Yanbo Fu [1,2], Wenjie Zhao [1,2,*] and Liu Liu [1]

[1] School of Aeronautics and Astronautics, Zhejiang University, Hangzhou 310027, China; 22124084@zju.edu.cn (Y.F.)
[2] Center for Unmanned Aerial Vehicles, Huanjiang Laboratory, Zhuji 311800, China
* Correspondence: zhaowenjie8@zju.edu.cn

Abstract: Ducted-fan tail-sitter unmanned aerial vehicles (UAVs) provide versatility and unique benefits, attracting significant attention in various applications. This study focuses on developing a safe reinforcement learning method for back-transition control between level flight mode and hover mode for ducted-fan tail-sitter UAVs. Our method enables transition control with a minimal altitude change and transition time while adhering to the velocity constraint. We employ the Trust Region Policy Optimization, Proximal Policy Optimization with Lagrangian, and Constrained Policy Optimization (CPO) algorithms for controller training, showcasing the superiority of the CPO algorithm and the necessity of the velocity constraint. The transition trajectory achieved using the CPO algorithm closely resembles the optimal trajectory obtained via the well-known GPOPS-II software with the SNOPT solver. Meanwhile, the CPO algorithm also exhibits strong robustness under unknown perturbations of UAV model parameters and wind disturbance.

Keywords: safe reinforcement learning; ducted fan; transition control; unmanned aerial vehicle (UAV)

Citation: Fu, Y.; Zhao, W.; Liu, L. Safe Reinforcement Learning for Transition Control of Ducted-Fan UAVs. *Drones* **2023**, *7*, 332. https://doi.org/10.3390/drones7050332

Academic Editor: Mostafa Hassanalian

Received: 15 April 2023
Revised: 12 May 2023
Accepted: 18 May 2023
Published: 22 May 2023

Copyright: © 2023 by the authors. Licensee MDPI, Basel, Switzerland. This article is an open access article distributed under the terms and conditions of the Creative Commons Attribution (CC BY) license (https://creativecommons.org/licenses/by/4.0/).

1. Introduction

In recent years, vertical take-off and landing (VTOL) unmanned aerial vehicles (UAVs) have gained considerable attention due to their unique advantages. Distinct from traditional fixed-wing UAVs, VTOL UAVs are capable of taking off and landing vertically, eliminating the need for a runway [1]. Moreover, when compared to multi-rotor UAVs, they offer several benefits, such as a larger payload capacity, higher cruise speed, and longer flight ranges [2]. This is primarily attributed to their reliance on wings for lift production, as opposed to the multiple rotors utilized by multi-rotor UAVs. Among the various VTOL UAV configurations, such as tilt rotor, tail sitter, and vectored thrust, the ducted-fan tail-sitter fixed-wing UAV stands out as a unique design. Notably, they eliminate the need for additional moving parts to achieve VTOL capabilities, leading to a simplified mechanical design that is both easier to maintain and less prone to mechanical failures [3]. Moreover, the incorporation of a ducted fan enhances the propulsion efficiency, allowing the UAVs to cover longer distances. The ducted fan also contributes to noise reduction, making these UAVs particularly suitable for noise-sensitive environments [4].

Ducted-fan tail-sitter fixed-wing UAVs integrate the characteristics of both traditional fixed-wing and multi-rotor UAVs, enabling them to perform level flight and hover operations. However, during the transition process, the UAV experiences aerodynamic instability due to a stall occurring at the fixed wing [5]. As a result, the control strategies for transition maneuvers between these two flight modes are of paramount importance, necessitating extensive investigation and research. Generally, two prevalent methods to address this issue can be found in the current literature [3].

The first approach conceptualizes the transition process as a trajectory optimization problem. In [6], the direct collocation method was employed to obtain the optimal transition trajectory by Li et al. An optimization approach using the interior point method was proposed that focused on altitude changes during transition in [7]. The optimal feed-forward control input for transition was computed via sequential quadratic programming by Kubo and Suzuki [8]. Banazadeh devised a gradient-based algorithm based on the classical Cauchy method to generate optimal transition trajectories [9]. Naldi and Marconi [10] utilized mixed-integer nonlinear programming to tackle the minimum-time and minimum-energy optimal transition problems.

The second approach is structured in two stages: the first stage entails devising the desired trajectory during the transition, and the second stage involves designing a controller to track the trajectory established in the first stage [3]. Jeong et al. proposed a continuous ascent transition trajectory consisting of the angle of attack and flight path angle and invoked dynamic inversion control for tracking [11]. Flores provided a desired velocity trajectory and implemented a recurrent neural network-based controller for feedback linearization [12]. Cheng and Pei established a transition corridor based on the constraint of maintaining a fixed altitude [13]. They planned the desired velocity in the corridor and utilized an adaptive controller [14].

Alternatively, reinforcement learning (RL) has emerged as a promising approach to address various challenges in the domain of UAVs, owing to its ability to learn and adapt to dynamic environments [15]. As a result, there has been growing interest in leveraging RL algorithms to tackle the transition problem of VTOL UAVs. In [16], an RL-based controller for hybrid UAVs was designed that can not only automatically complete the transition but can also be adapted to different configurations. Xu [17] proposed a soft landing control algorithm based on the RL method. Yuksek employed the deep deterministic policy gradient algorithm to address the transition flight problem for tilt-rotor UAVs [18].

In this paper, we solve the back-transition task of ducted-fan tail-sitter fixed-wing UAVs (i.e., from level flight mode to hover mode) using safe RL algorithms. While prior work [16] mainly emphasizes the successful execution of transition maneuvers, our focus is on minimizing altitude changes and the transition time, while adhering to velocity constraints. Our method, compared to [18], integrates trajectory optimization and control problems into an RL-based learning process, thereby reducing the complexity and computational effort. To the best of our knowledge, this is one of the first works in which the RL methodology is utilized to solve the back-transition control problem of ducted-fan tail-sitter UAVs.

The main contributions of our work are as follows.

1. We develop a mathematical model of ducted-fan fixed-wing UAV dynamics. Based on this model, we create a training environment for ducted-fan UAVs in OpenAI GYM [19] using the fourth-order Runge–Kutta method.
2. Taking into account the velocity constraint during the transition process, we train controllers using Trust Region Policy Optimization (TRPO) with fixed penalty, Proximal Policy Optimization with Lagrangian (PPOLag), and Constrained Policy Optimization (CPO). We assess the performance of these algorithms and demonstrate the superiority of the CPO algorithm.
3. Comparing the CPO algorithm with the optimal trajectory obtained via GPOPS-II [20], we find that the performance of CPO closely approximates the optimal trajectory. In addition, the CPO algorithm has robustness under unknown perturbations of UAV model parameters and wind disturbance, which is lacking in the GPOPS-II software.

This paper is organized as follows. In Section 2, a mathematical model of a ducted-fan fixed-wing UAV is described. In Section 3, the general structure of the RL transition controller is introduced, and the reward function, action, and observation are explained. In Section 4, comparisons between CPO and other RL algorithms are reported. We also compare the transition trajectory of CPO with GPOPS-II and verify the robustness of the CPO algorithm. In Section 5, concluding remarks and future works are reported.

2. Mathematical Modeling

In this section, we describe a three-degree-of-freedom (DOF) longitudinal model for ducted-fan fixed-wing UAVs. The model is derived from the Newton and Euler theorems and is simplified from the full 6-DOF dynamic model. This 3-DOF model is able to speed up the process of assessing the impact of different reward functions and hyper-parameter settings compared to the 6-DOF dynamic model [21]. It is equipped with four groups of control vanes as the main control surfaces, each consisting of three movable vanes. In addition, four groups of fixed vanes are situated above the main control surfaces, intended to balance the anti-torque generated by the rotor, with each fixed group consisting of two fixed vanes (see Figure 1). The four groups of control vanes are numbered 1, 2, 3, and 4 and are employed to change the attitude of the UAVs (see Figure 2). Each group of three movable vanes, controlled by a single servo, deflects by the same angle. We use δ_1, δ_2, δ_3, and δ_4 to represent the deflection of group numbers 1, 2, 3, and 4.

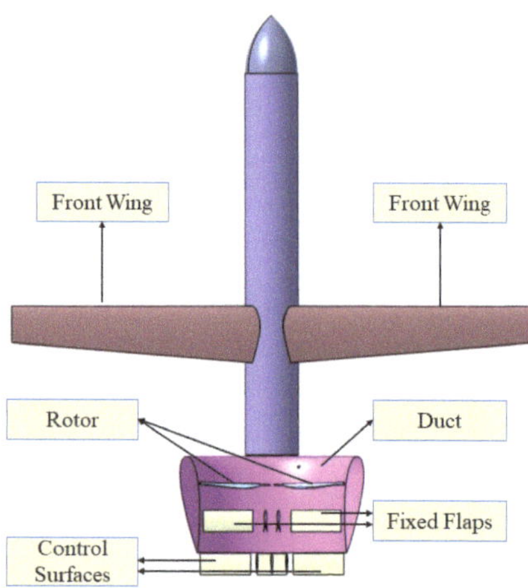

Figure 1. Ducted-fan UAV layout.

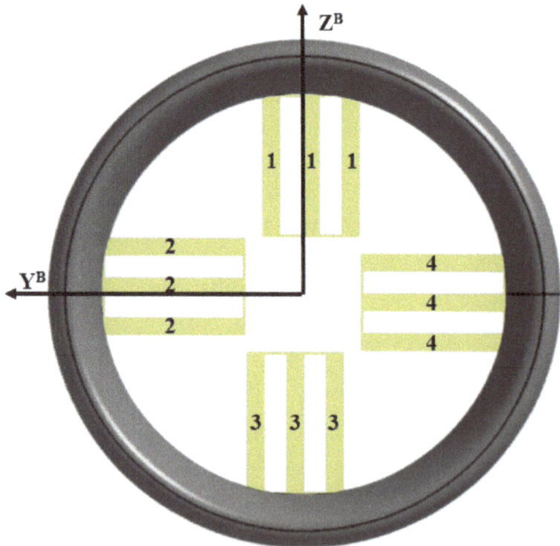

Figure 2. Control vanes.

2.1. Three-DOF Dynamics

For the ducted-fan VTOL UAV, two right-handed coordinate systems are applied to describe the states of the aircraft (see Figure 3). The inertial frame axes are denoted as $\{\Gamma^I{:}X^I,Y^I,Z^I\}$ and the body frame axes are denoted as $\{\Gamma^B{:}X^B,Y^B,Z^B\}$ with the origin located at the center of mass. Throughout this section, the superscripts $(.^I)$ and $(.^B)$ are utilized to specify whether a variable is formulated in the inertial or body frame. The position of the aircraft in Γ^I is described by $\xi = [x,y,z]^T$ and the velocity of the aircraft in Γ^B is defined as $V^B = [u,v,w]^T$. The Euler angle vector (i.e., roll, pitch, and yaw) is described by $\Theta = [\phi,\theta,\psi]^T$ and the angular velocity vector with respect to the body frame is denoted by $\Omega = [p,q,r]^T$. It should be noted that based on the above-mentioned definition, the pitch angle $\theta = 0°$ at the level flight condition, and the pitch angle $\theta = 90°$ at the landing and hover conditions (see Figure 4). Thus, the longitudinal dynamics of the aircraft are derived as follows:

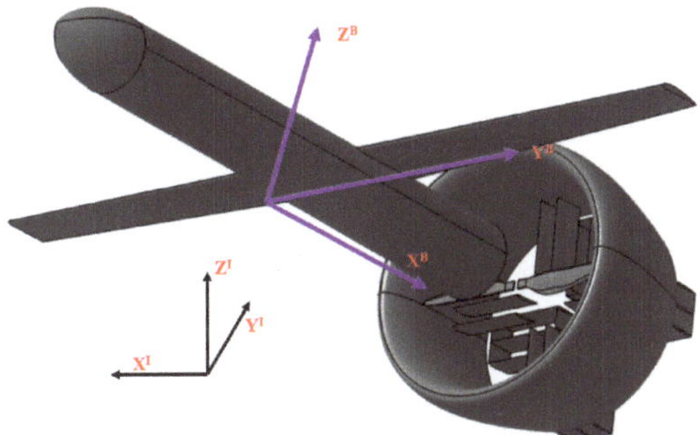

Figure 3. Inertial frame and body frame.

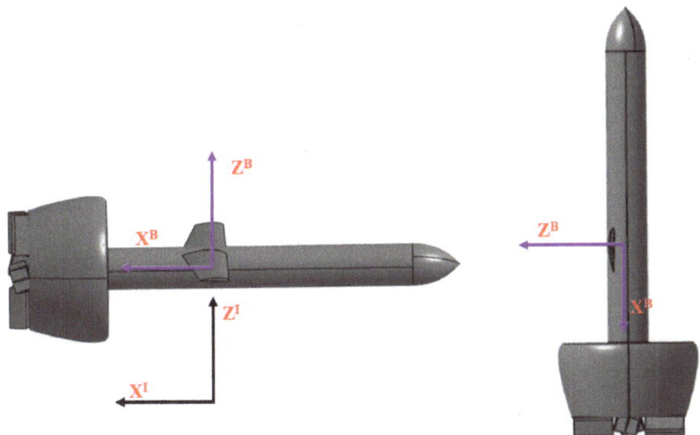

Figure 4. Left: level flight; **Right**: hover mode.

$$
\begin{cases}
\dot{u} = g\sin\theta + (D\cos\alpha - T - L\sin\alpha)/m - qw \\
\dot{w} = (F_{cw} + L\cos\alpha + D\sin\alpha + F_{cs})/m + qu - g\cos\theta \\
\dot{x} = w\cos\theta - u\sin\theta \\
\dot{z} = u\cos\theta + w\sin\theta \\
\dot{\theta} = q \\
\dot{q} = \left(M_{pitch} + M_{cw} + M_{cs}\right)/I_{yy}
\end{cases}
\tag{1}
$$

where g is gravity acceleration; I_{yy} is the moment of inertia on the Y^B axis; α represents the angle of attack. All the forces and moments are discussed below.

2.2. Rotor

In this subsection, the rotor model is discussed based on basic momentum theory and blade element theory [22]. Considering the airspeed along the X^B axis, the configuration of the blades, the airflow through the rotor, and the thrust of the rotor can be expressed as [23]

$$v_b = u + \frac{2}{3}\omega_r r \left(\frac{3}{4}K_{twist}\right) \tag{2}$$

$$T = \frac{1}{4}(v_b - v_i)\omega_r r^2 \rho_\infty a_0 b c_r \tag{3}$$

where $V^B = [u, v, w]^T$ is the velocity in the body frame, ρ_∞ is the air density, ω_r represents the angular velocity of the rotor, r is the radius of the rotor and K_{twist} is the twist of the blades, a_0 is the rotor lift curve slope, b is the number of blades, c_r is the chord of the rotor blade, and the induced velocity v_i and the far-field velocity v_f can be expressed as

$$v_f = \sqrt{(u - v_i)^2 + w^2 + v^2} \tag{4}$$

$$v_i = T/2\rho_\infty \pi r^2 v_f \tag{5}$$

The expressions for T and v_i can be solved iteratively through Equations (2)–(5) using the Newton–Raphson method.

2.3. Aerodynamics Model

The aerodynamic forces and moments include the lift L, the drag D, and the pitching moment M_{pitch}. They are primarily dependent on the fuselage, duct, and mostly on the wings. The angle of attack has a significant impact on the aerodynamic model, which can be expressed as follows:

$$\alpha = \begin{cases} 0 & \text{if } w = 0 \\ -sgn(w)\dfrac{\pi}{2} & \text{if } u = 0 \\ \arctan(\dfrac{w}{u}) & \text{if } u < 0 \\ \arctan(\dfrac{w}{u}) + \pi & \text{if } u > 0 \end{cases} \tag{6}$$

The ducted-fan UAV can be regarded as an entire lifting body, including the fuselage, wings, and duct [15]. This simplification introduces some inaccuracies in the aerodynamic data. To address these discrepancies, it is essential to incorporate compensation for the disturbances within the aerodynamic data. In order to account for the modeling errors, the lift and drag coefficients are multiplied by a perturbation factor, sampled from a uniform distribution ranging between 0.8 and 1.2.

During the transition mode, high angle of attack conditions can induce wing stall phenomena, leading to a significant reduction in lift. Traditional linear aerodynamic coefficient models are insufficient in accurately capturing the complexities of the aerodynamic behavior. As demonstrated in [24], an advanced aerodynamic model that integrates both the linear lift model and the effects of wing stall can be formulated as follows:

$$\begin{aligned} C_L(\alpha) &= (1 - \sigma(\alpha))\left[C_{L_0} + C_{L_\alpha}\alpha\right] \\ &\quad + \sigma(\alpha)[C_{L_{max}} \sin\alpha \cos\alpha] \end{aligned} \tag{7}$$

$$C_D(\alpha) = C_{D_p} + \frac{(C_{L_0} + C_{L_\alpha}\alpha)^2}{\pi e AR} \tag{8}$$

where

$$\sigma(\alpha) \;=\; \frac{1 + e^{-M(\alpha-\alpha_0)} + e^{M(\alpha+\alpha_0)}}{\left(1 + e^{-M(\alpha-\alpha_0)}\right)\left(1 + e^{M(\alpha+\alpha_0)}\right)} \tag{9}$$

$$C_{L_\alpha} \;=\; \frac{\pi AR}{1 + \sqrt{1 + (AR/2)^2}} \tag{10}$$

The lift and drag coefficients are shown in Figure 5. The pitching moment coefficients will be expressed by the linear model as

$$C_m(\alpha) = C_{m_0} + C_{m_\alpha}\alpha \tag{11}$$

Thus, the lift L, drag D, and pitching moment M_{pitch} can be written as

$$\begin{aligned}
L &= \frac{1}{2}\rho_\infty V^2 S C_L(\alpha) \\
D &= \frac{1}{2}\rho_\infty V^2 S C_D(\alpha) \\
M_{pitch} &= \frac{1}{2}\rho_\infty V^2 S C_m(\alpha)
\end{aligned} \tag{12}$$

V is the air speed of the UAV and S is the wing area.

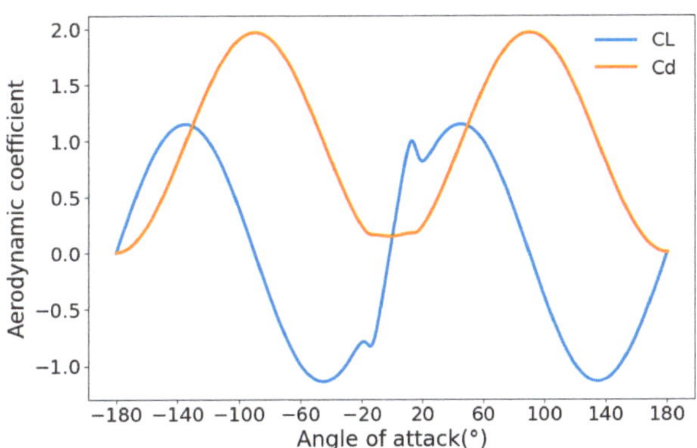

Figure 5. Lift and drag coefficient.

2.4. Momentum Drag

Due to the existence of crosswinds, the duct must generate a force to align the incoming airflow with its orientation. This results in a reaction force known as momentum drag. Moreover, crosswinds lead to the formation of a region with higher velocity over the near edge of the duct, as the air surrounding it is pulled into the duct by the rotor. This increased lift on the edge produces a moment that causes the vehicle to turn away from the crosswind, referred to as the momentum moment [22]. The formulas for the momentum drag and moment can be represented as follows:

$$F_{cw} = -v_i \rho_\infty \pi r^2 w \tag{13}$$

$$M_{cw} = C_{duct}\rho_\infty r |r| w \tag{14}$$

where w is the velocity of the Z-axis in the body frame, and v_i is determined by Equation (5).

2.5. Control Vanes

Each control vane can be modeled as an airfoil. In [3,14,15], the lift slope coefficient of vane is assumed to be constant. However, our Computational Fluid Dynamics simulations have shown that the lift coefficient of the vane remains virtually unchanged when subjected to a large angle of attack. Thus, based on the simple model in [25], the lift slope coefficient C_{lcs} of the vane can be expressed as

$$C_{lcs} = \begin{cases} 0.0625 & \text{if } |\alpha| \leq 16° \\ 0 & \text{if } |\alpha| > 16° \end{cases} \tag{15}$$

The dynamic pressure q_{cs} on each vane can be expressed as

$$q_{cs} = \frac{1}{2}\rho_\infty (u - v_i)^2 \tag{16}$$

where u is the velocity of the X-axis in the body frame. Based on the control allocation method [15], the equivalent vane deflection of pitch (Y-axis) is given as follows:

$$\delta_y = \delta_2 - \delta_4 \tag{17}$$

The drag forces of vanes can be neglected [15], and the vane's angle of attack only depends on the vane's deflection [3]. Thus, the force and moment generated by the control vanes are as follows:

$$F_{cs} = q_{cs} S_{cs} C_{lcs} \delta_y \tag{18}$$

$$M_{cs} = q_{cs} S_{cs} C_{lcs} l_e \delta_y \tag{19}$$

where S_{cs} represents the vane area and l_e is the arm of pitch moment.

3. Approach

3.1. Problem Formulation

In this section, we provide a mathematical expression for the UAV back-transition problem. In the back-transition process, the most straightforward strategy is that the UAV climbs at a large pitch angle while converting the kinetic energy into potential energy [26] (see Figure 6a). Although this approach offers a short transition time, it results in substantial altitude loss. In this paper, our desired trajectory, referred to as a "neat transition", aims to complete the transition as quickly as possible with minimal altitude loss [13] (see Figure 6b). To achieve this transition, we introduce an additional constraint on the Z-axis velocity, expressed as

$$\|v_z\| < 1$$

Thus, we pay attention to the altitude loss and transition time after the transition has been completed. The RL problem is defined as follows.

RL UAV Back-Transition Problem:

Minimize:

1. Terminal Velocity Error: $\|v\|$ at $t = t_f$
2. Terminal Attitude Error: $\|\theta - \pi/2\|$ at $t = t_f$
3. Terminal Angular Velocity Error: $\|\omega\|$ at $t = t_f$
4. Terminal Time: t_f
5. Height loss: $\left\|z_{t_f} - z_{t_0}\right\|$

Subject to:

1. Angular Velocity Constraint: pitch angle velocity rate $q < 180°$
2. Control Constraints: $|\delta_y| < 30°$ and $\omega_{rmin} < \omega_r < \omega_{rmax}$
3. Neat Transition Constraint: $\|v_z\| < 1$ m/s

4. Equations of Motion (set by the environment)

To primarily optimize the terminal altitude loss and transition time, we opt to provide error margins for the terminal velocity, terminal angle, and terminal angular velocity. When the error falls within these margins, the transition is deemed successful. The terminal error range can be expressed as follows:

$$\|v_{t_f}\| \leq 1 \text{ m/s}$$
$$\left\|\theta_{t_f} - \pi/2\right\| \leq 0.08 \text{ rad}$$
$$\|\omega_{t_f}\| \leq 0.08 \text{ rad/s}$$

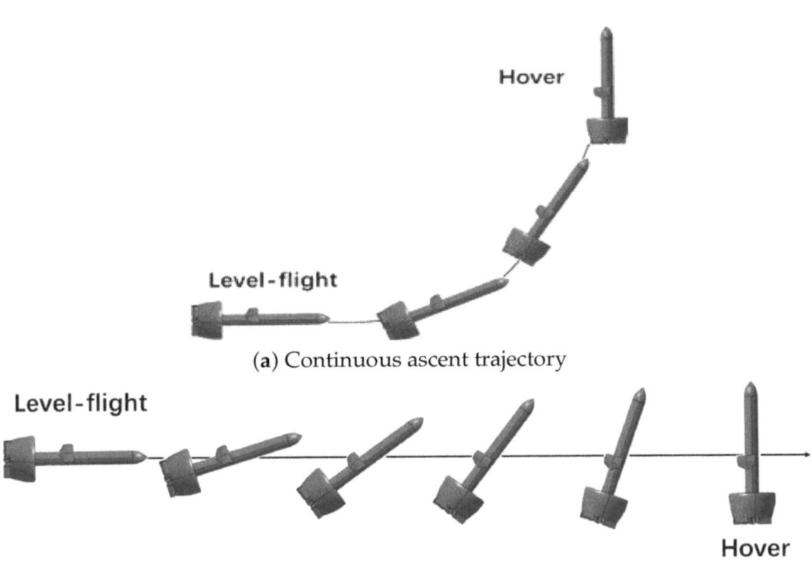

(a) Continuous ascent trajectory

(b) Neat transition trajectory

Figure 6. Different transition trajectories.

3.2. Algorithm

To address the neat transition constraint, traditional RL methods typically apply a fixed penalty through the reward function. However, if the penalty is too small, the agent may learn unsafe behavior, while an excessively severe penalty might result in the agent's inability to learn anything. In contrast, safe reinforcement learning algorithms employ the Lagrangian approach to tackle constraints, automatically balancing the weights between rewards and penalties. This ensures that the entire transition process adheres to the constraints while exploring the optimal transition performance. Among safe RL algorithms, we introduce the Constrained Policy Optimization (CPO) algorithm to solve the ducted-fan UAV back-transition problem [27].

3.2.1. Constrained Markov Decision Process (CMDP)

A Markov decision process (MDP) is a tuple, (S, A, P, γ, R), where S is the set of states, A is the set of actions, and $R : S \times A \times S \rightarrow \mathbb{R}$ is the reward function. $P : S \times A \times S \rightarrow [0, 1]$ is the transition probability function (where $P(s' \mid s, a)$ is the probability of transitioning to state s' under the previous state s and the action a). $\gamma \in [0, 1]$ is the discount factor for future rewards, which represents a trade-off between immediate and future rewards. π refers to a policy, which is a function that maps a state to an action. $\pi(a|s)$ denotes the probability of selecting action a in state s. A trajectory τ is a set of MDP tuples in one episode during which an agent interacts with the environment under policy π.

A constrained Markov decision process (CMDP) is an MDP subjected to some constraints. A set $C_1, \ldots, C_m$ (with each one as $C_i : S \times A \times S \to \mathbb{R}$) is referred to as cost functions, similar to the reward function [28]. The limits $d_1, \ldots, d_m$ represent the thresholds of the cost functions. For the policy gradient algorithm, value functions, action-value functions, and advantage functions are defined as V^π, Q^π, and A^π.

$$
\begin{aligned}
Q^\pi(s_t, a_t) &= \mathbb{E}_\pi\left[\sum_{k=0}^{\infty} \gamma^k R_{t+k}\right] \\
V^\pi(s_t) &= \mathbb{E}_\pi\left[\sum_{k=0}^{\infty} \gamma^k R_{t+k}\right] \\
A^\pi(s, a) &= Q^\pi(s, a) - V^\pi(s)
\end{aligned}
\tag{20}
$$

Similarly, we can replace reward R in the above equation with cost C to acquire V_C^π, Q_C^π, and A_C^π.

In a CMDP, the actor–critic framework consists of three parts: the actor, the critic, and the cost critic. The actor is responsible for selecting actions based on the current state of the agent, while the critic and cost critic estimate the expected discounted reward and cost based on the current state. Typically, these three parts are usually approximated by neural networks (NN). In the case of CMDP, let $J(\pi)$ denote the expected discounted reward:

$$
J(\pi) = \mathop{\mathrm{E}}_{\tau \sim \pi}\left[\sum_{t=0}^{\infty} \gamma^t R_t\right]
\tag{21}
$$

Similarly, the cumulative cost can be described as

$$
J_{C_i}(\pi) = \mathop{\mathrm{E}}_{\tau \sim \pi}\left[\sum_{t=0}^{\infty} \gamma^t C_i(s_t, a_t, s_{t+1})\right]
\tag{22}
$$

Thus, the set of feasible policies is

$$
\Pi_C \doteq \left\{\pi \in \Pi : \forall i, J_{C_i}(\pi) \leq d_i\right\}
\tag{23}
$$

The objective in a CMDP is to find a policy that maximizes $J(\pi)$ in Π_C:

$$
\pi^* = \arg\max_{\pi \in \Pi_C} J(\pi)
\tag{24}
$$

3.2.2. Constrained Policy Optimization

CPO adheres to the monotonic improvement theory proposed by Trust Region Policy Optimization (TRPO) [29]. By constraining the difference between the old and new strategies to a small step size, it becomes possible to approximate the lower bound of the objective function and the upper bound of the cost objective function between the old and new policies, resulting in the following expression [27]:

$$
J(\pi') - J(\pi) \approx \mathbb{E}_{\tau \sim \pi}\left[\sum_{t=0}^{\infty} A^\pi(s_t, a_t)\frac{\pi'(a_t|s_t)}{\pi(a_t|s_t)}\right]
\tag{25}
$$

$$
J_{C_i}(\pi') - J_{C_i}(\pi) \approx \mathbb{E}_{\tau \sim \pi}\left[\sum_{t=0}^{\infty} A_{C_i}^\pi(s_t, a_t)\frac{\pi'(a_t|s_t)}{\pi(a_t|s_t)}\right]
\tag{26}
$$

Due to the approximation of the difference and the requirement in importance sampling that the two strategy distributions must not be too far apart, a constraint is added on the average Kullback–Leibler (KL) divergence between the old and new policies:

$$\bar{D}_{KL}\left(\pi \| \pi'\right) \leq \delta \tag{27}$$

where δ is a step size. By employing the minorization-maximization (MM) algorithm, policy boosting can be guaranteed through maximizing an approximate lower bound on the difference between the old and new policies (often called the alternative objective function). Similarly, if the sum of the upper bound of the difference between the cost function of the old and the new strategy and the cost function of the old strategy satisfies the constraint, then the cost objective function of the new policy can also satisfy the constraint. As a result, this optimization problem can be expressed as

$$\max_{\theta} \mathbb{E}_{(s_t,a_t)\sim\pi} \left[\frac{\pi_\theta(a_t \mid s_t)}{\pi(a_t \mid s_t)} \widehat{A}^\pi(s_t, a_t) \right]$$

$$\text{s.t. } J_{C_i}(\pi) + \mathbb{E}_{(s_t,a_t)\sim\pi} \left[\frac{\pi_\theta(a_t \mid s_t)}{\pi(a_t \mid s_t)} \widehat{A}^\pi_{C_i}(s_t, a_t) \right] \leq d_i \quad \forall i \tag{28}$$

$$\bar{D}_{KL}\left(\pi \| \pi^\theta\right) \leq \delta$$

where θ is the parameter of the policy network and $\widehat{A}$ is the advantage value function estimated by generalized advantage estimation [30]. Since the KL divergence limits the difference in the policy distribution to a small step size, the objective and constraint functions in the above equation can be expanded in the first order, while the KL divergence constraint is expanded in the second order. Let the gradient of the objective function be g, the gradient of constraint i be b_i, the Hessian of KL divergence be H, and c be defined as $J_{C_i}(\pi) - d_i$. Thus, the approximate problem is

$$\theta_{k+1} = \arg\max_{\theta} g^T(\theta - \theta_k)$$

$$\text{s.t. } c_i + b_i^T(\theta - \theta_k) \leq 0 \quad i = 1, \ldots, m \tag{29}$$

$$\frac{1}{2}(\theta - \theta_k)^T H(\theta - \theta_k) \leq \delta$$

Due to the computational complexity of the Hessian matrix, CPO uses a Fisher information matrix to approximate H, which makes the problem above a convex optimization problem [27]. To solve this convex optimization problem, we denote the Lagrangian multipliers λ and ν. Then, when the original problem has a feasible solution, a dual to (29) can be expressed as

$$\max_{\substack{\lambda \geq 0 \\ \nu \geq 0}} \frac{-1}{2\lambda} \left(g^T H^{-1} g - 2\nu^T b^T H^{-1} g + \nu^T b^T H^{-1} b\nu \right) + \nu^T c - \lambda\delta \tag{30}$$

If λ^* and ν^* are the optimal solution to the dual problem, the policy can be updated as follows:

$$\theta^{k+1} = \theta^k + \frac{1}{\lambda^*} H^{-1}(g - b\nu^*) \tag{31}$$

When there is no feasible solution to the original problem, CPO will execute the recovery method, which will change the search direction. The recovery method is as follows:

$$\theta^{k+1} = \theta^k - \sqrt{\frac{2\delta}{b^T H^{-1} b}} H^{-1} b \tag{32}$$

Inspired by TRPO, the line search method is used to adjust the step size of the update to avoid any error caused by the approximation.

3.2.3. Implementation Details

In this section, we give the corresponding design of the state space, action space, and reward function in RL. The state and action spaces are described as

$$s = \left(\sin\theta, \cos\theta, v_x, v_z, q, t_{left}, I_{v_x}, I_{v_z}, I_\theta, z \right)^T \in R^{10},$$

$$a = \left(\delta_y, \omega_r \right)^T \in R^2. \tag{33}$$

where δ_y represents the vane deflection of pitch, ω_r is the angular velocity of the rotor, θ is the pitch angle, v_x and v_z are the velocity of the X-axis and Z-axis in the inertial frame, q is the angular velocity of the pitch angle, t_{left} is the remaining time during one episode (the maximum time for an episode is 10 s), and z is the height. $I_{v_x}, I_{v_z}, I_\theta$ is the integral error of v_x, v_z, θ, which can be expressed as

$$e_\theta = \theta - \pi/2, e_{v_x} = v_x, e_{v_z} = v_z \tag{34}$$

$$I_n = \eta I_{n-1} + e_n \tag{35}$$

where η is the integral coefficient, and the integration of the error is approximated by such an incremental approach as Equation (35). The design of the reward function is a key factor affecting the performance of the reinforcement learning algorithm. The reward function R and the cost function C are given by the following:

$$\Phi(s_t) = k_1\sqrt{v_x^2 + v_z^2} + k_2|\theta - \pi/2| \tag{36}$$

$$R = \kappa + (\Phi(s_{t+1}) - \Phi(s_t)) \times (0.5)^{t/10} - \chi I(s) - \beta(q > \pi) - \xi(|z| > 2) - \tau \tag{37}$$

$$C = 1 \text{ if } \|v_z\| \geq 1 \text{ else } 0 \tag{38}$$

where the various terms are described as follows.

1. κ is a bonus reward for successful transition, where the terminal velocity, pitch angle, and angular velocity are all within specified limits. In addition, different amounts of bonuses are also given based on the loss of height when a successful transition state is reached. We use Z_T to represent the terminal height loss.

$$\kappa = \begin{cases} 1000 & \text{if } Z_T \leq 0.125\,\text{m} \\ 100 - 100 \times \log_2 Z_T & \text{if } Z_T \leq 1\,\text{m} \\ 100 - 90 \times \log_2 Z_T & \text{if } Z_T \leq 2\,\text{m} \\ 10 & \text{else} \end{cases} \tag{39}$$

2. $\Phi(s_{t+1}) - \Phi(s_t)$ is in the form of reward shaping. The aim of reward shaping is to accelerate the learning process, where k1 and k2 denote the weights of the speed and angle rewards. The term $(0.5)^{t/10}$, which is gradually decaying over time, is designed to reduce the transition time.

3. χ is constant and $I(s) = \|0.02 \times I_{v_x}^2 + I_\theta^2 + I_{v_z}^2\|$, β is a penalty for an angle exceeding the limit, and ξ is a penalty when the UAV loses more than two meters in altitude. τ is a fixed time penalty.

The cost limit d is set at 5. The terminal condition of an episode is either meeting the termination condition specified in Section 3.1 or reaching the maximum time of 10 s.

4. Simulation and Results

In this section, we present the experimental results. In our experiments, the controller update and sensor data download frequency was set as 50 HZ. For our experiments, the initial flight mode was set to level flight, with the initial angle of attack sampled from a uniform distribution ranging between 6° and 15°. The initial pitch angle was set equal to the angle of attack, the initial height was set at 50 m, and the corresponding horizontal velocity could be determined using the following equation:

$$v_{level} = -\sqrt{\frac{mg}{\frac{1}{2}C_L(\alpha)\rho_\infty S}} \tag{40}$$

4.1. Comparing CPO and TRPO with Fixed Penalty

To demonstrate the superiority of safe reinforcement learning compared to other conventional reinforcement learning algorithms with a fixed penalty, we conducted three different sets of experiments, which were named CPO, TRPO with penalty 1, and TRPO with penalty 5. The experimental results are shown in Figure 7.

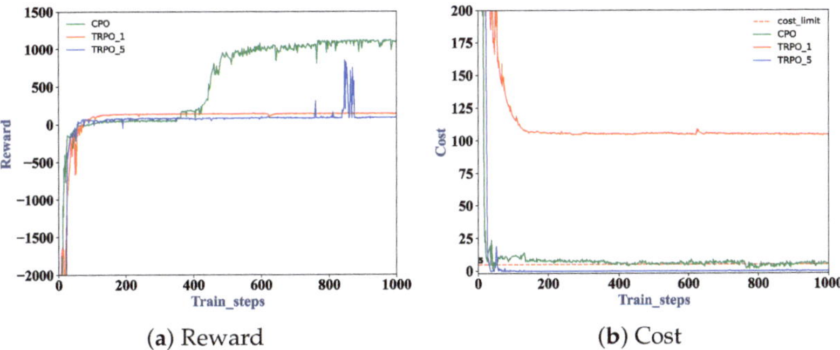

(a) Reward   (b) Cost

Figure 7. Comparison of learning curve.

During training, we took into account the randomization of the initial state and the perturbation of the aerodynamic coefficients. However, in evaluating the transition performance of CPO and TRPO with penalty, we have not considered these factors. Three sets of experiments with the same initial state (angle of attack at 6°) for the sake of comparison are shown in Figure 8, where the arrows represent the terminal state.

In Figures 7 and 8, we can observe that the trajectory corresponding to TRPO with penalty 1 converges to a locally optimal continuous ascending path, resulting in a fast transition time but significant altitude loss. In contrast, the trajectory corresponding to TRPO with penalty 5 consistently satisfies the constraints, but its terminal angle and speed exceed the predefined range, ultimately failing to complete the task. The CPO algorithm, however, effectively meets the constraints while accomplishing the transition with a minimal altitude loss of 0.1 m, which complies with the requirements of a neat transition.

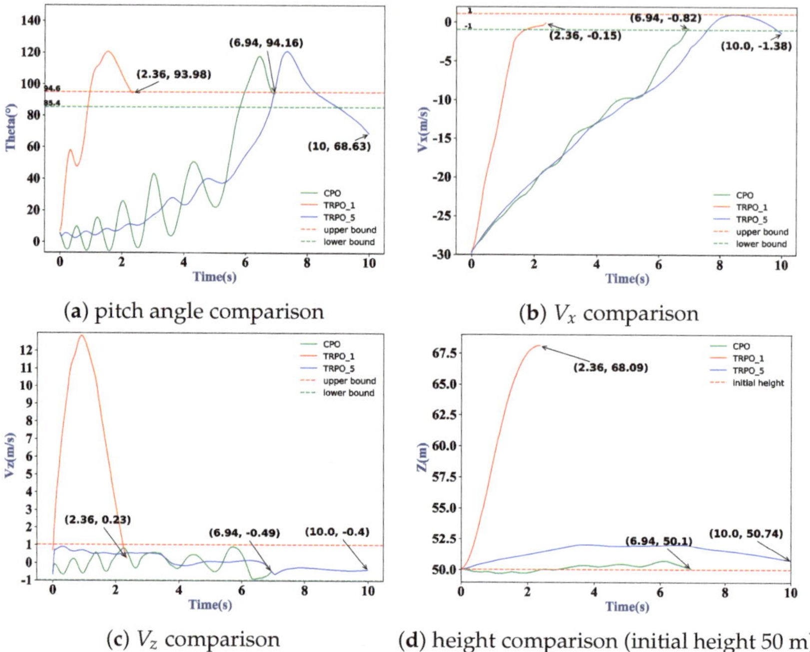

(**a**) pitch angle comparison

(**b**) V_x comparison

(**c**) V_z comparison

(**d**) height comparison (initial height 50 m)

Figure 8. Trajectory comparison.

4.2. Comparing CPO and PPOLag

In the domain of safe reinforcement learning, numerous algorithms can address constraint-related issues. For the sake of code development convenience, we have chosen to compare the CPO algorithm with Proximal Policy Optimization with Lagrangian, also called PPOLag (see Figure 9).

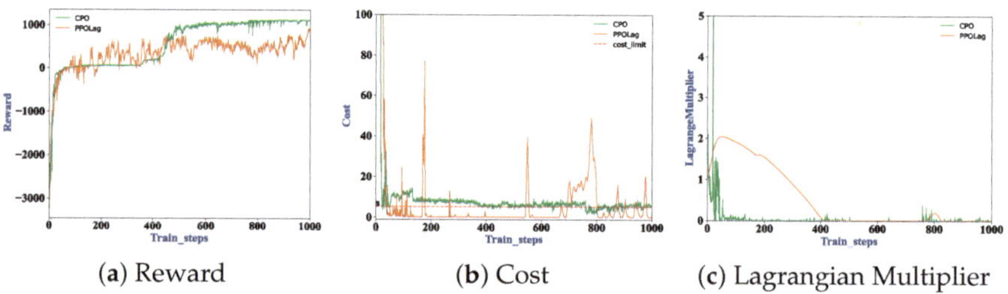

(**a**) Reward

(**b**) Cost

(**c**) Lagrangian Multiplier

Figure 9. CPO vs. PPOLag.

PPOLag is a variant of the Proximal Policy Optimization (PPO) algorithm, which employs a Lagrangian relaxation approach to handle constraints. Constraints are integrated into the objective function using a penalty term that is computed with the help of the Lagrange multiplier. The Lagrange multiplier is updated during the training process based on the discrepancy between the current constraint value and its cost limit, which serves as a weighting factor to balance the trade-off between the reward function and the constraints.

PPOLag is a simpler algorithm in principle and easier to implement. However, as shown in Figure 9b, this method may cause oscillations near the cost limit of the constraint, leading to poor performance of the agent (as shown in Figure 9a). Consequently, although PPOLag can complete the transition, its performance is inferior to that of CPO. The best-trained model with PPOLag (at an angle of attack of 6°) has a terminal height loss of 1.2 m and a transition time of 7.84 s, but CPO has a terminal height loss of 0.1 m and a transition time of 6.94 s.

4.3. Comparison with GPOPS-II

According to the problem formulation in Section 3.1, this problem can also be considered as an optimal control problem. As a result, we choose to employ GPOPS-II when computing the optimal trajectory. In this case, the control variables for GPOPS-II are the same as in our approach—namely, the angular velocity of the rotor and elevator deflection. The experimental results without perturbations at an initial angle of attack $10°$ are compared in Figures 10 and 11. The cost function of GPOPS-II is the following:

$$J = 0.5t_f + 0.2\left|Z_{t_f}\right| \tag{41}$$

where t_f is the transition time and Z_{t_f} is the terminal height loss.

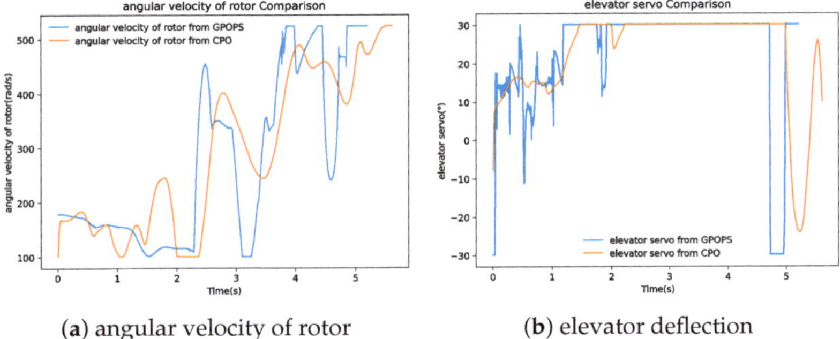

(**a**) angular velocity of rotor   (**b**) elevator deflection

Figure 10. Control input comparison for CPO and GPOPS-II.

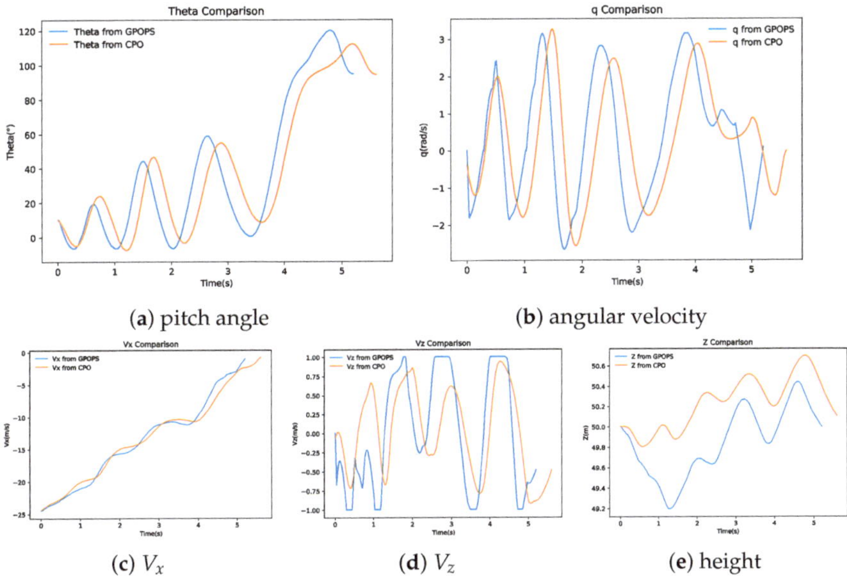

(**a**) pitch angle   (**b**) angular velocity

(**c**) V_x (**d**) V_z (**e**) height

Figure 11. State comparison for CPO and GPOPS-II.

The height loss of GPOPS-II is 0.01 m with a transition time of 5.21 s, while the height loss of CPO is 0.103 m with a transition time of 5.62 s. As can be seen from the figures, the performance of the CPO algorithm closely approximates that of GPOPS-II. However, GPOPS-II has two main drawbacks. Firstly, as a model-based approach, it provides only optimal feed-forward control input, which is an ideal solution. When the dynamics model of the UAV changes, it requires re-solving based on the altered model. In contrast, the CPO algorithm is highly robust and can still perform the transition task with high performance despite certain modeling errors and wind disturbance, as discussed later in Sections 4.4 and 4.5. Secondly, as micro-controllers onboard aerial robots generally only have limited computational power, the optimization of GPOPS-II can be only executed offline. However, RL algorithms such as CPO can solve the transition problem online after a policy model has been trained, which saves a great deal of computational resources.

4.4. Robustness Validation

During the training process, the randomness of the UAV is only determined by the randomness of the initial state and aerodynamic parameters. Therefore, in order to assess the robustness of the system, we chose to randomize various factors, including the mass, moment of inertia, and aerodynamic parameters of lift and drag (with a larger range), by employing a uniform distribution. Furthermore, we took sensor noise into account because measurement errors in the height can potentially lead to a significant decline in the system's performance, especially for terminal height loss. To validate the robustness of sensor noise, Gaussian noise was introduced into the height measurements. It is important to note that we did not consider these perturbations during training, meaning that the UAV had not been exposed to them before.

To enable a fair comparison, we only randomized the initial state and tested each perturbation individually. Specifically, we conducted 500 experiments for each perturbation, while maintaining the same transition success condition as in Section 3.1. We report the average performance of the UAV across all experiments (see Table 1). At the same time, the trajectories under each disturbance group (50 in total) are also provided. We chose to use the pitch angle and velocity of the UAV during the transition process to draw 2D plane curves (see Figures 12–15), where the different color simply indicates that these trajectories are different. Through the figures, we can find that the UAV can still complete the transition task with excellent performance despite the change in UAV model parameters. At the same time, we can observe that even when the UAV does not transition successfully, its terminal velocity and pitch angle mostly remain close to our desired range.

Table 1. Experimental results under different perturbations.

Perturbation Type	Parameter Range	Success Rate	Transition Time	Terminal Height Loss	Terminal Velocity	Terminal Pitch Angle
Mass	21%	94%	6.30 s	0.27 m	1.01 m/s	95.09°
Lift and Drag	40%	97.0%	6.09 s	0.109 m	0.966 m/s	93.76°
Inertia	40%	100%	6.02 s	0.072 m	0.80 m/s	93.42°
Sensor Noise	$\|\mu\| \leq 0.2, 0.005 \leq \sigma \leq 0.05$	100%	6.54 s	0.133 m	0.95 m/s	93.05°

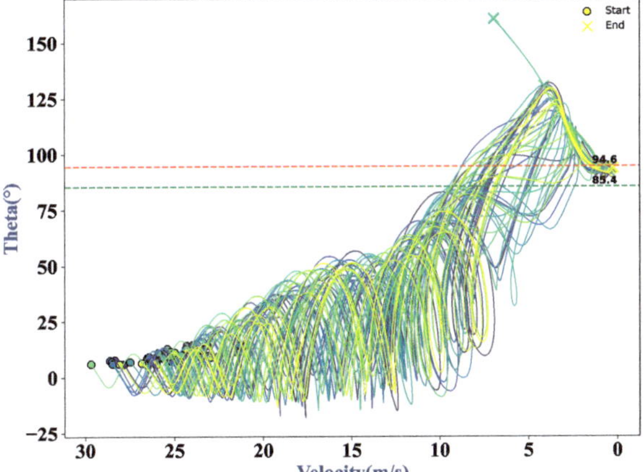

Figure 12. Mass perturbation.

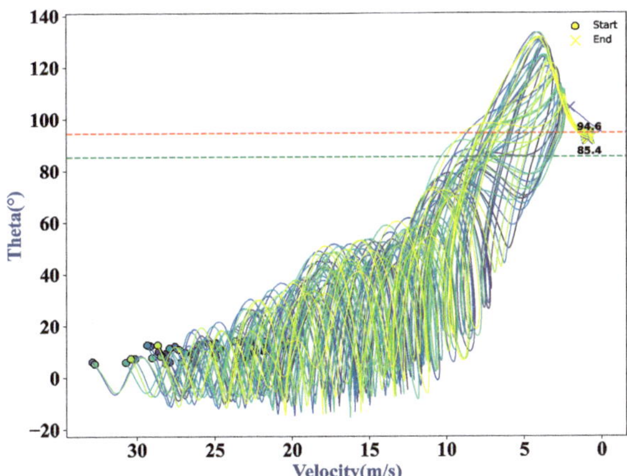

Figure 13. Lift and drag perturbation.

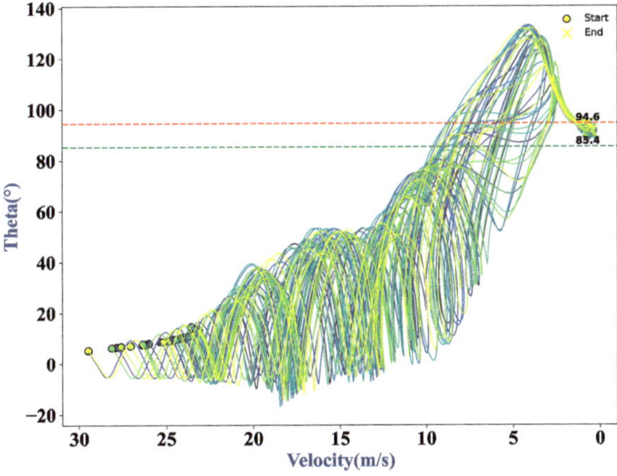

Figure 14. Inertia perturbation.

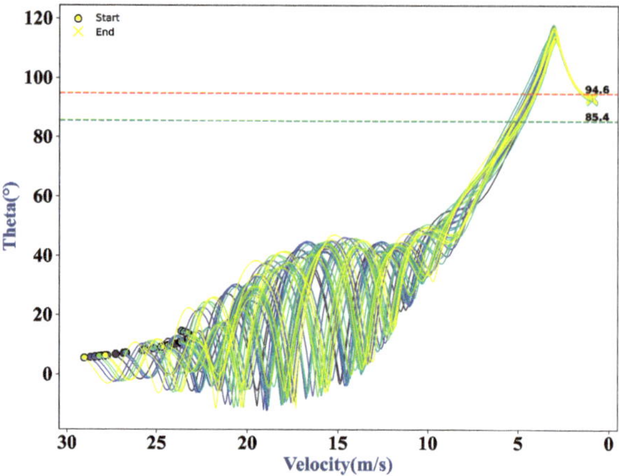

Figure 15. Sensor noise perturbation.

4.5. Transition under Wind Disturbance

In this section, we aim to verify the wind disturbance rejection ability of our method. In RL-based robotic control design, the sim-to-real gap is a challenging problem because there are inevitable mismatches between the simulator setting and real-world setting in the UAV control problem. This is not only due to the unknown perturbations of the UAV parameters but also the presence of adversarial disturbances such as wind in the real world. To account for wind disturbance, we introduce two different wind scenarios: constant-magnitude wind (ranging from -5 m/s to 5 m/s) and sinusoidal wind (with an amplitude of 5 m/s and time periods of 2 s, 3 s, 4 s, and 5 s), both along the horizontal direction, which is the X^I axis. Furthermore, the dynamics of ducted-fan fixed-wing UAVs in the presence of wind are described in detail in [3,14]. Therefore, we recreate the dynamic equations of the UAV using their modeling of wind.

To address the wind disturbance, we employ domain randomization [31] and retrain the RL agent under three different conditions: the two wind scenarios mentioned above and a no-wind scenario. Domain randomization is an approach used to overcome the sim-to-real gap by randomizing the simulator environment during the training of the RL control policy. By adapting the domain randomization approach, we aim to retrain a UAV controller that is robust against wind disturbance.

The trajectory comparison among the constant-magnitude wind (5 m/s), sinusoidal wind (time period 2 s), and no-wind conditions is shown in Figure 16, where the arrows represent the terminal state and all initialized from the angle of attack 6°.

From Figures 8 and 16, we can see that compared to the controller trained without domain randomization, the terminal height loss of the drone retrained with domain randomization remains nearly the same, while the time taken increases by 1.52 s. This phenomenon can be intuitively understood as follows: the RL algorithm selects actions with the highest expected returns under various wind disturbance scenarios, rather than actions that can achieve high returns in windless environments, but may fail to complete the task in windy conditions. Thus, a certain degree of performance loss is reasonable.

In order to better evaluate the resistance of our method to wind disturbances, we applied the UAV in 100 experiments under three conditions and from different initial states, in addition to considering the perturbation of the aerodynamic coefficients during the transition (a uniform distribution ranging between 0.8 and 1.2). The experimental results are shown in Figure 17 (success rate 90%), where the different color simply indicates that these trajectories are different.

In Figure 17, we observe that in most cases, the UAV is able to withstand wind disturbances and complete the transition task. However, for those trajectories that do not satisfy the terminal constraint in Section 3.1, we find that the trajectories generally terminate with an attitude close to vertical hover (90°) and a small velocity. This suggests that the UAV is less resistant to interference in the hovering state. In conventional approaches, a common method is to switch to the hover controller when the UAV transitions from level flight to near-hover. Wind disturbance can be resisted by switching between the two controllers and enhancing the hover controller's ability to handle wind disturbance. In RL, resistance to wind disturbance in hovering should be treated as a separate task, and the problem should be addressed using multi-task RL or meta-learning, which will be the focus of our future work.

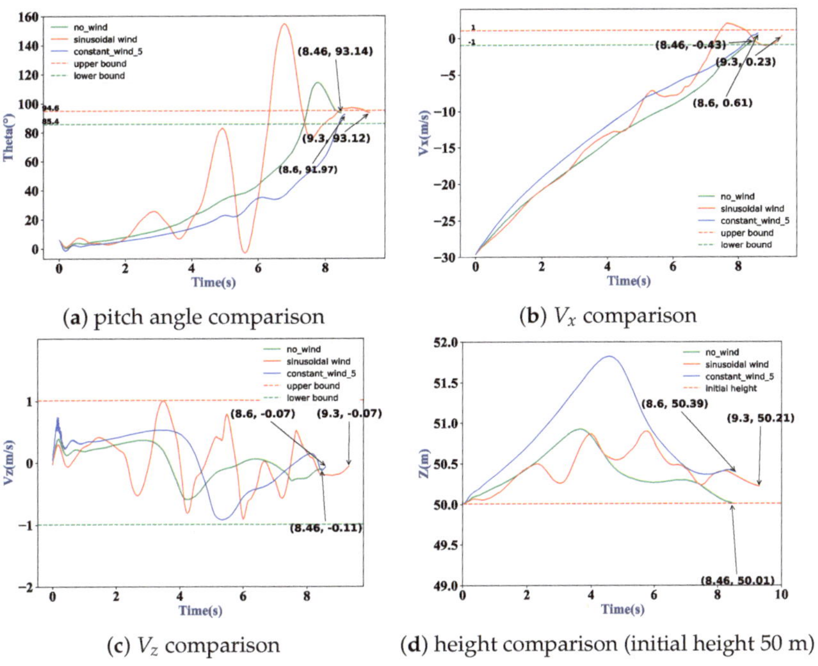

(**a**) pitch angle comparison

(**b**) V_x comparison

(**c**) V_z comparison

(**d**) height comparison (initial height 50 m)

Figure 16. Trajectory comparison under wind disturbance.

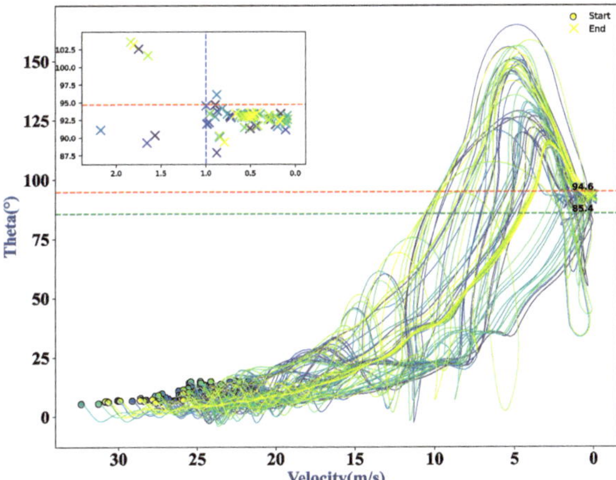

Figure 17. Trajectory under wind disturbance.

5. Conclusions

In this study, we have successfully developed a safe reinforcement learning-based approach for neat transition control during the back-transition process of ducted-fan fixed-wing UAVs. By constructing a three-degree-of-freedom longitudinal model and implementing the CPO algorithm, our method effectively addresses the challenges of integrating trajectory optimization and control methods. By comparison, we found that the introduction of a velocity constraint leads to better performance compared to adding a penalty to the reward. Furthermore, we also found that our method closely resembles GPOPS-II's performance without the need for prior knowledge. Additionally, we confirmed the robustness of the CPO algorithm and found that even when the transition was not successful, the terminal conditions remained close to our desired terminal range. Future research directions include enabling the UAV to complete multi-tasks (i.e., from hover to level flight, hover under wind disturbance, and level flight), ensuring robustness against wind disturbance, and validating the approach in the real world.

Author Contributions: Conceptualization, Y.F. and W.Z.; Data curation, Y.F. and L.L.; Formal analysis, Y.F.; Funding acquisition, W.Z.; Investigation, Y.F. and W.Z.; Methodology, Y.F.; Project administration, W.Z.; Resources, W.Z. and L.L.; Software, Y.F.; Supervision, W.Z.; Validation, Y.F.; Visualization, Y.F. and L.L.; Writing—original draft, Y.F.; Writing—review and editing, Y.F. and W.Z. All authors have read and agreed to the published version of the manuscript.

Funding: This work was supported by the 1912 project, the Key Research and Development Program of Zhejiang Province, China (Grant No. 2020C05001), and the Fundamental Research Funds for the Central Universities, China (Grant No. 2021QNA4030).

Data Availability Statement: The data presented in this study are available on request.

Conflicts of Interest: The authors declare no conflict of interest.

Nomenclature

Symbol	Description	Units
x, y, z	position of UAV	m
u, v, w	velocity in the body frame	m/s
ϕ, θ, ψ	roll, pitch, and yaw angle	rad
p, q, r	angular velocity	rad/s
g	gravity acceleration	m/s^2
α	angle of attack	rad
m	mass	kg
I_{yy}	moment of inertia	kg · m^2
L, D, T	aerodynamic lift, drag, and thrust	N
F_{cw}, F_{cs}	momentum drag and control vane force	N
$M_{pitch}, M_{cw}, M_{cs}$	pitch, momentum, and control vane moment	N · m
ρ_∞	air density	Kg/m^3
ω_r	angular velocity of the rotor	rad/s
r	radius of the rotor	m
K_{twist}	twist of blades	
a_0	rotor lift curve slope	
b	number of blades	
c_r	chord of the rotor blade	m
v_b, v_i, v_f	airflow through rotor, induced velocity, and far-field velocity	m/s
C_L, C_D, C_m	lift, drag, and pitch moment coefficient	
V	air speed	m/s
C_{lcs}	lift slope coefficient of vane	

q_{cs}	dynamic pressure	$\text{kg}/\text{m} \cdot \text{s}^2$
S_{cs}	vane area	m^2
l_e	arm of pitch moment	m
s	state space	
R	reward function	
C	cost function	
a	action space	
P	transition probability	
γ	discount factor	
V^π	value function	
A^π	advantage value function	
$J(\pi)$	expected discounted reward	

References

1. Ozdemir, U.; Aktas, Y.O.; Vuruskan, A.; Dereli, Y.; Tarhan, A.F.; Demirbag, K.; Erdem, A.; Kalaycioglu, G.D.; Ozkol, I.; Inalhan, G. Design of a commercial hybrid VTOL UAV system. *J. Intell. Robot. Syst.* **2014**, *74*, 371–393. [CrossRef]
2. Okulski, M.; Ławryńczuk, M. A Small UAV Optimized for Efficient Long-Range and VTOL Missions: An Experimental Tandem-Wing Quadplane Drone. *Appl. Sci.* **2022**, *12*, 7059. [CrossRef]
3. Argyle, M.E. Modeling and Control of a Tailsitter with a Ducted Fan. Ph.D. Thesis, Ira A. Fulton College of Engineering and Technology, Provo, UT, USA, 2016.
4. Graf, W.E. Effects of Duct Lip Shaping and Various Control Devices on the Hover and Forward Flight Performance of Ducted Fan UAVs. Ph.D. Thesis, Virginia Tech, Blacksburg, VA, USA, 2005.
5. Oosedo, A.; Abiko, S.; Konno, A.; Uchiyama, M. Optimal transition from hovering to level-flight of a quadrotor tail-sitter UAV. *Auton. Robot.* **2017**, *41*, 1143–1159. [CrossRef]
6. Li, B.; Sun, J.; Zhou, W.; Wen, C.Y.; Low, K.H.; Chen, C.K. Transition optimization for a VTOL tail-sitter UAV. *IEEE/ASME Trans. Mechatronics* **2020**, *25*, 2534–2545. [CrossRef]
7. Verling, S.; Stastny, T.; Bättig, G.; Alexis, K.; Siegwart, R. Model-based transition optimization for a VTOL tailsitter. In Proceedings of the 2017 IEEE International Conference on Robotics and Automation (ICRA), Singapore, 29 May–3 June 2017; pp. 3939–3944.
8. Kubo, D.; Suzuki, S. Tail-sitter vertical takeoff and landing unmanned aerial vehicle: transitional flight analysis. *J. Aircr.* **2008**, *45*, 292–297. [CrossRef]
9. Banazadeh, A.; Taymourtash, N. Optimal control of an aerial tail sitter in transition flight phases. *J. Aircr.* **2016**, *53*, 914–921. [CrossRef]
10. Naldi, R.; Marconi, L. Optimal transition maneuvers for a class of V/STOL aircraft. *Automatica* **2011**, *47*, 870–879. [CrossRef]
11. Jeong, Y.; Shim, D.; Ananthkrishnan, N. Transition Control of Near-Hover to Cruise Transition of a Tail Sitter UAV. In Proceedings of the AIAA Atmospheric Flight Mechanics Conference, Toronto, ON, Canada, 2–5 August 2010; p. 7508.
12. Flores, A.; Flores, G. Transition control of a tail-sitter UAV using recurrent neural networks. In Proceedings of the 2020 International Conference on Unmanned Aircraft Systems (ICUAS), Athens, Greece, 1–4 September 2020; pp. 303–309.
13. Cheng, Z.H.; Pei, H.L. Transition analysis and practical flight control for ducted fan fixed-wing aerial robot: Level path flight mode transition. *IEEE Robot. Autom. Lett.* **2022**, *7*, 3106–3113. [CrossRef]
14. Cheng, Z.; Pei, H.; Li, S. Neural-networks control for hover to high-speed-level-flight transition of ducted fan uav with provable stability. *IEEE Access* **2020**, *8*, 100135–100151. [CrossRef]
15. Zhang, W.; Quan, Q.; Zhang, R.; Cai, K.Y. New transition method of a ducted-fan unmanned aerial vehicle. *J. Aircr.* **2013**, *50*, 1131–1140. [CrossRef]
16. Xu, J.; Du, T.; Foshey, M.; Li, B.; Zhu, B.; Schulz, A.; Matusik, W. Learning to fly: Computational controller design for hybrid uavs with reinforcement learning. *ACM Trans. Graph. (TOG)* **2019**, *38*, 1–12. [CrossRef]
17. Xu, X.; Chen, Y.; Bai, C. Deep reinforcement learning-based accurate control of planetary soft landing. *Sensors* **2021**, *21*, 8161. [CrossRef] [PubMed]
18. Yuksek, B.; Inalhan, G. Transition Flight Control System Design for Fixed-Wing VTOL UAV: A Reinforcement Learning Approach. In Proceedings of the AIAA SCITECH 2022 Forum, San Diego, CA, USA, 3–7 January 2022; p. 0879.
19. Brockman, G.; Cheung, V.; Pettersson, L.; Schneider, J.; Schulman, J.; Tang, J.; Zaremba, W. Openai gym. *arXiv* **2016**, arXiv:1606.01540.
20. Patterson, M.A.; Rao, A.V. GPOPS-II: A MATLAB software for solving multiple-phase optimal control problems using hp-adaptive Gaussian quadrature collocation methods and sparse nonlinear programming. *ACM Trans. Math. Softw. (TOMS)* **2014**, *41*, 1–37. [CrossRef]
21. Gaudet, B.; Linares, R.; Furfaro, R. Deep reinforcement learning for six degree-of-freedom planetary landing. *Adv. Space Res.* **2020**, *65*, 1723–1741. [CrossRef]
22. Johnson, E.N.; Turbe, M.A. Modeling, control, and flight testing of a small-ducted fan aircraft. *J. Guid. Control Dyn.* **2006**, *29*, 769–779. [CrossRef]

23. Heffley, R.K.; Mnich, M.A. Minimum-complexity helicopter simulation math model. Technical Report; Manudyne Systems, Inc.: Los Altos, CA, USA, 1988.
24. Beard, R.W.; McLain, T.W. *Small Unmanned Aircraft: Theory and Practice*; Princeton University Press: Princeton, NJ, USA, 2012.
25. Puopolo, M.; Reynolds, R.; Jacob, J. Comparison of three aerodynamic models used in simulation of a high angle of attack UAV perching maneuver. In Proceedings of the 51st AIAA Aerospace Sciences Meeting including the New Horizons Forum and Aerospace Exposition, Grapevine, TX, USA, 7–10 January 2013; p. 242.
26. Kikumoto, C.; Urakubo, T.; Sabe, K.; Hazama, Y. Back-Transition Control with Large Deceleration for a Dual Propulsion VTOL UAV Based on Its Maneuverability. *IEEE Robot. Autom. Lett.* **2022**, *7*, 11697–11704. [CrossRef]
27. Achiam, J.; Held, D.; Tamar, A.; Abbeel, P. Constrained policy optimization. In Proceedings of the International Conference on Machine Learning, PMLR, Sydney, Australia, 6–11 August 2017; pp. 22–31.
28. Yang, T.Y.; Rosca, J.; Narasimhan, K.; Ramadge, P.J. Projection-based constrained policy optimization. *arXiv* **2020**, arXiv:2010.03152.
29. Schulman, J.; Levine, S.; Abbeel, P.; Jordan, M.; Moritz, P. Trust region policy optimization. In Proceedings of the International Conference on Machine Learning PMLR, Lille, France, 6–11 July 2015; pp. 1889–1897.
30. Schulman, J.; Moritz, P.; Levine, S.; Jordan, M.; Abbeel, P. High-dimensional continuous control using generalized advantage estimation. *arXiv* **2015**, arXiv:1506.02438.
31. Tobin, J.; Fong, R.; Ray, A.; Schneider, J.; Zaremba, W.; Abbeel, P. Domain randomization for transferring deep neural networks from simulation to the real world. In Proceedings of the 2017 IEEE/RSJ international conference on intelligent robots and systems (IROS), Vancouver, BC, Canada, 24–28 September 2017; pp. 23–30.

Disclaimer/Publisher's Note: The statements, opinions and data contained in all publications are solely those of the individual author(s) and contributor(s) and not of MDPI and/or the editor(s). MDPI and/or the editor(s) disclaim responsibility for any injury to people or property resulting from any ideas, methods, instructions or products referred to in the content.

Article

Distributed Multi-Target Search and Surveillance Mission Planning for Unmanned Aerial Vehicles in Uncertain Environments

Xiao Zhang [1,2], Wenjie Zhao [1,2,*], Changxuan Liu [1,2] and Jun Li [1,2]

1   School of Aeronautics and Astronautics, Zhejiang University, Hangzhou 310014, China; 12024061@zju.edu.cn (X.Z.)
2   Center for Unmanned Aerial Vehicles, Huanjiang Laboratory, Zhuji 311800, China
*   Correspondence: zhaowenjie8@zju.edu.cn

Abstract: In this paper, a distributed, autonomous, cooperative mission-planning (DACMP) approach was proposed to focus on the problem of the real-time cooperative searching and surveillance of multiple unmanned aerial vehicles (multi-UAVs) with threats in uncertain and highly dynamic environments. To deal with this problem, a time-varying probabilistic grid graph was designed to represent the perception of a target based on its a priori dynamics. A heuristic search strategy based on pyramidal maps was also proposed. Using map information at different scales makes it easier to integrate local and global information, thereby improving the search capability of UAVs, which has not been previously considered. Moreover, we proposed an adaptive distributed task assignment method for cooperative search and surveillance tasks by considering the UAV motion environment as a potential field and modeling the effects of uncertain maps and targets on candidate solutions through potential field values. The results highlight the advantages of search task execution efficiency. In addition, simulations of different scenarios show that the proposed approach can provide a feasible solution for multiple UAVs in different situations and is flexible and stable in time-sensitive environments.

Keywords: UAV swarm; cooperative search surveillance; mission planning; image pyramid

Citation: Zhang, X.; Zhao, W.; Liu, C.; Li, J. Distributed Multi-Target Search and Surveillance Mission Planning for Unmanned Aerial Vehicles in Uncertain Environments. *Drones* **2023**, *7*, 355. https://doi.org/10.3390/drones7060355

Academic Editor: Carlos Tavares Calafate

Received: 16 April 2023
Revised: 23 May 2023
Accepted: 26 May 2023
Published: 28 May 2023

Copyright: © 2023 by the authors. Licensee MDPI, Basel, Switzerland. This article is an open access article distributed under the terms and conditions of the Creative Commons Attribution (CC BY) license (https://creativecommons.org/licenses/by/4.0/).

1. Introduction

With the mission environment becoming increasingly complex and dynamic, multiple unmanned aerial vehicles (UAVs) have been used to form a cooperative combat system with complementary advantages and cooperation to enhance the overall combat capability of UAVs [1]. UAV swarm systems are inspired by swarm intelligence derived from the biological swarm behavior in nature such as the behavior of ants and bees, which is coordinated and controlled according to the swarm intelligence principle, making full use of local perception and interaction ability to complete relatively complex tasks. Meanwhile, as the number of UAVs increases, their computational and communication complexity will increase dramatically. Moreover, a UAV swarm has high requirements for system robustness, communication reliability, and capacity. Distributed control architecture employs an autonomous and cooperative method, breaking down complex issues into smaller sub-problems that can be addressed by individual nodes. This approach maximizes the independent abilities of each UAV and significantly enhances the computational efficiency and is widely used in search and rescue (SAR) [2], surveillance [3], civil security [4], searching [5–7], task allocation [8], and mapping [9] applications.

In the above tasks, the problem of path planning is crucial to ensure the safe and efficient completion of a particular task. In general, path planning issues associated with UAV flight can be categorized into two types: target-oriented issues and area coverage issues. In target-oriented problems, the objective is for the UAV to arrive at a designated target [10]. For instance, in a dynamic environment, UAVs may need to reach different targets for various purposes while resolving conflicts. In contrast, area coverage problems do not involve a specific destination, and the goal is typically to achieve full coverage of a designated area in the least amount of time possible [11]. Considering that in collaborative search and surveillance mission-planning problems, the UAV is not knowledgeable about the mission environment or the a priori state of the target, we mainly considered the study of the area coverage. In practical applications, monitoring missions that do not involve a ground control station can be quite complex. The control logic of an individual UAV must be meticulously designed to allow for smooth and autonomous switching between various operating modes such as search mode and continuous target-surveillance mode.

Existing research on multi-UAV collaborative search problems has focused on geometry based methods, probability-based methods, and other related methods. The geometry-based approach focuses on individual UAV planning, integrating area decomposition, and task assignment algorithms to accomplish a multi-UAV coverage search of the area. Xu et al. [12] put forward an optimal terrain coverage algorithm for UAV mission areas. This offline method segments the open space into several cells and calculates the full-coverage motion path that connects these regions, generating coverage of the deployed sensor by solving a linear programming formulation. The authors of [13] divided the task area into several equal sub-regions, each corresponding to a specific robot, to ensure complete coverage while minimizing the coverage path. Mansouri et al. [14] addressed the problem of infrastructure inspection tasks by developing a theoretical framework that slices the infrastructure through the XOY plane to facilitate the identification of branches and provide each UAV with a path to achieve full infrastructure coverage. Choi et al. [15] used this method to consider the energy required by a UAV in different phases of flight, and for the problem of covering multiple regions with regular shapes, it was reduced to a traveling sales problem (TSP) with the goal of completely covering all regions with the shortest path. Although the geometry-based approach to planning offers benefits such as no omissions, a low repetition rate, and high reconnaissance efficiency, its centralized and offline nature represents a limitation of this approach in dynamic environments. If a UAV fails or the detection environment shifts, the entire team must execute a replanning process, resulting in significant computational cost.

The probability-based approach can assist UAVs in developing cooperative search strategies. With a probabilistic model characterizing the uncertainty of the unknown environment, UAVs can continuously sense the environment and update the probability map, thereby effectively using real-time detection information for online decision making, which is applicable to dynamic search processes. Takahiro et al. [16] investigated time-constrained multi-intelligent search and action problems. They used probability density maps and a time–cost-based reward prediction function to evaluate actions. Multiple agents can make near-optimal decisions, leading to maximum gains using probabilistic reasoning. Zhen Ziyang [17] introduced a collaborative mission-planning scheme for multiple UAVs using a hybrid artificial potential field and ant colony optimization (HAPF-ACO) method for UAVs to search for and attack moving targets in uncertain environments. Zheng et al. [18] proposed a biogeographically based optimization (BBO) method to minimize the expected time for aid workers to reach a target in response to the problem of human–UAV cooperative search planning and conducted experiments showing that the proposed algorithm was superior to many popular algorithms. Duan [19] proposed a dynamic discrete pigeon-inspired optimization (D^2PIO) algorithm to handle cooperative search-attack mission planning for UAVs and achieved promising experimental results. However, probability-based path optimization methods are still subject to problems such as missing search targets and long search times. For example, the greedy strategy is short-sighted and only focuses on the largest value of the objective function in the current candidate raster (i.e., it only considers the information of a single raster or a single point) without considering the information of the whole region. Therefore, the frequently adopted greedy strategy generates duplicate paths and missing regions, and the whole search process may easily fall into local optimality, leading to decreased search efficiency as the task progresses. Furthermore, reinforcement learning (RL) has garnered increasing interest from researchers for search problems. By planning each UAV's navigation trajectory, the problem of autonomous UAV navigation in extensive, complex environments has been tackled, treating it as a Markov decision process, and the problem of autonomous UAV navigation in large-scale complex environments has been solved by planning the navigation trajectory of each UAV, which is described as a Markov decision process [20,21]. However, these methods require a priori determination of the degree of impact of previous policies in different training phases and often require a large number of calculations, which may not be suitable for real-time systems.

In addition, different strategies have been adopted in target reconnaissance missions according to specific tactical needs. Three typical applications of tactical intent in reconnaissance missions can be specified: (1) UAVs converge to the vicinity of the target in a short period of time for follow-on fire operations [17]; (2) UAV swarms chase and round up the target in a hunting manner [22]; (3) UAVs converge quickly to the vicinity of the target and maintain surveillance continuously until the target is no longer functional [3]. For the third tactical requirement, in this paper, we propose a distributed autonomous cooperative mission-planning (DACMP) method in the mission scenario of the search and reconnaissance of targets by a heterogeneous UAV swarm, which enables the UAV swarm to achieve the reconnaissance of targets while still maintaining the optimal search capability.

Therefore, the main contributions of this paper can be summarized as follows:

(1) A DACMP method is initially presented for the search and surveillance of time-sensitive moving targets in an uncertain dynamic environment by considering the constraints of maneuverability, collision avoidance, and threat avoidance. To the best of our knowledge, this DACMP method has not been explored in the literature to date.

(2) This distributed method allows heterogeneous UAVs to adaptively switch between area coverage and target surveillance missions, ensuring that each target is continuously monitored by a suitable UAV while still maintaining the optimal search capability of the entire UAV swarm for other potential targets or to handle special situations.

(3) Inspired by the successful use of image pyramids in computer vision, we developed a grid map pyramid to represent the environment. This grid graph pyramid-based heuristic enables a multiscale view of the environment, which can prevent the search process from getting stuck in local optima, thus enhancing the search efficiency of the UAV swarm.

2. Cooperative Search and Surveillance Problem Description

In this section, the cooperative search and surveillance mission is modeled, and the mission environment is established. The search and surveillance mission-planning problem is defined, and the constraints of the model are provided.

2.1. Hybrid Mission-Planning Architecture and Assumption

Multi-UAV cooperative search and surveillance mission planning is a complex optimization problem with the aim of discovering and monitoring as many targets as possible under various constraints. As previously stated, our research focused on a group of heterogeneous fixed-wing UAVs performing search and surveillance missions in an unknown region. Assume that:

1. UAVs are equipped with optical sensors that have a fixed detection range and are projected in a circle;
2. The UAVs do not have any a priori knowledge of the threat and target location;
3. Two different types of heterogeneous UAVs with different maximum speeds are employed. We believe that higher-speed UAVs have better search capabilities and are better-suited for search missions, whereas lower-speed UAVs are more suitable for continuous surveillance missions after the target has been detected;
4. Each target only needs to be monitored by one UAV, and other UAVs are encouraged to conduct more exploratory movements to find other potential targets;

5. The maximum speed of the target is lower than the speed of the UAV to ensure surveillance efficiency.

At the beginning of the mission, the UAVs are deployed in the mission area, and each UAV performs a coverage search of the area following a real-time online planned path, as shown in Figure 1. The aim is to discover as many targets as possible. Each UAV maintains a probabilistic grid map in the form of locally stored data. The UAVs then heuristically evaluate candidate solutions and make the next state-shifting decision within the performance constraints. Once the target is detected, UAVs within the target's sphere of influence are flexible in adaptive mission allocation, with low-speed UAVs maintaining continuous detection of the target, whereas high-speed UAVs are better-suited to continue searching for other potential targets. By planning the next waypoint of the UAVs, the search and surveillance efficiency of the UAV swarm is maximized during the mission, resulting in improved execution performance.

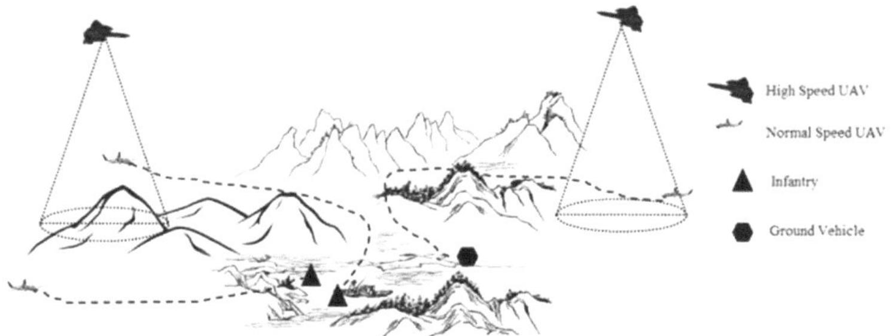

Figure 1. Illustration of a search and surveillance mission performed by the UAV swarm.

2.2. Cooperative Search and Surveillance Model of the UAV Swarm

The status of the UAV swarm is defined as:

$$\chi = \{x_1, x_2, \ldots, x_i, \ldots, x_N\} \tag{1}$$

$$x_i = [x \; y \; v \; \theta]^T \tag{2}$$

where x_i is the state of UAV i, and N is the number of UAVs. The purpose of cooperative search and surveillance of the UAV swarm is to cover the mission area and detect and monitor multiple targets while adhering to the given constraints. The objective function of this problem is defined as follows:

$$\begin{aligned} V^* &= \underset{V}{\mathrm{argmax}}(J(\chi)) \\ s.t. &\; C \leq 0 \\ J(\chi) &= J_S(\chi) + J_T(\chi) \end{aligned} \tag{3}$$

where V is the decision input, which represents the waypoints of the UAV swarm in the next iteration; C represents a set of constraint items; $J_S(\chi)$ represents the search benefit for the whole area of the UAV swarm; $J_T(\chi)$ represents the surveillance benefit of the UAV swarm, which can be defined as follows:

$$T(\chi) = \sum_{m=1}^{M} value_m \tag{4}$$

where $value_m$ is the value of the target m, and M is the number of targets monitored by the UAV swarm.

For the distributed control structure, each sub-UAV is equipped with a processor to build its own solution so that the centralized search-track mission planning indicator can be transformed into a distributed form:

$$V_i^* = \underset{V_i}{\mathrm{argmax}}(J_i(\chi))$$
$$s.t.\ C_i \leq 0$$
$$J_i(X_i) = \omega_i J_{Si}(x_i, \hat{x}_i) + (1 - \omega_i)J_{Ti}(x_i) \tag{5}$$

where $\hat{x}_i$ represents the set of neighboring UAVs under the same communication topology, and ω_i is a Boolean variable, $\omega_i = 1$ means that UAV i is performing the search task; otherwise, UAV i is performing a surveillance task. The global objective function $J(\chi)$ can be achieved by summing the local objective functions generated by each UAV:

$$J(\chi) = \sum_{i=1}^{N}(\omega_i J_{Si}(x_i, \hat{x}_i) + (1 - \omega_i)J_{Ti}(x_i)) \tag{6}$$

The specific form of $J_S(\chi)$ relies on the way the environment is modeled.

2.3. Uncertainty Map Model of The Environment

Due to the time-varying nature of the mission area, the characteristics of the target should be taken into consideration when modeling the environment. Considering that the target may move to the previously detected area, the environment was modeled using a time-varying uncertainty map [23] to ensure that the UAV swarm completes the tasks of search and tracking of all targets. Each UAV maintains an uncertainty map, which is divided into a two-dimensional probabilistic grid map $\{q_{11}, q_{12}, \ldots, q_{HW}\}$. According to the prior information of the mission, the uncertainty of a grid is expressed as follows:

$$q_{hw} = 1 - e^{-\tau_v t} \tag{7}$$

where t is the time elapsed since the last search. The uncertainty of a grid indicates the probability that it contains a hiding target, which depends on how long it has been since it was last detected and the prior mobility of the target. When $t = 0$, the grid is within the current detection radius of the UAV, and the uncertainty of the grid is zero. The uncertainty of the grid increases over time, and the rate of change is determined by the memory factor (τ_v), which depends on the prior velocity (v) of the target. The uncertainty map for a single UAV over time is shown in Figure 2. In the search process, the uncertainty of the area that has not been detected is high, and the uncertainty within the UAV detection radius is 0. The uncertainty of the searched area gradually increases with time to achieve effective return visits to the searched area by the UAV in the long search process and timely follow-up of the mission area situation. The aim of the UAV swarm searching the map is to increase the certainty of the map. The more uncertain the map's subtractions, the more aware the UAV swarm is of its environment, and the fewer places the target can hide. The surveillance coverage ratio indicates how certain the UAV swarm is about the mission area, which is calculated as the sum of the uncertainty of all grids [24]:

$$J_S(\chi) = 1 - \sum_{h=1}^{H}\sum_{w=1}^{W}\frac{q_{hw}}{HW} \tag{8}$$

where H, W indicates the size of the origin grid map.

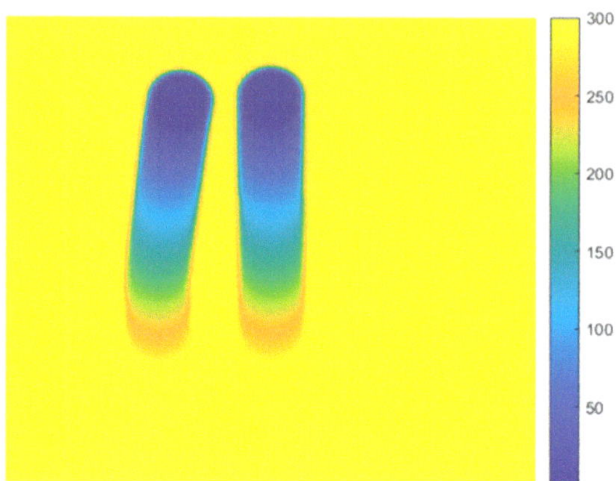

Figure 2. Uncertainty grid map of a single UAV.

2.4. Constraints of the Cooperative Planning Problem

The main constraints of multi-UAV flight path planning include dynamic constraints, collision avoidance constraints, threat avoidance constraints, and communication constraints, which are described as follows.

2.4.1. Dynamic Constraints

When carrying out a search mission, a fixed-wing UAV usually moves on a horizontal plane. The motion model can be simplified to a particle's motion on a two-dimensional plane without considering the size of the UAV.

$$\begin{bmatrix} \dot{x} \\ \dot{y} \end{bmatrix} = v \begin{bmatrix} \cos\theta \\ \sin\theta \end{bmatrix} \tag{9}$$

$$\begin{bmatrix} \dot{v} \\ \dot{\theta} \end{bmatrix} = \begin{bmatrix} a \\ \omega \end{bmatrix} \tag{10}$$

where $[x\ y]^T$ is the position of the fixed-wing UAV in flight profile, and θ is the heading angle. For fixed-wing UAVs, the linear acceleration (a) is restricted by engine performance, and its speed (v) is affected by air:

$$\begin{aligned} C_a &: |a| - a_{\max} \leq 0 \\ C_v &: v_{\min} - v \leq 0, \ v - v_{\max} \leq 0 \end{aligned} \tag{11}$$

The turning angular velocity is expressed by $\omega = \frac{v}{r}$, where the turning radius (r) should be larger than the minimum turning radius ($r_{\min}$).

$$C_r : r_{\min} - r \leq 0 \tag{12}$$

2.4.2. Collision Avoidance Constraints

Considering flight safety, the distance (d_{uav}) between UAVs should be longer than the minimum distance ($d_{\min}$) to avoid colliding:

$$C_{d_{uav}} : d_{\min} - d_{uav} \leq 0 \tag{13}$$

2.4.3. Threat Constraints

In general, the mission area contains numerous threats, which have a negative impact on the mission execution of UAVs. Therefore, UAVs must avoid threats while carrying out missions. The distance (d_{threat}) between a UAV and the threat should be longer than the threat radius (R_T), which can be expressed as:

$$C_c : R_T - d_{threat} \leq 0 \tag{14}$$

3. Design of DACMP

3.1. Model Solving

Inspired by the foraging behavior of birds, PSO is a technique with many key advantages that has been widely used in path planning for multiagent navigation [25–27]. Two important characteristics of PSO are related to swarm intelligence, namely, cognition and social coherence, which allow each particle of the population to search for solutions by following individual and group experiences rather than using traditional evolutionary operators such as mutation and crossover [25]. Therefore, compared with other algorithms, the PSO algorithm has advantages in terms of computational efficiency and stable solution convergence [28]. Moreover, PSO is not considerably affected by changes in the initial conditions and the objective function, and can adapt to various environmental constructions with a small number of parameters [29].

Suppose that N UAVs are performing a cooperative search and surveillance mission, and each UAV corresponds to a particle population composed of M particles, which represent a possible solution. Each particle can be regarded as a searching individual with a certain flight speed and direction in n-dimensional search space, and the current position of the particle is a candidate solution to the corresponding optimization problem. Particles are randomly initialized within the blue circle outside the detection radius (R_s), as shown in Figure 3. The particle moving to the optimal position changes its state by referring to its previous optimal experience and the experience of other individuals in the swarm.

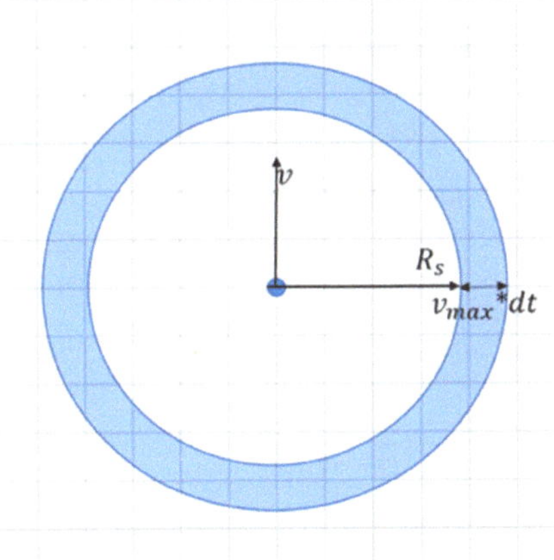

Figure 3. The distribution of particles.

To avoid early convergence to a local optimum and further improve the performance of the PSO, the algorithm is improved. First, the initial values of the control variables obey a normal distribution, traversing the allowed range. Second, according to the literature [30], inertia weight (ω) and cognitive and social coefficients (c_1, c_2) are introduced, which decrease linearly with each iteration. Finally, a random value (Q) is added to the position update formula, expressed by:

$$z_{kd}^{t+1} = z_{kd}^t + \Delta z_{kd}^t Q \tag{15}$$

where z_{kd}^t and Δz_{kd}^t are the position and position increment vectors of particle k in the d-th dimension, respectively.

3.2. State Transfer

The state transfer of a UAV is used to evaluate the quality and feasibility of each particle, and the objective value ($J(\chi)$) is maximized by a reasonable design of the state transfer function. We abstract the motion environment of the UAV swarm as a potential field, and the potential field value ($f(k)$) is used to measure particle k. In the search process, the higher the uncertainty in the region, the greater the search gain of the UAV and the more the UAV tends to move to that region, so the effect of the uncertainty map on particle k is modeled by the attraction potential value ($f_s(k)$). When the target appears within the detection range of the UAV, the target generates an attractive potential field for the particles corresponding to the UAV within its influence range, thereby converging the UAV's waypoints toward the target. In order to ensure that a target is monitored by only one UAV, the competing repellent potential values between UAVs are also considered so that the potential value received by particle k of a UAV within the target influence range is represented as $f_t(k)$. The total potential value ($f(k)$) of a particle can be described as:

$$f(k) = f_s(k) + f_t(k) \tag{16}$$

3.2.1. Construction of Pyramid Map

When a UAV performs a search mission, the goal is to reduce the uncertainty in the mission area to maximize the probability of the detection of the target. Therefore, UAVs tend to fly to locations with high uncertainty. However, a single level of an uncertainty map is not enough to represent the whole information of the mission area. Information at different scales also contributes to the understanding of the environment. With the aim of obtaining a larger global view, the image pyramids initially used in computer vision were a series of images arranged in a pyramid shape with gradually decreasing resolution from the same original image.

Inspired by its successful use in computer vision, a map pyramid was constructed for a better understanding of the mission area. The map pyramid is a multiscale representation of maps, which is an effective structure with simple concepts to interpret maps with multiple resolutions for a better understanding of the task area. The best-known hierarchical structures are Gaussian [31] and Laplacian pyramids [24]. The pyramid model is one of the most intuitive multiscale descriptions of the signal and generally consists of two steps: first, the map is smoothed by a Gaussian filter; then, the smoothed map is sampled or interpolated to obtain a sequence of scaled-down or scaled-up maps, as shown in Figure 4. Each level of the map in the sequence is the result of the sampling of every other row and column after Gaussian filtering of the previous level of the map. That is,

$$G_l(\alpha, \beta) = \sum_{m=-2}^{2} \sum_{n=-2}^{2} w(m, n) G_{l-1}(2\alpha + m, 2\beta + n) \tag{17}$$

where $G_l(\alpha, \beta)$ is the map of the l-layer Gaussian pyramid; G_0 is the original map as the lower layer of the Gaussian pyramid; $w(m, n) = h(m) \cdot h(n)$ is an $m \times n$ window function with low-pass property; $h(m)$ is the Gaussian density distribution function that satisfies the following constraints: normalization, symmetry, parity term, and other contribution terms. Then, a typical 5×5 window function $w(5, 5)$ can be expressed as follows:

$$w(5,5) = \frac{1}{256} \begin{bmatrix} 1 & 4 & 6 & 4 & 1 \\ 4 & 16 & 24 & 16 & 4 \\ 6 & 24 & 36 & 24 & 6 \\ 4 & 16 & 24 & 16 & 4 \\ 1 & 4 & 6 & 4 & 1 \end{bmatrix} \tag{18}$$

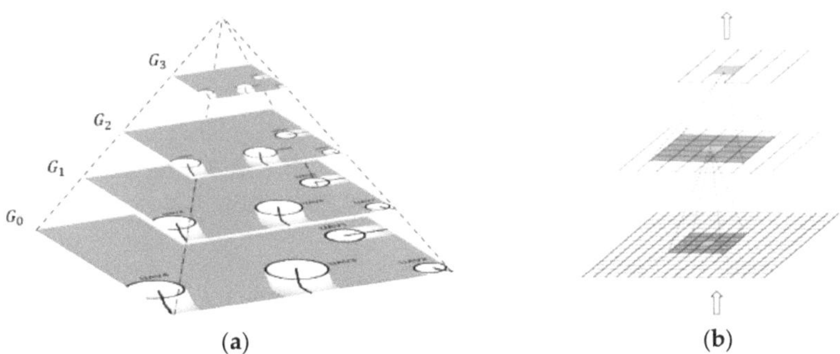

Figure 4. Illustration of the uncertainty gird map pyramid: (**a**) pyramid map; (**b**) generation process.

Thus, G_0, G_1, ..., G_L form the Gaussian pyramid of the uncertainty map.

To prevent the waypoints from falling into local optima during the search process, we used information from each layer of the map to construct an attractive potential function, which can be expressed as:

$$f_{S,l}(k) = \frac{\sum_{l=1}^{L} \lambda_l P(X_i^k(t), l)}{1 + \left(\frac{D(x_i(t), X_i^k(t)) - R_i^{search}}{R_i^{search}} \right)^2} \tag{19}$$

where $P(X_i^k(t), l)$ is the uncertainty that the potential target exists in the same region as particle k in the l-th level of the map pyramid at moment t. Regions with high uncertainty should have a higher search priority. λ_l is a weight used to balance the uncertainty of each layer, and R_i^{search} is the detection radius of UAV i. A Dubins path $(D(x_i(t), X_i^k(t)))$ is used to measure the distance from particle k to UAV i.

$f_{S,l}(k)$ describes the local potential field of the UAV; therefore, knowledge of the environment is still short-sighted. In order to further improve the global search capability of the UAV, an area with high uncertainty and that could be reached by the current UAVs in the next phase in the whole area was detected in the upper layer of the pyramid. Taking the seven-layer pyramid as an example, the seventh layer of the pyramid divided the map into 4 x 4 sub-areas. The center point of the area with the highest uncertainty among the eight areas adjacent to the current UAV was selected. The attractive field function of the global guidance point of particle k is expressed by:

$$f_{s,g}(k) = \frac{1}{1 + e^{\frac{\theta(P(x,y), x_i(t), x_i^k(t))}{\pi}}} \tag{20}$$

Figure 5 shows a larger view of the global guideline $(P(x,y))$ represented by the center of the nearby area. $\theta(P(x,y), x_i(t), x_i^k(t))$ is the angle among the position of the global guideline (P), UAV i, and particle k, which causes the optimized waypoint to be more inclined to the direction of the global guideline and guides the UAV to search in the direction of the high-uncertainty region, thereby improving the search efficiency.

$$f_S(k) = \eta f_{S,l}(k) + (1 - \eta) f_{s,g}(k) \tag{21}$$

where η is the balancing factor and is usually set to 0.5 to maintain a balance between the local and global information.

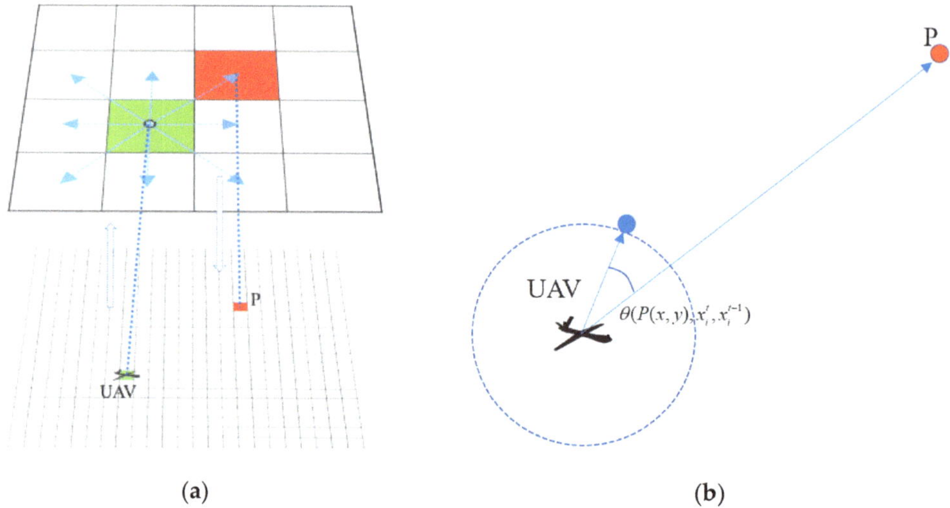

(a)   **(b)**

Figure 5. The use of the global guideline: (**a**) global guideline (*P*); (**b**) illustration of the angle.

3.2.2. Distributed Adaptive Target Allocation Approach

When the UAVs detect one or more targets, the targets have an attractive effect on the particles within their sphere of influence, guiding the UAVs toward the targets and maintaining surveillance on them. The attraction potential value of the target to particle k is expressed as follows:

$$f_{t,att}(k) = \begin{cases} \sum_{m=1}^{M} \dfrac{\omega_{t,i}}{d(X_i^k(t),g_m(t))v_i^{max}v_i^{max}} & if\,(d(X_i^k(t),g_m(t)) < D_i) \\ 0 & else \end{cases} \tag{22}$$

where $\omega_{t,i}$ is the weight of an attractive potential item; $g_m(t)$ is the state of target m at time t; $d(X_i^k(t),g_m)$ denotes the distance between the particle k and target m; v_i^{max} is the maximum velocity of UAV i. D_i is the maximum radius of influence of the target on UAV i. As shown in Figure 6, both the high-speed UAV1 and the low-speed UAV2 were within the range of influence of the target. Because the high-speed UAV is more suitable for area exploration and the low-speed UAV is more suitable for maintaining continuous surveillance of the target, the low-speed UAV had a larger range of influence and was more vulnerable to the target than the high-speed UAV.

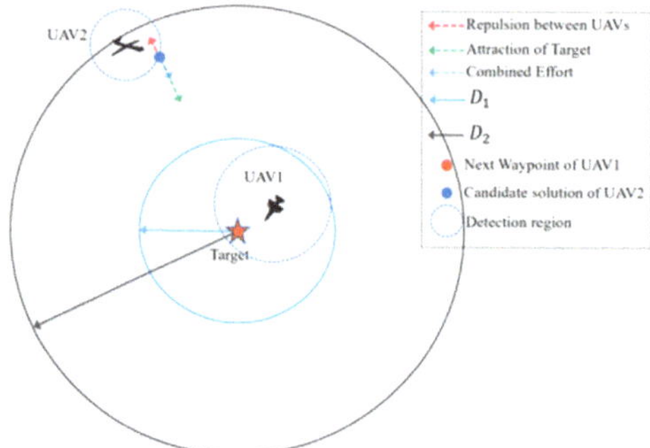

Figure 6. Virtual force diagram of the candidate solution.

When more than one UAV is within the scope of the target, the exclusion behavior between UAVs is introduced. The UAV swarm is encouraged to perform more exploration actions while ensuring that one target is scouted by one UAV, preventing UAVs from clustering in a small group, therefore searching for all targets faster. The corresponding repulsive field value of particle k is expressed as:

$$f_{t,rep}(k) = \begin{cases} -\sum_{j=1,j\neq i}^{N} \dfrac{\omega_{r,i}}{d(X_i^k(t),x_j(t+1))v_i^{max}v_j^{max}} & if\ both\ in\ the\ scope\ of\ target \\ 0 & else \end{cases} \tag{23}$$

where $\omega_{r,i}$ is the weight of the repulsive potential item, and $d(X_i^k(t), x_j(t+1))$ denotes the distance between particle k of UAV i and the waypoint of UAV j at the next moment $(t+1)$. Therefore, the potential energy function combining the target and other UAV effects is:

$$f_t(k) = f_{t,att}(k) + f_{t,rep}(k) \tag{24}$$

According to the different motion characteristics of the UAVs, we can adjust the values of $f_{t,att}$ and $f_{t,rep}$ to meet the task assignment requirements. Figure 7 depicts a case of two heterogeneous UAVs competing for a target. Assuming that a high-speed UAV with greater search capability detects the target first, if there is no low-speed UAV within the target range, the high-speed UAV maintains surveillance of the target, and if a low-speed UAV appears within the target range, the integrated potential energy field of the nearby low-speed UAV changes at this time. To make the high-speed UAV with higher search capability continue to explore the potential target and the lower-speed UAV maintain surveillance of the target, the attractive potential value of the target to the particle of the low-speed UAV should be greater than the repulsive potential value of the high-speed UAV to the particle of the low-speed UAV. Then, the integrated potential field guides the low-speed UAV toward the target; this inequality can be achieved by adjusting ω_t and ω_r as follows:

$$\frac{\omega_{t,low}}{v_{max}^{low}} > \frac{\omega_{r,low}}{v_{max}^{high}} \tag{25}$$

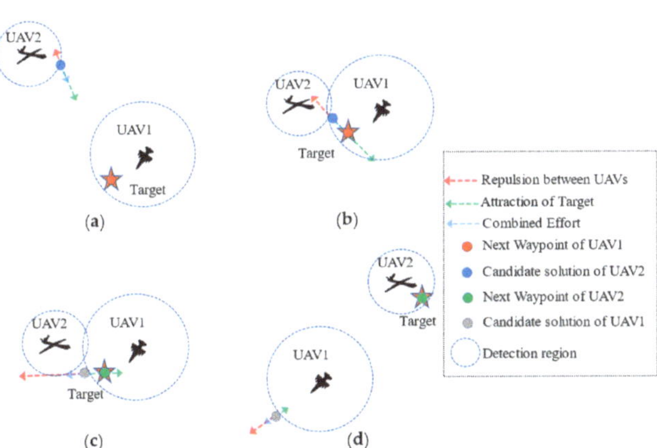

Figure 7. Analysis of the motion and force situation of a low-speed UAV and a high-speed UAV during mission exchange: (**a**) UAV 1 detects the target and maintains surveillance; (**b**) the particle swarm of UAV 2 converges to the position of the target; (**c**) after UAV 2 moves to the vicinity of the target and UAV 1, the repulsive value on the particles of UAV 1 is greater than the attractive value, and its particle swarm no longer converges to the position of the target, thereby relieving surveillance of the target; (**d**) after a period, UAV 1 maintains surveillance of the target, and UAV 2 leaves the target area and continues area exploration.

When the low-speed UAV moves near the target, the attractive potential value of the target to the particle of the high-speed UAV should be lower than the repulsive potential value of the low-speed UAV to the particle of the high-speed UAV:

$$\frac{\omega_t, high}{v^{low}_{max}} < \frac{\omega_r, high}{v^{high}_{max}} \tag{26}$$

Then, the integrated force field leads the particle swarm of the high-speed UAV away from the target and back to the search state. The result is the same when the low-speed UAV finds the target first.

Assuming that two UAVs are isomorphic, to prevent multiple UAVs from tracking a target at the same time, the repulsive potential value between the particles of the isomorphic UAVs should be greater than the attractive potential value of the target to them:

$$\omega_{t,i} < \omega_{r,i} \tag{27}$$

3.2.3. Local Planning Module Based on APF

Appropriate control sequences should be provided to the UAV to ensure that it can reach the next optimal track point quickly and safely. In a dynamic and uncertain environment, the main challenge associated with this problem is to plan and execute an inflation-free path. In recent years, the method based on APF has shown considerable potential in path planning in highly dynamic environments due to its rapidity and simplicity.

The next optimized waypoint generates a gravitational field (U_{att}) with the APF method, which causes the UAV to approach the next waypoint by applying a gravitational force (F_{att}) to the UAV. Based on the real-time waypoints ($f_i(X)$) of UAV i, the attraction field is expressed as:

$$U_{att,i} = k_{att} \cdot f_i(X) \tag{28}$$

where $k_{att} > 0$ represents the gravitational coefficient. The gravitational force is the gradient of the gravitational potential field, which can be expressed as:

$$F_{att,i}(X) = \nabla U_{att,i}(X) \tag{29}$$

The UAV may encounter no-fly zones that need to be avoided to ensure safety. Assuming that the threat is a circle, based on the APF method, the threat area creates a repulsive force field, keeping the UAV away from the threat area. The repulsive force of threats can be obtained by:

$$F^T_{rep,i} = \begin{cases} k_{rep} \cdot \sum_{e \in S} \left(\frac{1}{\|x_{ie}\|^2} - \frac{1}{(d^t_{max,e}-d_0)^2} \right) \cdot \hat{x}_{ie}, & \|x_{ie}\|^2 \le d^t_{max,e} \\ 0, & \|x_{ie}\|^2 > d^t_{max,e} \end{cases} \tag{30}$$

where k_{rep} is the repulsion coefficient, and S is the threat set. x_{ie} is the position vector of the detected e-th threat pointing to the i-th UAV, $\hat{x}_{ie} = x_{ie}/\|x_{ie}\|$. $d^t_{max,e}$ represents the influence radius of the e-th threat's repulsion field. d_0 is the minimum safe distance between the UAV and the threat center.

As the number of UAVs in the same airspace increases, the trajectories between UAVs may appear to be close or even overlap in time and space during mission execution. To ensure the safety of UAVs, we introduced repulsive fields between the UAVs [17]:

$$U^V_{rep,i} = \begin{cases} \sum\limits_{j=1,j\neq i}^{N} \frac{b}{e^{\frac{\|x_{ij}\|}{c}} - e^{\frac{\|x_{ij}\|_{min}}{c}}}, & if\ d_{uav} \in [d_{min}, d^v_{max}] \\ 0 & else \end{cases} \tag{31}$$

where b and c are adjustable parameters that govern the magnitude and rate of change of the repulsive field, respectively; $\|x_{ij}\|$ is the distance from UAV i to the UAV j; d_{max}

represents the maximum operating distance of the repulsion field between the UAVs. The repulsion force between the UAVs is:

$$F_{rep,i}^{V} = -\nabla U_{rep,i} = \sum_{j=1,j\neq i}^{N} \frac{b}{c} \cdot \frac{1}{(e^{\frac{\|x_{ij}\|}{c}} - e^{\frac{\|x_{ij}\|_{min}}{c}})^2} e^{\frac{\|x_{ij}\|}{c}} \cdot \hat{x}_{ij} \tag{32}$$

where $\hat{x}_{ij} = x_{ij}/\|x_{ij}\|$ is the unit vector pointing from UAV j to UAV i.

4. Experimental Analysis

To provide a comprehensive analysis of the DACMP algorithm presented in this paper, a series of simulation experiments was conducted. The simulation platform consisted of a desktop computer equipped with a 64-bit Windows 10 operating system, an Inter(R) Core (TM) i5-8265u 1.6 GHz CPU, and 8 G of RAM. The experiment was conducted using C++14 programming language, and MATLAB was employed for the analysis of the experimental results. In accordance with the information provided in Section 3, the procedural flow of the simulation is shown in Figure 8.

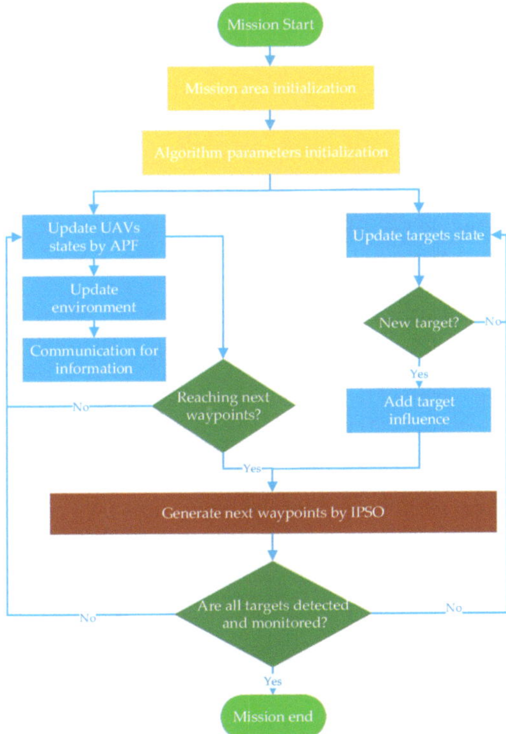

Figure 8. Flow chart of the DACMP algorithm.

4.1. Experimental Parameters

The mission area was 50 km × 50 km, inside which both the UAVs and targets moved. A seven-layer map pyramid with a 100 m × 100 m bottom resolution was used to model the mission area. To verify the performance of the algorithm for heterogeneous UAVs in a multi-target search and tracking scenario, two different types of UAVs were adopted. We used the same kinetic model of the UAV to simulate the movement of the target and a Wiener random process of acceleration. The information on the UAVs and targets is shown in Table 1.

Table 1. Parameter setup of the UAVs and targets.

Parameter	UAV		Target
Type	A	B	/
Speed	30~50 m/s	60–90 m/s	0–20 m/s
Linear acceleration	±0.4 m/s^2	±0.6 m/s^2	±0.4 m/s^2
Maximum bank angle	30°	20°	/
Detection radius	1.5 km	2 km	/
ω_t	5.0	2.0	/
ω_r	5.1	2.1	/
D	10 km	4 km	

4.2. Comparison with the Search Model

Given that our method is an online, real-time, and distributed approach, it is suitable for comparison with other probability-based methods. For this purpose, we selected the basic but effective greedy strategy as the benchmark for comparison. To analyze the performance of the DACMP algorithm in the search process, we conducted a set of experiments and compared our heuristic strategy with the greedy strategy. The advantages of the two methods were analyzed by comparing two metrics: the coverage of the region and the number of targets searched. We used average values to estimate the performance of the two strategies after 30 executions.

Assuming that six type-B UAVs search for 14 targets in the mission area without considering threats, the initial positions of the UAVs were (0, 20) km, (0, 22) km, (0, 24) km, (0, 26) km, (0, 28) km, and (0, 30) km. We believed that the target information was captured without further surveillance if the target appeared in the UAV's field of view, and the UAV then continues to search for the remaining targets.

The trajectories of the UAV swarm for a regional search task guided by two methods are shown in Figure 9 under the same initial conditions. The colored curves and dots represent the path and current position of each UAV, respectively. The black dashed line indicates the path of the target. Figure 9a shows the coverage path of multi-UAVs using the greedy strategy. The multiple UAVs can complete the target search of the region, but only the uncertainty of the current position of the candidate solution is considered in the optimization process, which can easily cause the whole search task to fall into the local optimum and lead to the repeated search of UAVs in a small area. Figure 9b shows the coverage path of multiple UAVs based on the DACMP heuristic strategy, which considers not only the uncertainty of the current candidate raster but also the uncertainty of the region near the raster as well as the directional guidance of the urgent region to be searched. The pursuit of long-term optimality enables the UAV to choose a better path to search the region, thereby improving the overall search coverage. The path of the UAV is evenly distributed in any part of the region.

Figure 10 displays the mission area coverage rates and the numbers of searched targets under the two methods. The proposed DACMP algorithm demonstrated better mission execution efficiency compared to the greedy strategy. As shown in Figure 10a, the mission area coverage rates of the greedy strategy and DACMP algorithm were roughly the same in the early stage. With an increase in the number of iterations, the advantage of the DACMP became increasingly apparent.

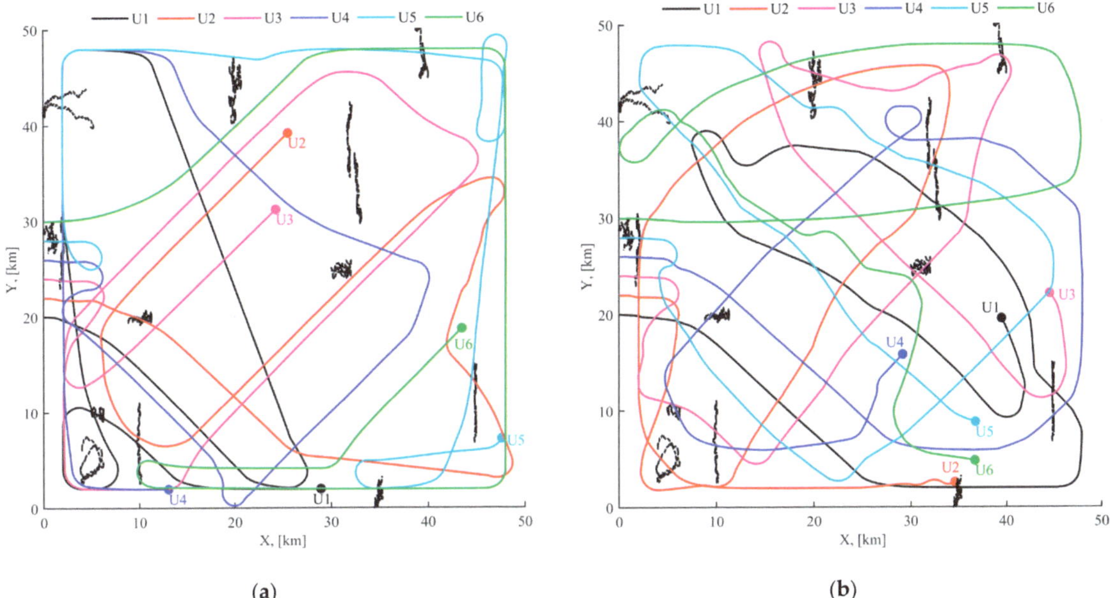

(**a**)   (**b**)

Figure 9. The paths of multiple UAVs after 2000 simulation steps: (**a**) greedy strategy; (**b**) DACMP strategy.

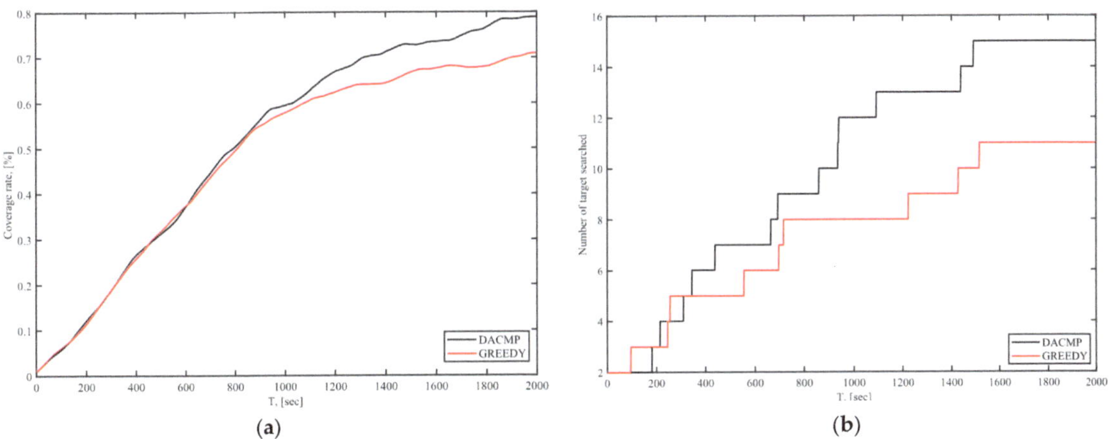

(**a**)   (**b**)

Figure 10. Comparison of the results of the two methods: (**a**) coverage rate; (**b**) number of targets searched.

To ensure reliability, we carried out 30 simulations of the two methods to better analyze the performance. As shown in Figure 11, the simulation time of each method was 2000 s. The results showed that the average coverage area under the greedy strategy was 0.712, with 12.1 targets searched. The DACMP algorithm could search for 13.7 targets, and the average area coverage was 0.763. Therefore, our method has a significant advantage in terms of the number of searched targets and the coverage region.

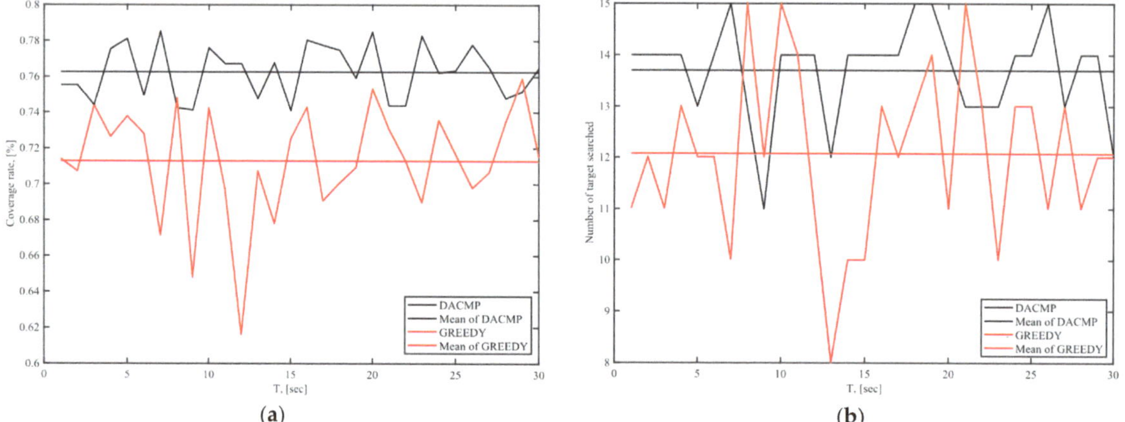

Figure 11. Average coverage rate of stationarity in 30 simulation times: (**a**) coverage rate; (**b**) number of targets searched.

4.3. Mission Execution Analysis

Organizing UAVs for collaborative search and surveillance missions under various constraints is a challenging task, particularly in complex environments with threats. To assess the feasibility and effectiveness of the proposed method, three cases were established. Case 1 and Case 2 were designed as non-threatening and threatening scenarios, respectively, to verify the feasibility and reliability of the proposed method. Case 3 builds on Case 2 by adding a new UAV in the middle of the mission to illustrate the scalability and adaptability of the solution.

Case 1: Three UAVs were deployed into the mission area, which contained two targets dispersed in a threat-free environment. UAV 1 and UAV 2 are high-speed UAVs with initial positions of (20, 22) km and (36, 22) km, with a low-speed UAV (3) located at (10, 22) km. Figure 12a–c represents multiple UAVs performing missions at different moments. As shown in Figure 12a, UAV 1 detected target 1 first, at which time the particles gradually converged to the position of target 1 under the effect of the attractive field, thereby guiding UAV 1 to fly toward target 1. Figure 12b shows that the low-speed UAV (3) within the sphere of influence of target 1 flew toward target 1 because its particles were more attracted to the target than repelled by UAV 1. When UAV 3 reached the vicinity of target 1, the particles of UAV 1 were subjected to a repulsive potential value greater than the attractive potential value, and UAV 1 automatically left target 1 to undertake the search task. As shown in Figure 12c, UAV 1 passed through the range of influence of target 2 while searching, but because UAV 2 maintained surveillance of target 2, UAV 1 was not disturbed by the attraction of the target and continued to search because the repulsive potential value among the isomorphic UAVs to which the particles of UAV 1 were subjected was greater than the attractive potential value of the target to the particles, so the nearest UAV could prioritize its own surveillance relationship with the target and avoid the entanglement of multiple isomorphic UAVs to one target, leading to a reduction in the search capability of the UAV swarm. Figure 12d shows the distance calculation results of multiple UAVs and target 1. This experiment shows that the proposed algorithm can make the UAVs adaptively switch between search and surveillance tasks according to their own capabilities so that the whole UAV swarm can reach the optimal search and surveillance state. In this case, the proposed algorithm can accomplish the distributed target assignment task quickly and efficiently without threats, which validates the basic feasibility of the proposed approach.

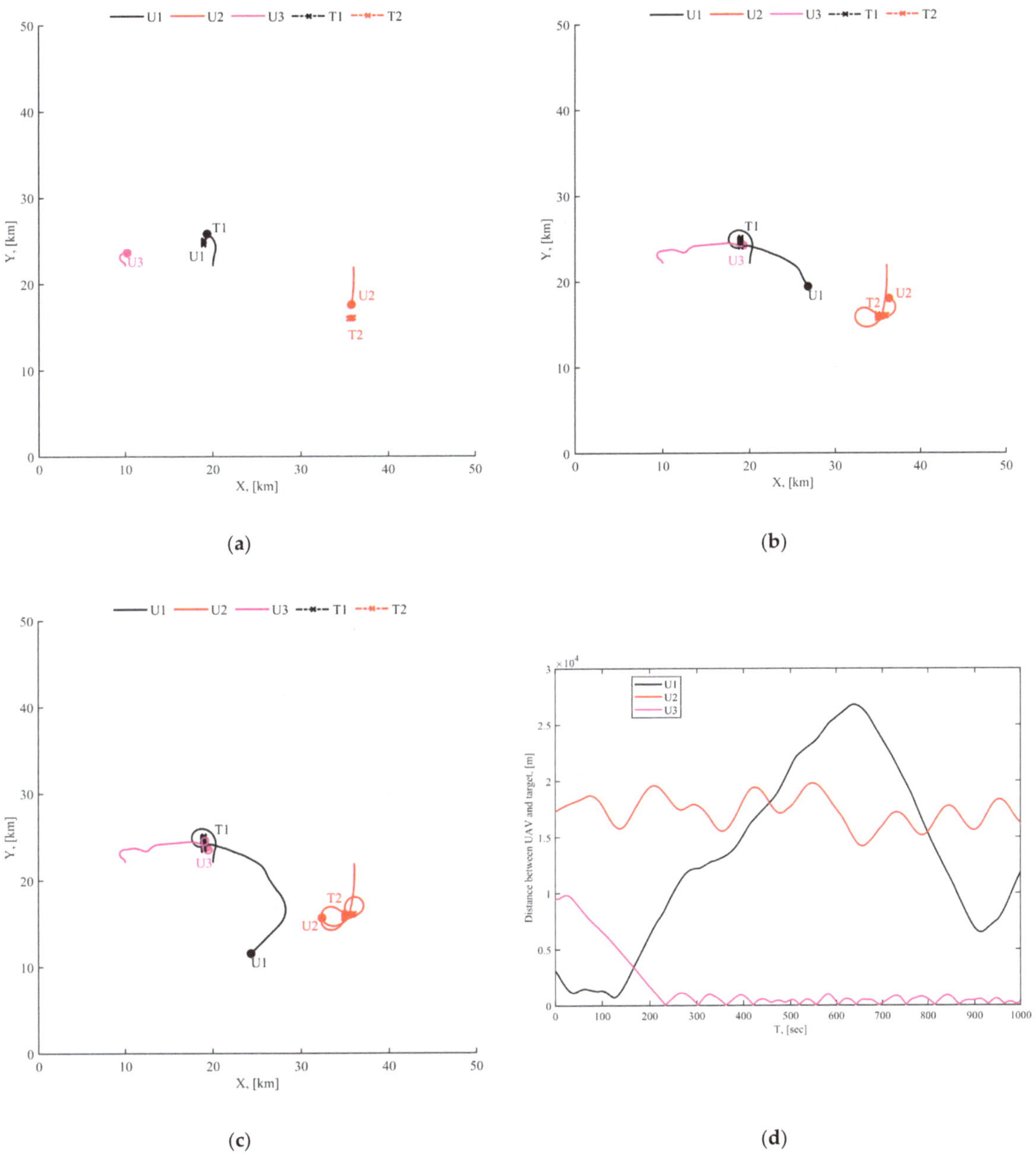

Figure 12. Cooperative search and surveillance mission-planning results in Case 1: (**a**) sub-mission at second 50; (**b**) sub-mission at second 250; (**c**) sub-mission at second 350; (**d**) distance between UAVs and target 1.

Case 2: Ten UAVs were deployed into the mission area, with six targets scattered in the environment and four threats. UAVs 1–5 are high-speed UAVs with initial positions of (10, 1) km, (29, 1) km, (30, 49) km, (15, 49) km, and (39, 49) km, and UAVs 6–10 are low-speed vehicles located at (20, 1) km, (35, 1) km, (5, 49) km, (23, 49) km, and (45, 49) km. Figure 13 represents multiple UAVs performing missions at different moments. Targets 4, 6, 1, 3, and

2 were the first to be detected and were temporarily kept under surveillance by high-speed UAVs 1, 5, 4, 3, and 2, respectively. Target 5 was detected and kept under continuous surveillance by low-speed UAV 7. After the high-speed UAVs detected the targets, they shared the target information with other UAVs through information interaction, and the particles of the low-speed UAVs near the targets started to converge toward the targets under the effect of the target attraction field. As shown in Figure 13b,c, UAV 9 flew toward target 3, and the particles of UAV 3 were repelled by UAV 9, which canceled the surveillance of the target and returned to search the area. The remaining high-speed UAVs (5, 4, and 1) also completed the mission transition with low-speed UAVs (10, 8, and 6, respectively). In this process, the low-speed UAVs replaced the high-speed UAVs to monitor the target, and the high-speed UAVs continued to search the area. The response result of the search and surveillance task is shown in Figure 14. Figure 15 illustrates the task execution of UAVs during the whole search and surveillance task-planning period. Figure 15 shows the coverage rate during the execution of the task. In this case, the UAVs completed the search and surveillance tasks while avoiding threats and adaptively completed the target assignment, illustrating the practicality of the proposed method in complex environments.

Case 3: To further verify the scalability and adaptability of the DACMP algorithm to the UAV swarm system, we added a new UAV in the process of mission execution in Case 2. We assumed that a new type-B UAV (UAV11) was initially located at position (50, 25) km and was added to the UAV swarm system at 300 s. The response result of the search and surveillance task is shown in Figure 16, and the task allocation of the UAV swarm is shown in Figure 17. As shown by the comparison of the coverage rate (Figure 18), the coverage increased after the addition of the new UAV, and the new UAV quickly adapted to the UAV swarm system and elevated the efficiency. Therefore, the multi-UAV system based on the DACMP algorithm is flexible, stable, and expandable.

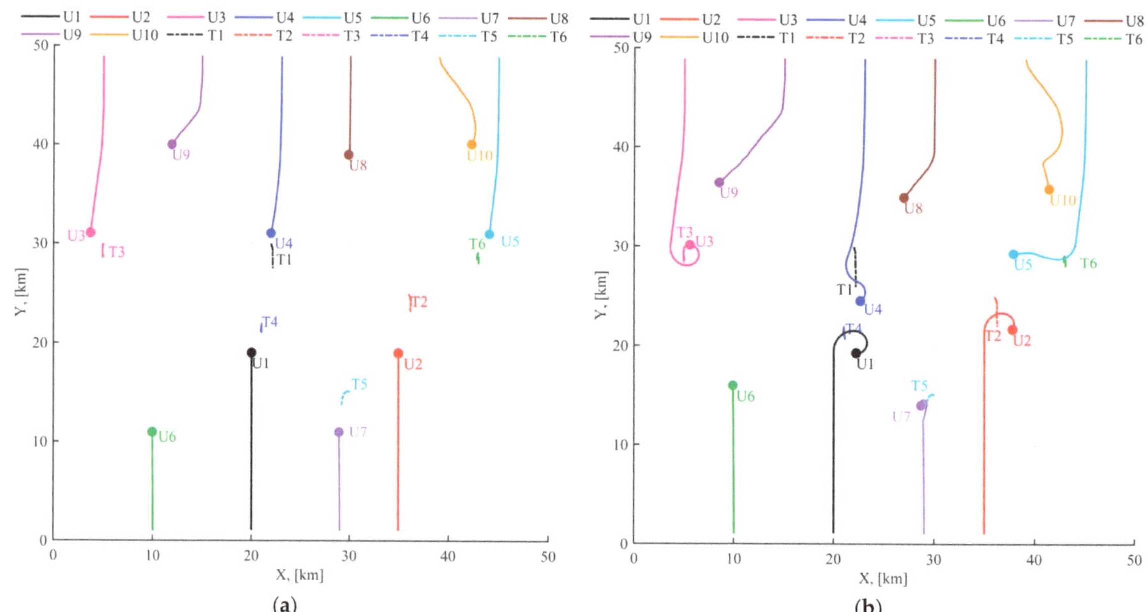

Figure 13. *Cont.*

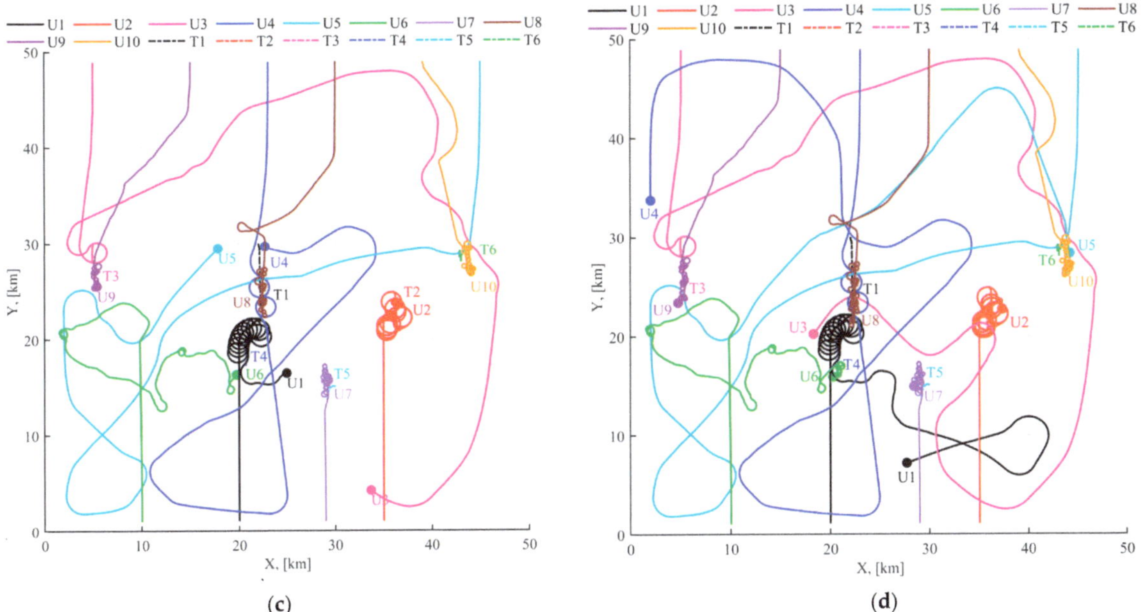

Figure 13. Cooperative search and surveillance mission planning results in Case 2: (**a**) sub-mission in second 200; (**b**) sub-mission in second 300; (**c**) sub-mission in second 1500; (**d**) sub-mission in second 2000.

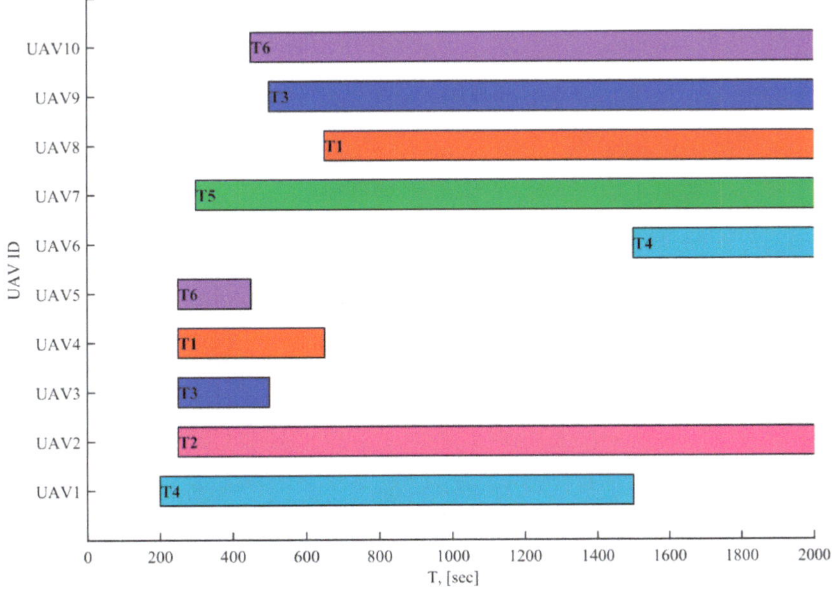

Figure 14. Task loads of the UAV swarm.

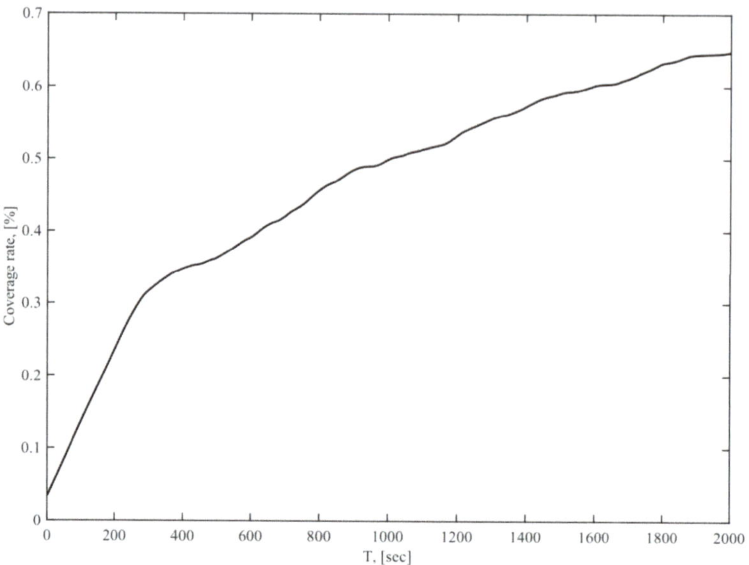

Figure 15. Coverage rate of the UAV swarm.

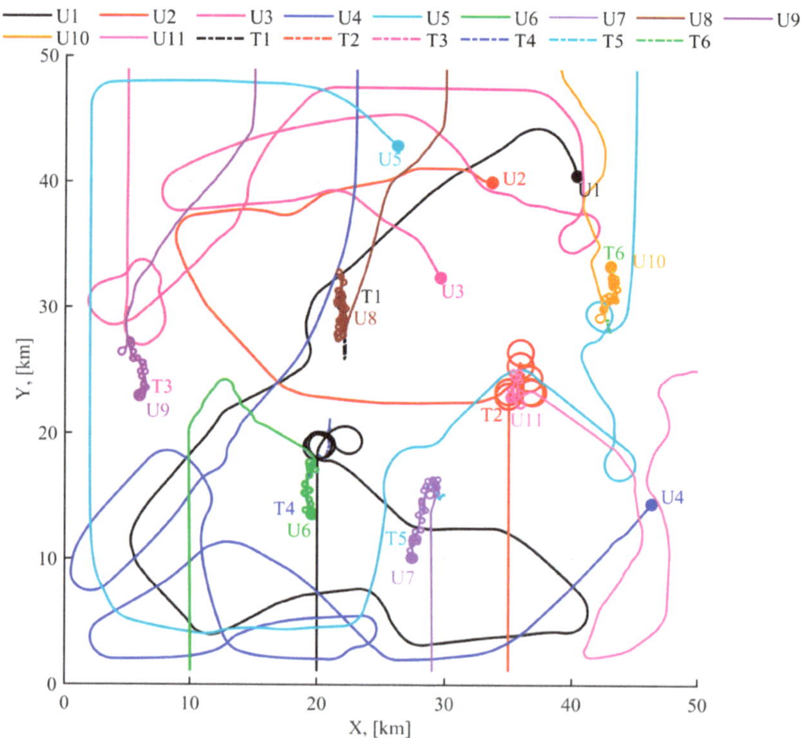

Figure 16. Cooperative search and surveillance mission-planning results in Case 3.

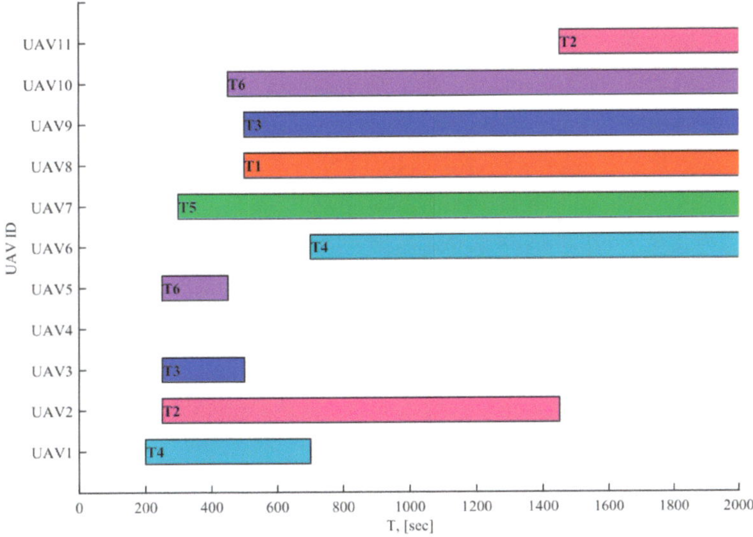

Figure 17. Task allocation of the UAV swarm.

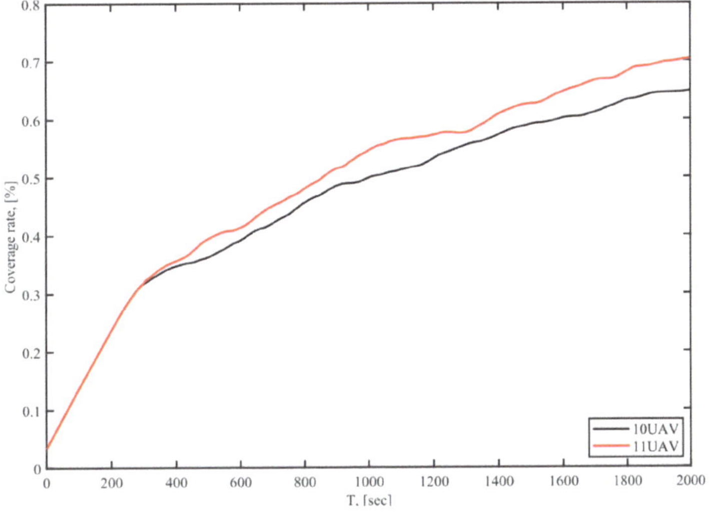

Figure 18. Comparison of the coverage rate between scenarios with 10 UAVs and 11 UAVs.

5. Conclusions

In this paper, we proposed a novel distributed autonomous collaborative mission-planning method for the multi-UAV search and surveillance mission-planning problem. Satisfactory performance was achieved using grid pyramid time-varying uncertainty maps to simulate environmental cognition and construct heuristic search strategies. By modeling the effects of uncertain maps and targets on candidate solutions as potential field values, an adaptive distributed mission assignment was enabled for multiple UAVs, maximizing the area coverage capability of heterogeneous UAV swarms. Numerical simulation results and analyses demonstrated that the proposed method could achieve fast area coverage and dynamic task assignment under multiple constraints and is robust to the dynamic topology of the UAV swarm.

However, only two types of UAVs with area coverage search strategies were analyzed in this paper. How to cope with more UAV types and more target types is a problem to be considered in the future, and experiments will be conducted on the performance of the algorithm in real scenarios.

Author Contributions: Conceptualization, X.Z. and W.Z.; methodology, X.Z.; software, X.Z.; validation, X.Z.; formal analysis, X.Z.; investigation, X.Z. and W.Z.; resources, W.Z.; data curation, X.Z.; writing—original draft preparation, X.Z.; writing—review and editing, W.Z. and C.L.; visualization, X.Z.; supervision, W.Z.; project administration, W.Z. and J.L.; funding acquisition, W.Z. and J.L. All authors have read and agreed to the published version of the manuscript.

Funding: This research was funded by the 1912 project; Key Research and Development Program of Zhejiang Province, China (grant No. 2020C05001); the Fundamental Research Funds for the Central Universities, China (Grant No. 2021QNA4030).

Data Availability Statement: Data available on request from the authors.

Conflicts of Interest: The authors declare no conflict of interest.

References

1. Ollero, A.; Kondak, K. 10 years in the cooperation of unmanned aerial systems. In Proceedings of the RSJ International Conference on Intelligent Robots and Systems, Vilamoura, Algarve, Portugal, 7–12 October 2012.
2. Du, Y.-C.; Zhang, M.-X.; Ling, H.-F.; Zheng, Y.-J. Evolutionary Planning of Multi-UAV Search for Missing Tourists. *IEEE Access* **2019**, *7*, 73480–73492. [CrossRef]
3. Wang, Y.; Bai, P.; Liang, X.; Wang, W.; Zhang, J.; Fu, Q. Reconnaissance Mission Conducted by UAV Swarms Based on Distributed PSO Path Planning Algorithms. *IEEE Access* **2019**, *7*, 105086–105099. [CrossRef]
4. Wu, Y.; Wu, S.; Hu, X. Multi-constrained cooperative path planning of multiple drones for persistent surveillance in urban environments. *Complex Intell. Syst.* **2021**, *7*, 1633–1647. [CrossRef]
5. Berger, J.; Happe, J. Co-evolutionary search path planning under constrained information-sharing for a cooperative unmanned aerial vehicle team. In Proceedings of the IEEE Congress on Evolutionary Computation, Barcelona, Spain, 18–23 July 2010.
6. Li, C.; Yang, C. Cooperative search of multiple robots with a distributed algorithm. In Proceedings of the 44th Annual Conference of the IEEE Industrial Electronics Society, Washington, DC, USA, 21–23 October 2018.
7. Zhen, Z.; Xing, D.; Gao, C. Cooperative search-attack mission planning for multi-UAV based on intelligent self-organized algorithm. *Aerosp. Sci. Technol.* **2018**, *76*, 402–411. [CrossRef]
8. Sujit, P.B.; Sousa, J.B. Multi-UAV task allocation with communication faults. In Proceedings of the 2012 American Control Conference (ACC), Montreal, QC, Canada, 27–29 June 2012.
9. Newaz, A.A.R.; Jeong, S.; Lee, H.; Ryu, H.; Chong, N.Y. UAV-based multiple source localization and contour mapping of radiation fields. *Robot. Auton. Syst.* **2016**, *85*, 12–25. [CrossRef]
10. Liu, H.; Li, X.; Fan, M.; Wu, G.; Pedrycz, W.; Suganthan, P.N. An Autonomous Path Planning Method for Unmanned Aerial Vehicle Based on a Tangent Intersection and Target Guidance Strategy. *IEEE Trans. Intell. Transp. Syst.* **2020**, *23*, 3061–3073. [CrossRef]
11. Arribas, E.; Mancuso, V.; Cholvi, V. Coverage Optimization with a Dynamic Network of Drone Relays. *IEEE Trans. Mob. Comput.* **2019**, *19*, 2278–2298. [CrossRef]
12. Xu, A.; Viriyasuthee, C.; Rekleitis, I. Optimal complete terrain coverage using an Unmanned Aerial Vehicle. In Proceedings of the 2011 IEEE International Conference on Robotics and Automation, Shanghai, China, 9–13 May 2011.
13. Kapoutsis, A.C.; Chatzichristofis, S.A.; Kosmatopoulos, E.B. DARP: Divide Areas Algorithm for Optimal Multi-Robot Coverage Path Planning. *J. Intell. Robot. Syst.* **2017**, *86*, 663–680. [CrossRef]
14. Mansouri, S.S.; Kanellakis, C.; Fresk, E.; Kominiak, D.; Nikolakopoulos, G. Cooperative coverage path planning for visual inspection. *Control. Eng. Pract.* **2018**, *74*, 118–131. [CrossRef]
15. Choi, Y.; Choi, Y.; Briceno, S.; Mavris, D.N. Energy-Constrained Multi-UAV Coverage Path Planning for an Aerial Imagery Mission Using Column Generation. *J. Intell. Robot. Syst.* **2020**, *97*, 125–139. [CrossRef]
16. Miki, T.; Popovic, M.; Gawel, A.; Hitz, G.; Siegwart, R. Multi-Agent Time-Based Decision-Making for the Search and Action Problem. In Proceedings of the 2018 IEEE International Conference on Robotics and Automation (ICRA), Brisbane, Australia, 21–25 May 2018.
17. Zhen, Z.; Chen, Y.; Wen, L.; Han, B. An intelligent cooperative mission planning scheme of UAV swarm in uncertain dynamic environment. *Aerosp. Sci. Technol.* **2020**, *100*, 105826–105841. [CrossRef]
18. Zheng, Y.J.; Du, Y.C.; Sheng, W.G.; Ling, H.F. Collaborative Human–UAV Search and Rescue for Missing Tourists in Nature Reserves. *INFORMS J. Appl. Anal.* **2019**, *49*, 371–383. [CrossRef]
19. Duan, H.; Zhao, J.; Deng, Y.; Shi, Y.; Ding, X. Dynamic Discrete Pigeon-Inspired Optimization for Multi-UAV Cooperative Search-Attack Mission Planning. *IEEE Trans. Aerosp. Electron. Syst.* **2021**, *57*, 706–720. [CrossRef]

20. Lakshmanan, A.K.; Mohan, R.E.; Ramalingam, B.; Le, A.V.; Veerajagadeshwar, P.; Tiwari, K.; Ilyas, M. Complete coverage path planning using reinforcement learning for Tetromino based cleaning and maintenance robot. *Autom. Constr.* **2020**, *112*, 103078. [CrossRef]
21. Pham, H.; La, H.; Feil-Seifer, D.; Nguyen, L. Cooperative and Distributed Reinforcement Learning of Drones for Field Coverage. *arXiv* **2018**, arXiv:1803.07250.
22. Liu, G.; Shu, C.; Liang, Z.; Peng, B.; Cheng, L. A Modified Sparrow Search Algorithm with Application in 3d Route Planning for UAV. *Sensors* **2021**, *21*, 1224. [CrossRef]
23. Jin, Y.; Liao, Y.; Minai, A.; Polycarpou, M. Balancing search and target response in cooperative unmanned aerial vehicle (UAV) teams. *IEEE Trans. Syst. Man Cybern. Part B (Cybernetics)* **2006**, *36*, 571–587. [CrossRef]
24. Burt, P.; Adelson, E. The Laplacian Pyramid as a Compact Image Code. *IEEE Trans. Commun.* **1983**, *31*, 532–540. [CrossRef]
25. Lee, K.Y.; Park, J.B. Application of Particle Swarm Optimization to Economic Dispatch Problem: Advantages and Disadvantages. In Proceedings of the 2006 IEEE PES Power Systems Conference and Exposition, Atlanta, GA, USA, 29 October–1 November 2006.
26. Mohammadi, V.; Ghaemi, S.; Kharrati, H. PSO tuned FLC for full autopilot control of quadrotor to tackle wind disturbance using bond graph approach. *Appl. Soft Comput.* **2018**, *65*, 184–195. [CrossRef]
27. Niknam, T.; Narimani, M.R.; Jabbari, M. Dynamic optimal power flow using hybrid particle swarm optimization and simulated annealing. *Int. Trans. Electr. Energy Syst.* **2012**, *23*, 975–1001. [CrossRef]
28. Gaing, Z.-L. Particle swarm optimization to solving the economic dispatch considering the generator constraints. *IEEE Trans. Power Syst.* **2003**, *18*, 1187–1195. [CrossRef]
29. Eberhart, R.C.; Shi, Y. Comparison between genetic algorithms and particle swarm optimization. In Proceedings of the in Evolutionary Programming VII, Berlin, Heidelberg, 25–27 March 2005.
30. Wu, X.; Bai, W.; Xie, Y.; Sun, X.; Deng, C.; Cui, H. A hybrid algorithm of particle swarm optimization, metropolis criterion and RTS smoother for path planning of UAVs. *Appl. Soft Comput.* **2018**, *73*, 735–747. [CrossRef]
31. Burt, P.J.; Adelson, E.H. A multiresolution spline with application to image mosaics. *ACM Trans. Graph.* **1983**, *2*, 217–236. [CrossRef]

Disclaimer/Publisher's Note: The statements, opinions and data contained in all publications are solely those of the individual author(s) and contributor(s) and not of MDPI and/or the editor(s). MDPI and/or the editor(s) disclaim responsibility for any injury to people or property resulting from any ideas, methods, instructions or products referred to in the content.

Article

A Robust Disturbance-Rejection Controller Using Model Predictive Control for Quadrotor UAV in Tracking Aggressive Trajectory

Zhixiong Xu [1], Li Fan [1,2,*], Wei Qiu [1], Guangwei Wen [2] and Yunhan He [1]

[1] College of Control Science and Engineering, Zhejiang University, Hangzhou 310027, China; 22232005@zju.edu.cn (Z.X.); cyanilius@yeah.net (W.Q.); heyunhan@zju.edu.cn (Y.H.)
[2] Huzhou Institute, Zhejiang University, Huzhou 313000, China; wengw@hizju.org
* Correspondence: fanli77@zju.edu.cn

Abstract: A robust controller for the waypoint tracking of a quadrotor unmanned aerial vehicle (UAV) is proposed in this paper, in which position control and attitude control are effectively decoupled. Model predictive control (MPC) is employed in the position controller. The constraints of motors are imposed on the state and input variables of the optimization equation. This design effectively mitigates the nonlinearity of the attitude loop and enhances the planning efficiency of the position controller. The attitude controller is designed using a nonlinear and robust control law based on $SO(3)$ space, which enables continuous control on the $SO(3)$ manifold. By extending the differential flatness of the quadrotor-UAV to the angular acceleration level, the mapping of the control reference from the position controller to the attitude controller is achieved. Simulations are carried out to demonstrate the capability of the proposed controller. In the simulations, multiple aggressive flight trajectories and severe external disturbances are designed. The results show that the controller is robust, with superior accuracy in tracking aggressive trajectories.

Keywords: waypoint tracking; model predictive control; nonlinear attitude controller; differential flatness

Citation: Xu, Z.; Fan, L.; Qiu, W.; Wen, G.; He, Y. A Robust Disturbance-Rejection Controller Using Model Predictive Control for Quadrotor UAV in Tracking Aggressive Trajectory. *Drones* **2023**, *7*, 557. https://doi.org/10.3390/drones7090557

Academic Editor: Vaios Lappas

Received: 30 June 2023
Revised: 10 August 2023
Accepted: 21 August 2023
Published: 29 August 2023

Copyright: © 2023 by the authors. Licensee MDPI, Basel, Switzerland. This article is an open access article distributed under the terms and conditions of the Creative Commons Attribution (CC BY) license (https://creativecommons.org/licenses/by/4.0/).

1. Introduction

Nowadays, quadrotor UAVs have been widely employed in various applications, such as agricultural plant protection [1], transportation and logistics distribution [2], and post-disaster rescue operations [3]. To meet the demands of different missions and stay robust under challenging environmental conditions, the control methods for quadrotor UAVs have been continuously developed.

In a general task, the controller is responsible for accurately tracking a time-dependent trajectory designated by the planner. As an underactuated system, a quadrotor-UAV has far more state variables than input variables, and the nonlinear characteristics of the system make the control problem challenging. The cascaded PID control method is commonly utilized for quadrotor-UAV control, in which the position of the next time step is considered a reference input. The inner loop controls the attitude, and the outer loop takes charge of the position. The mapping from the position loop to the attitude loop is completed using the differential-flatness attribute of quadrotor UAVs [4–6]. However, this error-based control method often performs well while tracing a smooth and slow trajectory, but struggles in tracking an aggressive trajectory.

To tackle the problem, a variety of advanced control algorithms for quadrotor UAVs have been proposed, among which MPC has received significant attention. The MPC method constructs an optimization problem based on the quadrotor-UAV model and applies constraints on the states and inputs of the system, and it generates control commands over the predicted time horizon by solving the optimization problem. References [7–9] have demonstrated the exceptional control efficiency of this approach. Furthermore, linearized model predictive control (LMPC) and nonlinear model predictive control (NMPC) have been developed and compared in many studies. In the former method, Taylor expansion is generally conducted around the hover point to linearize the quadrotor-UAV model. Linearization saves the computation cost, but downgrades the performance in terms of aggressive trajectory tracking, as the higher-order terms are truncated in the model. On the contrary, the nonlinear model is employed in NMPC for predictive control, resulting in superior tracking accuracy and robustness. However, the optimization offered by this method is usually non-convex, which leads to longer computational time and risks of solver non-convergence. In addition, the reinforcement learning-based control methods have been demonstrated to be effective for quadrotor-UAV control [10–13]. Han et al. proposed a robust controller based on a hierarchical control framework and reinforcement learning [14]. A robust control performance is achieved without prior knowledge of quadrotor-UAV dynamics. Kaufmann et al. showcased the application of learning-based approaches in acrobatic quadrotor-UAV flights [12]. However, the end-to-end training approach limits the reusability and generalization of the method.

As the quadrotor-UAV is an underactuated system, the state variables are coupled with each other, which makes it difficult to accurately track all the states. The accuracy of attitude tracking is directly related to the accuracy of position tracking. Therefore, attitude control for quadrotor UAVs has been widely investigated in the literature. Traditional attitude control methods for quadrotor-UAVs often rely on Euler angles to represent the attitude, where each Euler angle is individually tracked to achieve attitude control. However, Euler angles reach singularity at specific attitudes. To avoid these issues, quaternion is employed to represent the attitude of quadrotor UAVs, as can be seen in references [15–18], resulting in improved attitude control performance. In geometric theory, the changes in attitude are correlated to the evolution of the $SO(3)$ manifold. Therefore, Lee et al. and Yu et al. designed attitude controllers based on Lie groups on $SO(3)$ space [19–21], which enables continuous control on $SO(3)$ space. Also, with the representation of the attitude using this method, singularities are avoided during the control process. What is more, many researchers integrated the classical robust control algorithms into attitude control for robustness. For instance, Lee et al. [22] came up with an adaptive control algorithm. An adaptive law is employed to provide robust control with bounded errors even in the presence of unknown quadrotor-UAV internal parameters. Considering that quadrotor-UAVs are often subjected to external disturbances during flight, sliding-mode control algorithms are employed for attitude control in references [23–25]. By incorporating sliding-mode-control terms into the control laws, the system disturbance rejection capability is, therefore, enhanced, and asymptotic convergences are ensured.

In most of the missions for a quadrotor-UAV, the objective of control is accurately tracking the position. The control of the attitude loop is essentially to fulfill the demands of the position loop. Most control approaches usually employ the entire dynamics model of quadrotor UAVs in the optimization process for predictive control, which introduces challenges due to the nonlinear nature of the attitude loop. However, if the desired values can be accurately traced in the attitude loop, the attitude loop can be decoupled from the position loop in the MPC optimization process. The nonlinearity in the attitude loop can, therefore, be avoided. Based on this idea, this paper aims to develop a quadrotor-UAV controller that can accurately track waypoints with temporal information. Contributions to this paper are listed as follows:

(1)   A robust controller for quadrotor-UAV waypoint tracking is proposed, in which position control and attitude control are decoupled. The control efficiency is enhanced while maintaining robustness.

(2)   The LMPC algorithm is used in position control, where the motor constraints are incorporated into both the input and state variables of the optimization problem. This approach effectively mitigates the nonlinear characteristics of the attitude loop and enhances the planning efficiency of the position controller.

(3)   For the attitude controller, a nonlinear robust control law is designed based on the $SO(3)$ space, which guarantees global asymptotic stability in the sense of Lyapunov.

(4)   Extensive simulations have verified the excellent control performance of the proposed controller, which can achieve high-precision tracking of the desired trajectory even under severe external disturbances.

The paper is divided into five sections. In Section 2, the model of the quadrotor-UAV as well as the coordinate systems used in current research are introduced. In Section 3, the framework of the proposed controller is elaborated. Descriptions of each module are provided in detail. In Section 4, control performances that have been demonstrated by multiple numerical simulations are provided. Finally, summaries and conclusions are provided in Section 5.

2. Model

In this article, variables are defined in a unified format. Matrices and vectors are denoted by bold non-italic capital letters and bold non-italic lowercase letters, respectively. Scalars are written in italic letters. The subscripts of a vector represent the components of that vector. For example, v_{wx}, v_{wy} and v_{wz} are the components of the velocity vector along the three axes in the world coordinate system, respectively.

The coordinate systems employed in this paper are the world system $\mathcal{W}$ and the body system $\mathcal{B}$. For the body coordinate system, the origin is located at the center of mass of the vehicle. The x-axis and y-axis are parallel to the two arms of the quadrotor-UAV as shown in Figure 1 and it is a right-handed system. For the world coordinate system, the z-axis points in the opposite direction of Earth's gravity. Similar to the paper by Mellinger and Kumar [26], an intermediate coordinate system $\mathcal{C}$ between the system $\mathcal{W}$ and system $\mathcal{B}$ is defined, it shares the same z-axis as the world coordinate system. The numbering of the four motors is illustrated in Figure 1.

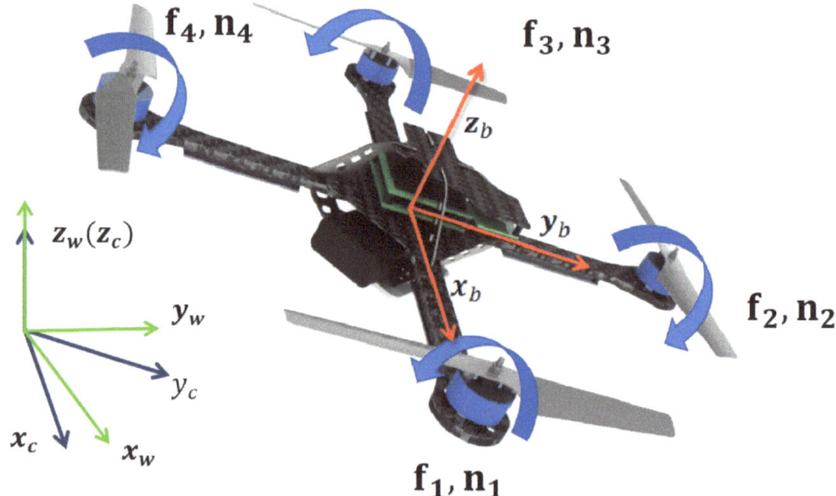

Figure 1. Coordinate system definition and motor numbering.

The propulsion system of the quadrotor-UAV is modeled as a second-order model. Following the discussion in reference [26], it is assumed that the force f_i and torque n_i produced by each motor are proportional to the square of its rotational speed ω_i, as indicated in Equation (1). Within the equation, both k_f and k_n are constants, while k_m stands for the motor response constant. In addition, considering the issue of motor saturation, there exists a maximum rotational speed constraint for each motor.

$$\dot{\omega}_i = k_m(\omega_i^d - \omega_i), \quad f_i = k_f\omega_i^2, \quad n_i = k_n\omega_i^2 \tag{1}$$

$$0 \leq \omega_i \leq \omega_{\max} \tag{2}$$

In the body coordinate system, the thrust T and the three-axis torque τ produced by the four motors are

$$\begin{bmatrix} \tau \\ T \end{bmatrix} = \mathbf{D}\omega, \quad \tau = \begin{bmatrix} \tau_x \\ \tau_y \\ \tau_z \end{bmatrix} \tag{3}$$

$$\mathbf{D} = \begin{bmatrix} 0 & k_f l & 0 & -k_f l \\ -k_f l & 0 & k_f l & 0 \\ -k_n & k_n & -k_n & k_n \\ k_f & k_f & k_f & k_f \end{bmatrix}, \quad \omega = \begin{bmatrix} \omega_1^2 \\ \omega_2^2 \\ \omega_3^2 \\ \omega_4^2 \end{bmatrix} \tag{4}$$

where l represents the length from the motor's rotational axis to the center of mass of the quadrotor-UAV.

Following the convention, all angular velocities are represented in the body frame, while position and velocity vectors are represented in the world frame. The quadrotor-UAV is subjected to the forces of gravity and the thrust generated by the motors along the z-axis of the body frame. Newton's second law related to the position control is

$$m\mathbf{a} = -mg\mathbf{z_w} + T\mathbf{z_b} \tag{5}$$

in which $\mathbf{z_w}$ and $\mathbf{z_b}$ indicate the unit vectors of the z-axis for the world and body frames. The mass of the quadrotor-UAV is m and the gravitational acceleration is g. The Euler equation related to the attitude control is

$$\mathbf{J\dot{w}_b} + \mathbf{w_b} \times \mathbf{Jw_b} = \tau \tag{6}$$

where $\mathbf{J}$ denotes the inertia tensor. The roll, pitch, and yaw angles (ϕ, θ and ψ) based on the Z-X-Y Euler angle sequence are employed to represent the attitude. The relation between the first derivative of Euler angles and the angular velocities in the body frame is

$$\begin{bmatrix} w_{bx} \\ w_{by} \\ w_{bz} \end{bmatrix} = \begin{bmatrix} \cos\theta & 0 & -\cos\phi\sin\theta \\ 0 & 1 & \sin\phi \\ \sin\theta & 0 & \cos\phi\cos\theta \end{bmatrix} \begin{bmatrix} \dot{\phi} \\ \dot{\theta} \\ \dot{\psi} \end{bmatrix} \tag{7}$$

3. Methodology

For the proposed controller, a sequence of reference temporal waypoints is the control input. The information of each waypoint mainly consists of the three-dimensional (3D) position and yaw angle $\begin{bmatrix} p_{wx}^d & p_{wy}^d & p_{wz}^d & \psi^d \end{bmatrix}$. Figure 2 provides the framework of the controller. The LMPC algorithm is employed in the position loop. The relevant model constraints are implemented on the state and input variables, thus planning the reference trajectory within a future time domain. Based on the differential flatness property of the quadrotor-UAV, the planned trajectory in the position loop can be mapped to the desired input for the attitude controller. In addition, an attitude control law is designed based on the $SO(3)$ space. The nonlinear robust control law achieves accurate tracking of the desired attitude angles. Regarding the control allocation of the quadrotor-UAV, an inverse kinematics solution is obtained based on the optimization approach. Each module will be described in detail in the following sections.

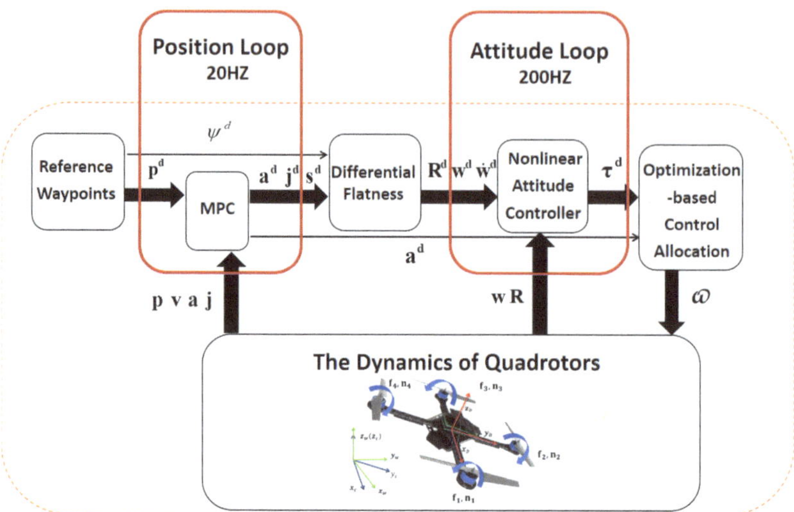

Figure 2. Framework of proposed the quadrotor-UAV controller based on LMPC.

3.1. Linear Model Predictive Control

The ability to predict over a wide time domain and consider the constraints of an actual model are the reasons why MPC is widely employed. In this paper, the LMPC is employed in the position controller, and the model constraints primarily incorporate considerations of motor speed limitations and motor response constraints. At each time step, the optimal control sequence for the entire prediction horizon is computed, but only the first control input is applied to the system. This process is iterated in a receding horizon manner, where optimization is performed repeatedly at each time step.

As shown in Figure 2, the selected state variables are the position, velocities, accelerations, and jerks along three axes. These variables are denoted by their respective initials.

$$
\mathbf{x} = \begin{bmatrix} \mathbf{p} \\ \mathbf{v} \\ \mathbf{a} \\ \mathbf{j} \end{bmatrix}, \mathbf{y} = [\mathbf{p}], \mathbf{u} = [\mathbf{s}], \mathbf{r} = [\mathbf{p^d}] \tag{8}
$$

The output of the system $\mathbf{y}$ is the 3D position, and the control input is the snap, which is the second derivative of acceleration. The reference input $\mathbf{r}$ corresponds to the desired position. Among these variables, $\mathbf{p}, \mathbf{v}, \mathbf{a}, \mathbf{j}, \mathbf{s} \in \mathbb{R}^3, \mathbf{y} \in \mathbb{R}^3, \mathbf{u} \in \mathbb{R}^3, \mathbf{r} \in \mathbb{R}^3$. The reasons for optimizing for the snap will be thoroughly analyzed in the subsequent sections.

The corresponding difference equation can be formulated based on the aforementioned input and state variables.

$$
\begin{aligned}
\mathbf{x}(k+1) &= \mathbf{A}\mathbf{x}(k) + \mathbf{B}\mathbf{u}(k) \\
\mathbf{y}(k+1) &= \mathbf{C}\mathbf{x}(k+1)
\end{aligned} \tag{9}
$$

where $\mathbf{A}$, $\mathbf{B}$, and $\mathbf{C}$ are defined as

$$
\mathbf{A} = \begin{bmatrix}
\mathbf{I}_{3\times3} & Ts*\mathbf{I}_{3\times3} & \frac{1}{2}Ts^2*\mathbf{I}_{3\times3} & \frac{1}{6}Ts^3*\mathbf{I}_{3\times3} \\
0 & \mathbf{I}_{3\times3} & Ts*\mathbf{I}_{3\times3} & \frac{1}{2}Ts^2*\mathbf{I}_{3\times3} \\
0 & 0 & \mathbf{I}_{3\times3} & Ts*\mathbf{I}_{3\times3} \\
0 & 0 & 0 & \mathbf{I}_{3\times3}
\end{bmatrix}, \quad
\mathbf{B} = \begin{bmatrix}
\frac{1}{24}Ts^4*\mathbf{I}_{3\times3} \\
\frac{1}{6}Ts^3*\mathbf{I}_{3\times3} \\
\frac{1}{2}Ts^2*\mathbf{I}_{3\times3} \\
Ts*\mathbf{I}_{3\times3}
\end{bmatrix} \tag{10}
$$

$$
\mathbf{C} = \begin{bmatrix} \mathbf{I}_{3\times3} & 0 & 0 & 0 \end{bmatrix}
$$

in which Ts represents the discrete time interval, and $\mathbf{I}_{3\times3}$ denotes the identity matrix. Given the prediction horizon N, we define the state vector $\mathbf{X}$, the reference input vector $\mathbf{\Gamma}$, the output vector $\mathbf{Y}$, and the control input vector $\mathbf{U}$ over the predicted time horizon.

$$
\mathbf{X} = \begin{bmatrix} \mathbf{x}(k+1) \\ \mathbf{x}(k+2) \\ \mathbf{x}(k+3) \\ \vdots \\ \mathbf{x}(k+N) \end{bmatrix} \quad \mathbf{U} = \begin{bmatrix} \mathbf{u}(k) \\ \mathbf{u}(k+1) \\ \mathbf{u}(k+2) \\ \vdots \\ \mathbf{u}(k+N-1) \end{bmatrix} \quad \mathbf{Y} = \begin{bmatrix} \mathbf{y}(k+1) \\ \mathbf{y}(k+2) \\ \mathbf{y}(k+3) \\ \vdots \\ \mathbf{y}(k+N) \end{bmatrix} \quad \mathbf{\Gamma} = \begin{bmatrix} \mathbf{r}(k+1) \\ \mathbf{r}(k+2) \\ \mathbf{r}(k+3) \\ \vdots \\ \mathbf{r}(k+N) \end{bmatrix} \tag{11}
$$

The relationship between them can be expressed as follows:

$$
\mathbf{X} = \begin{bmatrix} \mathbf{B} & 0 & 0 & \cdots & 0 \\ \mathbf{AB} & \mathbf{B} & 0 & \cdots & 0 \\ \mathbf{A}^2\mathbf{B} & \mathbf{AB} & \mathbf{B} & \cdots & 0 \\ \vdots & \vdots & \vdots & \ddots & \vdots \\ \mathbf{A}^{N-1}\mathbf{B} & \mathbf{A}^{N-2}\mathbf{B} & \mathbf{A}^{N-3}\mathbf{B} & \cdots & \mathbf{B} \end{bmatrix} \mathbf{U} + \begin{bmatrix} \mathbf{A} \\ \mathbf{A}^2 \\ \mathbf{A}^3 \\ \vdots \\ \mathbf{A}^N \end{bmatrix} \mathbf{x}(k) = \mathbf{PU} + \mathbf{Mx}(k) \tag{12}
$$

$$
\mathbf{Y} = \begin{bmatrix} \mathbf{C} & 0 & 0 & 0 & 0 \\ 0 & \mathbf{C} & 0 & 0 & 0 \\ 0 & 0 & \mathbf{C} & 0 & 0 \\ 0 & 0 & 0 & \ddots & 0 \\ 0 & 0 & 0 & 0 & \mathbf{C} \end{bmatrix} \mathbf{X} = \mathbf{EX} \tag{13}
$$

$$
\Delta\mathbf{U} = \begin{bmatrix} \mathbf{I}_{3\times3} & 0 & 0 & \cdots & 0 \\ -\mathbf{I}_{3\times3} & \mathbf{I}_{3\times3} & 0 & \cdots & 0 \\ 0 & -\mathbf{I}_{3\times3} & \mathbf{I}_{3\times3} & \cdots & 0 \\ \vdots & \vdots & \vdots & \ddots & \vdots \\ 0 & 0 & 0 & -\mathbf{I}_{3\times3} & \mathbf{I}_{3\times3} \end{bmatrix} \mathbf{U} + \begin{bmatrix} -\mathbf{I}_{3\times3} \\ 0 \\ 0 \\ 0 \\ 0 \end{bmatrix} \mathbf{u}(k-1) = \mathbf{SU} + \mathbf{Hu}(k-1) \tag{14}
$$

where $\mathbf{x}(k)$ represents the state values at time step k, and $\mathbf{u}(k-1)$ represents the control input at the previous time step.

Considering the limitations of the control inputs and the tracking accuracy, a problem aiming to minimize tracking errors and control input variations can be formulated as

$$
\min_{\mathbf{U}} (\mathbf{\Gamma} - \mathbf{Y})^T \mathbf{G}_1 (\mathbf{\Gamma} - \mathbf{Y}) + \Delta\mathbf{U}^T \mathbf{G}_2 \Delta\mathbf{U} \tag{15}
$$

where $\mathbf{G}_1$ is the weight matrix related to tracking errors, and $\mathbf{G}_2$ is the weight matrix related to the control input. Both $\mathbf{G}_1$ and $\mathbf{G}_2$ are positive definite diagonal matrices.

The model constraints considered in the controller are the motor speed constraints and motor response constraints. Combining Equations (1) and (5), we obtain

$$
m\mathbf{a} = \mathbf{z_b} \sum_{i=1}^{4} k_f \omega_i^2 - mg\mathbf{z_w} \tag{16}
$$

The motor's maximum speed constraint is imposed on the acceleration term of the state variables. $[\cdot]_{\max}$ and $[\cdot]_{\min}$ represent the maximum and minimum values of each component in the vector, respectively. The attitude of quadrotor-UAV is assumed to be unchanged within a single time step. This assumption may lead to motor speed saturation in some extreme cases. However, the optimization-based control allocation approach significantly reduces the impact of motor saturation.

$$\mathbf{a}_{\min} = \left[\frac{\mathbf{z_b}}{m} \sum_{i=1}^{4} k_f \omega_i^2 - g\mathbf{z_w} \right]_{\min} \quad \mathbf{a}_{\max} = \left[\frac{\mathbf{z_b}}{m} \sum_{i=1}^{4} k_f \omega_i^2 - g\mathbf{z_w} \right]_{\max} \tag{17}$$

By taking the second derivative of both sides of Equation (5) and applying the motor response constraints to the snap term of the input variables, we have

$$m\mathbf{s} = 2\mathbf{z_b} \sum_{i=1}^{4} k_f(\omega_i \ddot{\omega}_i + \dot{\omega}_i^2) \tag{18}$$

$$\mathbf{s}_{\min} = \frac{2\mathbf{z_b}}{m} \sum_{i=1}^{4} k_f(\omega_i \ddot{\omega}_{i,\min} + \dot{\omega}_i^2), \quad \mathbf{s}_{\max} = \frac{2\mathbf{z_b}}{m} \sum_{i=1}^{4} k_f(\omega_i \ddot{\omega}_{i,\max} + \dot{\omega}_i^2) \tag{19}$$

$$\ddot{\omega}_{i,\min} = k_m(\omega_{\min}^d - \omega_i), \quad \ddot{\omega}_{i,\max} = k_m(\omega_{\max}^d - \omega_i) \tag{20}$$

where $\ddot{\omega}_{i,\min}$ and $\ddot{\omega}_{i,\max}$ represent the minimum and maximum motor responsiveness, respectively. Additional constraints, such as velocity limitations, can also be incorporated into the current algorithm. By substituting the variables in Equation (15), an optimization problem is formulated as

$$\min_{\mathbf{U}} \mathbf{U}^T(\mathbf{P}^T\mathbf{E}^T\mathbf{G_1}\mathbf{E}\mathbf{P} + \mathbf{S}^T\mathbf{G_2}\mathbf{S})\mathbf{U} + 2(\mathbf{x}(k)^T\mathbf{M}^T\mathbf{E}^T\mathbf{G_1}\mathbf{E}\mathbf{P} - \mathbf{\Gamma}^T\mathbf{G_1}\mathbf{E}\mathbf{P} + \mathbf{u}(k-1)^T\mathbf{H}^T\mathbf{G_2}\mathbf{S})\mathbf{U}$$
$$s.t. \mathbf{a}_{\min} \leq \mathbf{a} \leq \mathbf{a}_{\max} \tag{21}$$
$$\mathbf{U}_{\min} \leq \mathbf{U} \leq \mathbf{U}_{\max}$$

It is worth mentioning that the aforementioned optimization problem can be simplified into a quadratic programming problem due to the absence of nonlinearity in the attitude loop. As a result, it can be solved quickly in real-time. Through solving this optimization equation, the desired position, velocity, acceleration, jerk, and snap can be effectively planned within the predicted time horizon. The first control sequence of the optimization results can then be applied to the controller in the next time step.

3.2. Nonlinear Attitude Controller Based on $SO(3)$

As errors in attitude significantly undermine the position tracking accuracy, an attitude controller that can achieve high-precision tracking is very important to the overall controller. In this paper, a nonlinear and robust attitude control law is designed based on the $SO(3)$ space, which ensures the attitude control is continuous on the $SO(3)$ manifold. Furthermore, the proposed attitude controller is proved to be global asymptotically stable in the sense of Lyapunov.

The Euler equation for the attitude control is

$$\mathbf{J}\dot{\mathbf{w}}_\mathbf{b} + \mathbf{w}_\mathbf{b} \times \mathbf{J}\mathbf{w}_\mathbf{b} = \boldsymbol{\tau} + \boldsymbol{\tau}_\mathbf{dis} \tag{22}$$

where $\boldsymbol{\tau}_\mathbf{dis}$ represents external disturbances and $\boldsymbol{\tau}$ is the input moment. Equation (22) can be rewritten as

$$\dot{\mathbf{w}}_\mathbf{b} = \mathbf{J}^{-1}\boldsymbol{\tau} + \mathbf{J}^{-1}(-\mathbf{w}_\mathbf{b} \times \mathbf{J}\mathbf{w}_\mathbf{b}) + \mathbf{J}^{-1}\boldsymbol{\tau}_\mathbf{dis} = \eta u + \xi + \mathbf{h}$$
$$\eta = \mathbf{J}^{-1}, \xi = \mathbf{J}^{-1}(-\mathbf{w}_\mathbf{b} \times \mathbf{J}\mathbf{w}_\mathbf{b}), \mathbf{h} = \mathbf{J}^{-1}\boldsymbol{\tau}_\mathbf{dis} \tag{23}$$

where $\mathbf{h}$ represents the external perturbation term, satisfying $||\mathbf{h}||_\infty \leq h_{max}$.

According to the Euler rotation theorem, any rotation matrix of $SO(3)$ can be equivalently achieved by rotating around a specific axis by a certain angle. Here, ${}^w_b\mathbf{R}$ denotes the rotation matrix of the current body coordinate system with respect to the world coordinate system, while ${}^w_b\mathbf{R}^d$ signifies the desired rotation matrix of the body coordinate system with respect to the world coordinate system. Without causing any ambiguity, they are denoted as $\mathbf{R}$ and $\mathbf{R}^\mathbf{d}$ for simplicity, respectively. The error of the rotation matrix can be obtained as

$$\mathbf{R_e} = \mathbf{R}^T\mathbf{R}^\mathbf{d} \tag{24}$$

Following the definition of the error of the rotation matrix, the angle error can be calculated as

$$\theta_e = cos^{-1}\left(\frac{tr(\mathbf{R_e} - 1)}{2}\right) \tag{25}$$

The angular velocity error is defined as the difference between the desired angular velocity $\mathbf{w_b^d}$ and the actual angular velocity $\mathbf{w_b}$:

$$\mathbf{w_e} = \mathbf{w_b^d} - \mathbf{w}_b \tag{26}$$

Hence, the attitude control law can be designed based on the angle error and the angular velocity error:

$$\mathbf{u} = \eta^{-1}\left[\dot{\mathbf{w}}_\mathbf{b}^\mathbf{d} - \xi + h_{max}\mathrm{sgn}(\mathbf{w_e}) + K_1\mathbf{w_e} + \frac{|\theta_e|\,||\mathbf{w_b^d}|| + K_2\theta_e^2}{||\mathbf{w_e}||^2}\mathbf{w_e} + \log(\mathbf{R_e})\right] \tag{27}$$

Referring to the exponential mapping and algebraic mapping between Lie groups and Lie algebras [27], the calculation of the desired rotation axis based on the error of the rotation matrix can be carried out as follows:

$$\log(\mathbf{R}_e) = \frac{\theta_e}{2\sin\theta_e}(\mathbf{R_e} - \mathbf{R_e^T})^\vee \in so(3) \tag{28}$$

where the operator $\wedge$ denotes a conversion of a three-dimensional vector into a 3×3 skew-symmetric matrix, while the operator $\vee$ represents the inverse operation of the operator $\wedge$.

$$\mathbf{c} \times \mathbf{d} = \mathbf{c}^\wedge\mathbf{d}, \quad \forall \mathbf{c}, \mathbf{d} \in \mathbb{R}^3 \tag{29}$$

The term $h_{max}\mathrm{sgn}(\mathbf{w_e})$ in the control law corresponds to the sliding mode control term for angular velocity control, and it effectively mitigates the influence of external disturbances. Moreover, this term exhibits negligible numerical values, thus ensuring the stability of the controller without inducing any instability concerns.

To demonstrate the robustness of the control law, a Lyapunov function is conducted as follows:

$$V = \frac{1}{2}\theta_e^2 + \frac{1}{2}\mathbf{w_e^T}\mathbf{w_e} \tag{30}$$

By taking the derivative of the Lyapunov function, we have

$$\dot{V} = \dot{\theta}_e\theta_e + \mathbf{w}_e^T\dot{\mathbf{w}}_e = \dot{V}_1 + \dot{V}_2 \tag{31}$$

The derivative of the angle term is expanded and the calculation is derived as

$$\begin{aligned}
\dot{V}_1 &= \dot{\theta}_e\theta_e \\
&= \theta_e\frac{d(cos^{-1}(\frac{tr(\mathbf{R_e}-1)}{2}))}{dt} \\
&= -\theta_e\frac{1}{2\sqrt{1-cos^{-1}(\frac{tr(\mathbf{R_e}-1)}{2})^2}}\frac{dtr(\mathbf{R_e})}{dt} \\
&= -\frac{\theta_e}{2\sin\theta_e}\frac{dtr(\mathbf{R}^T\mathbf{R}^d)}{dt} \\
&= -\frac{\theta_e}{2\sin\theta_e}tr\left(\frac{d\mathbf{R}^T}{dt}\mathbf{R}^d + \mathbf{R}^T\frac{d\mathbf{R}^d}{dt}\right) \\
&= -\frac{\theta_e}{2\sin\theta_e}tr(\hat{\mathbf{w}}_\mathbf{b}^T\mathbf{R}^T\mathbf{R}^d + \mathbf{R}^T\mathbf{R}^d\hat{\mathbf{w}}_\mathbf{b}^\mathbf{d}) \\
&= \log(\mathbf{R_e})^T\mathbf{w_e}
\end{aligned} \tag{32}$$

Substituting the above equation and the control law into Equation (31) and scaling it appropriately, we have

$$
\begin{aligned}
\dot{V} &= \dot{\theta}_e \theta_e + \mathbf{w}_e^T \dot{\mathbf{w}}_e \\
&= \log\left(\mathbf{R_e}\right)^T \mathbf{w_e} + \mathbf{w_e}^T \left(\dot{\mathbf{w}}_\mathbf{b}^\mathbf{d} - \zeta - \mathbf{h} - \dot{\mathbf{w}}_\mathbf{b}^\mathbf{d} + \zeta - h_{max}\mathrm{sgn}(\mathbf{w}_e) \right. \\
&\quad \left. - K_1\mathbf{w_e} - \frac{|\theta_e| \|\mathbf{w}_\mathbf{b}^\mathbf{d}\| + K_2\theta_e^2}{\|\mathbf{w_e}\|^2}\mathbf{w_e} - \log(\mathbf{R_e}) \right) \\
&= -K_1 \mathbf{w}_e^T \mathbf{w_e} - \left(h_{max}\mathrm{sum}(|\mathbf{w}_e|) + \mathbf{h}^T \mathbf{w}_e\right) - \|\|\mathbf{w}_\mathbf{b}^\mathbf{d}\|\|\theta_e| - K_2\theta_e^2 \\
&\leq -K_1\mathbf{w}_e^T\mathbf{w_e} - K_2\theta_e^2 \\
&\leq -4\min(K_1, K_2)V
\end{aligned}
\tag{33}
$$

where $\mathrm{sum}(|\mathbf{w}_e|)$ represents the sum of the absolute values of each element in $\mathbf{w_e}$. It is evident that the constructed Lyapunov function is positive definite, and its derivative is negative definite. Consequently, it can be concluded that the control law guarantees the global asymptotic stability of the system in the sense of Lyapunov. By adjusting the coefficients K_1 and K_2, the convergence rate of the system can be tailored.

The attitude control law is designed based on the $SO(3)$ space to avoid singularities. Concurrently, the control outputs are more efficient on the $SO(3)$ manifold. Simulations presented in Section 4 have demonstrated the exceptional performance of the proposed attitude control law, even in the presence of disturbances and observation uncertainties. The precise tracking of the attitude loop ensures its capability to fulfill the demands of the position loop. Although the attitude loop is excluded from the MPC optimization, its nonlinearity can still be compensated by the attitude controller, thus the overall control performance remains superior.

3.3. Differential Flatness

Given the above control scheme, the position and attitude controller can be designed independently. By leveraging the differential flatness properties of the quadrotor-UAV, a mapping from the position loop to the attitude loop is achieved. Differential flatness is referred to as the selection of appropriate variables from the state space and describing the entire state space using these variables and their derivatives. Similar to most articles in the literature, the following flat outputs are selected, namely the 3D position and the yaw angle:

$$
\sigma = [p_{wx}, p_{wy}, p_{wz}, \psi]^T
\tag{34}
$$

The position is planned using the LMPC method within the position loop, and the yaw angle information is pre-defined. A direct mapping from the flat outputs to the desired attitude, angular rate, and angular acceleration is obtained based on these variables and their derivatives.

According to Mellinger and Kumar [26], the desired attitude is calculated through acceleration. As mentioned in Section 2, the coordinate system $\mathcal{C}$ is an intermediate coordinate system obtained by rotating the world coordinate system around the z-axis with the yaw angle, and $\mathbf{x_c}$ indicates the unit vectors of the x-axis for this coordinate system.

$$
\begin{aligned}
\mathbf{z_b} &= \frac{\mathbf{a}^\mathbf{d} + g\mathbf{z_w}}{\|\mathbf{a}^\mathbf{d} + g\mathbf{z_w}\|} \\
\mathbf{x_c} &= [\cos\psi, \sin\psi, 0]^T \\
\mathbf{y_b} &= \frac{\mathbf{z_b} \times \mathbf{x_c}}{\|\mathbf{z_b} \times \mathbf{x_c}\|} \\
\mathbf{x_b} &= \mathbf{y_b} \times \mathbf{z_b}, \quad {}_b^w\mathbf{R}^d = [\mathbf{x_b} \quad \mathbf{y_b} \quad \mathbf{z_b}]
\end{aligned}
\tag{35}
$$

By differentiating both sides of Equation (5), the jerk planned in the position loop can be mapped into the desired angular velocity.

$$
\begin{aligned}
m\dot{\mathbf{a}} &= \dot{T}\mathbf{z_b} + T(\mathbf{w_w} \times \mathbf{z_b}) \\
\dot{T} &= \mathbf{z_b} \cdot m\dot{\mathbf{a}} \\
\mathbf{h} &= \mathbf{w_w} \times \mathbf{z_b} = \tfrac{m}{T}(\mathbf{j^d} - (\mathbf{z_b} \cdot \mathbf{j^d})\mathbf{z_b}) \\
w_{bx} &= -\mathbf{h} \cdot \mathbf{y_b}, \quad w_{by} = \mathbf{h} \cdot \mathbf{x_b}
\end{aligned}
\tag{36}
$$

Also in paper [26], however, the z-axis angular velocity is taken as zero for a constant yaw setpoint, which is actually nonzero. In this paper, the z-axis angular velocity is calculated through Equation (7). By taking the inverse of Equation (7), we obtain

$$
\begin{bmatrix} \dot{\phi} \\ \dot{\theta} \\ \dot{\psi} \end{bmatrix} =
\begin{bmatrix}
\cos\theta & 0 & \sin\theta \\
\tan\phi\sin\theta & 1 & -\tan\phi\cos\theta \\
-\sec\phi\sin\theta & 0 & \sec\phi\cos\theta
\end{bmatrix}
\begin{bmatrix} w_{bx} \\ w_{by} \\ w_{bz} \end{bmatrix}
\tag{37}
$$

By solving Equation (38), the z-axis angular velocity can be determined based on the derivative value of the yaw angle.

$$
\dot{\psi}\cos\phi = -w_{bx}\sin\theta + w_{bz}\cos\theta
\tag{38}
$$

In the attitude controller, both angular velocity and angular acceleration are required as feed-forward terms to enhance tracking accuracy in attitude control. Thus, the differential flatness property is extended to the angular acceleration to calculate its desired value. Similarly, by taking the second derivative of Equation (5), we have

$$
m\mathbf{s^d} = \ddot{T}\mathbf{z_b} + 2\dot{T}(\mathbf{w_w} \times \mathbf{z_b}) + T\frac{\delta(\mathbf{w_w} \times \mathbf{z_b})}{\delta t}
\tag{39}
$$

The equation for calculating the second derivative of thrust T is

$$
\ddot{T} = (\mathbf{w_w} \times \mathbf{z_b}) \cdot m\dot{\mathbf{a}} + \mathbf{z_b} \cdot m\mathbf{s^d}
\tag{40}
$$

Expanding Equation (39), the cross product of the angular acceleration with the z-axis of the world coordinate system can be calculated. By projecting this quantity onto the x-axis and y-axis of the body coordinate system, the values of $\dot{w}_{bx}$ and $\dot{w}_{by}$ can be derived as

$$
\begin{aligned}
\frac{\delta(\mathbf{w_w} \times \mathbf{z_b})}{\delta t} &= (\hat{\mathbf{w}}_\mathbf{w}{}_b^w\mathbf{R}\mathbf{w_b} + {}_b^w\mathbf{R}^{-1}\tfrac{\delta\mathbf{w_b}}{\delta t}) \times \mathbf{z_b} + \mathbf{w_w} \times (\mathbf{w_w} \times \mathbf{z_b}) \\
\boldsymbol{\zeta} &= \dot{\mathbf{w}}_\mathbf{w} \times \mathbf{z_b} = {}_b^w\mathbf{R}^{-1}\tfrac{\delta\mathbf{w_b}}{\delta t} \times \mathbf{z_b} \\
\dot{w}_{bx} &= -\boldsymbol{\zeta} \cdot \mathbf{y_b}, \quad \dot{w}_{by} = \boldsymbol{\zeta} \cdot \mathbf{x_b}
\end{aligned}
\tag{41}
$$

Furthermore, we take the derivative of Equation (38):

$$
\ddot{\psi}\cos\phi = \dot{w}_{bz}\cos\theta - \dot{w}_{bx}\sin\theta - \dot{\theta}\dot{\phi} + \dot{\phi}\dot{\psi}\sin\phi
\tag{42}
$$

By solving Equation (42), the angular acceleration of the z-axis in the body coordinate system can be calculated. Based on differential flatness, the information in the position controller, a^d, j^d, s^d, can be mapped to the inputs of the attitude controller, $R^d, w^d, \dot{w}^d$. Therefore, the second derivative of acceleration, i.e., the snap variable, must be considered in the position controller. Apart from better handling of the constraint of motor responsiveness, another crucial reason for this is to calculate the desired angular acceleration for the attitude controller.

3.4. Optimization-Based Control Allocation

The desired thrust can be calculated based on the desired acceleration from the position controller. When combined with the desired torque from the attitude controller, the desired control inputs can be obtained. Under normal conditions, it is sufficient to compute the desired rotational speeds for each motor by solving the system of linear Equations (3). Nevertheless, considering potential inaccuracies in the applied constraints and the possibility of excessive torque required during extreme flight conditions, the desired motor rotational speed might still exceed the actual speed limitation. To minimize the effects of motor saturation, the final module employs an optimization-based approach for control allocation. The algorithm is expressed as

$$\min_{\omega} \left(\begin{bmatrix} \tau \\ T \end{bmatrix} - \mathbf{D}\omega \right)^T \mathbf{Q} \left(\begin{bmatrix} \tau \\ T \end{bmatrix} - \mathbf{D}\omega \right)$$
$$s.t. \omega_{\min} \leq \omega \leq \omega_{\max}$$

(43)

where $\mathbf{Q}$ represents the weight matrix. The desired rotational speeds for the four motors are determined by solving the aforementioned optimization problem. Subsequently, these desired speeds are translated into the corresponding PWM signals and transmitted to the motors.

4. Results

The simulations were performed within the MATLAB software, and the LMPC optimization equa+tions were solved using the OSQP solver. All modules were individually implemented on a HP OMEN8 Plus laptop, which is equipped with an Intel Core i7-12700H processor and 16 GB RAM. Under this specific hardware configuration, the average computation time for solving the MPC optimization problem is 4 ms.

A series of simulation scenarios were conducted to validate the tracking performance of the proposed controller. The relevant parameters of the quadrotor-UAV are provided in Table 1, which is referred to [28]. The performance of the proposed attitude controller in terms of attitude angle tracking is demonstrated in Section 4.1. The trajectory tracking performance of the entire controller is presented in Section 4.2.

Table 1. Parameters of the quadrotor-UAV in simulation.

Variables	Description	Values
m	The mass of the quadrotor-UAV (kg)	0.98
J	The inertia tensor of the quadrotor-UAV	$Diag\{2.64 \times 10^{-3}, 2.64 \times 10^{-3}, 4.96 \times 10^{-3}\}$
k_f	The force scaling factor of motor	8.98132×10^{-9}
k_n	The moment scaling factor of motor	1.1694×10^{-10}
k_m	Motor Response Constant	10
l	The arm length of the quadrotor-UAV(m)	0.26
ω_{max}	Maximum motor speed (RPM)	42.000

4.1. Nonlinear Attitude Controller Based on $SO(3)$

It is crucial to substantiate the control performance and robustness of the attitude controller, given its direct relevance to the overall controller. In the attitude control law as shown in Equation (27), both K_1 and K_2 were set as 10. The desired tracking attitude is given as

$$\phi(t) = \frac{\pi}{9} \sin(\pi t), \theta(t) = \frac{\pi}{9} \cos(\pi t), \psi(t) = 0$$

(44)

Firstly, the simulation was carried out without the external disturbances. In Figure 3, the actual and desired attitude and angular rate in three axes are plotted over time. As can be observed in the graph, even with an initial deviation between the actual and desired pitch angle, the system stabilized within 2 seconds. Subsequently, the system accurately tracked the specified attitude angles and angular velocities throughout the ensuing timeframe.

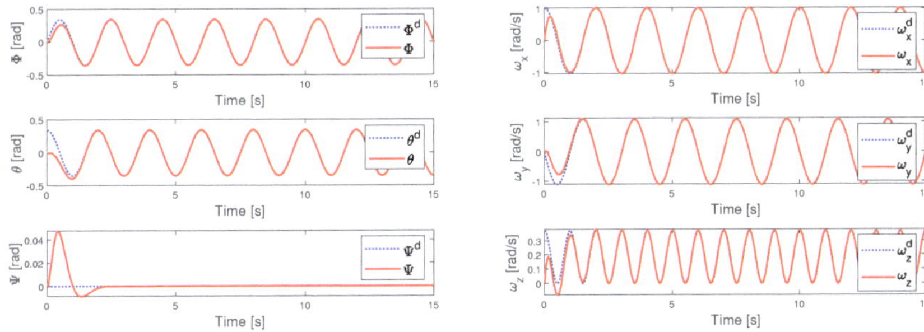

Figure 3. Actual and desired attitude and angular rate without external disturbances.

Next, observation errors with an amplitude of 0.1 were added to the measured values of attitude and angular rate, while white noise sampled every 0.2 s with an amplitude of 20% of the controller normal operating conditions was employed to simulate the external disturbances. The actual and desired attitude and angular rate are plotted over time in Figure 4. Despite such a large initial pitch angle deviation and a massive amount of disturbances, the system stabilized and converged to the desired states quickly. It is important to note that there were several significant peaks in the actual angular velocities. This was caused by the presence of large random disturbances and observation errors at that particular moment. The attitude controller compensated for tracking errors by outputting a large torque. Actually, the angular velocity remained stable, but there were certain instances where it exhibits higher rates of variation.

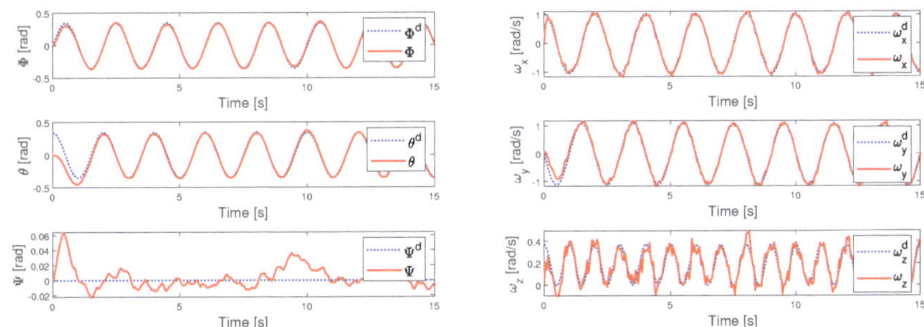

Figure 4. Actual and desired attitude and angular rate with external disturbances.

The results shown above have proved that the proposed nonlinear attitude controller based on $SO(3)$ space has a robust control performance even under considerable external disturbances.

4.2. Trajectory Tracking

In order to validate the tracking performance of the proposed LMPC-based robust controller, four simulations were conducted: (a) aggressive trajectory tracking, (b) tracking lemniscate of Bernoulli, (c) acrobatic flight trajectory tracking, and (d) aggressive trajectory tracking under observation errors and extreme external disturbances. In these simulations, other controllers were employed for the purpose of comparison.

In the simulations, the prediction horizon for the position control was set to 50, which means that the LMPC optimizes the reference trajectory of the position loop for the next 50 time steps. Relevant parameters in the LMPC algorithm are presented in Table 2. The planning frequency of the position controller was set as 20 Hz, while the control frequency of the attitude controller was set as 200 Hz.

Table 2. MPC parameter settings in simulation.

Variables	Description	Values
N	The prediction horizon	50
$\mathbf{G_1}$	The weight matrix of tracking error	$Diag\{20, 20, \cdots, 20, 200\}$
$\mathbf{G_2}$	The weight matrix of control input	$Diag\{0.01, 0.005, \cdots, 0.005\}$
$\mathbf{Q}$	The weight matrix for control allocation	$Diag\{5, 5, 5, 10\}$
K_1	The proportional coefficient in attitude control law	10
K_2	The proportional coefficient in attitude control law	10

4.2.1. Aggressive Trajectory Tracking

In the first simulation scenario, the initial position for the quadrotor-UAV was at $[0, 0, 0]$. Both velocity and acceleration were configured as zero at the initial time. The given reference waypoints are

$$
\begin{aligned}
p^d_{x,k} &= 3\cos\left(\tfrac{2\pi}{5}t_k\right) \\
p^d_{y,k} &= 2\sin\left(\tfrac{2\pi}{5}t_k\right) \\
p^d_{z,k} &= 4 - 2\cos\left(\tfrac{2\pi}{5}t_k\right)
\end{aligned}
\tag{45}
$$

which requires aggressive attitude changes in both roll and pitch angles. A differential flatness-based cascaded PID controller with velocity and acceleration feedforward [24] was employed for comparison. Furthermore, a controller that combines MPC for position control with PID for attitude control was conducted as a comparative experiment to validate the performance of our attitude controller. The tracking performance is evaluated by comparing the desired and actual dynamics states (i.e., position, velocity, acceleration, attitude, and angular rate), as shown in Figure 5.

The mean square error (MSE) of position for the aforementioned three controllers is shown in Table 3.

Table 3. The MSE in tracking an aggressive trajectory.

Algorithms/Time	0–10 s	10–20 s	0–20 s
Proposed	$[0.4670, 0.0373, 0.2753]$	$[6.507 \times 10^{-6}, 2.917 \times 10^{-6}, 3.559 \times 10^{-6}]$	$[0.2335, 0.0186, 0.1376]$
PID + FF	$[0.8527, 0.1652, 0.2536]$	$[0.2183, 0.1099, 0.0098]$	$[0.5355, 0.135, 0.1317]$
MPC + PID	$[0.4632, 0.0507, 0.2781]$	$[0.0008, 0.0097, 0.0007]$	$[0.2320, 0.0302, 0.1394]$

In this scenario, substantial initial deviations occurred in both the x and z-axes. Moreover, there were also significant attitude changes in roll and pitch angles with an amplitude of 1 radius during flight. As shown in Figure 5 and Table 3, the PID controller with feedforward failed to track the trajectory, leading to observable phase lag in tracking. This shortfall could be attributed to the presence of initial errors, impeding the provision of precise feedforward information. For the controller that combines MPC and PID, although it exhibited a good tracking performance overall, there was a significant tracking error in the y-axis position compared to the proposed controller. This disparity became particularly conspicuous during peaks in the y-axis position, where substantial tracking errors manifest. On the contrary, the proposed controller tracked the given trajectory with high spatial and temporal precision.

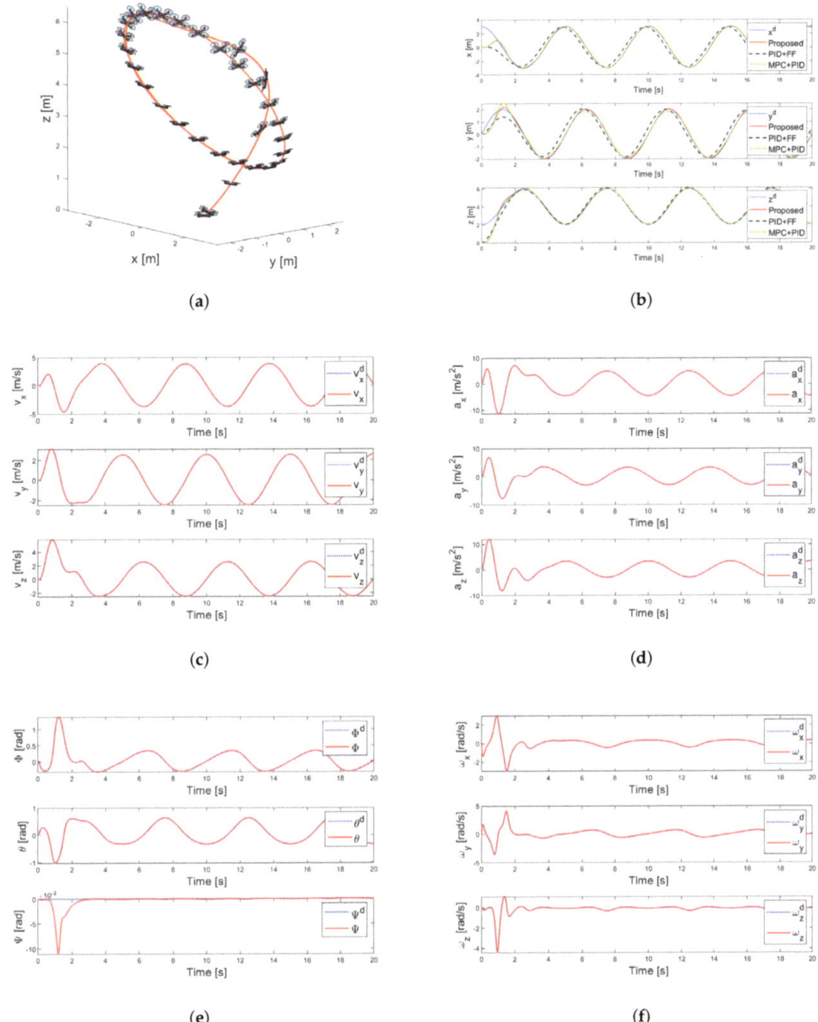

Figure 5. Aggressive trajectory tracking without disturbances. (**a**) Aggressive trajectory tracking. (**b**) Desired position (blue curve), actual position by the proposed controller (red curve), actual position by the controller that combines MPC for position control with PID for attitude control (green curve), and actual position by the PID controller with feedforward (dashed black curve). (**c**) Actual and desired velocity. (**d**) Actual and desired acceleration. (**e**) Actual and desired attitude. (**f**) Actual and desired angular rate.

4.2.2. Comparison of Different Controllers in Tracking Lemniscate of Bernoulli

To examine the performance of the proposed controller in waypoint tracking, a simulation was conducted utilizing the Bernoulli lemniscate as the reference trajectory. Similarly, the above two controllers (the PID controller with velocity and acceleration feedforward and the controller that combines MPC for position control with PID for attitude control) were implemented for comparison. The given reference waypoints are as follows. The initial position of the quadrotor-UAV was set at $[0, 4, 1]$, and both velocity and acceleration were set to zero at the initial time.

$$
\begin{aligned}
R &= 4, w = \frac{\pi}{3} \\
p_{x,k}^d &= -\frac{R\cos(wt_k)\sin(wt_k)}{\sin^2(wt_k)+1} \\
p_{y,k}^d &= \frac{R\cos(wt_k)}{\sin^2(wt_k)+1} \\
p_{z,k}^d &= 1
\end{aligned}
\tag{46}
$$

The waypoints and the actual flight path by the three controllers are illustrated in Figure 6. The comparisons between desired and actual values of position, velocity, acceleration, attitude, and angular rate are plotted in separate subplots of Figure 6.

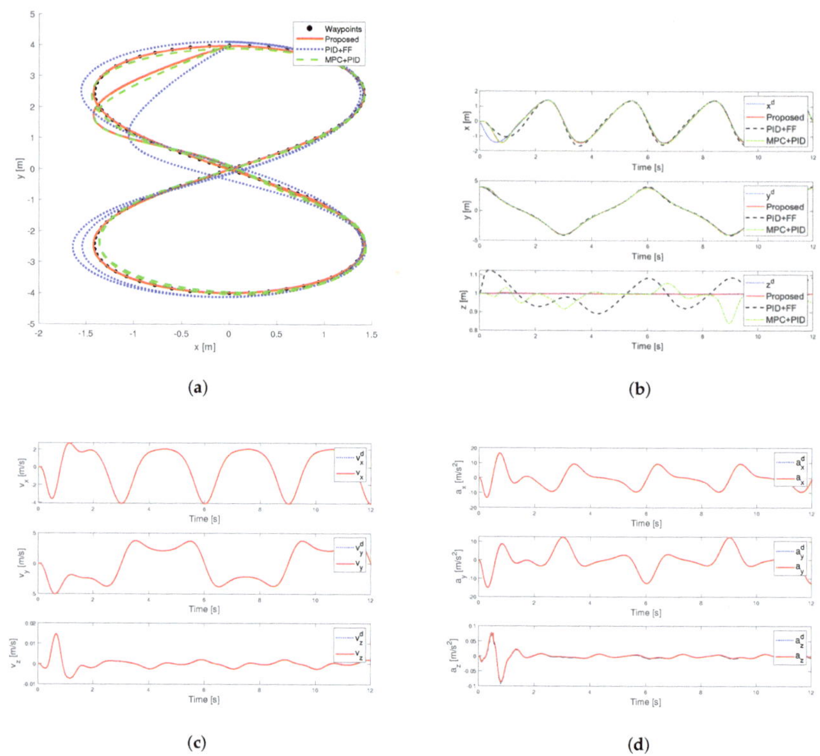

Figure 6. *Cont.*

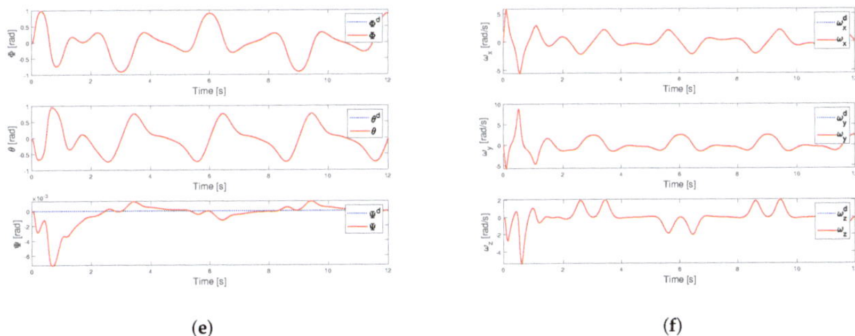

(e) (f)

Figure 6. Lemniscate of Bernoulli tracking. (**a**) Trajectory of different controllers. (**b**) Actual and desired position. (**c**) Actual and desired velocity. (**d**) Actual and desired acceleration. (**e**) Actual and desired attitude. (**f**) Actual and desired angular rate.

The MSE of position for the aforementioned three controllers is presented in Table 4.

Table 4. The MSE in tracking Lemniscate of Bernoulli.

Algorithms/Time	0–6 s	6–12 s	0–12 s
Proposed	$[0.0420, 0.0074, 1.4256 \times 10^{-6}]$	$[5.027 \times 10^{-5}, 5.128 \times 10^{-5}, 4.127 \times 10^{-7}]$	$[0.0210, 0.0037, 9.191 \times 10^{-7}]$
PID + FF	$[0.0832, 0.0049, 0.0045]$	$[0.0079, 0.0075, 0.0032]$	$[0.0455, 0.0062, 0.0038]$
MPC + PID	$[0.0451, 0.0070, 0.0008]$	$[0.0013, 0.0025, 0.0020]$	$[0.0232, 0.0047, 0.0014]$

In this simulation, the PID feedforward controller accurately provided the velocity and acceleration feedforward information. However, its effectiveness was limited by the position loop, which only output the desired acceleration. Consequently, it failed to provide additional feedforward information (desired angular velocity and desired angular acceleration) to the attitude control loop. As a result, significant tracking errors could be observed. The MPC combined with the PID controller showcased a satisfying tracking performance. However, as presented in Figure 6 and Table 4, this controller achieved precise tracking along straight paths, but exhibited larger tracking errors during aggressive turns. This discrepancy arose due to the requirement of the quadrotor-UAV for substantial adjustments in attitude during the turns, which the PID attitude controller failed to track accurately. In contrast, the proposed controller exhibited superior tracking performance. It accurately traced the desired trajectory after the initial adjustments.

4.2.3. Acrobatic Flight Trajectory Tracking

Since the design of the attitude controller is based on the $SO(3)$ space, it possesses the capacity to track dynamic trajectories encompassing substantial attitude maneuvers. An acrobatic flying trajectory in the X-Z plane was prescribed in Figure 7 for testing the proposed controller. The controller that combines MPC for position control with PID for attitude control was added as a comparative experiment. The given reference waypoints are

$$
\begin{aligned}
w_k^d &= \frac{\sqrt{v_{\min}^2 + 2gR(1 + \cos\beta_k)}}{R} \\
\beta_{k+1} &= \beta_k + w_k^d (t_{k+1} - t_k) \\
p_{x,k}^d &= R\sin\beta_k \\
p_{y,k}^d &= 0 \\
p_{z,k}^d &= R(1 - \cos\beta_k) + 1 \\
v_{\min} &\geq \sqrt{gR}
\end{aligned}
\tag{47}
$$

where R was set to 2 m, and v_{min} was set to 6 m/s. The initial position for the quadrotor-UAV was established as $[0, 0, 1]$, and both velocity and acceleration were set to zero at the initial time. In order to track the given waypoints accurately, the pitch angle of the quadrotor-UAV was required to complete a full attitude maneuver from 0 to 2π. Additionally, the large steps in the velocity and acceleration terms at the initial time made it more challenging for the controller.

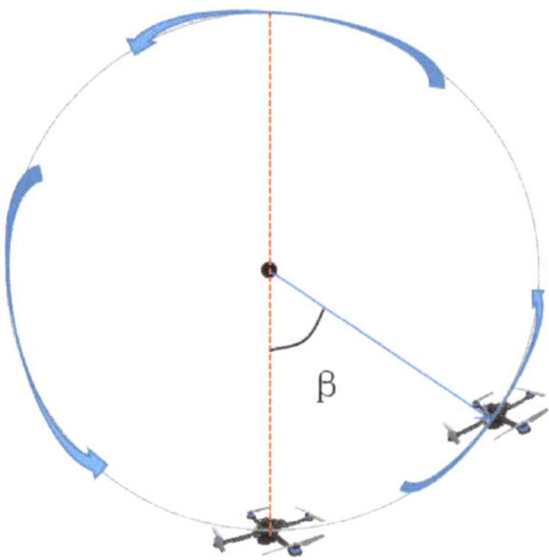

Figure 7. Acrobatic flight trajectory.

The performance in terms of tracking the acrobatic flight trajectory is evaluated by comparing the desired position, velocity, acceleration, attitude, and angular rate to the actual values and they are plotted in Figure 8.

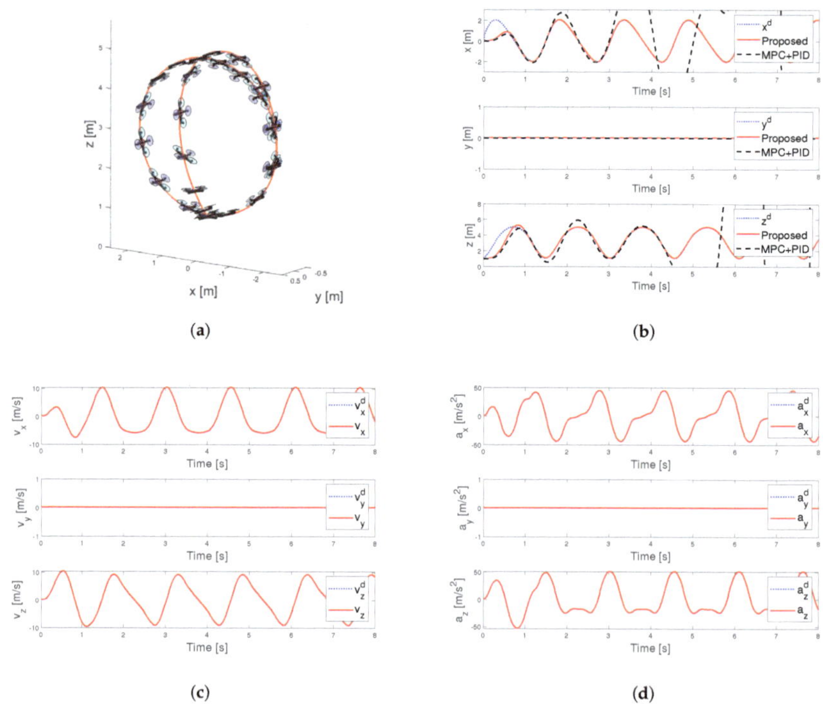

(a)　(b)　(c)　(d)

Figure 8. *Cont.*

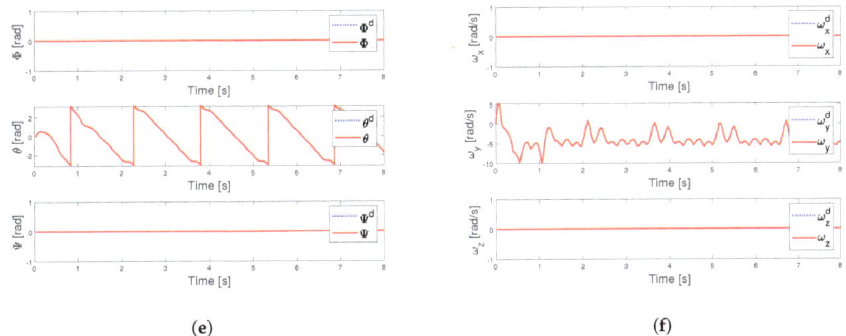

(e)   (f)

Figure 8. Acrobatic flight trajectory tracking of the quadrotor-UAV without disturbances. (**a**) Acrobatic flight trajectory tracking. (**b**) Actual and desired position. (**c**) Actual and desired velocity. (**d**) Actual and desired acceleration. (**e**) Actual and desired attitude. (**f**) Actual and desired angular rate.

The MSE of position for tracking the acrobatic flight trajectory is shown in Table 5.

Table 5. The MSE in tracking acrobatic flight trajectory.

Algorithms/Time	0–4 s	4–8 s	0–8 s
Proposed	[0.2486, 0, 0.3217]	$[2.404 \times 10^{-4}, 0, 3.26 \times 10^{-4}]$	[0.1244, 0, 0.1610]

In the simulation, given the trajectory involving full attitude maneuvers, our attitude controller achieved stable tracking without encountering any singularities. Despite the presence of significant errors in both velocity and acceleration at the initial time, good accuracy in tracking the desired waypoints could be observed in Figure 8 and Table 5. As a comparative experiment, the controller that combines MPC with PID failed to track the given waypoints. This inadequacy could be attributed to the PID attitude controller's limited ability to accurately trace the desired attitude, particularly when dealing with aggressive attitude changes.

4.2.4. Aggressive Trajectory Tracking under Observation Errors and Extreme External Disturbances

Considering the complex environmental factors the quadrotor-UAV may encounter during the actual flight, in the simulation, the artificial state observation errors were introduced for simulating the real-world conditions. Specifically, the random values were added to the state variables to emulate the measurement values. The amplitudes of the random values associated with each state variable are presented in Table 6.

Table 6. Observation errors in state variables.

Variables	Observation Error Amplitude
Displacement	0.05 [m]
Velocity	0.05 [m/s]
Acceleration	0.03 [m/s^2]
Attitude angle	0.02 [rad]
Angular rate	0.01 [rad/s]

The reference waypoints in this case are defined by Equation (45). The initial position for the quadrotor-UAV was set at $[0, 0, 0]$, with both velocity and acceleration initialized to zero at the initial time. Moreover, a significant external disturbance was introduced during the 9th to 11th seconds as shown in Figure 9. During this time interval, a constant force was

115

exerted on the quadrotor-UAV to simulate external disturbances, with magnitudes of 2.5N, 2N, and 1.5N along the three axes, respectively. Given the desired trajectory, observation errors, and external disturbances, the tracking performance of the controller is evaluated by comparing the desired position, velocity, acceleration, attitude, and angular velocity to the actual values in Figure 9.

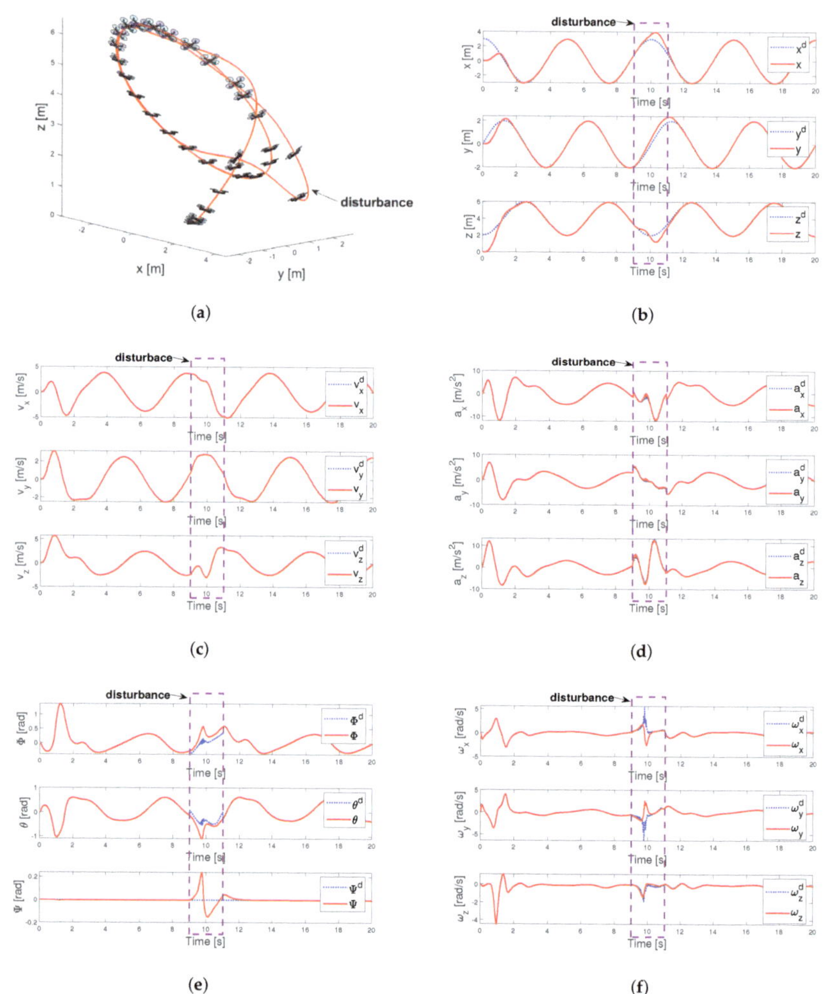

Figure 9. Aggressive trajectory tracking under observation errors and external disturbances. (**a**) Trajectory of proposed controller. (**b**) Actual and desired position. (**c**) Actual and desired velocity. (**d**) Actual and desired acceleration. (**e**) Actual and desired attitude. (**f**) Actual and desired angular rate.

The MSE of position for tracking aggressive trajectory under observation errors and extreme external disturbances is shown in Table 7.

Table 7. The MSE in tracking aggressive trajectory under disturbances.

Algorithms/Time	0–8 s	8–12 s	12–20 s
Proposed	[0.5846, 0.0469, 0.3450]	[0.2929, 0.0951, 0.1592]	$[1.518 \times 10^{-4}, 1.1 \times 10^{-4}, 4.923 \times 10^{-5}]$

The results presented in Figure 9 and Table 7 demonstrated that the proposed controller exhibited strong disturbance rejection capabilities. Even in the presence of observation errors and substantial external disturbances, the controller was able to maintain robust control. It is worth mentioning that the replanning time of our controller in the position loop is less than 0.01 s, which endows our controller with online control capability.

5. Conclusions

In the context of waypoint tracking tasks for quadrotor UAVs, a robust controller is introduced in this paper. The LMPC algorithm is employed in the position loop and a nonlinear control law based on the $SO(3)$ space is designed for the attitude control. The accurate tracking of the attitude loop enables the decoupling of the attitude control from the optimization within the MPC. The motor model constraints are taken into account for wide-time-domain prediction and feasible trajectory optimization. Furthermore, the nonlinear characteristics can be effectively compensated by the attitude controller, therefore allowing the controller to maintain its superior performance while reducing the computational expenses. Multiple aggressive trajectories and external disturbances were designed and simulations were conducted to test the controller. The proposed scheme showed high precision in tracking performance. Since the controller requires low computational resources, it is estimated to be implementable for real-time control for an actual quadrotor-UAV.

Author Contributions: Conceptualization, Z.X.; methodology, Z.X.; software, Z.X.; validation, Z.X. and W.Q.; formal analysis, Z.X., W.Q. and G.W.; investigation, Z.X.; resources, L.F.; data curation, Z.X.; writing—original draft preparation, Z.X.; writing—review and editing, G.W. and W.Q.; visualization, Z.X. and W.Q.; supervision, L.F., G.W., W.Q. and Y.H.; project administration, L.F., G.W., W.Q. and Y.H.; funding acquisition, L.F. All authors have read and agreed to the published version of the manuscript.

Funding: This work is supported by the Intelligent Aerospace System Team (2022R01003) of Zhejiang Provincial Leading Innovative Teams and the project of "Key Technology and Verification of Intelligent Piloted Electric Aircraft" (HYI21005) from Huzhou Institute of Zhejiang University.

Data Availability Statement: Not applicable.

Conflicts of Interest: The authors declare no conflict of interest.

References

1. Xu, H.; Yi, L.; Li, C.; Sun, Y.; Hou, L.; Bai, J.; Kong, F.; Han, X.; Lan, Y. Design and Experiment of Ecological Plant Protection UAV Based on Ozonated Water Spraying. *Drones* **2023**, *7*, 291. [CrossRef]
2. Liang, X.; Fang, Y.; Sun, N.; Lin, H. Nonlinear hierarchical control for unmanned quadrotor transportation systems. *IEEE Trans. Ind. Electron.* **2017**, *65*, 3395–3405. [CrossRef]
3. Xiong, T.; Liu, F.; Liu, H.; Ge, J.; Li, H.; Ding, K.; Li, Q. Multi-Drone Optimal Mission Assignment and 3D Path Planning for Disaster Rescue. *Drones* **2023**, *7*, 394. [CrossRef]
4. Li, J.; Li, Y. Dynamic analysis and PID control for a quadrotor. In Proceedings of the 2011 IEEE International Conference on Mechatronics and Automation, Beijing, China, 7–10 August 2011; pp. 573–578.
5. Bouabdallah, S.; Noth, A.; Siegwart, R. PID vs. LQ control techniques applied to an indoor micro quadrotor. In Proceedings of the 2004 IEEE/RSJ International Conference on Intelligent Robots and Systems (IROS) (IEEE Cat. No. 04CH37566), Sendai, Japan, 28 September–2 October 2004; Volume 3, pp. 2451–2456.
6. Greeff, M.; Schoellig, A.P. Flatness-based model predictive control for quadrotor trajectory tracking. In Proceedings of the 2018 IEEE/RSJ International Conference on Intelligent Robots and Systems (IROS), Madrid, Spain, 1–5 October 2018; pp. 6740–6745.
7. Lu, G.; Xu, W.; Zhang, F. On-manifold model predictive control for trajectory tracking on robotic systems. *IEEE Trans. Ind. Electron.* **2022**, *70*, 9192–9202. [CrossRef]
8. Romero, A.; Sun, S.; Foehn, P.; Scaramuzza, D. Model predictive contouring control for time-optimal quadrotor flight. *IEEE Trans. Robot.* **2022**, *38*, 3340–3356. [CrossRef]
9. Sun, S.; Romero, A.; Foehn, P.; Kaufmann, E.; Scaramuzza, D. A comparative study of nonlinear mpc and differential-flatness-based control for quadrotor agile flight. *IEEE Trans. Robot.* **2022**, *38*, 3357–3373. [CrossRef]
10. Zhao, W.; Liu, H.; Lewis, F.L. Robust formation control for cooperative underactuated quadrotors via reinforcement learning. *IEEE Trans. Neural Netw. Learn. Syst.* **2020**, *32*, 4577–4587. [CrossRef]
11. Wang, Q.; Namiki, A.; Asignacion, A., Jr.; Li, Z.; Suzuki, S. Chattering Reduction of Sliding Mode Control for Quadrotor UAVs Based on Reinforcement Learning. *Drones* **2023**, *7*, 420. [CrossRef]

12. Kaufmann, E.; Loquercio, A.; Ranftl, R.; Müller, M.; Koltun, V.; Scaramuzza, D. Deep drone acrobatics. *arXiv* **2020**, arXiv:2006.05768.

13. Li, G.; Tunchez, A.; Loianno, G. Learning model predictive control for quadrotors. In Proceedings of the 2022 International Conference on Robotics and Automation (ICRA), Philadelphia, PA, USA, 23–27 May 2022; pp. 5872–5878.

14. Han, H.; Cheng, J.; Xi, Z.; Yao, B. Cascade Flight Control of Quadrotors Based on Deep Reinforcement Learning. *IEEE Robot. Autom. Lett.* **2022**, *7*, 11134–11141. [CrossRef]

15. Serrano, F.; Castillo, O.; Alassafi, M.; Alsaadi, F.; Ahmad, A. Terminal sliding mode attitude-position quaternion based control of quadrotor unmanned aerial vehicle. *Adv. Space Res.* **2023**, *71*, 3855–3867. [CrossRef]

16. Fresk, E.; Nikolakopoulos, G. Full quaternion based attitude control for a quadrotor. In Proceedings of the 2013 European Control Conference (ECC), Zurich, Switzerland, 17–19 July 2013; pp. 3864–3869.

17. Meradi, D.; Benselama, Z.A.; Hedjar, R.; Gabour, N.E.H. Quaternion-based Nonlinear MPC for Quadrotor's Trajectory Tracking and Obstacles Avoidance. In Proceedings of the 2022 2nd International Conference on Advanced Electrical Engineering (ICAEE), Constantine, Algeria, 29–31 October 2022; pp. 1–6.

18. Choutri, K.; Lagha, M.; Dala, L.; Lipatov, M. Quadrotors trajectory tracking using a differential flatness-quaternion based approach. In Proceedings of the 2017 7th International Conference on Modeling, Simulation, and Applied Optimization (ICMSAO), Sharjah, United Arab Emirates, 4–6 April 2017; pp. 1–5.

19. Lee, T. Global Exponential Attitude Tracking Controls on SO(3). *IEEE Trans. Autom. Control* **2015**, *60*, 2837–2842. [CrossRef]

20. Yu, Y.; Yang, S.; Wang, M.; Li, C.; Li, Z. High performance full attitude control of a quadrotor on SO(3). In Proceedings of the 2015 IEEE International Conference on Robotics and Automation (ICRA), Seattle, WA, USA, 26–30 May 2015; pp. 1698–1703.

21. Lee, T.; Leok, M.; McClamroch, N.H. Geometric tracking control of a quadrotor UAV on SE(3). In Proceedings of the 49th IEEE Conference on Decision and Control (CDC), Atlanta, GA, USA, 15–17 December 2010; pp. 5420–5425.

22. Lee, T. Robust Adaptive Attitude Tracking on $SO(3)$ With an Application to a Quadrotor UAV. *IEEE Trans. Control Syst. Technol.* **2012**, *21*, 1924–1930.

23. Lian, S.; Meng, W.; Shao, K.; Zheng, J.; Zhu, S.; Li, H. Full Attitude Control of a Quadrotor Using Fast Nonsingular Terminal Sliding Mode With Angular Velocity Planning. *IEEE Trans. Ind. Electron.* **2022**, *70*, 3975–3984. [CrossRef]

24. Tian, B.; Liu, L.; Lu, H.; Zuo, Z.; Zong, Q.; Zhang, Y. Multivariable finite time attitude control for quadrotor UAV: Theory and experimentation. *IEEE Trans. Ind. Electron.* **2017**, *65*, 2567–2577. [CrossRef]

25. Islam, S.; Faraz, M.; Ashour, R.; Cai, G.; Dias, J.; Seneviratne, L. Adaptive sliding mode control design for quadrotor unmanned aerial vehicle. In Proceedings of the 2015 International Conference on Unmanned Aircraft Systems (ICUAS), Denver, CO, USA, 9–12 June 2015; pp. 34–39.

26. Mellinger, D.; Kumar, V. Minimum snap trajectory generation and control for quadrotors. In Proceedings of the 2011 IEEE International Conference on Robotics and Automation, Shanghai, China, 9–13 May 2011; pp. 2520–2525.

27. Murray, R.M.; Li, Z.; Sastry, S.S.; Sastry, S.S. *A Mathematical Introduction to Robotic Manipulation*; CRC Press: Boca Raton, FL, USA, 1994.

28. Zhou, X.; Wang, Z.; Ye, H.; Xu, C.; Gao, F. Ego-planner: An esdf-free gradient-based local planner for quadrotors. *IEEE Robot. Autom. Lett.* **2020**, *6*, 478–485. [CrossRef]

Disclaimer/Publisher's Note: The statements, opinions and data contained in all publications are solely those of the individual author(s) and contributor(s) and not of MDPI and/or the editor(s). MDPI and/or the editor(s) disclaim responsibility for any injury to people or property resulting from any ideas, methods, instructions or products referred to in the content.

 **drones**

MDPI

Article

A Multi-Regional Path-Planning Method for Rescue UAVs with Priority Constraints

Lexu Du [1,2], Yankai Fan [3], Mingzhen Gui [1,2] and Dangjun Zhao [1,2,*]

1 School of Automation, Central South University, Changsha 410083, China; dulx3364@csu.edu.cn (L.D.); guimingzhen@csu.edu.cn (M.G.)
2 Hunan Provincial Key Laboratory of Optic-Electronic Intelligent Measurement and Control, Changsha 410083, China
3 Academy of Astronautics, Nanjing University of Aeronautics and Astronautics, Nanjing 211106, China; yankai.fan@nuaa.edu.cn
* Correspondence: zhao_dj@csu.edu.cn

Abstract: This study focuses on the path-planning problem of rescue UAVs with regional detection priority. Initially, we propose a mixed-integer programming model that integrates coverage path planning (CPP) and the hierarchical traveling salesman problem (HTSP) to address multi-regional path planning under priority constraints. For intra-regional path planning, we present an enhanced method for acquiring reciprocating flight paths to ensure complete coverage of convex polygonal regions with shorter flight paths when a UAV is equipped with sensors featuring circular sampling ranges. An additional comparison was made for spiral flight paths, and second-order Bezier curves were employed to optimize both sets of paths. This optimization not only reduced the path length but also enhanced the ability to counteract inherent drone jitter. Additionally, we propose a variable neighborhood descent algorithm based on K-nearest neighbors to solve the inter-regional access order path-planning problem with priority. We establish parameters for measuring distance and evaluating the priority order of UAV flight paths. Simulation and experiment results demonstrate that the proposed algorithm can effectively assist UAVs in performing path-planning tasks with priority constraints, enabling faster information collection in important areas and facilitating quick exploration of three-dimensional characteristics in unknown disaster areas by rescue workers. This algorithm significantly enhances the safety of rescue workers and optimizes crucial rescue times in key areas.

Keywords: UAV coverage path planning; traveling salesmen problem; priority constraints; path optimization

Citation: Du, L.; Fan, Y.; Gui, M.; Zhao, D. A Multi-Regional Path-Planning Method for Rescue UAVs with Priority Constraints. *Drones* **2023**, *7*, 692. https://doi.org/10.3390/drones7120692

Academic Editors: Jihong Zhu, Heng Shi, Zheng Chen and Minchi Kuang

Received: 21 October 2023
Revised: 25 November 2023
Accepted: 27 November 2023
Published: 29 November 2023

Copyright: © 2023 by the authors. Licensee MDPI, Basel, Switzerland. This article is an open access article distributed under the terms and conditions of the Creative Commons Attribution (CC BY) license (https://creativecommons.org/licenses/by/4.0/).

1. Introduction

Unmanned aerial vehicles (UAVs) have been widely used in many domains due to their small size, sensitive control, and high scalability. Especially in the civil field, UAVs equipped with cameras, infrared, LiDAR, or other sensors can conduct various missions, including personnel searches [1], field monitoring [2], and terrain detection [3]. As UAVs can survey target areas without causing any damage, they are well-suited to performing information sensing tasks in remote or hazardous environments and complex terrains. By reaching a rescue site before rescue personnel, drones provide timely and precise target information, greatly enhancing the efficiency of rescue operations. To this end, a reasonable path should be planned such that the mobile sensor carried by a UAV can cover a region in a finite time, giving rise to a coverage path-planning (CPP) problem with various constraints, including energy consumption, time consumption, etc.

In recent years, extensive research on the CPP problem for covering single regions with energy constraints and photography constraints has been carried out [4–8]. A covered route primarily exhibits various shapes, including round-trip and spiral patterns. However, the UAV's motion capabilities in actual flight are limited, making it challenging to precisely follow sharp corners in simulated routes at turning points. Therefore, smoothing out the sharp corners of the flight path becomes essential to save UAV flight time and reduce jitter during turning. This challenge is extensively studied in the realm of UAV obstacle avoidance in flight and is also applicable in UAV path-planning route optimization. The Bezier curve was initially widely employed in robot motion planning. In recent years, its application has expanded to include the field of UAVs. Machmudah et al. [9] studied the incline and turn flight trajectory optimization of fixed-wing UAVs at a fixed altitude. Utilizing the Bezier curve as the maneuvering path, the speed change reduces the load coefficient of the inclined steering mechanism, and a simultaneous on-arrival target mission has also been successfully conducted when the turning radius was small. However, in practical scenarios of large-scale search and rescue missions, it is often not feasible to consider the entire disaster area as a single region for coverage. Instead, the area is divided into multiple areas of interest (AOIs) based on disaster information. Multiple disaster locations are then selectively surveyed to effectively obtain post-disaster information. Therefore, the challenge of path planning to cover multiple regions becomes a compelling research topic.

In general, the multi-regional path-planning problem can be converted to a combination of two subproblems: a traveling salesmen problem (TSP) and a CPP problem [10], constituting a TSP-CPP problem. A two-step path-planning method [11] is proposed to cover multiple disjoint regions: (1) the access order of UAVs between regions is determined in the first step by using genetic algorithms, and (2) the coverage path inside a region is determined in the second step by using the rotating caliper algorithm [12]. Xie et al. [10] planned the coverage path for multiple two-dimensional rectangular areas based on the grid approach and dynamic programming methods. The target area is first split into mesh grids according to the sensor's sampling range; thus, the original TSP-CPP problem is converted to a TSP problem, which is then solved by a dynamic programming method. Further, the authors proposed a heuristic algorithm based on NN-2Opt [13] for efficiently covering all regions even when there is a large number of regions. In [14], the minimum distance strategy was considered, and an improved simulated annealing algorithm was proposed to determine the access sequence of multiple regions, after which a back-and-forth (BF) path is generated to cover multiple convex polygonal regions. Ko et al. [15] proposed a novel UAV trajectory-planning method to optimize location-dependent visual coverage. In this method, the UAV dynamically adjusts its altitude to meet varying image-resolution requirements. Comprising three components, the approach effectively minimizes task completion time.

Indeed, in numerous disaster relief missions, the significance of regions is determined by factors such as severity, distance from the disaster center, and population density. While the aforementioned methods have effectively tackled the traditional TSP-CPP problem, they have failed to consider the diverse priorities of multi-target regions inherent in many large-scale rescue missions. To date, there has been scarce research on path-planning problems that accommodate distinct priorities for different regions. Miao et al. [16] introduced UAV-assisted moving edge computing (MEC) using UAVs as MEC nodes and proposed a multi-UAV-assisted MEC unloading algorithm based on global and local path planning. The approach takes into account the priority of monitoring sites but focuses on optimizing drone swarm scheduling, distribution, and communication coverage to minimize flight length and energy consumption. In [17], the access order of multiple regions is manually prescribed, and a heuristic algorithm is used to generate the sequence of regions without considering the different priority levels of different regions.

In order to deal with priority constraints, the planned access order for multiple regions depends on the prescribed priority levels of each region; thus, the original TSP should be extended as a hierarchical traveling salesman problem (HTSP) [18]. In [19], regions with the same priority are clustered into one cluster; correspondingly, regions with different priorities are clustered into different clusters, whose access sequences are determined by their priority levels. However, this planning approach ensures that the accesses are planned in order of priority, but it invariably results in a significant degree of path-length redundancy. In some application scenarios, this strategy may not be the most efficient and effective. Panchamgam et al. [18] proposed a d-relaxed priority model, in which priority was adjusted to a certain extent during the planning process while taking path length into account. The rule of this model is as follows: if p is the highest priority of all unvisited locations, the vehicle is allowed to access one of the positions whose priority is p, $p + 1, \ldots, p + d$. The value of the positive integer d can be flexibly controlled to realize the trade-off between path cost and emergency degree. Hà et al. [20] established a d-relaxed priority integer programming model based on [18] and introduced a metaheuristic method based on the framework of iterated local search with problem-tailored operators to find approximate solutions.

In the realm of drone path planning for real-life rescue scenarios, a comprehensive and systematic approach for multi-regional path-planning tasks with priority constraints is lacking. Such a method should have the ability to determine priority sequences and path lengths tailored for evaluating emergency rescue tasks effectively. This paper presents the following contributions: building upon the work presented in [18], we formally define the HTSP-CPP problem and formulate it as a mixed-integer programming model with d-relaxed constraints. In the realm of intra-regional path planning, we present an enhanced BF path coverage method that leverages the minimum width of polygonal regions. This method ensures complete coverage of convex regions by sensors equipped with circular sampling ranges. And simulate a comparison with the spiral path, optimizing both paths using Bezier curves. To optimize inter-regional access order planning, we introduce two different strategies for generating initial solutions, enabling efficient determination of the access sequence for multiple regions, and utilizing the RVND algorithm to optimize the initial solution. Additionally, we propose a distance–priority evaluation rule to assess the trade-off between distance and priority with respect to the solutions.

The structure of this paper is as follows. In Section 2, we present a mixed-integer programming model based on the HTSP-CPP. In Section 3, we discuss the related algorithms and explain the specific algorithm design and process for intra-regional and inter-regional path planning. We also propose a result evaluation index during the experimental design stage, design a specific simulation scheme, and present the simulation and experiment results in Sections 4 and 5.

2. Mathematical Model of the HTSP-CPP

The sensors integrated into UAVs mainly consist of LiDARs, RGB cameras, NIR cameras, and others. Studies by Salach et al. [21] and Domingo et al. [22] have highlighted LiDARs' superiority in terrain detection and 3D modeling compared to other sensors. However, concerning search and rescue missions, NIR cameras and RGB cameras demonstrate more significant potential. In complete disaster relief operations, UAVs equipped with LiDAR are initially utilized to gather terrain data in affected regions, aiding in disaster severity assessment and the formulation of relief strategies. Subsequently, UAVs outfitted with infrared or optical sensors conduct a secondary search in areas where individuals may be trapped, precisely identifying their locations for efficient rescue efforts. Therefore, the path-planning method should be applicable to various sensor types. However, due to the unique sampling shape of LiDAR, this paper focuses on improving the coverage method based on circular sampling ranges. Table 1 shows the main symbols used in this article.

Table 1. Summary table of important symbols.

Symbol	Definition	Symbol	Definition
A_i	The region's parameter set	d	The coefficient of relaxation
C	The set of accessed regions	d_L	The distances from the strip's sides to the base edge
C_i	The central point coordinates of the region	e_p^i	Decision variable, whether the waypoint, p, is the entrance of the region
$\mathbf{D}$	The distance matrix	g	The number of priorities
E	The set of unvisited regions	m_i	The number of waypoints in the region
G	The priority set	n_i	The number of vertices in the region
L_i	The sides of the strip proximal to the bottom	o	The path overlap rate of the UAV
L_i'	The sides of the strip further to the bottom	p_i	The priority of the region
N	The region's number set	t_p^i	Decision variable, whether the waypoint, p, is the export of the region
N_0	The region's number set including the depot	u_i	The position of the region in the access sequence
N_b	The number of stripes in the region	v_{ij}	The vertex coordinates of the region
$\mathbf{P}$	The priority matrix	w_{ij}	The coordinate of the waypoints
R_k	The set of regions with the same priority	x_{ij}	Decision variable, whether there is a connecting path between two regions, i, j
S	The width of the convex region	y_{pq}^i	Decision variable, whether there is a connecting path between two waypoints, p, q, in the region
V_i	The vertex set of the region	ω	The sensor's sampling diameter
W_i	The waypoints set of the region	Δd	The distance between the UAV's scan lines

2.1. Problem Description

We assume that there are n convex polygonal regions with different sizes and shapes to be covered, and these regions are denoted by $A_i = <V_i, C_i, p_i>$, where $V_i = \{v_{i1}, v_{i2}, \ldots, v_{in_i}\}$ is the vertex coordinates of the clockwise arrangement of polygons, v_{ij} is the coordinates of the j-th vertex of the i-th region, n_i is the number of vertices in the i-th region, C_i is the central point coordinate of the i-th region, and p_i is the priority of the i-th region, where $p_i \in G = \{1, \ldots, g\}$.

Now, a UAV equipped with a sensor (LiDAR) is utilized to initiate a comprehensive coverage detection of these n areas starting from the base for acquiring the height and obstacle information of each region. Upon completion of the coverage task, the UAV returns to the base. The ground sampling range of the sensor is circular in shape, as depicted in Figure 1.

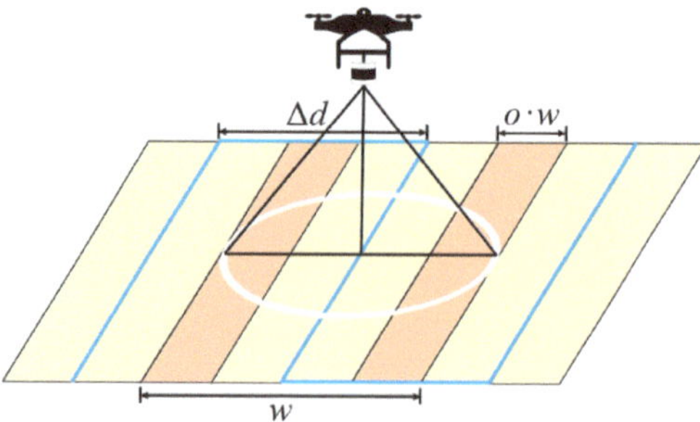

Figure 1. Ground sampling range of sensor.

The sampling diameter of the sensor is influenced not only by the sensor's performance, including the field of view angle, the angular resolution, and the maximum detection radius, but also the flight parameters, such as the flight height and velocity. As the flight altitude increases, the sampling range expands while the range that satisfies the required sampling precision decreases. Therefore, there exists a maximum flight altitude that ensures adequate sampling precision. Similarly, there exists a minimum flight altitude that guarantees the minimum required sampling range. When the UAV operates within this permissible altitude range, minor fluctuations in altitude will not impact sampling effectiveness. This allowable range is determined by both task-specific sampling precision requirements and sensor parameters. To simplify matters, we treat the sampling range necessary for achieving the desired precision as a fixed value and introduce a concept of "sampling overlap rate" to ensure that UAV altitude changes during flight do not compromise the task's overall quality of data collection. The sampling diameter is w, and the distance between scan lines is given as:

$$\Delta d = \omega(1 - o) \tag{1}$$

Since UAVs maintain a constant flight height regardless of the terrain's ups and downs, this problem can be considered a two-dimensional HTTP-CPP problem. Assuming that the sensor's sampling footprint can completely cover the target region, the aggregate flight path length of the UAV is taken as the flight cost, and the access sequence between regions and the flight path within each region are the decision variables to be optimized for minimizing the flight cost.

2.2. Problem Modelling

The path of a UAV is generated by a set of a series of waypoints $W_i = \{w_{i1}, w_{i2}, \cdots, w_{im_i}\}$, $i \in \{1, 2, \cdots, n\}$, where n is the number of target areas and m_i represents the number of waypoints in the i-th region, indicating the number of elements in the point set W_i. Each waypoint signifies a change in direction for the UAV. It is important to note that the UAV maintains a straight-line trajectory between any two consecutive waypoints. We divide the flight paths into two types: (1) intra-regional paths in a single region and (2) inter-regional paths connecting different regions.

a.   Intra-regional paths

Let y^i_{pq} denote the access order of waypoints $p, q \in \{1, 2, \cdots, m_i\}$ of the intra-regional path for the i-th region and $y^i_{pq} = 1$ indicate that the UAV flies from point p to point q, while $y^i_{pq} = 0$ indicates that there is no connecting path between the waypoints p and q. Then, let e^i_p and t^i_p denote the import and export of the i-th region, respectively. When $e^i_p = 1$ means that the UAV flies into region i from the waypoint p, $e^i_p = 0$ means that the waypoint p is not the entry point in region i. Similarly, when $t^i_p = 1$ represents that the UAV flies out of region i from the waypoint p, $t^i_p = 0$ represents that the waypoints p is not the exit point in region i.

b.   Inter-regional paths

The priority constraints should be imposed on the inter-regional path planning to obtain the optimal access sequence of multiple regions. Intending to cover high-priority regions as extensively as possible while minimizing flight costs, it becomes necessary to slightly relax the priority of individual regions. In this paper, we employ d-relaxed priority to model this problem. The d-relaxed priority approach ensures that during the planning of an inter-regional access sequence, if the current region has a priority of $k \in G = \{1, \cdots, g\}$, the priority of the subsequent region to be accessed should not exceed $k + d$. When $d = 0$, regional access should strictly adhere to the order of priority from high to low. Conversely, when $d = g - 1$, the problem is degraded into an ordinary TSP problem without any priority constraints. By selecting an appropriate value for d, a suitable compromise can be achieved between the distance traveled and the priority of regions.

Suppose the target regions are classified into different groups, R_k, according to their priorities, which represent the set of regions with priority level k. When planning the sequence of inter-regional access, an access sequence *Order* is generated. The value of u_i represents the sequence number in which the i-th region is accessed, that is, the index value of region i in *Order*. Let $N = \{1, 2, \cdots, n\}$ be the set of all area numbers to be accessed and $N_0 = \{0, 1, 2, \cdots, n\}$ be the set of numbers containing the depot, where 0 is the number of the depot. A decision variable x_{ij} $(i, j \in N_0)$ is introduced to represent the inter-regional access order, and $x_{ij} = 1$ if the UAV flies from region i to region j, while $x_{ij} = 0$ if there is no connection path between region i and region j. Therefore, the constraints on u_i and x_{ij} can be expressed as follows.

$$u_i + 1 - M(1 - x_{ij}) \leq u_j, \ \forall i, j \in N_0, j \neq 0 \tag{2}$$

$$u_i + 1 < u_j, \ \forall i \in R_k, j \in R_l; \ k, l \in G, l > k + d \tag{3}$$

The first constraint states that for any two regions i and j, when region j is not the depot and there exists a path from i to j, then the access order of region j should be after region i. Alternatively, if there is no path from i to j, this inequality is also satisfied when a sufficiently large M is used. The second constraint states that when the priority of region j does not meet the d-relaxed constraint, the access order of region j should be after region i and not in the immediate subsequent access position after region i.

c. Integer programming model of the HTSP-CPP Consider the following specific scenarios:

1. In the case where the sensor's sampling range can only cover region i, i.e., $m_i = 1$, meaning there is only one waypoint, w_{i1}, in region i, which coincides with the center of mass of region i. The UAV enters and exits region i from w_{i1} simultaneously. In other cases where $m_i > 0$, it is necessary to ensure that the UAV enters and exits region i from different points to avoid redundant path lengths.
2. When there are multiple regions and each region has only one waypoint, i.e., $N > 1$, then the problem simplifies to the TSP.
3. When there is only one region and its area exceeds the sampling range of the sensor, i.e., $N = 1$ and $m_1 > 0$, the problem becomes a CPP problem with a starting point and an ending point.

Taking into account the above scenarios, the objective function for this problem is as follows.

$$\begin{aligned}
J = &\sum_{i=1}^{n} \sum_{j=1, j \neq i}^{n} \sum_{p=1}^{m_i} \sum_{q=1}^{m_j} x_{ij} t_p^i e_q^j d(w_{ip}, w_{jq}) + \sum_{i=1}^{n} \sum_{p=1}^{m_i} \sum_{q=1, q \neq p}^{m_i} y_{pq}^i d(w_{ip}, w_{iq}) \\
&+ \sum_{i=1}^{n} \sum_{p=1}^{m_i} x_{0i} e_p^i d(c_0, w_{ip}) + \sum_{i=1}^{n} \sum_{p=1}^{m_i} x_{0i} t_p^i d(w_{ip}, c_0)
\end{aligned} \tag{4}$$

where $d(a, b)$ represents the Euclidean distance from a to b and c_0 is the coordinates of the depot. In Equation (4), the first term is the path length between regions, the second term represents the sum of path lengths within all regions, the third term represents the path length from the base to the entrance of the first region, and the fourth term represents the path length from the exit of the last region back to the base.

In addition, the following constraints should be imposed on Equation (4):

$$\sum_{j=0, j \neq i}^{n} x_{ij} = 1, \ \forall i \in N_0 \tag{5}$$

$$\sum_{i=0, i \neq j}^{n} x_{ij} = 1, \ \forall j \in N_0 \tag{6}$$

$$u_i + 1 - M(1 - x_{ij}) \leq u_j, \ \forall i, j \in N_0, \ j \neq 0 \tag{7}$$

$$u_i + 1 < u_j, \ \forall i \in R_k, \ j \in R_l; \ k, l \in G, \ l > k + d \tag{8}$$

$$\sum_{i \in R_k} \sum_{j \in R_l} x_{ji} = 0, \forall k, l \in G, l > k + d \tag{9}$$

$$\sum_{i \in R_k} x_{0i} = 0, \ \forall k \in G, \ k > 1 + d \tag{10}$$

$$\sum_{i \in R_k} x_{i0} = 0, \ \forall k \in G, \ k < g - d \tag{11}$$

$$\sum_{q=1, q \neq p}^{m_i} y_{pq}^i = 1 - t_p^i, \ \forall i \in N, \ p \in \{1, 2, \cdots, m_i\} \tag{12}$$

$$\sum_{p=1, p \neq q}^{m_i} y_{pq}^i = 1 - e_p^i, \ \forall i \in N, \ q \in \{1, 2, \cdots, m_i\} \tag{13}$$

$$\sum_{p=1}^{m_i} e_p^i = 1, \ \sum_{p=1}^{m_i} t_p^i = 1, \ \forall i \in N \tag{14}$$

$$\sum_{i,j \in M_1} x_{ij} \leq |M_1| - 1, \ \forall M_1 \subset \{0, 1, 2, \cdots, N\}, \ 2 \leq |M_1| \leq N - 1 \tag{15}$$

$$\sum_{p,q \in M_2} y_{pq}^i \leq |M_2| - 1, \ \forall i \in \{1, 2, \cdots, N\}, \ M_2 \subset \{1, 2, \cdots, m_i\}, \ 2 \leq |M_2| \leq m_i - 2 \tag{16}$$

$$x_{ij} \in \{0, 1\}, \ \forall i, j \in N_0 \tag{17}$$

$$u_i, u_j \in N_0, \ \forall i, \ j \in N_0, \ j \neq 0 \tag{18}$$

$$y_{pq}^i, e_p^i, t_p^i \in \{0, 1\}, \ \forall i \in N, \ p, q \in \{1, 2, \cdots, m_i\} \tag{19}$$

$$e_p^i + t_p^i \leq 1, \ \forall i \in N, \ m_i > 1, \ p \in \{1, 2, \cdots, m_i\} \tag{20}$$

Equations (5) and (6) specify that the UAV enters and exits each region, including the depot, only once. This implies that each region can be visited only once. Equation (7) represents the relationship between the position variable, u_i, and the decision variable, x_{ij}. Equations (8)–(11) describe the constraint conditions of d-relaxed priority: Equation (8) defines the relationship between regional positions under the d-relaxed priority constraint and Equation (9) specifies that, when the d-relaxed priority constraint is not satisfied, a region with lower priority cannot be directly transferred to a region with higher priority. Equations (10) and (11) are the constraints of the d-relaxed priority rule when leaving the depot and returning to the depot, respectively. Equations (12) and (13) indicate that, apart from the entrance and exit of each region, for each other waypoint, the UAV will fly from one waypoint to another, ensuring that each waypoint can be accessed only once and avoiding the redundant path length caused by repeated access to waypoints. Equation (14) states that each region has one and only one entrance and exit. Equations (15) and (16) ensure the continuity of paths to eliminate subloops within and between regions, and $|M|$ represents the cardinality of set M. Equations (17)–(20) specify the value ranges of the decision variables to ensure their effectiveness.

3. Algorithm Design

Based on the aforementioned analysis, the resolution of the HTSP-CPP problem articulated by Equations (4)–(19) can be solved through the proposed algorithm, which encompasses the following three key steps:

(1) Calculation of the inter-regional path and parameters: The initial step involves determining the optimal flight direction for the UAV within each convex polygonal region. This is achieved by calculating the width of the region. Simultaneously, the distance between flight paths is determined based on predefined sampling requirements. This information facilitates the generation of parallel flight lines within the region, subsequently yielding four potential candidate flight entry points;

(2) Construction of the priority-constrained TSP: The algorithm designates both the center of each region and the depot as "cities" to be visited. This lays the foundation for formulating a traveling salesman problem with priority constraints. To compute the most efficient order of access, a heuristic algorithm is employed. This step aids in identifying the optimal sequence for visiting the designated cities;

(3) Selection of optimal entry points and path generation: Utilizing the determined optimal access order, the algorithm proceeds to select the best entry points within each region. By amalgamating all selected waypoints, a coherent and comprehensive UAV path is formulated, which represents the final output of the algorithm.

In the following subsections, we will delve into the principles and code logic of this algorithm, conducting a comprehensive analysis of its effectiveness in solving HTSP-CPP problems. Additionally, we introduce a region order optimization algorithm and showcase the optimization impact on region access order using distance and priority evaluation criteria.

3.1. Calculation of Inter-Regional Paths and Parameters

In this paper, we employ a BF path pattern for comprehensive coverage of the designated region. Notably, executing turns with a UAV entails heightened energy consumption and concomitant augmentation of the overall path length. Consequently, diminishing the frequency of turns stands as a pivotal means of curtailing drone flight costs. Guided by the imperative of turn reduction, this paper embarks on a quest to ascertain the optimal width of convex polygons [23]. Subsequently, the UAV's BF path aligns with the vertical direction of the width of these polygons, a strategic alignment that serves to minimize path length across the targeted expanse.

a. Calculate the width of the convex polygon to determine the best flight direction of the UAV

The authors of [23] proposed a method for determining the width of a convex polygon, which is defined as the minimum span of the polygon. This characteristic is evident specifically in polygon configurations featuring vertex–edge patterns. Notably, the span between two parallel edges can also represent the width of the polygon, which is recognized as a distinct instance of the vertex–edge scenario.

To calculate the width of a convex polygon, a comprehensive procedure is followed for each of its sides. This involves the calculation of the distances from all vertices to the selected edge. The longest calculated distance corresponds to the height of the chosen edge, thereby identifying the vertex associated with this particular height. By performing this calculation for all sides of the convex polygon and subsequently comparing the computed lengths, the minimum height emerges as the polygon's width, denoted as S. Remarkably, the edge linked with this minimum height establishes the optimal flight direction for the UAV [12].

The optimal flight direction is shown in Figure 2. Herein, the red edge and the green vertex are indicative of the edge and vertex corresponding to the width of the polygon, respectively. Consequently, the optimal flight orientation for the UAV is one that parallels the identified red edge.

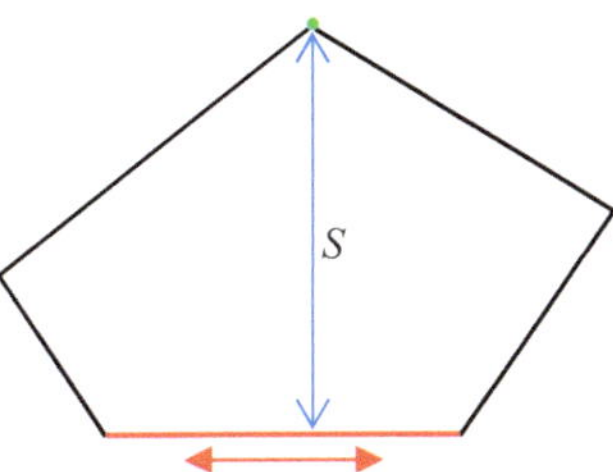

Figure 2. Optimum flight direction.

b.   Dividing the region into strips along the width direction according to the distance between flight lines

Due to the circular sampling range of the sensor system, the scenarios depicted in Figure 3a,b are likely to arise when employing the conventional convex polygon coverage method [12]. In Figure 3, the red-shaded regions within the red boxes illustrates the uncovered area. The blue arrows denote the flight path, the green dots indicate the starting point of the flight path, and the orange area represents the coverage achieved by the sensor. To address this limitation and ensure comprehensive coverage of both vertices or edges, this paper employs a strategy involving the division of a polygon into multiple strip-like subregions along the width direction [24]. The resultant coverage effect is illustrated in Figure 3c. It becomes evident that this approach facilitates the complete coverage of all corners within the designated area, while concurrently adhering to the requisite standards for sampling quality.

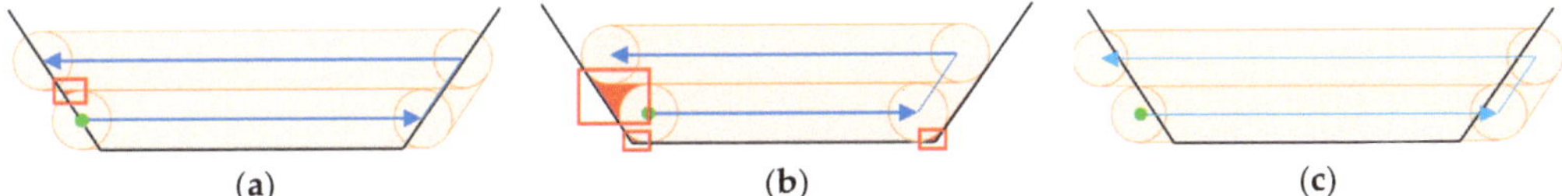

(a)   (b)   (c)

Figure 3. Effects of different coverage methods. (**a**) The flight line starts on the edge of the polygon. (**b**) The edge of the sensor is tangent to the edge of the polygon. (**c**) The method proposed in this paper.

The width of a strip corresponds to the scanning diameter, w, of the sensor. As shown in Figure 4, the median line that runs parallel to the optimal flight direction in the strip serves as the designated UAV flight trajectory. Furthermore, the intersection points between this median line and the strip delineate the specific path points for the UAV. Notably, the length of the strip plays a pivotal role in dictating the extent of the flight trajectory, thereby ensuring that the sensor can effectively encompass regions at the vertex or edge of the polygon. This strategic arrangement safeguards against scenarios akin to those depicted in Figure 3a,b.

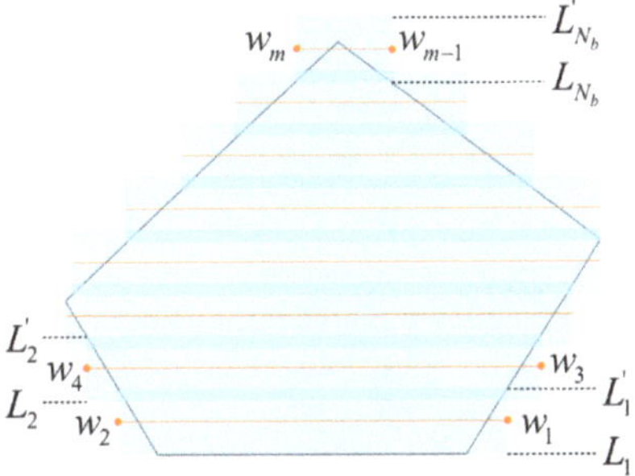

Figure 4. Strip division method.

For the i-th strip, which aligns with the best flight direction, the two parallel sides are denoted as L_i and L_i', respectively. Among these, L_i corresponds to the side proximal to the bottom, while L_i' pertains to the side situated further from the bottom. The determination of the strip's position is achieved by calculating the distances, d_{L_i} and $d_{L_i'}$, from the strip's sides to the base edge. This calculation is performed according to the following equation:

$$\begin{cases} d_{L_i} = \Delta d \cdot (i-1) \\ d_{L_i'} = d_{L_i} + \omega \end{cases} \tag{21}$$

where $\Delta d = \omega \cdot (1 - o), i \in \{1, 2, \cdots, N_b\}$. N_b denotes the overall quantity of stripes, and its calculation is delineated as follows:

$$N_b = \begin{cases} \left\lfloor \frac{S}{\Delta d} \right\rfloor, \ if \ S\backslash\Delta d \leq \omega \cdot o \\ \left\lceil \frac{S}{\Delta d} \right\rceil, \ if \ S\backslash\Delta d > \omega \cdot o \end{cases} \tag{22}$$

The extent of a strip's length is determined by the minimal measure required for each strip to precisely encompass the polygon. This entails considering three distinct scenarios:

(1) When both edges of the strip intersect the polygon, generating two intersections, as illustrated in Figure 5a, the strip's length is essentially the greater of the distances between the two intersections. Mathematically, the length of the strip is represented by $d(I_3, I_4)$;

(2) In instances where the strip is defined by a single edge that intersects with the polygon at two points, depicted in Figure 5b, the strip's length equates to the distance between these two intersection points. This is succinctly expressed as $d(I_1, I_2)$;

(3) If a strip encompasses vertices within its scope, as illustrated in Figure 5c, the strip's length is the shortest distance capable of covering the given vertex. This length is symbolized by $d(I_5, I_6)$.

c. Generate candidate paths based on the four entry points

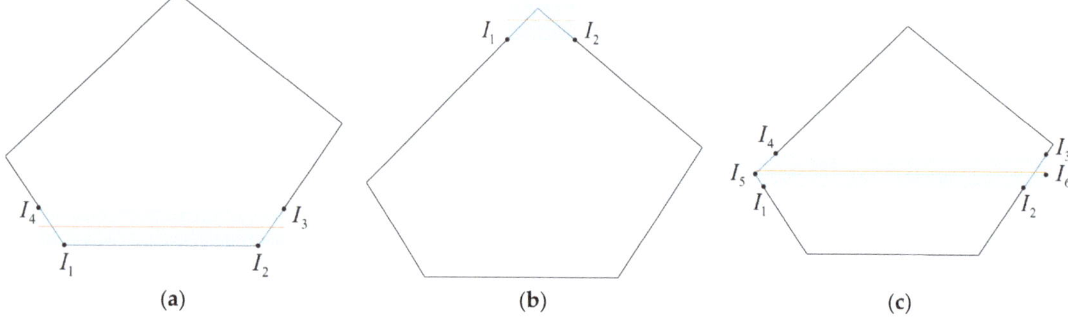

Figure 5. Strip length analysis. (**a**) Two edges, two points. (**b**) One edge, two points. (**c**) Strip containing vertices.

Once N_b parallel scan lines have been determined, the initial and terminal points of the first scan line and the N_b-th scan line inherently constitute the four potential flight entry points for the current designated region. Subsequently, the UAV embarks on its coverage mission by entering the region through these four flight entry points. This sequential process engenders the creation of four distinct, complete paths confined within the region's bounds, as visually depicted in Figure 6.

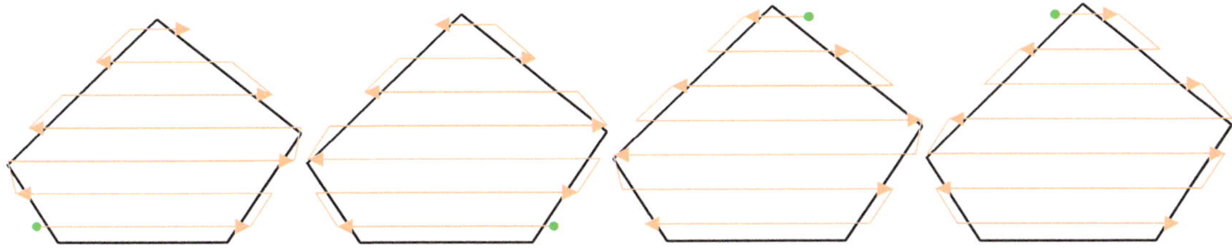

Figure 6. Different entry points of the path.

With the inter-regional access sequence firmly established, the judicious arrangement of entry points assumes a paramount role in minimizing path lengths. Therefore, subsequent to the formulation of the optimal inter-regional access sequence, a pragmatic selection is made among these four paths. This selection serves as a strategic step toward ultimately deriving the most optimal overall path configuration.

The pseudocode for generating path points within the region is presented in Algorithm 1. Lines 1–6 encompass the computation of the region's width and determine the optimal flight direction for the UAV. Subsequently, lines 7–25 provide a comprehensive description of how strip-related parameters are calculated along with the determination of endpoints for parallel scanning lines of the UAV. Moving forward, lines 26–30 ascertain both the set of path points and their corresponding lengths at each of the four inlet points. Ultimately, this function yields four distinct paths and their respective lengths within a given region.

Algorithm 1: *getIntraWay* (A_i, ω, o)

Input: Regional parameter A_i, sensor parameter ω, o
Output: Complete collection of waypoints *path*, the path length in the region *dist*

1:       // Get the width of the region and the best flight direction
2:     **For** each j in 1 to n_i **do**
3:        Calculate the distance between the vertex v_{ij} and all edges of the region, and take its maximum value, denoted as h_j, and the corresponding edge denoted as $edge_j$
4:     **End for**
5:     $S \leftarrow \min([h_1, h_2, \cdots, h_{n_i}])$
6:     $EDGE \leftarrow$ the edge corresponding to S
7:     // Get the scan lines by calculating strip parameters
8:     $\Delta d \leftarrow \omega * (1 - o)$
9:     **If** $S \backslash \Delta d \leq \omega * o$ **then**
10:    $N_b \leftarrow \lfloor S \backslash \Delta d \rfloor$
11:    **Else**
12:    $N_b \leftarrow \lceil S \backslash \Delta d \rceil$
13:    **End if**
14:    **Dor** each k in 1 to N_b **do**
15:       $d_{L_k} \leftarrow \Delta d(k - 1)$
16:       $d_{L_k'} \leftarrow d_{L_k} + \omega$
17:       The length of the strip *len* $\leftarrow$ the maximum value generated by the intersection of the two sides and the middle line of the strip with the region
18:       $L_k \leftarrow$ the coordinates of the two endpoints on one side of the strip are determined by d_{L_k} and *len*
19:       $L_k' \leftarrow$ the coordinates of the two endpoints on the other side of the strip are determined by $d_{L_k'}$ and *len*
20:    **End for**

21:    $m_i \leftarrow 2 * N_b$
22:    **For** each l in 1 to m_i **do**
23:        $w_{il} \leftarrow$ the coordinates of the midpoints of the lines connecting corresponding points in L_k and L'_k
24:        $W_i \leftarrow \{W_i, w_{il}\}$
25:    **End for**
26:    // Get the candidate paths based on the four entry points
27:    **For** each t in 1 to 4 **do**
28:        *path(t)* $\leftarrow$ Sort the waypoints in W_i with the t-th candidate point as the entry point
29:        *dist(t)* $\leftarrow$ the length of the *path(t)*
30:    **End for**
31:    **Return** *path, dist*

In addition to BF coverage, the spiral coverage method can avoid the frequent acceleration and deceleration of the drone during turning, reducing drone jitter. It is an effective method for obtaining stable terrain data. To generate a spiral flight trajectory, for any convex polygonal region, start by selecting adjacent sides among all edges to form perpendicular bisectors, resulting in the same number of intersection points (coordinates may be identical). Take the average of these coordinates, and the resulting point is marked as the reference point for the polygon. Subsequently, connect the vertices of the polygon to the reference point, calculate the lengths of the resulting line segments, and then divide these lengths by the sensor diameter to obtain the segment count. Assuming the UAV enters from the reference point of the polygon, it first flies toward the farthest vertex from that point for a distance equal to the sensor diameter. Then, it flies in the direction of the second-farthest vertex and continues flying in that direction until it exits the polygonal region, as illustrated in Figure 7. The green lines in Figure 7 represent the connections between reference point and vertices, the blue points denote flight waypoints, and the red lines depict the flight path.

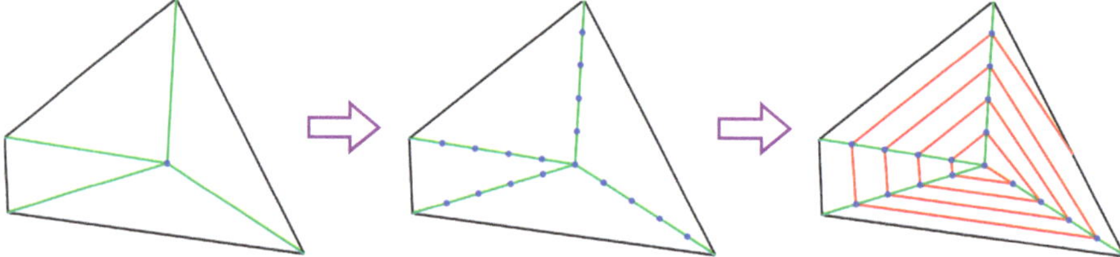

Figure 7. Spiral-curve generation method.

However, the UAV's motion capabilities in actual flight are limited, making it challenging to precisely follow sharp corners in simulated routes at turning points. Therefore, smoothing out the sharp corners of the flight path becomes essential to save UAV flight time and reduce jitter during turning.

If there are two points, P0 and P2, around the turning point P1 in the UAV flight path, these three points can be used to form a second-order Bezier curve. This is illustrated in Figure 8a, wherein:

$$\begin{cases} p_{01}(t) = (1-t)P_0 + tP_1 \\ p_{12}(t) = (1-t)P_1 + tP_2 \end{cases} \Rightarrow p_2(t) = (1-t)p_{01} + tp_{12}, \ t \in [0,1]$$

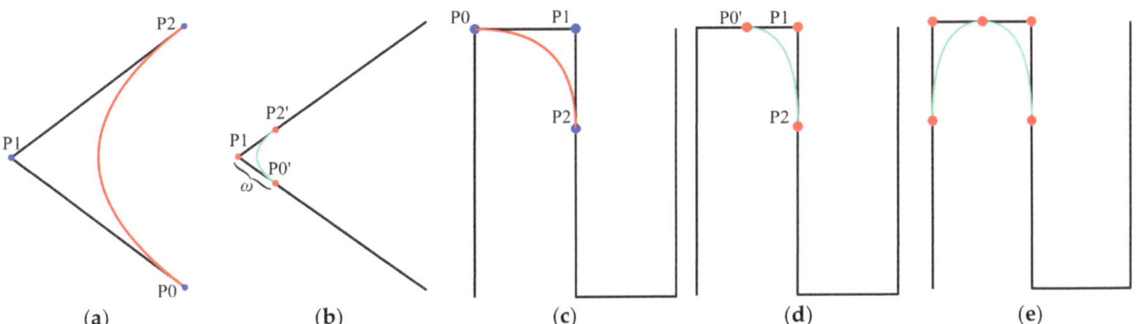

Figure 8. Bezier curve and optimized path. (**a**) A second-order Bezier curve. (**b**) Determine control points according to parameter ω. (**c**) A special case. (**d**) The solution. (**e**) Two consecutive paths.

Therefore, the turning point of the UAV flight path is considered as P1. Two points, denoted as P0′ and P2′, are selected on the two adjacent edges of point P1 at a sensor sampling radius of w. These three points, P0, P1, and P2, act as control points to create a Bezier curve, resulting in a smoothed UAV flight path, as depicted in Figure 8b. This approach helps reduce the jitter during UAV flight. Additionally, the length of the flight path P0→P0′→P2′→P2 is shorter than the straight flight path from P0→P1→P2.

In the method described above, a special case arises when the distance between P0 (or P2) and P1, determined by the sensor sampling radius, w, is not less than the distance between P1 and the adjacent vertices. More precisely, the distance between the two vertices is less than twice the sensor radius, resulting in an incorrect position for point P0 (or P2), as illustrated in Figure 8c. To address this issue, in such cases, the original point P0 (or P2) can be replaced by the midpoint of the adjacent vertices, as shown in Figure 8d. This optimization scheme ensures the correct placement of the point. Two continuous Bezier curves established by the two vertices of this short edge are then depicted in Figure 8e.

When the UAV utilizes a circular sensor for flight with an improved back-and-forth path, the Bezier curve can optimize the flight path, as depicted in Figure 9a, to the improved path shown in Figure 9b, thus optimizing two consecutive turning movements into a single U-turn. When the drone adopts a spiral trajectory, Bezier curves can optimize the flight path shown in Figure 9c to resemble that shown in Figure 9d, smoothing the turning points in the path. Bezier curves are applicable to both of these coverage methods to reduce the length of the flight path and mitigate the jitter phenomenon caused by a too-small turning radius, almost without compromising the coverage effect.

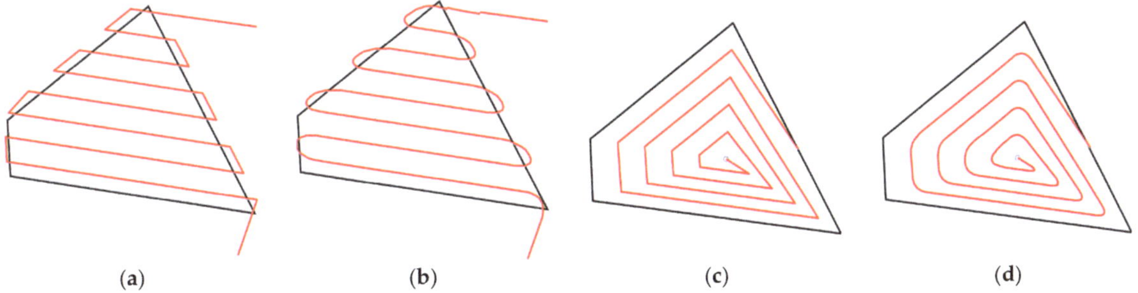

Figure 9. Optimization effects of Bezier curves on BF and spiral paths. (**a**) BF path without Bezier. (**b**) BF path with Bezier. (**c**) Spiral path without Bezier. (**d**) Spiral with Bezier.

3.2. Construction of the Priority-Constrained TSP

The formulation of the inter-regional access order entails addressing a TSP that accommodates the imposition of priority constraints. In this paper, a meta-heuristic algorithm rooted in the random variable neighborhood descent (RVND) framework [25] is harnessed for this purpose. The algorithmic approach commences by generating an initial solution via the K-nearest neighbors (KNN) heuristic, subsequently adopting the variable neighborhood descent (VND) technique for local search operations. The combination of these strategies effectively yields an approximate optimal solution for the complex problem.

a.   Two methods of initial solution generation based on PKNN

KNN is a well-established classification algorithm within the realm of machine learning predominantly employed for categorizing samples characterized by similar features. An extension of this, the priority K-nearest neighbors (PKNN) algorithm, draws inspiration from the principles of the KNN algorithm. In the PKNN algorithm, during each iteration, the highest-priority region is chosen from the K-nearest neighboring regions with respect to the current region, serving as the subsequent target for access. The value of K in the PKNN algorithm functions as a limiting factor for the search scope. This algorithm encompasses the capacity to effectively balance priority considerations and path lengths to a reasonable degree.

In this paper, two distinct strategies are advanced for the selection of K-nearest neighbor regions. The following discussion initially outlines the procedural steps intrinsic to the implementation of the first strategy, which is named PKNN-Full, focused on identifying K-neighboring regions across all regions.

Step 1: Introduce a set denoted as C serving as a repository for region numbers that have undergone access alongside a collection E designated to accommodate region numbers that remain unexplored. Simultaneously, establish a collection termed *Order* designed to retain the region numbers, which are systematically arranged according to the computed access order. Given that UAVs are required to initiate coverage operations from a depot, the initialization phase incorporates the inclusion of the depot within the *Order* collection. Proceed to formulate a region matrix denoted as **D** and devise a corresponding priority matrix named **P**. Within **D**, each i-th row systematically arranges the sequence numbers of regions in an ascending order, signifying their proximity to region i. Simultaneously, **P**, having the same dimensions as **D**, ascribes each individual element to represent the priority attributed to regions occupying corresponding positions in **D**.

Step 2: The last element in the *Order* represents the current region. By consulting matrix **D**, the K-nearest neighboring regions pertaining to the current region are discerned, and their presence in set C is evaluated—this assessment essentially determines whether they have been visited or not. Subsequently, the regions that are yet unvisited are compiled in E, effectively emerging as candidate regions. The respective priorities corresponding to these unvisited regions are simultaneously acquired from matrix **P**.

Step 3: Select the region with the highest priority from collection E as the subsequent target for access. Incorporate its sequence number into the *Order* while concurrently removing the region from the set C. In instances where the highest-priority regions exhibit a non-unique presence, the region that is proximate to the current region is elected for access.

Step 4: Iterate through the execution of Step 2 and Step 3 until all the regions within set C have been visited, leading to the eventual emptiness of C. This iterative process culminates in the attainment of an inter-regional access sequence *Order*.

The ensuing scenarios are anticipated during the implementation of Step 2 and Step 3, each warranting its own resolution strategy, as outlined below:

(1)   If all K-neighboring regions of the current region have been visited, continue to select the subsequent set of K-neighboring regions for the Step 2 operation; if it is found that all such regions have been visited as well, repeat this process until an unvisited region emerges within any group of K-neighboring regions. This unvisited region is

then cataloged within the collection in E, subsequently triggering the commencement of Step 3 operations;

(2) In instances where the final set of neighboring regions adjacent to the current region consists of fewer elements than the stipulated value K, it is still treated as a valid set for Step 2 operations;

(3) When all neighboring regions in the current region have been visited, the current region is designated as the last region in the sequence of visits. It is subsequently appended to the *Order* collection, thereby culminating in the aforementioned iterative loop.

The pseudocode for PKNN-Full is presented in Algorithm 2, where lines 1–5 correspond to Step 1, mentioned above. Lines 6–18 constitute a loop body that corresponds to the loop in Steps 2–4, described earlier. Finally, the function outputs the initial solution of the access order.

Algorithm 2: *PKNN-Full* (A, c_0, K)

Input: All regional parameter set A, when $A \leftarrow \{A_1 \cdots A_n\}$, the coordinate of depot c_0, nearest neighbor parameter K
Output: The initial solution of the inter-regional access sequence *Order*
1:    Initialize: Create a set C for the unvisited region set where all region numbers are stored, an empty candidate region set E, and an empty set *Order* for the access sequence
2:    **For** each i in N_0 **do**
3:       $\mathbf{D}(i, :) \leftarrow$ region numbers in ascending order of distance from region i
4:       $\mathbf{P}(i, :) \leftarrow$ the priority of the region corresponding to the location in $\mathbf{D}$
5:    **End for**
6:    $Order \leftarrow$
7:    **While** $C \neq \varnothing$
8:       $cr \leftarrow$ the last element in *Order*
9:       $Item = 1$
10:   **While** $E \cup C = \varnothing$
11:     $E \leftarrow \mathbf{D}(cr, (item - 1) \cdot K + 1 : item \cdot K)$
12:     $Item \leftarrow item + 1$
13:   **End while**
14:   $ar \leftarrow$ the number of the area with the highest priority in E according to $\mathbf{P}$
15:   $Order \leftarrow \{Order, ar\}$
16:   $C \leftarrow C$ set without region ar
17:   $E \leftarrow \varnothing$
18: **End while**
19: **Return** *Order*

The second KNN search strategy, referred to as PKNN-Excluded, involves seeking out the K-nearest regions among the unselected regions. In contrast to the aforementioned procedure, the divergence lies solely within the implementation of Step 2. The modified Step 2 operations are delineated as follows:

Step 2: The last element in the set *Order* denotes the current region. Subsequently, matrix $\mathbf{D}$ is consulted to identify the K-nearest regions that remain unvisited in relation to the current region. These K regions are stored in the collection E as candidate regions. Their corresponding priorities are simultaneously extracted from matrix $\mathbf{P}$.

The pseudocode of PKNN-Excluded is presented in Algorithm 3, where only lines 9–13 exhibit variations from Algorithm 2 due to the distinct approach employed for neighborhood region selection.

Algorithm 3: *PKNN-Excluded (A, c_0, K)*

Input: All regional parameter set A, when $A \leftarrow \{A_1 \cdots A_n\}$, the coordinate of depot c_0, nearest neighbor parameter K

Output: The initial solution of the inter-regional access sequence *Order*

1: Initialize: Create a set C for the unvisited region set where all region numbers are stored, an empty candidate region set E, and an empty set Order for the access sequence

2: **For** each i in N_0 **do**

3: $\mathbf{D}(i, :) \leftarrow$ region numbers in ascending order of distance from region i

4: $\mathbf{P}(i, :) \leftarrow$ the priority of the region corresponding to the location in $\mathbf{D}$

5: **End for**

6: $Order \leftarrow \{Order, 0\}$

7: **While** $C \neq \varnothing$

8: $cr \leftarrow$ the last element in *Order*

9: **While** $E \cup C = \varnothing$

10:   $F \leftarrow \mathbf{D}(cr, :)$

11:   $F \leftarrow F$ set without region in *Order*

12:   $E \leftarrow$ the first K elements of the set F

13: **End while**

14: $ar \leftarrow$ the number of the area with the highest priority in E according to $\mathbf{P}$

15: $Order \leftarrow \{Order, ar\}$

16: $C \leftarrow C$ set without region ar

17: $E \leftarrow \varnothing$

18: **End while**

19: **Return** *Order*

The primary objective of the KNN algorithm is to expedite the visitation of regions with higher priority by increasing the value of K. As K varies within the range from 1 to the total number of regions $n-1$, the search scope progressively extends from nearby regions to encompass all available regions. In situations where the count of searchable regions equals 1, it signifies that, exclusively, the region closest to the current region qualifies for selection as the next target. This approach indeed facilitates ensuring shorter travel distances to a certain extent. However, as the K value gradually increases towards $N-1$, the scope of searchable regions encompasses the entirety of available regions. In such instances, the next target region is attributed to the region characterized by both the highest priority and proximity to the current location. Ultimately, this methodology guarantees the attainment of optimal priority sequencing. Furthermore, when dealing with regions of identical priority, the principle of minimum distance governs the arrangement of access paths, thereby underlining a comprehensive optimization approach.

Within the context of the two KNN search strategies, PKNN-Full incorporates the regions that have been accessed within the process of identifying K neighbors. This strategy proves advantageous when dealing with a limited number of regions in proximity to the current region. Instead of expending additional distance to uncover regions of higher priority, this strategy endeavors to locate regions in closer proximity, thereby optimizing resource utilization. PKNN-Excluded involves identifying K neighbors by excluding regions that have already been accessed. The advantage of this strategy lies in its immunity to disruption from regions already accessed. This prevents regions of higher priority from being overlooked, ensuring consistent access to the region with higher priority. Both strategies demonstrate flexibility in their application. Therefore, the KNN algorithm incorporating these two search strategies avoids the necessity of rigidly defining a standardized or constant K-value selection logic throughout the research process. Instead, it prioritizes adaptability, dynamically employing the most effective strategy to achieve the highest total score for comprehensive coverage of the target region. This approach is centered on optimizing conditions in response to changing variables. In essence, it strives to generate the finest quality and most gratifying search path, thereby offering invaluable, comprehensive assistance in time-sensitive relief operations functioning under tight time constraints.

b.   Local search strategy

Random variable neighborhood descent (RVND) stands as a meta-heuristic algorithm framework proposed by Mladenovi et al. [26]. This framework leverages a diverse range of neighborhood structures, each comprising distinct actions, to facilitate alternating searches, thereby striving for optimal results. Let t denote the number of neighborhood structures and $\{N_1, N_2, \cdots, N_t\}$ denote the set of neighborhood structures. Within the RVND approach, when the present neighborhood structure fails to improve the current optimal solution, the algorithm seamlessly transitions to the subsequent neighborhood to continue the search. The search process concludes once all neighborhoods exhaustively fail to yield improvements to the optimal solution.

In this paper, the approach of the variable neighborhood search (VNS) strategy [20] with a d-relaxed priority constraint is employed to solve the intricacies of access order planning between regions governed by priority constraints. The configuration of the neighborhood structure set, denoted as N_i, is underpinned by the d-relaxed priority constraint. The corresponding requirements to satisfy the d-relaxed criteria are outlined below. Importantly, the current operation is performed solely when the constraints corresponding to the pertinent neighborhood structure are successfully satisfied.

(1)   Relocated (1)—N_1: Reallocate $Order[i]$ to a position succeeding the j-th index in the access sequence Order contingent upon the fulfillment of any of the following conditions: ① If $j < i$,

$$p^{min}_{j(i-1)} \geq p_{Order[i]} - d \tag{23}$$

② If $i < j - 1$,

$$p^{max}_{(i+1)(j-1)} \leq p_{Order[i]} + d \tag{24}$$

(2)   Relocated (2)—N_2: Reallocate $Order[i]$, $Order[i+1]$ to a position succeeding the j-th index in the access sequence $Order$ contingent upon the fulfillment of any of the following conditions: ① If $j < i$,

$$p^{min}_{j(i-1)} \geq \max\left(p_{Order[i]}, p_{Order[i+1]}\right) - d \tag{25}$$

② If $j > i + 2$,

$$p^{max}_{(i+2)(j-1)} \leq \min\left(p_{Order[i]}, p_{Order[i+1]}\right) + d \tag{26}$$

(3)   Swap (1-1)—N_3: Swap $Order[i]$ and $Order[j]$ contingent upon the fulfillment of any of the following conditions: ① If $i < j$,

$$p^{min}_{i(j-1)} \geq p_{Order[j]} - d \text{ and } p^{max}_{(i+1)j} \leq p_{Order[i]} + d \tag{27}$$

② If $j < i$,

$$p^{min}_{i(j-1)} \geq p_{Order[j]} - d \text{ and } p^{max}_{(i+1)j} \leq p_{Order[i]} + d \tag{28}$$

(4)   Swap (2-1)—N_4: Swap two adjacent regions $Order[i]$, $Order[i+1]$ and another region $Order[j]$ in the access sequence $Order$ contingent upon the fulfillment of any of the following conditions: ① If $j > i + 1$,

$$p^{min}_{i(j-1)} \geq p_{Order[j]} - d \text{ and } p^{max}_{(i+2)j} \leq \min\left(p_{Order[i]}, p_{Order[i+1]}\right) + d \tag{29}$$

② If $j < i$,

$$p^{min}_{j(i-1)} \geq \max\left(p_{Order[i]}, p_{Order[i+1]}\right) - d \text{ and } p^{max}_{(j+1)(i-1)} \leq p_{Order[j]} + d \tag{30}$$

(5)   Swap (2-2)—N_5: Swap the two adjacent regions $Order[i]$, $Order[i+1]$ and the other two adjacent regions $Order[i]$, $Order[i+1]$ in the access sequence $Order$ contingent upon the fulfillment of the following conditions: If $i + 1 < j$,

$$p^{min}_{j(i-1)} \geq \max\left(p_{Order[i]}, p_{Order[i+1]}\right) - d \ \text{ and } \ p^{max}_{(i+2)j} \leq \min\left(p_{Order[i]}, p_{Order[i+1]}\right) + d \tag{31}$$

where p^{min}_{ij} and p^{max}_{ij} respectively represent the minimum and maximum values of priorities from region i to region j in the access sequence *Order*.

c.   Disturbance

When the local search cannot improve the current solution, a perturbation operator comes into play, introducing a random perturbation to guide the current solution away from the local optimality. This study uses a straightforward yet efficacious perturbation strategy encompassing two core operations: Relocated (1) and Swap (1-1). These operations align with the d-relaxed constraints and are executed with distinct selection probabilities of p and $(1 - p)$, respectively.

The primary section of the pseudocode for access order optimization is illustrated in Algorithm 4. Lines 1–2 depict the generation of the initial solution, while the subsequent lines focus on RVND optimization. The function incorporates the five neighborhood structures of d-relaxed constraints as predefined parameters.

Algorithm 4: *getOrder* (A, K)

Input: All regional parameter set A, when $A = \{A_1 \cdots A_n\}$, the coordinate of depot c_0, nearest neighbor parameter K
Output: The inter-regional access sequence *Order*
1:      *Order* ← *PKNN-Excluded* (A, c_0, K) or *Order* ← *PKNN-Full* (A, c_0, K)
2:      *Dist* ← the length of the *Order* is obtained from the center point of the region
3:      *Order'* ← *Order*
4:      **For** each t in 1 to 5 **do**
5:        **For** each i in 1 to $n + 1$ **do**
6:        **For** each j in 1 to n + 1 *do*
7:           *Order'* ← use the neighborhood structure N_t to operate on *Order*
8:           *Dist'* ← the length of the *Order'* is obtained from the center point of the region
9:           **If** *Dist'* < *Dist* **then**
10:          *Order* ← *Order'*
11:         **End if**
12:       **End for**
13:      **End for**
14:     **End for**
15:     **Return** *Order*

3.3. Selection of Optimal Entry Points and Path Generation

In this paper, drawing inspiration from the principles of the greedy algorithm, a method is devised to determine the entry points for each region based on the inter-regional access order derived in Section 3.2. Specifically, from the four prospective candidate entry points identified in Section 3.1 for each region, the entry point that is closest to the flight point of the previous region is selected as the entry point of the current region. Subsequently, a meticulous arrangement of endpoints for the flying scan lines is orchestrated in alignment with the designated entry points. This arrangement is integral to facilitating the formation of coherent BF scanning path trajectories, effectively constituting the waypoints in the region. This process culminates in the procurement of a comprehensive collection of waypoints spanning multiple regions. These waypoints encompass both the depot and region-based waypoints, organized according to the order of internal access.

The pseudocode for the function that generates the complete path is illustrated in Algorithm 5. This function takes the outputs of Algorithms 1 and 4 as inputs and sequentially connects the intra-regional paths based on regions by selecting appropriate fly-in points. It should be noted that while Algorithm 1 generates intra-regional paths for a single region, Algorithm 5 requires intra-regional paths for all regions, necessitating running Algorithm 1 separately for each region before executing it.

Algorithm 5: *getWaypoint* (c_0, *Order*, *path*)

Input: Depot coordinates c_0, access sequence *Order*, intra-regional waypoints *path*
Output: Complete collection of waypoints *PATH*
1:    $PATH \leftarrow \{PATH; c_0\}$
2:    **For** each i in 1 to $n + 1$ **do**
3:    **For** each j in 1 to 4 **do**
4:       $d_j \leftarrow$ the distance between the last coordinate in the *PATH* and the j-th entry point in the region *Order (i)*
5:    **End for**
6:    *Path'* $\leftarrow$ the entry point corresponding to the minimum value in $d_1 \cdots d_4$ and the *path* within region *Order (i)* starting from that point
7:    $PATH \leftarrow \{PATH; Path'\}$
8:    **End for**
9:    **Return** *PATH*

3.4. Path Evaluation Criteria Based on Priority and Distance

Given that the paths generated in individual regions adhere to specific length criteria, the evaluation metric in question exclusively assesses the effectiveness of the inter-regional access sequence planning algorithm under the ambit of priority constraints. In this paper, we produce two distinct categories of solutions: distance reference solutions without factoring priority and priority reference solutions without factoring distance. These solutions are generated through specific strategies tailored for each approach. Subsequently, the solution derived from the proposed algorithm—which incorporates both distance and priority considerations—is juxtaposed against these two aforementioned solutions. This comparative process yields the distance score and priority score separately. By summing these scores, the overall score for the current solution is computed. A higher score is indicative of superior optimization outcomes. The subsequent sections outline the methodologies employed for distance scoring and priority scoring.

a.   Distance scoring strategy

As the strategy overlooks the intra-regional path and the inherent regional priority, a simplification is employed, treating each region as a singular particle and regarding the central point's coordinates as the representative location of the region. Furthermore, the task of planning the paths connecting every region to the depot is regarded as a conventional TSP.

The realm of the general TSP problem is well-established and has been extensively studied. Given reasonable constraints on the number and dimensions of regions, numerous algorithms have been developed to uncover the optimal solution. In this paper, a genetic algorithm is employed to obtain the reference solution $Order_{Ref}$ along with its corresponding distance, $Dist_{Ref}$. Comparatively, the optimization algorithm proposed herein yields $Order$ as the solution, accompanied by its respective distance, $Dist$. However, as the algorithm introduced in this paper considers both distance and priority, it necessitates a trade-off, which is manifest as a certain sacrifice in distance length to harmonize the prioritization sequence. As a result, the distance score of the current solution can be expressed as follows:

$$SC_D = \frac{Dist_{Ref}}{Dist} \tag{32}$$

It can be seen that the smaller the distance from $Dist$, the higher the distance score SC_D.

b.   Priority scoring strategy

In this paper, we employ the strategy of sequential penalty accumulation for priority scoring. This approach aims to assign a penalty factor, denoted as Pf, to each level of priority. In cases where the priority level is defined as $G = \{1, 2, \cdots, g\}$, the respective penalty factor is designated as $Pf = \{g, g-1, \cdots, 1\}$. This assignment illustrates the relationship between priority and penalty factors, with higher priority levels receiving higher penalty factors in accordance with their importance. For a given access sequence *Order*, let us consider the i-th region, whose priority is represented as $p_{Order[i]} = k \in G$, and it corresponds to a specific penalty factor denoted as Pf_k. In the context of sequential penalty accumulation, the penalty accumulation process for the *Order* can be defined as follows:

$$T = \sum_{i=1}^{n} i \cdot Pf_k, \; k = p_{Order[i]} \tag{33}$$

Evidently, when the pathway sequence meticulously adheres to the prioritization levels—meaning that regions endowed with higher priorities are addressed foremost—an optimal outcome marked by the minimal sequence penalty accumulation value, T_{Ref}, can be achieved. The algorithm expounded within this paper considers both distance and priority assignment, which consequently leads to the emergence of heightened penalty accumulation. This augmented penalty accumulation acts as a counterbalance to the distance aspect when weighed against a solution hewing strictly to priority constraints. Consequently, the formulation for the priority score of the prevailing solution can be articulated as follows:

$$SC_P = \frac{T_{Ref}}{T} \tag{34}$$

c.   Comprehensive scoring strategy

Designating the significance of both spatial distance and priority within the solution, we introduce the distance weight, w_D, and the priority weight, w_P. In fact, these weights serve to quantify the respective contributions of distance and priority in the solution's evaluation process. Consequently, the comprehensive appraisal of the current solution is encapsulated in the integrated score, which is precisely the weighted mean of the distance score and the priority score:

$$SC = \frac{w_D \cdot SC_D + w_P \cdot SC_P}{w_D + w_P} \tag{35}$$

Within the context of this paper, a balanced consideration between the significance of spatial distance and priority is established by fixing both w_D and w_P at 0.5. This deliberate choice underscores the equivalence of importance attributed to both factors within the solution evaluation framework. Nevertheless, it is important to acknowledge that, in real-world applications, the assignment of the distance weight and the priority weight can be tailored to the precise demands of the scenario at hand. The flexibility to adjust these weights allows for a customized approach that aligns more closely with the specific requirements of the problem under investigation.

4. Simulation and Experiment

To commence, the initial step involved generating a collection of 20 convex quadrilaterals, each encircled by an external circle with a radius of 50 m. These quadrilaterals were designated as the target regions necessitating access. Subsequently, a random assignment of priority values, ranging from 1 to 3, was applied to each distinct region. This distribution is visually represented in Figure 10, where priority 1 regions are demarcated by the red zones, priority 2 regions are indicated by the yellow zones, and priority 3 regions are highlighted in green.

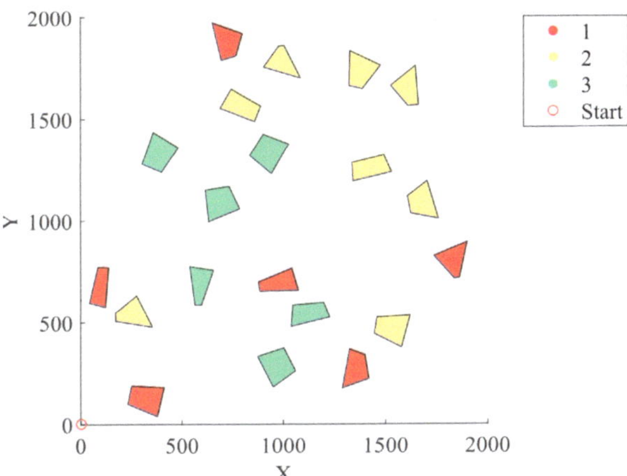

Figure 10. Original area.

Within this spatial setup, the point of origin (0, 0) was established as the starting point for the UAV's coverage task. Similarly, the termination point of this coverage task coincided with the same origin. To govern the UAV's sensory reach, a sensor coverage radius spanning 20 m was employed. Moreover, to ensure a judicious sampling process along the UAV's paths, an overlap rate of 0.1 was instituted.

These defined parameters collectively formed the foundation for the simulation. They served as the bedrock upon which the algorithm's efficacy was scrutinized and validated in a simulated environment.

4.1. Obtainment of Three Different Initial Solutions

The simulation entailed the execution and analysis of path-planning results for two distinct initial solutions, employing the set of randomly generated regions. Through this simulated process, an intricate evaluation was performed to dissect both the merits and drawbacks of the outcomes. Moreover, a careful examination of the trajectory of scores was undertaken to discern evolving trends.

By subjecting these initial solutions to rigorous simulation, a comprehensive understanding emerged regarding their practical implications. This scrutiny not only dissected the strengths and weaknesses exhibited in the generated paths but also traced the trajectory of scores across the simulation. Such insights contributed significantly to appraising the effectiveness of the path-planning algorithm under scrutiny.

a.   The PKNN-Full Strategy

The evaluation methodology followed the PKNN-Full strategy to derive distinct metrics: the distance score, the priority score, and the cumulative total score for varying K values, as well as the relationship between these scores and the corresponding K values, which were meticulously examined and are visualized in Figure 11.

As illustrated in the figure, a notable trend emerged where the path's length score exhibited a rapid reduction accompanied by a gradual ascent of the priority score towards full realization coinciding with increasing values of K. This pattern can be attributed to the amplified tendency of the path to preferentially select regions of higher priority as target destinations within the process of augmenting K values. This inclination was further substantiated by the positive correlation between higher K values and an increased availability of alternative regions, thereby ensuring earlier selection of regions with elevated priority status.

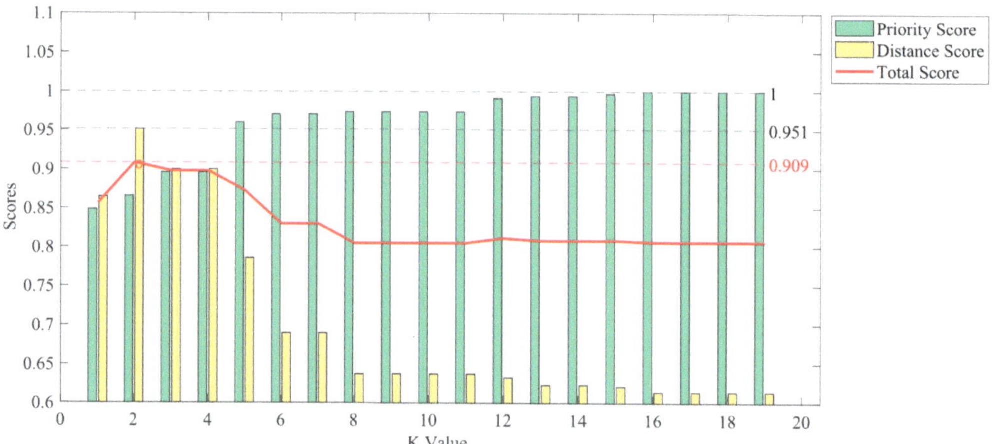

Figure 11. PKNN-Full score trend.

Examining the distinctive instances highlighted in the graphical representation, a pivotal observation arises: a zenith in the total score can be observed at K = 2. This specific point signifies the pinnacle performance achievable through the proposed planning methodology. This outcome is indicative of the path length closely approximating the optimal trajectory length while concurrently maintaining a relatively intact priority hierarchy.

Upon closer inspection, for K = 3 and K = 4, a sustained elevation characterizes both scores, manifesting a harmonious equilibrium. This equilibrium proves especially pertinent in scenarios demanding a balanced consideration of both priority and path dynamics, thus adhering to the requisites of balanced solution paradigms.

Upon entering the domain of K = 8, a notable transition can be observed. Here, while the priority score strictly adheres to the higher-priority sequence delineated by distance-based constraints in the selection of a larger set of neighbors, there is a conspicuous decline in the distance score. Consequently, an overall evaluation slightly inferior to the preceding cases ensued.

Beginning at K = 16, a distinctive pattern emerges. The trajectory of region selection becomes stabilized, culminating in the attainment of an optimal priority configuration. Subsequent path planning unfolds meticulously, aligning exactly with the priority order specified by distance-based constraints. This configuration finds particular relevance in environments where priority considerations hold substantial weight.

Guided by the principles underpinning the KNN algorithm, it is evident that, as K approaches $n - 1$, the path progressively attains an optimal priority orientation. To provide a visual portrayal, the comprehensive path corresponding to K = 2 is depicted in Figure 12. The red line in the Figure 12 represents the flight path of the UAV from the depot to the first waypoint, while the remaining flight paths are depicted in black. Various colored polygonal areas represent different priority levels.

b. The PKNN-Excluded Strategy

The evaluation methodology followed the PKNN-Full strategy to derive distinct metrics: the distance score, the priority score, and the cumulative total score for varying K values, as well as the relationship between these scores and the corresponding K values, which were meticulously examined and are visualized in Figure 13.

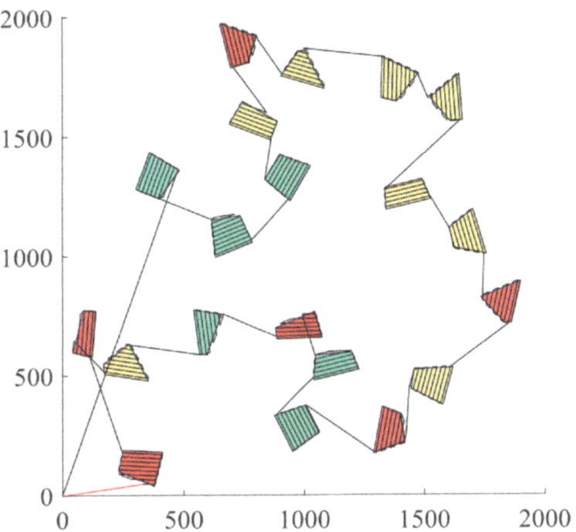

Figure 12. Completed path with PKNN-Full.

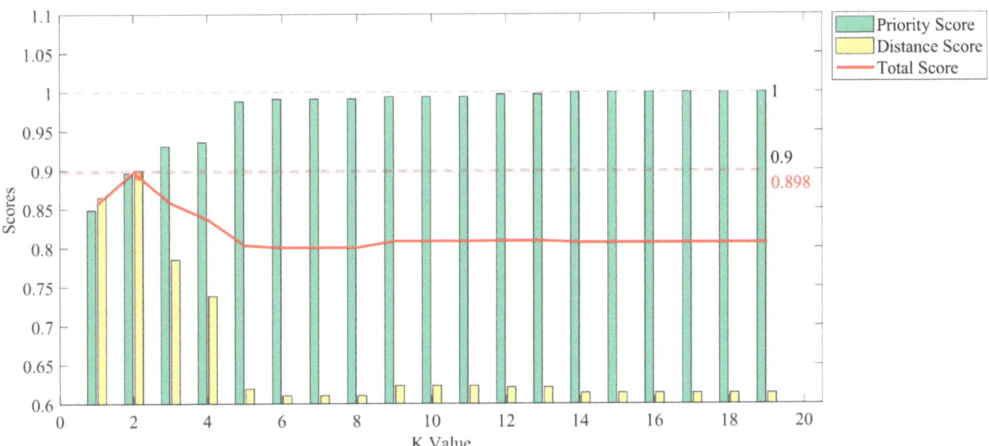

Figure 13. PKNN-Excluded score trend.

Figure 13 showcases the outcomes of the KNN algorithm implemented with the adapted search strategy, and, notably, it attained the highest overall score when K = 2. This algorithm variant yielded a path length marginally shorter than that of the previously discussed search methodologies. Nevertheless, it significantly approximated the optimal path length while concurrently elevating the priority, resulting in similar total scores that signified commendable equilibrium. As K = 3, a conspicuous decline in distance becomes apparent. By the onset of K = 5, the priority score experiences a gradual stabilization at an elevated threshold, which also coincides with a deceleration in the descent of the distance score. Echoing the trend, from K = 16 onwards, the trajectory of region selection stabilizes, and its priority reaches its apex configuration. Illustrating this empirically, the comprehensive path for K = 2 is visually presented in Figure 14. The meanings of the colors in Figure 14 are the same as those in Figure 12.

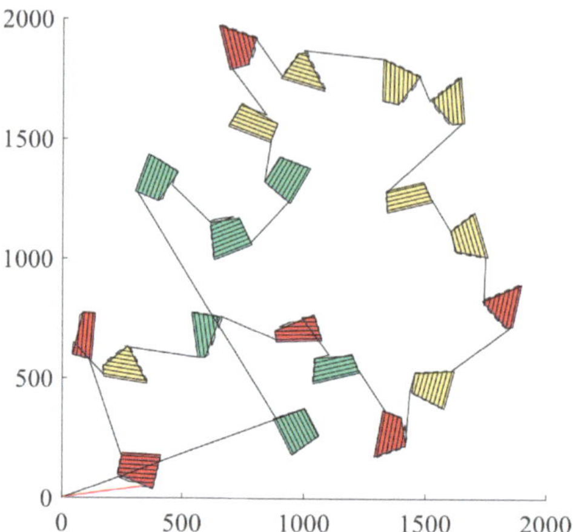

Figure 14. Completed path with PKNN-Excluded.

4.2. Optimizing Initial Solutions by RVND

Building upon the initial solution, we employed the RVND algorithm for the purpose of optimizing the path. By comparing the outcomes of this optimization process with those of the initial solution, we substantiated the efficacy of the RVND algorithm in curtailing path length while upholding the assured priority score. This comparative analysis served to validate the algorithm's capability in achieving path length reduction without compromising the stipulated priority constraints. The influence of the RVND algorithm on the optimization of the path was examined within the context of the PKNN-Full strategy. Initially, the trend of the total scores was computed across all K values, encompassing the range from 0 to $g - 1$, during d-relaxation. Subsequently, this trend was juxtaposed with the total score progression exhibited by the initial solution, visualized through a line chart, as depicted in Figure 15a. Analogously, the trend of the optimization's total scores for the PKNN-Excluded strategy was also obtained and illustrated in Figure 15b. This comparative analysis served to shed light on the impact of the RVND algorithm on the optimization efficacy of these strategies.

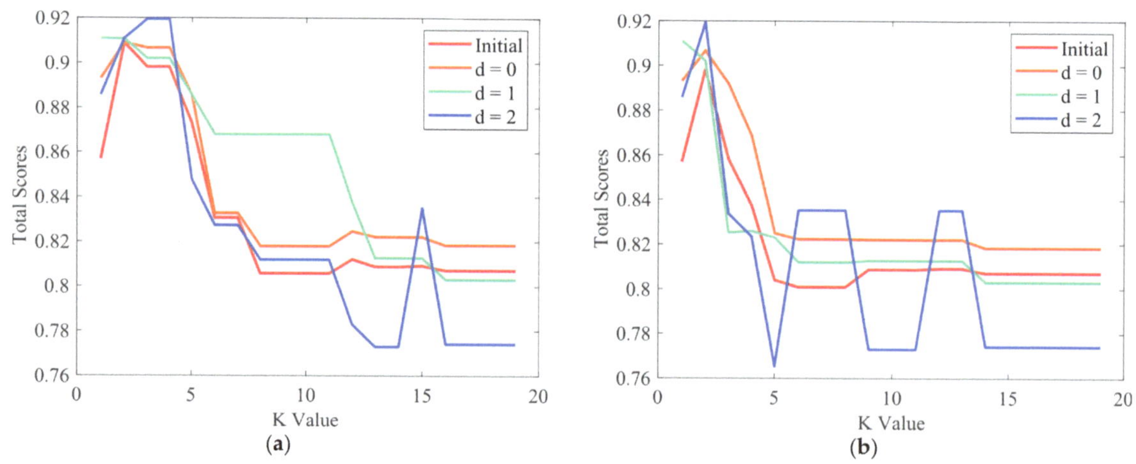

Figure 15. Different results of the two strategies: (**a**) d-value comparison using PKNN-Full; (**b**) d-Value comparison using PKNN-Excluded.

The simulation results distinctly illustrate the discernible efficacy of the RVND algorithm in enhancing the initial solutions generated by the PKNN-Full strategy. Particularly noteworthy was its performance in scenarios where $d = 1$, showcasing a pronounced enhancement range across several outcomes and thereby exhibiting conspicuous optimization capabilities. However, in instances where $d = 2$, the magnitude of priority relaxation was substantial, leading to the compromise of priority's significance in favor of a strictly distance-optimized approach. Consequently, given the prevailing distribution of regional priorities, the total score experienced a reduction due to the abrupt depreciation of the priority score.

Similarly, in cases where $d = 0$ and the priority order remained unchanged, the RVND algorithm still displayed discernible path-optimization capabilities. Broadly, the RVND algorithm demonstrably possesses the capacity to optimize path outcomes and is capable of ascertaining relatively optimal pathways tailored to specific requirements. This underscores the algorithm's adaptive capabilities in tailoring solutions in accordance with distinct priorities and demands. The impact of the RVND algorithm on optimizing the initial solution generated by the PKNN-Excluded strategy was notably pronounced. Upon juxtaposition with Figure 15, it becomes evident that this strategy is inherently more inclined to prioritize priority performance. Consequently, its total score exhibits a marginal reduction due to its lower distance score, a characteristic that sets it apart from the Full strategy.

In this context, a notable observation emerges: the disparity between the two algorithms in terms of the maximum value of the initial solution at K = 2 was effectively diminished. Additionally, the value of K derived from the initial solution was optimized towards its maximum value. This phenomenon underscores that an algorithm with modest baseline performance can achieve the same peak outcome after undergoing optimization by a superior algorithm. This manifestation illuminates the robust optimization prowess of the RVND algorithm.

When K = 2 and $d = 1$, the RVND algorithm was used to optimize the two initial solutions, and the complete paths obtained are shown in Figure 16a,b. The meanings of the colors in Figure 16 are the same as those in Figure 12.

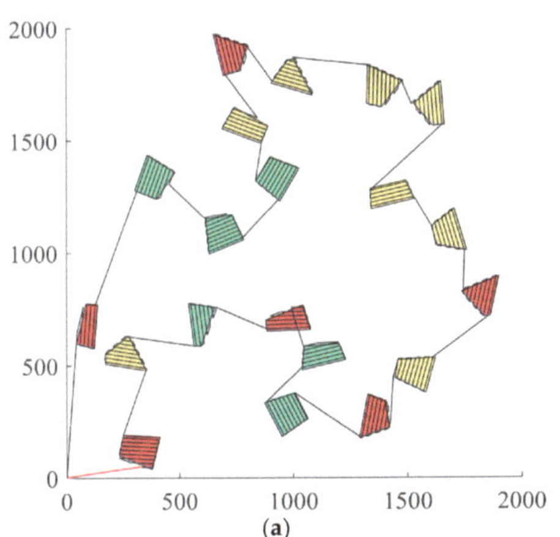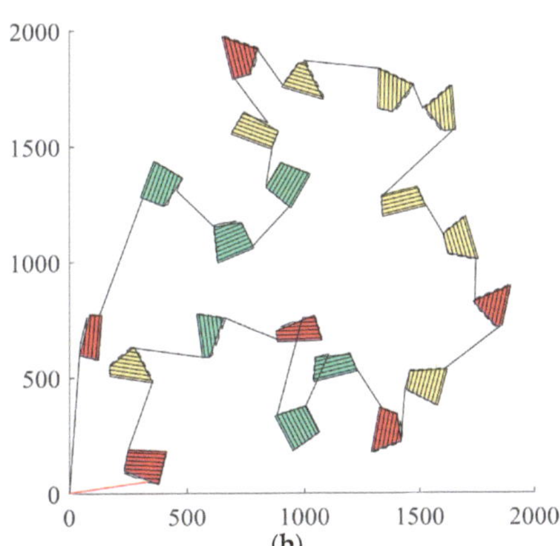

Figure 16. Complete paths obtained with PKNN-Full-RVND and PKNN-Excluded-RVND: (**a**) PKNN-Full-RVND; (**b**) PKNN-Excluded-RVND.

4.3. Analysis and Comparison of Optimization Results

a.   Intra-regional path optimization analysis

At the turning points of the intra-regional flight path, replacing the original polyline with Bezier curves and calculating the overall path length as shown in Table 2, through comparative analysis of path length calculations, it can be inferred that Bezier curves can optimize approximately 5% of the path length without affecting the coverage effectiveness within the region. This indicates a notable energy-saving efficiency. Similarly, applying the research conclusions about Bezier curves mentioned in the introduction, several optimization effects can be observed in the overall UAV path after incorporating Bezier curves:

(1)   Smooth Trajectory: Bezier curves contribute to smoothing the turning angles, reducing the drone's jitter and oscillation during turns, thereby improving flight stability.
(2)   Energy Saving: Bezier curves effectively reduce motion energy consumption in aspects such as path length and motion control, resulting in energy savings for UAV operations
(3)   Ease of Control: The control method is simple and easy to implement, leading to improved operational efficiency for the UAV.

Table 2. Path length optimization rate of Bezier curve.

Path	PKNN-Full	PKNN-Exclude	PKNN-Full RVND	PKNN-Exclude RVND
Original	29,178.3249	29,486.1863	28,617.9472	28,726.4042
Bezier	27,553.6353	27,862.4824	27,003.0889	27,121.3444
Opt. (%)	5.5681	5.5067	5.6428	5.5874

b.   Comparison of two optimized coverage methods

Comparing the BF and SP coverage methods, common experimental parameters were set as follows: the minimum circumscribed circle radius of the polygon: 50 m; the sensor scanning radius: 8 m. The number of vertices increased from 4 to 15, with the experiment repeated 20 times for each set of vertices. The average path length for both methods was calculated, and the variation in the average path length is shown in Figure 17. From the figure, it can be observed that, with a small number of polygon edges, the BF path demonstrated better coverage. However, as the number of polygon edges increased, the growth in path length for BF became significantly higher than that for SP. At a vertex count of eight, the two paths were closest, but after reaching nine vertices, the effectiveness of SP surpassed that of BF.

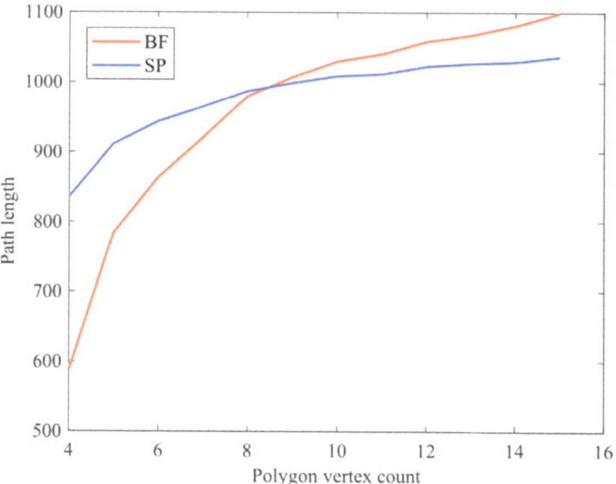

Figure 17. Trend: Path length with the number of polygon vertices.

As stated in Section 3.1, both coverage methods can achieve complete coverage within a region, but there will be a certain amount of redundant area. Continuing the analysis of the simulation results, by calculating the coverage area and the polygon area, the redundancy rate of the sensor coverage area for both methods under different vertex conditions was obtained, as shown in Figure 18. It can be observed that the redundancy rate of the BF path remained stable, while the redundancy rate of the SP path, although substantial when the number of polygon vertices was low, significantly decreased as the vertex count increased. Vertex counts of eight and nine were also critical points for the superiority or inferiority of the two methods in terms of redundancy rate.

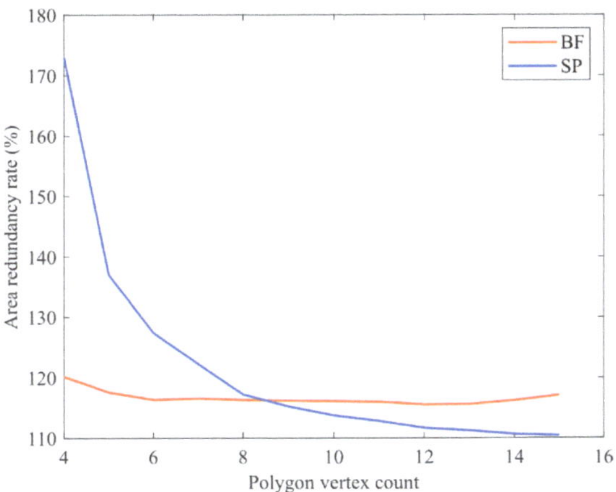

Figure 18. Trend: Area redundancy rate with the number of polygon vertices.

From the two aforementioned analytical approaches, it can be observed that if the polygon vertices are evenly distributed around the center of the minimum circumscribed circle of the polygon and when the number of polygon vertices is small, the BF path is likely to have better coverage advantages. It can reduce coverage-area redundancy while obtaining a shorter flight path. However, as the number of polygon vertices increases and the polygon shape gradually becomes smoother, with mostly large internal angles, the UAV is more suitable for using the SP path for coverage flight.

c.   Inter-regional path optimization analysis

Through simulating different numbers of regions and priorities, the optimization performance of the RVND algorithm was analyzed, as presented in Table 3. In this table, min*Dist.* represents the length of the shortest generated path, while $maxSC_P$, $maxSC_D$, and $maxSC$ represent the maximum priority score, the maximum distance score, and the maximum total score, respectively. The term Time denotes the computation time under specific computer performance conditions. Additionally, Opt. signifies the optimization rate of the algorithm towards improving initial solution scores, while Gap indicates the difference in scores between two optimal optimization results.

According to the simulation results recorded in Table 3, when the number of regions and priorities is small, the optimization algorithm exhibits limited effectiveness. This can be attributed to the sufficiency of the generation algorithm for initial solutions in this task, resulting in significantly reduced computation time compared to the optimization algorithm. Hence, when dealing with a small number of regions and priorities, employing two PKNN algorithms can yield higher computational speed. Conversely, when confronted with a large number of priorities, the priority factor plays a more pronounced and crucial role in planning outcomes, thereby highlighting the impact of the optimization algorithm. Considering real-world path considerations, it is recommended to utilize the optimization algorithm for path optimization under conditions involving numerous priorities and regions to achieve superior planning results.

Table 3. Summary table of important symbols.

No	n	g	d	PKNN-Full Min Dist.	Max SC_P	Max SC_D	Max SC	Time (s)	Opt. (%)	PKNN-Excluded Min Dist.	Max SC_P	Max SC_D	Max SC	Time (s)	Opt. (%)	Gap F-E (%)
1	10	2	-	7271.9	1.000	0.995	0.937	0.003		7271.9	1.000	0.995	0.937	0.008		
			0	7271.9	1.000	0.995	0.937	0.028	1.17	7271.9	1.000	0.995	0.937	0.029	1.17	0
			1	7232.9	0.929	1.000	0.948	0.028		7232.9	0.929	1.000	0.948	0.029		
2	10	2	-	7244.8	1.000	0.899	0.928	0.003		7244.8	1.000	0.899	0.863	0.003		
			0	6553.4	1.000	0.994	0.944	0.025	3.23	6553.4	1.000	0.994	0.910	0.028	5.79	+4.93
			1	6515.4	0.938	1.000	0.958	0.029		6515.4	0.826	1.000	0.913	0.029		
3	10	2	-	7468.4	1.000	0.927	0.918	0.003		7468.4	1.000	0.927	0.918	0.003		
			0	7468.4	1.000	0.927	0.918	0.028	0.87	7468.4	1.000	0.927	0.918	0.025	0.87	0
			1	6920.1	0.897	1.000	0.927	0.032		6920.1	0.897	1.000	0.927	0.027		
4	10	3	-	8323.2	1.000	0.858	0.895	0.003		8646.7	1.000	0.826	0.895	0.003		
			0	7904.6	0.986	0.904	0.914	0.025	6.25	8400.7	0.973	0.850	0.899	0.261	2.23	+3.93
			1	7144.2	0.973	1.000	0.951	0.029		7144.2	0.890	1.000	0.915	0.026		
			2	7144.2	0.830	1.000	0.915	0.029		7144.2	0.901	1.000	0.915	0.032		
5	10	3	-	6948.2	1.000	0.879	0.876	0.003		6948.2	1.000	0.879	0.878	0.003		
			0	6735.9	0.989	0.907	0.903	0.036	8.90	6859.0	0.989	0.891	0.890	0.028	8.66	0
			1	6108.1	0.908	1.000	0.954	0.030		6108.1	0.908	1.000	0.954	0.050		
			2	6108.1	0.908	1.000	0.954	0.033		6108.1	0.908	1.000	0.954	0.031		
6	10	3	-	7332.5	1.000	0.945	0.859	0.003		7332.5	1.000	0.945	0.869	0.003		
			0	7332.5	0.967	0.945	0.873	0.024	7.92	7332.5	0.989	0.945	0.869	0.027	6.67	0
			1	7332.5	0.937	0.945	0.927	0.026		7332.5	0.927	0.945	0.927	0.029		
			2	7136.5	0.927	0.971	0.927	0.031		7136.5	0.918	0.971	0.903	0.029		
7	20	3	-	8335.3	1.000	0.951	0.909	0.033		11,165	1.000	0.784	0.836	0.013		
			0	7930.2	1.000	0.951	0.909	0.293	1.10	10,827	1.000	0.808	0.851	0.280	1.79	+7.99
			1	7946.9	0.966	0.998	0.911	0.338		10,082	0.868	0.863	0.834	0.300		
			2	7930.2	0.848	1.000	0.919	0.303		9518.7	0.819	0.919	0.842	0.303		
8	20	3	-	10417.0	1.000	0.840	0.843	0.009		8812.0	1.000	0.9	0.898	0.011		
			0	9942.3	1.000	0.880	0.875	0.275	5.46	8582.7	1.000	0.924	0.906	0.287	2.34	−3.37
			1	9856.1	0.886	0.888	0.875	0.325		7946.9	0.894	0.998	0.911	0.292		
			2	8749.9	0.823	1.000	0.889	0.293		8008.8	0.848	0.990	0.919	0.314		
9	20	3	-	8870.3	1.000	0.943	0.887	0.008		9043.1	1.000	0.925	0.853	0.011		
			0	8657.8	0.997	0.967	0.897	0.284	2.71	8772.1	0.997	0.954	0.872	0.281	6.21	+0.56
			1	8369.3	0.926	1.000	0.911	0.286		8369.3	0.924	1.000	0.894	0.293		
			2	8456.6	0.895	0.990	0.911	0.315		8555.6	0.895	0.978	0.906	0.306		
10	20	5	-	9301.3	1.000	0.871	0.852	0.008		9301.3	1.000	0.871	0.852	0.011		
			0	9048.7	0.995	0.895	0.864	0.275		9048.7	1.000	0.895	0.864	0.277		
			1	8643.9	0.941	0.937	0.890	0.291	6.57	8643.9	0.964	0.937	0.887	0.281	6.57	0
			2	8175.1	0.943	0.991	0.908	0.294		8175.1	0.912	0.991	0.908	0.292		
			3	8118.3	0.890	1.000	0.908	0.305		8175.1	0.916	0.991	0.908	0.298		
			4	8175.1	0.849	0.991	0.908	0.288		8175.1	0.857	0.991	0.908	0.302		
11	20	5	-	9144.6	1.000	0.821	0.786	0.007		9144.6	1.000	0.821	0.775	0.013		
			0	8940.2	0.993	0.840	0.802	0.315		8940.2	0.993	0.840	0.802	0.287		
			1	8473.3	0.947	0.886	0.824	0.300	7.00	8473.3	0.933	0.886	0.850	0.295	9.68	−1.07
			2	8320.6	0.912	0.903	0.841	0.307		8320.6	0.896	0.903	0.841	0.293		
			3	8285.5	0.866	0.906	0.828	0.322		8285.5	0.830	0.906	0.828	0.299		
			4	8113.4	0.874	0.926	0.830	0.318		8113.4	0.874	0.926	0.826	0.318		
12	20	5	-	10,569.0	1.000	0.855	0.848	0.007		11,034	1.000	0.817	0.791	0.012		
			0	10,132.0	0.992	0.890	0.865	0.264		10,403	1.000	0.866	0.821	0.275		
			1	9938.9	0.955	0.907	0.864	0.286	2.00	10,266	0.955	0.878	0.851	0.290	8.97	+0.35
			2	10,100.0	0.911	0.892	0.863	0.289		10,266	0.911	0.878	0.862	0.290		
			3	9790.2	0.842	0.921	0.863	0.304		10,278	0.867	0.877	0.858	0.298		
			4	9790.2	0.801	0.921	0.846	0.293		9941.1	0.850	0.907	0.837	0.304		

Although there is no fixed optimal K value due to the influence of regional characteristics, an optimal *d* value can generally be identified from the simulation results: when facing few priorities, a larger *d* value brings about greater distance benefits, thus making it preferable as an optimal *d* value; however, when both region and priority numbers are slightly high simultaneously, the optimal *d* value often appears around the median among all *d* values.

4.4. UAV Path-Planning Simulation Platform Based on Unity3D

In order to simulate the flight of a UAV in a realistic environment, we have developed a UAV path-planning simulation system using the Unity3D virtual engine. We import a realistic 3D terrain and drone model and add rigid body components to both the terrain and drone. Additionally, for better realism in simulating the environment, it is necessary to incorporate environmental components, such as wind direction, wind speed, lighting, etc.

Unity3D relies on scripts to implement the operational logic of each object within the virtual environment. Therefore, several scripts need to be added, including region drawing, UAV attitude control, motion trajectory display, sensor data access, etc., along with design of the UI interface and control scripts for UI components. All these scripts work together to ensure the smooth operation of the simulation system. The UI interface of this system is depicted in Figure 19.

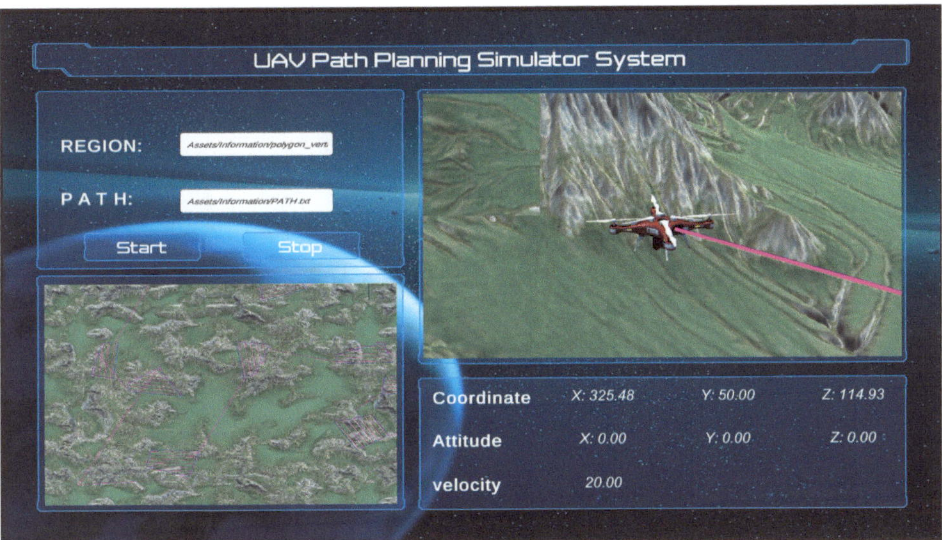

Figure 19. UI interface of UAV path-planning simulation system.

The input data for this simulation system consist of path points and area information obtained from MATLAB planning. Upon running the system, it first generates an area range for the terrain that needs coverage; subsequently, the drone flies through this scene based on waypoints while displaying its real-time trajectory. Furthermore, the position and attitude of the drone can also be observed in real time via the UI panel.

4.5. Flight-Path Experiment

The experimental section of this study is based on the MATLAB simulation analysis results discussed earlier. Through the use of a UAV for actual coverage flights in a specified area, the effectiveness of adaptive flight trajectories in practical applications was validated, as shown in Figure 20. We selected a spacious environment near the laboratory, as shown in Figure 20a, designating two nearby open areas for overall coverage. These areas were used to delineate internal regions, and a random location within was chosen as the base position, as illustrated in Figure 20b. In the figure, the region enclosed by the red line is the experimental area, the gray polygon represents the coverage area, and the yellow points indicate the depot of the UAV. The flight path was imported into the UAV remote controller to guide the UAV in flying along the planned path. The resulting flight path in the simulation program is depicted as blue lines in Figure 20c. After the flight experiment, the flight route from the UAV's flight log was exported and is shown as green lines in Figure 20d.

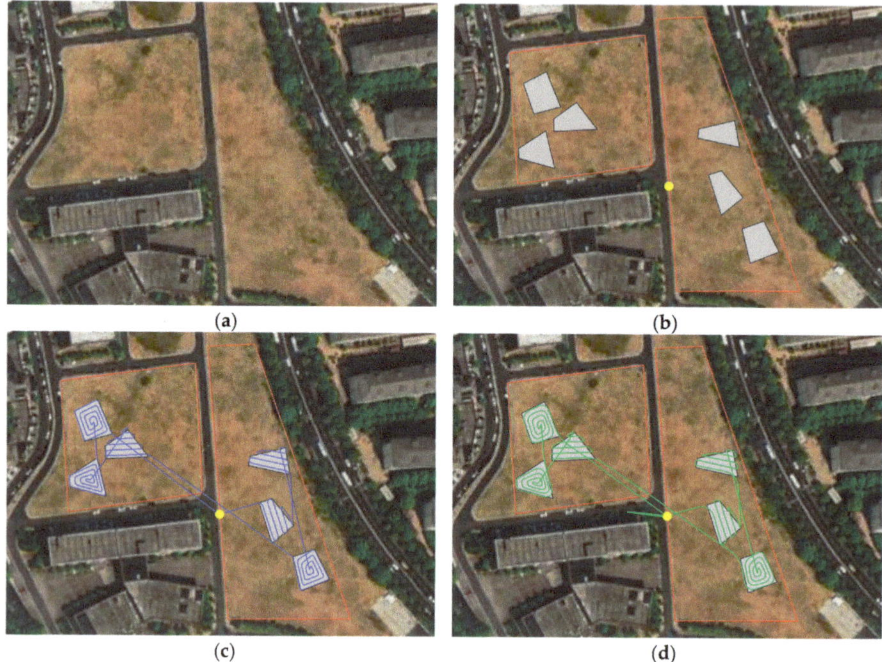

Figure 20. Experimental environment and flight experiments. (**a**) The experimental area. (**b**) The drone base, scanning area, and polygonal region. (**c**) The simulated trajectory. (**d**) The experimental trajectory.

Comparative analysis of simulation paths in real environments and paths from actual flight experiments demonstrates that the path-planning algorithm used in this study is well-suited for practical applications in real environments. It effectively achieves UAV path planning within and between areas. The algorithm's performance in real-world applications is thus confirmed. Hence, the flight experiments were successful, as the UAV accurately tracked the theoretical and simulated flight trajectories, validating the practicality of the path-planning method proposed in this study.

5. Conclusions

This paper introduces a comprehensive and easily implementable solution to the UAV path-planning problem under priority constraints. We enhance the coverage approach by employing a BF path, ensuring complete coverage within circular sensor sampling ranges and employing Bezier curves to optimize both the round-trip path and the spiral path. Furthermore, we introduce two initial solution generation techniques for priority paths based on the KNN algorithm. These methods are employed to devise the access sequence between regions, incorporating priority considerations. Through comparisons with planning algorithms lacking priority planning capabilities, our approach demonstrates its ability to intelligently plan paths based on priority orders.

The results of evaluation metrics demonstrate that our proposed method can quickly find high-quality solutions in terms of distance and priority. Furthermore, by optimizing both initial solutions using the RVND algorithm, we enhanced the optimization capabilities of the paths. The simulation results demonstrate the algorithm's strong performance in both distance and priority, indicating its ability to refine solutions from initial states. These outcomes validate the algorithm's effectiveness. Based on our real-world experiments, the algorithm has been demonstrated to exhibit favorable practical prospects in actual application environments. Consequently, the path-planning method presented in this paper holds significant potential for widespread application in the realm of emergency rescue.

Future research will focus on employing intelligent optimization algorithms, such as genetic algorithms, differential evolution, and reinforcement learning, to further enhance the optimization capabilities. It will also work on optimizing CPP path generation strategies to better enable UAVs to cope with external interference. In addition, we plan to explore the use of multi-drone cooperation to simulate more regions and priorities to accomplish complex missions more efficiently.

Author Contributions: Conceptualization, L.D. and Y.F.; methodology, L.D.; validation, L.D. and Y.F.; formal analysis, D.Z. and M.G.; investigation, L.D.; resources, D.Z.; data curation, L.D. and Y.F.; writing—original draft preparation, L.D. and Y.F.; writing—review and editing, D.Z. and M.G.; visualization, L.D. and M.G.; supervision, D.Z. and M.G; project administration, D.Z. All authors have read and agreed to the published version of the manuscript.

Funding: This research was funded by grants from the National Key Research and Development Program of China (grant no. 2021YFC3090401) and the Open Fund of the Laboratory of Pinghu.

Data Availability Statement: Partial data is already included in the charts of the article. The remaining portion of the raw/processed data cannot be shared temporarily as it is part of ongoing research.

Acknowledgments: The authors would like to thank the editors and the reviewers for their constructive comments and to thank Chenggong Li and Xingda Zhu for their support during the data collection and analysis.

Conflicts of Interest: The authors declare no conflict of interest.

References

1. Jiong, D.; Ota, K.; Mianxiong, D. UAV-based Real-time Survivor Detection System in Post-disaster Search and Rescue Operations. *IEEE J. Miniat. Air Space Syst.* **2021**, *2*, 209–219. [CrossRef]
2. Naidoo, Y.; Stopforth, R.; Bright, G. Development of an UAV for search & rescue applications. In Proceedings of the IEEE Africon '11, Victoria Falls, Zambia, 13–15 September 2011; pp. 1–6.
3. Verykokou, S.; Doulamis, A.; Athanasiou, G.; Ioannidis, C.; Amditis, A.; Instrumentat, I.; Measurement, S. UAV-Based 3D Modelling of Disaster Scenes for Urban Search and Rescue. In Proceedings of the IEEE International Conference on Imaging Systems and Techniques (IST)/IEEE International School on Imaging, Chania, Greece, 4–6 October 2016; pp. 106–111.
4. Cabreira, T.M.; Di Franco, C.; Ferreira, P.R.; Buttazzo, G.C. Energy-Aware Spiral Coverage Path Planning for UAV Photogrammetric Applications. *IEEE Robot. Autom. Lett.* **2018**, *3*, 3662–3668. [CrossRef]
5. Jensen-Nau, K.R.; Hermans, T.; Leang, K.K. Near-Optimal Area-Coverage Path Planning of Energy-Constrained Aerial Robots With Application in Autonomous Environmental Monitoring. *IEEE Trans. Autom. Sci. Eng.* **2021**, *18*, 1453–1468. [CrossRef]
6. Wang, H.P.; Zhang, S.Y.; Zhang, X.Y.; Zhang, X.B.; Liu, J.T. Near-Optimal 3-D Visual Coverage for Quadrotor Unmanned Aerial Vehicles Under Photogrammetric Constraints. *IEEE Trans. Ind. Electron.* **2022**, *69*, 1694–1704. [CrossRef]
7. Luna, M.A.; Isaac, M.S.A.; Ragab, A.R.; Campoy, P.; Peña, P.F.; Molina, M. Fast Multi-UAV Path Planning for Optimal Area Coverage in Aerial Sensing Applications. *Sensors* **2022**, *22*, 2297. [CrossRef] [PubMed]
8. Cabreira, T.M.; Brisolara, L.B.; Paulo, R.F. Survey on Coverage Path Planning with Unmanned Aerial Vehicles. *Drones* **2019**, *3*, 4. [CrossRef]
9. Machmudah, A.; Shanmugavel, M.; Parman, S.; Abd Manan, T.S.; Dutykh, D.; Beddu, S.; Rajabi, A. Flight Trajectories Optimization of Fixed-Wing UAV by Bank-Turn Mechanism. *Drones* **2022**, *6*, 69. [CrossRef]
10. Xie, J.F.; Carrillo, L.R.G.; Jin, L. An Integrated Traveling Salesman and Coverage Path Planning Problem for Unmanned Aircraft Systems. *IEEE Control. Syst. Lett.* **2019**, *3*, 67–72. [CrossRef]

11. Irving Vasquez-Gomez, J.; Herrera-Lozada, J.-C.; Olguin-Carbajal, M. Coverage Path Planning for Surveying Disjoint Areas. In Proceedings of the International Conference on Unmanned Aircraft Systems (ICUAS), Dallas, TX, USA, 12–15 June 2018; pp. 899–904.
12. Torres, M.; Pelta, D.A.; Verdegay, J.L.; Torres, J.C. Coverage path planning with unmanned aerial vehicles for 3D terrain reconstruction. *Expert Syst. Appl.* **2016**, *55*, 441–451. [CrossRef]
13. Xie, J.F.; Carrillo, L.R.G.; Jin, L. Path Planning for UAV to Cover Multiple Separated Convex Polygonal Regions. *IEEE Access* **2020**, *8*, 51770–51785. [CrossRef]
14. Chen, X.; Chen, J.; Du, C.; Xu, Y. Region Coverage Path Planning of Multiple Disconnected Convex Polygons Based on Simulated Annealing Algorithm. In Proceedings of the 2021 IEEE 4th International Conference on Computer and Communication Engineering Technology (CCET), Beijing, China, 13–15 August 2021; pp. 238–242.
15. Ko, Y.C.; Gau, R.H. UAV Velocity Function Design and Trajectory Planning for Heterogeneous Visual Coverage of Terrestrial Regions. *IEEE Trans. Mob. Comput.* **2023**, *22*, 6205–6222. [CrossRef]
16. Miao, Y.M.; Hwang, K.; Wu, D.; Hao, Y.X.; Chen, M. Drone Swarm Path Planning for Mobile Edge Computing in Industrial Internet of Things. *IEEE Trans. Ind. Inform.* **2023**, *19*, 6836–6848. [CrossRef]
17. Khanam, Z.; Saha, S.; Ehsan, S.; Stolkin, R.; McDonald-Maier, K. Coverage Path Planning Techniques for Inspection of Disjoint Regions With Precedence Provision. *IEEE Access* **2021**, *9*, 5412–5427. [CrossRef]
18. Panchamgam, K.; Xiong, Y.P.; Golden, B.; Dussault, B.; Wasil, E. The hierarchical traveling salesman problem. *Optim. Lett.* **2013**, *7*, 1517–1524. [CrossRef]
19. Ahmed, Z.H. The Ordered Clustered Travelling Salesman Problem: A Hybrid Genetic Algorithm. *Sci. World J.* **2014**, *2014*, 258207. [CrossRef]
20. Hà, M.H.; Phuong, H.N.; Nhat, H.T.N.; Langevin, A.; Trépanier, M. Solving the clustered traveling salesman problem with d-relaxed priority rule. *Int. Trans. Oper. Res.* **2022**, *29*, 837–853. [CrossRef]
21. Salach, A.; Bakula, K.; Pilarska, M.; Ostrowski, W.; Górski, K.; Kurczynski, Z. Accuracy Assessment of Point Clouds from LiDAR and Dense Image Matching Acquired Using the UAV Platform for DTM Creation. *ISPRS Int. J. Geo Inf.* **2018**, *7*, 342. [CrossRef]
22. Domingo, D.; Orka, H.O.; Næsset, E.; Kachamba, D.; Gobakken, T. Effects of UAV Image Resolution, Camera Type, and Image Overlap on Accuracy of Biomass Predictions in a Tropical Woodland. *Remote Sens.* **2019**, *11*, 948. [CrossRef]
23. Yu-Song, J.; Xin-Min, W.; Hai, C.; Yan, L. Research on the Coverage Path Planning of UAVs for Polygon Areas. In Proceedings of the 2010 5th IEEE Conference on Industrial Electronics and Applications (ICIEA 2010), Taichung, Taiwan, 15–17 June 2010; pp. 1467–1472. [CrossRef]
24. Berger, C.; Wzorek, M.; Kvarnström, J.; Conte, G.; Doherty, P.; Eriksson, A. Area Coverage with Heterogeneous UAVs using Scan Patterns. In Proceedings of the 14th IEEE International Symposium on Safety, Security, and Rescue Robotics (SSRR), Lausanne, Switzerland, 23–27 October 2016; pp. 342–349.
25. Silva, M.M.; Subramanian, A.; Vidal, T.; Ochi, L.S. A simple and effective metaheuristic for the Minimum Latency Problem. *Eur. J. Oper. Res.* **2012**, *221*, 513–520. [CrossRef]
26. Mladenovic, N.; Hansen, P. Variable neighborhood search. *Comput. Oper. Res.* **1997**, *24*, 1097–1100. [CrossRef]

Disclaimer/Publisher's Note: The statements, opinions and data contained in all publications are solely those of the individual author(s) and contributor(s) and not of MDPI and/or the editor(s). MDPI and/or the editor(s) disclaim responsibility for any injury to people or property resulting from any ideas, methods, instructions or products referred to in the content.

drones

Article

Reducing Oscillations for Obstacle Avoidance in a Dense Environment Using Deep Reinforcement Learning and Time-Derivative of an Artificial Potential Field

Zhilong Xi [1,†], **Haoran Han** [1,†], **Jian Cheng** [1,*] **and Maolong Lv** [2]

[1] School of Information and Communication Engineering, University of Electronic Science and Technology of China, Chengdu 611731, China; xijohnston@gmail.com (Z.X.); hanadam@163.com (H.H.)

[2] Air Traffic Control and Navigation College, Air Force Engineering University, Xi'an 710051, China; maolonglv@163.com

[*] Correspondence: chengjian@uestc.edu.cn

[†] These authors contributed equally to this work.

Abstract: Obstacle avoidance plays a crucial role in ensuring the safe path planning of quadrotor unmanned aerial vehicles (QUAVs). In this study, we propose a hierarchical framework for obstacle avoidance, which combines the use of artificial potential field (APF) and deep reinforcement learning (DRL) for training low-level motion controllers. Unlike traditional potential field methods, our approach modifies the state information received by the motion controllers using the outputs of the APF path planner. Specifically, the assumed target position is pushed away from obstacles, resulting in adjustments to the perceived position errors. Additionally, we address path oscillations by incorporating the target's velocity information, which is calculated based on the time-derivative of the repulsive force. Experimental results have validated the effectiveness of our proposed framework in avoiding collisions with obstacles and reducing oscillations.

Keywords: quadrotor unmanned aerial vehicle (QUAV); obstacle avoidance; artificial potential field (APF); deep reinforcement learning (DRL); oscillations

Citation: Xi, Z.; Han, H.; Cheng, J.; Lv, M. Reducing Oscillations for Obstacle Avoidance in a Dense Environment Using Deep Reinforcement Learning and Time-Derivative of an Artificial Potential Field. *Drones* **2024**, *8*, 85. https://doi.org/10.3390/drones8030085

Academic Editors: Jihong Zhu, Heng Shi, Zheng Chen and Minchi Kuang

Received: 10 January 2024
Revised: 26 February 2024
Accepted: 27 February 2024
Published: 1 March 2024

Copyright: © 2024 by the authors. Licensee MDPI, Basel, Switzerland. This article is an open access article distributed under the terms and conditions of the Creative Commons Attribution (CC BY) license (https:// creativecommons.org/licenses/by/ 4.0/).

1. Introduction

Due to their high flexibility, quadrotor unmanned aerial vehicles (QUAVs) have gained significant popularity in various applications, including parcel delivery [1], precision agriculture [2–4], search and rescue [5,6], and surveillance [7]. In these scenarios, the QUAV is typically required to autonomously navigate to a target position. However, in dense environments, unexpected obstacles can obstruct the path and lead to collisions. This task is even more challenging in a multi-agent system [8,9]. Therefore, obstacle avoidance becomes a crucial task for ensuring safe path planning. The QUAV must find an unobstructed path to the target while considering its physical limitations. Conventional approaches to obstacle-free path planning include Dijkstra, probabilistic roadmap (PRM), rapidly-exploring random trees (RRT), and artificial potential field (APF) [10–13]. Since these approaches do not assume the physical realization of the agent to be of a certain type, they are applicable to QUAV navigation. APF faces a critical challenge in dense environments. When multiple obstacles narrow the passageway to the target, the agent may experience oscillations due to the repulsive force created by the obstacles that conflicts with the attractive force towards the target. As a result, reaching the target smoothly becomes difficult.

Most existing approaches to reducing APF oscillations either rely on second-order optimization theory techniques or employ an escaping mechanism when the agent detects oscillations. However, these approaches are not fully compatible with a hierarchical path planning framework because they directly modify the forces applied to the QUAV without considering its often nonlinear dynamics. Additionally, they commonly attribute oscillations to local minima, which is not always the case. One possible solution to this problem is to utilize global obstacle information to bend an imaginary rubber band path [14]. However, this approach poses challenges for real-time decision-making, as it requires prior knowledge of obstacle positions.

Based on the symmetrical motion controller design in [15], our previous work [16] enabled the utilization of an APF in a hierarchical framework, utilizing only local information provided by sensors mounted on the QUAV. In this framework, an APF path planner generates a state that incorporates modified position errors, while low-level motion controllers are responsible for both position and attitude control. In [17], we introduced the Hessian matrix of the APF as a damping term, which effectively dissipates system energy, and we proved its stability using an energy-based Lyapunov function. In this current work, we utilize the time-derivative of the APF to further reduce oscillations in QUAV path planning within a hierarchical control framework.

Considering that the QUAV dynamic system is nonlinear and strongly coupled, we use deep reinforcement learning (DRL) to train the low-level motion controllers [18]. DRL not only alleviates the labor-intensive process of parameter tuning, as compared to the traditional proportional–integral–derivative (PID) method, but also demonstrates remarkable fitting capability with the aid of deep neural networks. The key contributions of this study are outlined as follows.

1. We incorporate the time-derivatives of the potential function to account for the velocities of intermediate target points, thereby resolving the conflict between the fixed-point chasing logic of the DRL motion controller and the potentially erratic movements of the intermediate target points.
2. To enhance the control performance of a high-dimensional nonlinear system, we reconfigure the states of the DRL motion controller to eliminate any asymmetry and ensure stability.
3. We conducted comparative simulation experiments to validate the effectiveness of the proposed method in reducing oscillations and preventing overshooting.
4. The complex dynamics of the QUAV system are considered, which is different from other approaches that either do not discuss agent dynamics or simplify the dynamics as a simple second-order point mass model.

The structure of this paper is outlined as follows. In Section 2, we provide a comprehensive overview of the existing research on obstacle-free path planning. In Section 3, we introduce the necessary background knowledge, including the principles of conventional artificial potential fields, drone dynamics, and deep reinforcement learning. Section 4 elaborates on the methodology that we have employed in this study. We present the experiments conducted in Section 5 to validate the effectiveness of the proposed method in reducing oscillation and overshooting. Section 6 discusses limitations of this work and its applicability to dynamic and irregular environments. Finally, in Section 7, we conclude our findings.

2. Related Work

We identify three major approaches to achieve obstacle-free path planning. Since they do not set restrictions on the type of agents being used, these approaches are applicable to QUAV obstacle avoidance. The first approach is grid search, exemplified by algorithms such as Dijkstra, A*, and D* [10,19,20]. These algorithms are guaranteed to find a viable least-cost path when the environment is finite. However, they suffer from the curse of dimensionality, as the number of vertices to be explored grows exponentially with the dimensionality of the working space. Additionally, the need for higher control resolutions further increases the computational burden. The second approach is sampling-based search, with PRM [11] and RRT [12] being typical examples. For example, Farooq et al. [21] guided the QUAV to avoid dangerous zones in a dynamic environment by computing the PRM. Compared to grid search, these algorithms are capable of quickly producing paths in high-dimensional spaces. However, since sampling-based methods rely on global information, such as environment boundaries, they are not well-suited for real-time control where sensor data are used to generate actions for the next step. The third approach is APF, which assumes that both the target and obstacles generate potential fields that influence the movement of the agent. By leveraging the gradient of the summed potential field, the agent is attracted to the target while being repelled from nearby obstacles. For example, Ma'arif et al. [22] used APF to guide a single QUAV to reach the target and avoid collisions with dynamic obstacles. APF is widely used in QUAV navigation due to its ease of implementation and ability to guide the QUAV using only local information.

Despite its numerous advantages, APF has certain inherent limitations [23]. One recognizable limitation is that the agent can easily become trapped in local minima, where the attraction and repulsion forces cancel each other out. Algorithms designed to address this problem can generally be classified into three categories. The first category is local minimum removal. For example, Kim et al. [24] employed harmonic functions to construct potential fields, allowing for the selection of singularity locations and the elimination of local minima in free space. The second category is local minimum escape. For instance, Park et al. [25] combined APF with simulated annealing, which introduces randomness to the agent's actions, enabling it to escape from local minima. Wang et al. [26] proposed the Left Turning scheme, which effectively handles U-shaped obstacles and helps the agent escape local minima. Lai et al. [27] proposed a dynamic step-size adjustment method to help the multi-UAV escape the local minima. The third category is local minimum avoidance. For instance, Doria et al. [28] utilized the deterministic annealing approach to expand the repulsive area and shrink the attractive area. This allows the agent to avoid local minima at the beginning, when the potential function is convex due to a high initial temperature. Additionally, Ge et al. [29] addressed a specific case of the local minimum problem known as goals non-reachable with obstacles nearby (GNRON). They considered the relative distance between the agent and the target when designing the repulsive potential function. As the agent approaches the target, this function approaches zero, thereby reducing the repulsive force in the target's vicinity and overcoming the local minimum problem. Overall, these different approaches highlight the efforts made to overcome the limitations of APF and improve its performance in various scenarios.

APF faces another dilemma, which is the occurrence of oscillations in narrow passages with densely distributed obstacles nearby. Previous research on oscillation reduction is relatively limited and mainly draws inspiration from optimization theory techniques. For instance, Ren et al. [30] proposed the use of Levenberg's modification of Newton's method as a solution to oscillation problems. This approach incorporates second-order information by utilizing the Hessian matrix. Additionally, they adjusted the control law to maintain a constant speed, ensuring smooth movement of the agents. Biswas et al. [31] further compared first-order gradient descent methods with two second-order approaches and concluded that Levenberg–Marquardt is more effective in generating smoother trajectories and improving convergence speeds.

The second branch of oscillation reduction algorithms introduces the concept of virtual obstacles or targets. Similar to LME used in tackling local minima, methods belonging to this branch often employ escaping techniques once oscillations are detected. For instance, Zhao et al. [32] enhanced the manipulator's predictive ability by incorporating dynamic virtual target points and utilized an extreme point jump-out function to escape oscillations. Zhang et al. [33] employed tangent APF to avoid local oscillations and introduced the back virtual obstacle setting strategy-APF algorithm, which enables the agent to return to previous steps and withdraw from concave obstacles. In a rule-based fashion, Zheng et al. [34] specified the condition for adding obstacles, compelling the resultant force to deflect when its angle to the obstacle center is too small. The dynamic step-size adjustment method proposed by Lai et al. [27] is also able to escape the jitter area where a local minimum is encountered, but it does not address oscillations in other cases, such as narrow passageways.

Oscillations can also be mitigated by modifying the repulsive forces in a certain manner. Tran et al. [35] estimated the projection of the repulsive force vector onto the attractive force vector and subtracted this component to prevent the agent from moving in the opposite direction of the attractive force. Martis et al. [36] introduced vortex potential fields to achieve seamless cooperative collision avoidance between mobile agents, and these were validated using Lyapunov stability analysis. For larger obstacles with irregular shapes, Szczepanski [37] combined the benefits of repulsive APF and vortex APF by defining multiple layers around the surface area of the obstacles. This approach surpassed traditional APF and pure vortex APF in terms of path smoothness.

There are also works specifically designed for QUAV path planning and obstacle avoidance. For example, Meradi et al. [38] proposed a nonlinear model predictive control method based on quaternions for QUAVs' obstacle avoidance. Valencia et al. [39] constructed a QUAV obstacle detection and avoidance system using a monocular camera. Gageik et al. [40] discussed the use of complementary low-cost sensors in QUAV collision avoidance. However, the problem of oscillation reduction in QUAV obstacle-free path planning based on APF is largely under-explored.

With the advancement in DRL technology, motion controllers based on DRL have gained widespread usage in QUAV path planning and obstacle avoidance by virtue of their strong fitting capability. In this context, the APF can be seen as an upper layer that indirectly or directly influences the agent's current pursuit position, while the low-level motion control task is handled by the DRL agent. For instance, the RL environment may employ convolutional neural networks to receive information about the surrounding potential energy and generate estimated rewards for various actions [41]. Xing et al. [42] combined the enhanced APF method with deep deterministic policy gradient to navigate autonomous underwater vehicles. In our previous research, we developed a hierarchical framework where the APF path planner was utilized to modify the position errors perceived by the DRL motion controllers, effectively altering the target position [16]. In this study, we further address oscillations by considering the velocity of the virtual target point, which has been validated in a collision-free formation control task [17]. Experimental results demonstrate that this improved algorithm reduces oscillations compared to approaches that do not incorporate second-order information, such as velocities.

3. Preliminaries

3.1. Drone Dynamics

The simulation environment is built based on gym-pybullet-drones [43]. In this work, a QUAV with an "X" configuration is considered. Similar to many fixed-wing UAVs [44], it has six degrees of freedom. Its position and Euler angles are defined as $\mathbf{p} = [x, y, z]^T$ and $\mathbf{\Theta} = [\phi, \theta, \psi]^T$, respectively. The Euler angles as shown in Figure 1 are obtained following the intrinsic z-y-x sequence (or, equivalently, extrinsic x-y-z sequence). Therefore, the rotation matrix from the body-fixed frame $\{\mathbf{x}_b, \mathbf{y}_b, \mathbf{z}_b\}$ to the earth frame $\{\mathbf{x}_e, \mathbf{y}_e, \mathbf{z}_e\}$ is calculated by

$$\mathbf{R} = [\mathbf{x}_b, \mathbf{y}_b, \mathbf{z}_b] = \begin{bmatrix} c_\theta c_\psi & c_\psi s_\theta s_\phi - s_\psi c_\phi & c_\psi s_\theta c_\phi + s_\psi s_\phi \\ c_\theta s_\psi & s_\psi s_\theta s_\phi + c_\psi c_\phi & s_\psi s_\theta c_\phi - c_\psi s_\phi \\ -s_\theta & s_\phi c_\theta & c_\phi c_\theta \end{bmatrix}, \tag{1}$$

where the coordinates of $\mathbf{x}_b$, $\mathbf{y}_b$, and $\mathbf{z}_b$ are in the earth frame, $\mathbf{z}_b = [z_b^x, z_b^y, z_b^z]^T$, c^ξ and s^ξ ($\xi = \phi, \theta, \psi$) are short for $\cos \xi$ and $\sin \xi$, respectively, and $\mathbf{R} \in SO(3)$.

In this work, we use the frame of SE(3) control [45] to avoid using a transformation matrix $\mathbf{W}$ between the Euler angles $\mathbf{\Theta}$ and the angular velocity $\boldsymbol{\omega} = [\omega^x, \omega^y, \omega^z]^T$, which suffers from singularity as θ approaches $\frac{\pi}{2}$. This technique overcomes the small angle restriction. The hat map $\hat{} : \mathbb{R}^3 \to \mathfrak{so}(3)$ is defined as

$$\hat{\boldsymbol{\alpha}} = \begin{bmatrix} 0 & -\alpha_3 & \alpha_2 \\ \alpha_3 & 0 & -\alpha_1 \\ -\alpha_2 & \alpha_1 & 0 \end{bmatrix}, \tag{2}$$

where $\boldsymbol{\alpha} = [\alpha_1, \alpha_2, \alpha_3]^T$ is a three-dimensional vector. Using the differential property of Lie Group, the QUAV's dynamic system can be formally described by

$$
\begin{aligned}
\dot{\mathbf{p}} &= \mathbf{v}, \\
m\dot{\mathbf{v}} &= \mathbf{RF} - m\mathbf{g}, \\
\dot{\mathbf{R}} &= \mathbf{R}\hat{\omega}, \\
\mathbf{J}\dot{\omega} &= -\omega \times \mathbf{J}\omega + \boldsymbol{\tau},
\end{aligned}
\tag{3}
$$

where the velocity $\mathbf{v} = [v^x, v^y, v^z]^T$ and the gravity $\mathbf{g} = [0, 0, g]^T$ are in the earth frame, external thrust $\mathbf{F} = [0, 0, f]^T$ and torque $\boldsymbol{\tau} = [\tau^\phi, \tau^\theta, \tau^\psi]^T$ are in the body-fixed frame, m is the mass, and $\mathbf{J} = diag(I^\phi, I^\theta, I^\psi)$ is the inertia matrix.

The four rotors are controlled by the pulse width modulation (PWM) signal $\mathbf{u}$, and the rotation speed $\boldsymbol{\Omega} = [\Omega_1, \Omega_2, \Omega_3, \Omega_4]^T$ is described by

$$
T_\mathrm{m}\dot{\boldsymbol{\Omega}} = k_\mathrm{m}\mathbf{u} + \mathbf{b}_\mathrm{m} - \boldsymbol{\Omega},
\tag{4}
$$

where $T_\mathrm{m}, k_\mathrm{m}$, and $\mathbf{b}_\mathrm{m}$ are the motors' coefficients, which remain constant. The thrust $\mathbf{F}$ and torque $\boldsymbol{\tau}$ are then generated by the four rotors according to

$$
\begin{bmatrix} f \\ \tau^\phi \\ \tau^\theta \\ \tau^\psi \end{bmatrix} = \begin{bmatrix} C_T & C_T & C_T & C_T \\ -\frac{L}{\sqrt{2}}C_T & -\frac{L}{\sqrt{2}}C_T & \frac{L}{\sqrt{2}}C_T & \frac{L}{\sqrt{2}}C_T \\ -\frac{L}{\sqrt{2}}C_T & \frac{L}{\sqrt{2}}C_T & \frac{L}{\sqrt{2}}C_T & -\frac{L}{\sqrt{2}}C_T \\ -C_D & C_D & -C_D & C_D \end{bmatrix} \begin{bmatrix} \Omega_1^2 \\ \Omega_2^2 \\ \Omega_3^2 \\ \Omega_4^2 \end{bmatrix},
\tag{5}
$$

where L is the arm length and C_T, C_D are the thrust and torque coefficients, respectively.

The environment contains N obstacles denoted by $\mathcal{O} = [1, 2, \cdots, N]$. The geometric relationship between the QUAV, the target, and obstacle i is shown in Figure 2, in which we define $\mathbf{p} = [x, y, z]^T$ as the QUAV's current position, $\mathbf{p}_\mathrm{d}$ as the target position, and $\mathbf{o}_i$ as the closest point of obstacle i to the QUAV. Moreover, $\mathbf{e}_i = \mathbf{p} - \mathbf{o}_i = [e_i^x, e_i^y, e_i^z]^T$ and $\mathbf{e}_\mathrm{d} = \mathbf{p} - \mathbf{p}_\mathrm{d} = [e^x, e^y, e^z]^T$ are error vectors to obstacle i and the target, respectively.

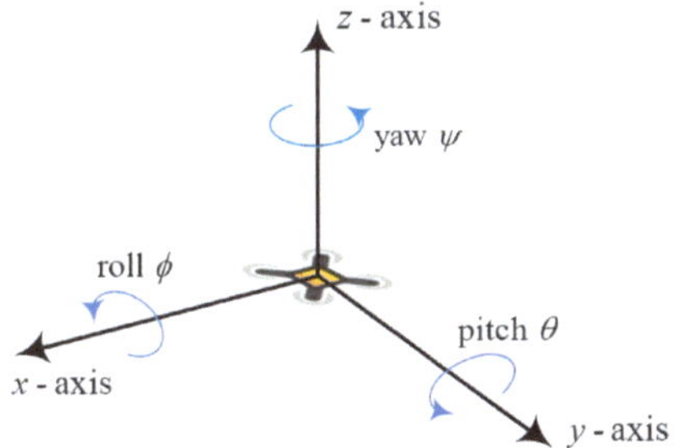

Figure 1. The body-fixed frame and the corresponding Euler angles.

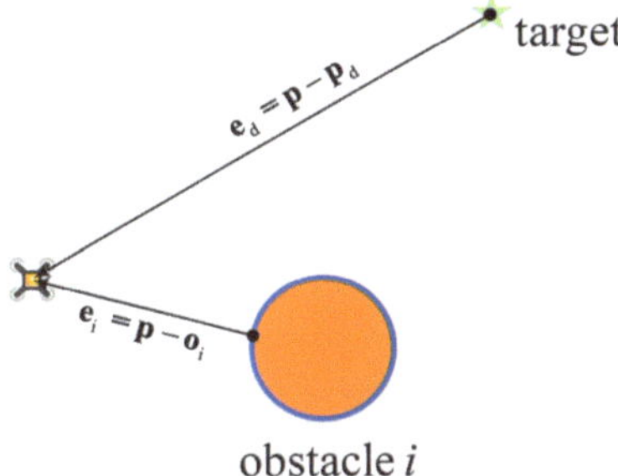

Figure 2. Geometric relationship of the QUAV, the target, and obstacle i.

3.2. Classical APF

The concept of APF was initially introduced by Khatib [13] with the aim of distributing collision avoidance across different levels of control. This approach involves creating a potential field, similar to an electrostatic field, within the agent's working space. In this field, the target location is represented as a valley, while obstacles are represented as peaks. By computing the negative gradient of the total potential function, the agent can determine a feasible path to the target while avoiding collisions with obstacles. From another perspective, the target exerts an attractive force on the agent, while the obstacles exert a repulsive force. To ensure the existence of these attractive and repulsive forces, the potential functions must be continuous and differentiable, except at the target position or within the obstacles. The calculation of the attractive and repulsive forces is performed by

$$\mathbf{F}_{\text{att}} = -\nabla_{\mathbf{p}} U_{\text{att}}(\|\mathbf{e}_{\text{d}}\|),$$
$$\mathbf{F}_{\text{rep}_i} = -\nabla_{\mathbf{p}} U_{\text{rep}_i}(\|\mathbf{e}_i\|), \tag{6}$$

where $U_{\text{att}}(\cdot)$ and $U_{\text{rep}_i}(\cdot)$ are potential functions that receive scalars instead of vectors. This maintains the isotropy of the overall system, since the norms of the APF forces are only determined by the radial distances and not by the relative angles to the attraction or repulsion sources.

Figure 3 depicts the attractive and repulsive forces in the APF algorithm. In this figure, the green star represents the target position, while the orange circles symbolize obstacles with specific regions of influence. The repulsive force generated by obstacle i is denoted as F_{rep}, and the attractive force generated by the target is denoted as F_{att}. The resultant force exerted on the agent is represented as F_{r}. It is worth noting that APF can be applied to both global path planning and local path planning, as illustrated in Figure 3a and Figure 3b, respectively. In the global case, the obstacles have a maximum influence distance ρ_0, and their positions and shapes are known in advance. However, in the local case, which is employed in this study, the QUAV solely relies on the relative distances and velocities obtained from sensors to calculate the APF forces. The maximum sensing range is denoted as d.

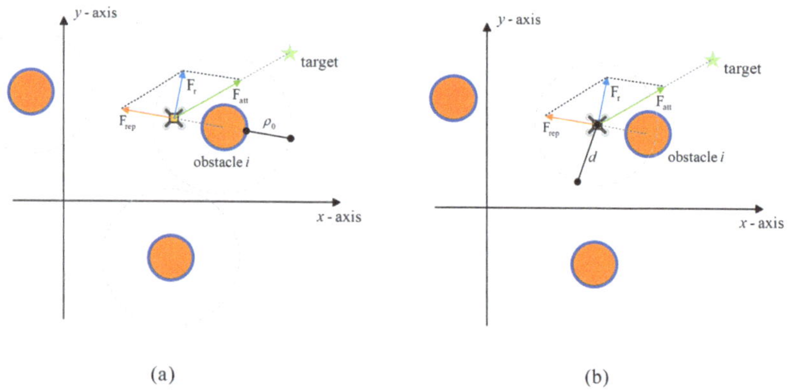

Figure 3. Schematic diagrams of APF forces. (**a**) The forces are generated based on the information returned by the QUAV's sensors. (**b**) The forces are generated using global geometric information.

3.3. Deep Reinforcement Learning

RL is a commonly adopted methodology in machine learning. An RL problem is usually characterized as a Markov decision process with the state initialized by a distribution $\mathcal{P}(s_1)$. In this work, we assume that the Markov property is satisfied. In other words, the transition probability for the next state is only determined by the current state and action, which is formalized as $\mathcal{P}(s_{t+1}|s_1, a_1, \ldots, s_t, a_t) = \mathcal{P}(s_{t+1})|s_t, a_t)$. At time step t, the agent receives the current state s_t and, depending on whether the policy is stochastic or deterministic, takes an action a_t according to its policy $\pi_\alpha(\cdot|s) : \mathcal{S} \to \mathcal{P}(\mathcal{A})$ or $\mu_\alpha(s) : \mathcal{S} \to \mathcal{A}$, where α denotes policy parameters. The environment will then give a scalar reward $r_t \sim \mathcal{R}(s_t, a_t)$ and the next state $s_{t+1} \sim \mathcal{P}(s_t, a_t)$. The aim of the agent is to maximize the expected return $G_t = \sum_{i=t}^{\infty} \gamma^{i-t} r_i$, where $\gamma \in (0,1)$ is the discount factor that prevents G_t from approaching infinity and controls how myopic the agent is.

To train the motion controller, we employ twin-delayed deep deterministic policy gradient (TD3), which is a model-free actor-critic DRL algorithm featuring three distinct characteristics. Firstly, it utilizes two critic networks, Q_{w_1} and Q_{w_2}, to tackle the issue of action value overestimation. Secondly, it incorporates smooth regularization to minimize the variance of the target update, which means the action is given by Equation (7), ensuring a more stable and reliable training process, which is crucial for achieving optimal performance:

$$\tilde{a}_{t+1} = \mu_{\alpha'}(s_{t+1}) + \epsilon, \quad \epsilon \sim \text{clip}(\mathcal{N}(0, \tilde{\sigma}), -c, c), \tag{7}$$

where $\mu_{\alpha'}$ is the target actor network, ϵ is the Gaussian noise, $\tilde{\sigma}$ is the standard deviation, and c is the noise bound. The loss function for the i-th critic network is represented by

$$L_{\text{TD3}}(w_i) = \mathbb{E}_{s_t, a_t, r_t, s_{t+1} \sim D}[(r_t + \gamma \min_{j=1,2} Q_{w_j'}(s_{t+1}, \tilde{a}_{t+1}) - Q_{w_i}(s_t, a_t))^2], \tag{8}$$

where D is the replay buffer and $Q_{w_j'}(j = 1, 2)$ represent the target critic networks.

The third feature of TD3 is that actor networks are updated only after a fixed number of updates d to the critic. The actor network's updates follow

$$\nabla_\alpha J(\alpha) = \mathbb{E}_{a_t = \mu_\alpha(s_t)}[\nabla_\alpha \mu_\alpha(s_t) \nabla_{w_1} Q_{w_1}(s_t, a_t)]. \tag{9}$$

4. Methodology

4.1. Motion Controller Training

As is shown in Figure 4, motion control is divided into two hierarchies: translational control and angular control. Due to the presence of gravity, vertical control differs inherently from horizontal control. Hence, we further subdivide translational control into vertical and horizontal control. As the three angular controllers adhere to the same control logic, and likewise the two horizontal controllers, only three networks are required. The actor networks' parameters for angular, vertical, and horizontal control are denoted by $\alpha_a, \alpha_v,$ and α_h, respectively. Their states are designed as

$$\begin{aligned}
s^\zeta &= [e^\zeta, \omega^\zeta]^T, \quad (\zeta = \phi, \theta, \psi) \\
s^x &= [e^x, v^x, \text{sign}(e^x)e^z, \text{sign}(e^x)v^z, \bar{\theta}, \dot{\theta}]^T, \\
s^y &= [e^y, v^y, \text{sign}(e^y)e^z, \text{sign}(e^y)v^z, \bar{\phi}, \dot{\phi}]^T, \\
s^z &= [e^z, v^z]^T,
\end{aligned} \tag{10}$$

in which $\bar{\theta}$ and $\bar{\phi}$, variables that characterize the pitch and the roll, are defined as

$$\begin{aligned}
\bar{\theta} &= \arctan \frac{z_b^x}{z_b^z} \\
\bar{\phi} &= \arctan \frac{z_b^y}{z_b^z}.
\end{aligned} \tag{11}$$

There are a few points to note here. First, $x_d, y_d, z_d,$ and ψ_d are given by external sources, while ϕ_d and θ_d are calculated from the outputs of the translational controllers. Second, horizontal states not only consider the corresponding position and velocity errors but also take into account height and angular information. Third, angular errors e_ζ are defined using the rotation matrix instead of the angular differences, and they are calculated by

$$\mathbf{e}_R = [e^\phi, e^\theta, e^\psi]^T = \frac{1}{2}(\mathbf{R}_d^T \mathbf{R} - \mathbf{R}^T \mathbf{R}_d)^\vee,$$ (12)

in which the vee map $\vee : \mathfrak{so}(3) \to \mathbb{R}^3$ is the inverse of the hat map, and $\mathbf{R}_d$ is the desired rotation matrix uniquely determined by ϕ_d, θ_d, and ψ_d.

To obtain ϕ_d and θ_d, the translational controllers' networks first generate desired forces along each axis following

$$\begin{aligned}
f_d^x &= mc_t\mu_{\alpha_h}(s^x),\\
f_d^y &= mc_t\mu_{\alpha_h}(s^y),\\
f_d^z &= m(c_t\mu_{\alpha_v}(s^z) + g),
\end{aligned}$$ (13)

where the networks' outputs are bounded by $(-1, 1)$, and c_t is the maximum translational acceleration. Then, the desired roll and pitch can be calculated as [15]

$$\begin{aligned}
\phi_d &= \arcsin \frac{s_{\psi_d}f_d^x - c_{\psi_d}f_d^y}{||[f_d^x, f_d^y, f_d^z]^T||},\\
\theta_d &= \arctan \frac{c_{\psi_d}f_d^x + s_{\psi_d}f_d^y}{f_d^z}.
\end{aligned}$$ (14)

The ultimate desired thrust and torques are generated following Equation (15), where c_r is the maximum angular acceleration. We can find the desired steady-state rotor speed $\Omega_{ss} = k_m\mathbf{u} + \mathbf{b}_m$ by taking the inverse matrix in Equation (5), and the PWM signal $\mathbf{u}$ will eventually be obtained.

$$\begin{aligned}
f_d &= \frac{f_d^z}{c_\phi c_\theta},\\
\tau_d^\xi &= I^\xi c_r\mu_{\alpha_a}(s^\xi). \quad (\xi = \phi, \theta, \psi)
\end{aligned}$$ (15)

Equation (16) defines the reward function for the angular, vertical, and horizontal reinforcement learning environments. In this equation, e_t^ξ represents the error at the current time, and ξ indicates the specific controller to which the error is applied. The agent receives a reward when the error approaches zero and is penalized otherwise. One notable advantage of this reward design is the absence of hyperparameters that require manual fine-tuning.

$$r_t^\xi = |e_t^\xi| - |e_{t+1}^\xi|$$ (16)

During the training process, the reinforcement learning environments receive the agent's action a_t^ξ and output the next state s_{t+1}^ξ as well as the reward r_t^ξ based on the current state s_t^ξ. In this work, the pitch and x-axis controllers are chosen as the prototypes for the angular and horizontal controllers, meaning that $\xi = \theta, x, z$ during training. The training follows a certain order. First, the roll and vertical controllers are trained independently. When the pitch controller is being trained, we set f_d, τ_d^ϕ, and τ_d^ψ as zero, and $\tau_d^\theta = I^\theta c_r\mu_{\alpha_a}(s^\theta)$, while when the vertical controller is being trained, we set f_d as $m(c_t\mu_{\alpha_h}(s^z) + g)/c_\phi c_\theta$ and the desired torques as zero. The x-axis controller is trained only after α_a and α_h are obtained because s^x contains height and pitch information, which is affected by the vertical and pitch controllers. No obstacles or external disturbances are considered during training.

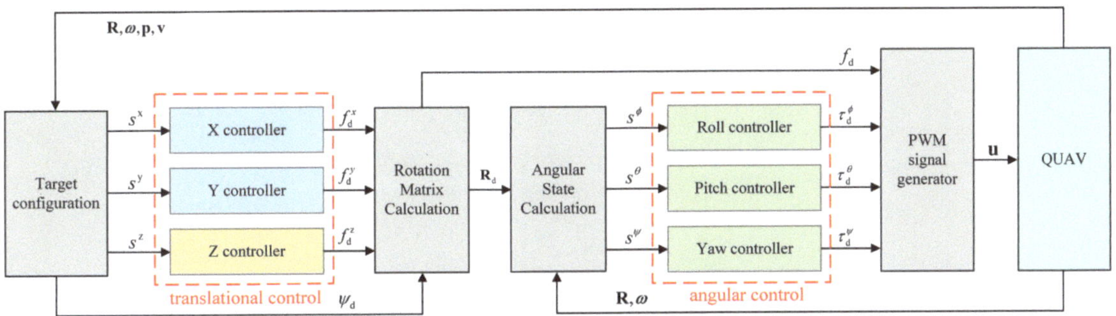

Figure 4. The general framework of the motion controller.

4.2. State Transformation

The trained motion controllers have the ability to navigate the QUAV in the corresponding dimension. However, when the states in the testing phase are too large, it becomes challenging for the networks to generate reasonable outputs based on the limited training samples. This issue typically arises in translational control, where the position errors are not bounded, except by the simulator. Since it is not feasible to gather information for position errors of arbitrary magnitude within a finite training period, using the policy networks in a large or boundless environment can result in uncontrollable behavior, which is inherently dangerous. To mitigate the impact of large horizontal position errors, we employ a hard-clipping technique on the variables e^x and e^y, with a clipping bound of d_c. Although no hard-clipping is applied to e^z, its range is constrained due to the presence of a floor and a ceiling in the working space of this study.

As observed in [18], the trained DRL controllers will decelerate to avoid overshooting, indicating that there exists a natural velocity upper bound $\tilde{v}$, since $\mu_\alpha(s) > 0$ when the velocity is smaller than $\tilde{v}$ and $\mu_\alpha < 0$ when it exceeds the bound. Considering that $\tilde{v}$ is sometimes too large to guarantee safety in the testing phase, we manually set an upper bound $\bar{v}$ to the velocity in the testing phase to prevent the QUAV from moving too quickly in a free space. To increase data efficiency, we multiply e^{v^x} and e^{v^y} by $\tilde{v}/\bar{v}$, which can map the horizontal velocity errors from $[0, \bar{v}]$ to $[0, \tilde{v}]$ and therefore force the actual velocity inputs to have a similar scale as the training samples.

In addition to the aforementioned issues, it is necessary to address the problem of anisotropic control performance. As depicted in Figure 5a, we denote the force exerted on the QUAV as $\mathbf{F}_1$ when it is located at position A and as $\mathbf{F}_2$ when it is at position B. Ideally, if positions A and B are equidistant from the target at a distance r, we expect the two force vectors to have identical magnitudes and angles relative to the corresponding position error vectors. This implies that the control performance should exhibit isotropy. However, the current motion controllers we have obtained do not guarantee isotropy due to two reasons: (1) the x-axis and y-axis controllers independently calculate the force components along each axis based on s^x and s^y, and (2) the policy networks are non-linear. To preserve this property, we transform the horizontal coordinate system such that the force vector components are calculated radially and tangentially to the horizontal error vector $\mathbf{e}_h = [e^x, e^y, 0]^T$. As shown in Figure 5b, the unit radial and tangential vectors are defined as

$$\mathbf{h} = \frac{\mathbf{e}_h}{||\mathbf{e}_h||}, \quad T(\mathbf{h}) = \mathbf{h} \times [0, 0, -1]^T. \tag{17}$$

The horizontal states in the testing phase are so far modified as

$$
\begin{aligned}
s^{\text{rad}} &= [e^{\text{rad}}, (\tilde{v}/\bar{v})\mathbf{v}^T\mathbf{h}, \text{sign}(e^{\text{rad}})e^z, \text{sign}(e^{\text{rad}})v^z, \tilde{\theta}, \dot{\tilde{\theta}}]^T, \\
s^{\text{tan}} &= [e^{\text{tan}}, (\tilde{v}/\bar{v})\mathbf{v}^T T(\mathbf{h}), \text{sign}(e^{\text{tan}})e^z, \text{sign}(e^{\text{tan}})v^z, \tilde{\phi}, \dot{\tilde{\phi}}]^T,
\end{aligned}
\tag{18}
$$

in which $e^{\mathrm{rad}}, e^{\mathrm{tan}}, \tilde{\theta}$, and $\tilde{\phi}$ are defined as

$$e^{\mathrm{rad}} = \mathrm{clip}(\mathbf{e}_{\mathrm{d}}^{T}\mathbf{h}, -d_c, d_c),$$
$$e^{\mathrm{tan}} = \mathrm{clip}(\mathbf{e}_{\mathrm{d}}^{T}T(\mathbf{h}), -d_c, d_c),$$
$$\tilde{\theta} = \arctan\frac{\mathbf{z}_{\mathrm{b}}^{T}\mathbf{h}}{z_{\mathrm{b}}^{z}}, \tag{19}$$
$$\tilde{\phi} = \arctan\frac{\mathbf{z}_{\mathrm{b}}^{T}T(\mathbf{h})}{z_{\mathrm{b}}^{z}}.$$

Similarly, the horizontal forces defined in Equation (13) should be modified following

$$[f_{\mathrm{d}}^{x}, f_{\mathrm{d}}^{y}, 0]^{T} = mc_t\mu_{\alpha_{\mathrm{h}}}(s^{\mathrm{rad}})\mathbf{h} + mc_t\mu_{\alpha_{\mathrm{h}}}(s^{\mathrm{tan}})T(\mathbf{h}). \tag{20}$$

It should be noted that, up to now, $e^{\mathrm{rad}} = \sqrt{(e^x)^2 + (e^y)^2} \geq 0$ and $e^{\mathrm{tan}} = 0$, and there is no reason for the QUAV to move along $T(\mathbf{h})$. This is because no obstacles are taken into consideration yet. As we shall see in the next section, they will be further modified by the repulsive potential fields.

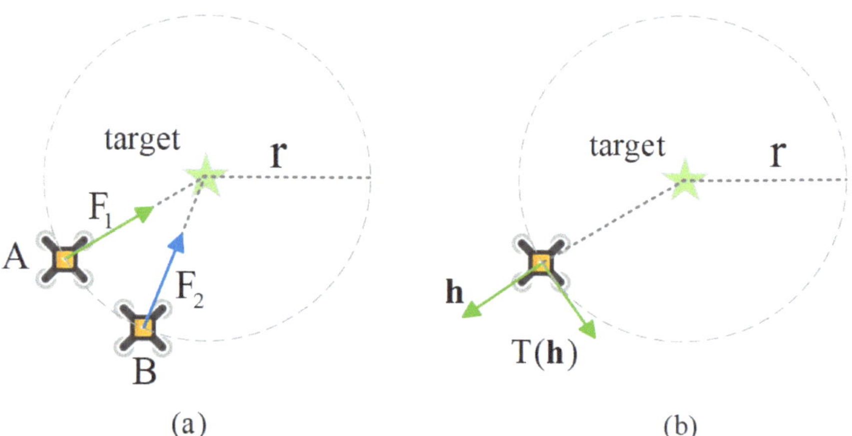

Figure 5. Isotropic control performance. (**a**) For the property of isotropy to be maintained, the norms of $\mathbf{F}_1$ and $\mathbf{F}_2$ should be the same. (**b**) The horizontal coordinate system is transformed such that horizontal axes are either parallel or tangential to the horizontal position error vector.

4.3. APF for Network Controller

In a hierarchical framework, RL motion controllers are solely responsible for the navigation of the QUAV in open spaces and do not address higher-level tasks such as obstacle avoidance. In this study, we incorporate the use of APF to guide the RL motion controllers by influencing the state inputs, thereby indirectly altering the QUAV's current pursuit position. Figure 6 provides an illustration of how the target is updated, with the saturated target being the direct result of the clipping operation along the radial direction. Similar to traditional APF methods, any obstacle that falls within the QUAV's sensing range generates a repulsive force. However, instead of directly applying this force to the agent, it is utilized to further modify the target's position. This is accomplished by providing the motion controllers with modified position errors in the following manner:

$$e^{\mathrm{rad}} = \mathrm{clip}(\mathbf{e}_{\mathrm{d}}^{T}\mathbf{h}, 0, d_c) - \mathbf{F}_{\mathrm{rep}}^{T}\mathbf{h},$$
$$e^{\mathrm{tan}} = -\mathbf{F}_{\mathrm{rep}}^{T}T(\mathbf{h}). \tag{21}$$

It is important to note that distance clipping is performed prior to the modification of errors by $\mathbf{F}_{\text{rep}}$, as depicted in Figure 6a. This implies that the radial and tangential position errors, e^r and e^t, are not strictly limited by the distance d. This is done to ensure that the repulsive forces exerted by nearby obstacles have a significant impact on the movements of the QUAV. Without this consideration, the actual influence of repulsion on the target would be insignificant, as shown in Figure 6b, particularly when the original target is located far away.

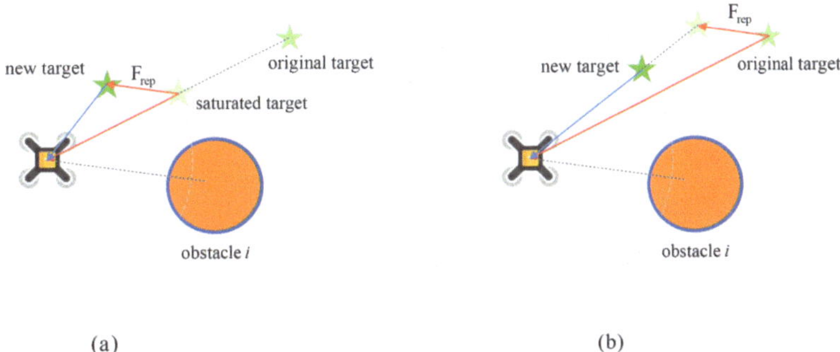

Figure 6. The target position is further adjusted based on the repulsive force. (**a**) In this study, the order of operations involves conducting the clipping operation before the target is pushed by the APF. (**b**) The clipping operation is performed at the end, which minimizes the impact of the repulsion.

The newly-introduced repulsive forces are dynamic, as the geometric relationship between the QUAV and adjacent obstacles changes. This implies that the position of the target also changes dynamically. Therefore, it is not enough to solely consider the agent's current velocity in the motion controllers' states, as the target was assumed to be stationary during the training process. In this study, we calculate the derivative of the repulsive force with respect to time to determine the target's velocity. Using the chain rule, we can obtain that

$$\mathbf{v}^{\text{tar}} = \dot{\mathbf{F}}_{\text{rep}} = -\sum_{i:d_i \leq d} \nabla_{\mathbf{p}}^2 U_{\text{rep}_i}(||\mathbf{e}_i||)\dot{\mathbf{p}} = -\sum_{i:d_i \leq d} \mathbf{H}_i \mathbf{v}, \tag{22}$$

in which $\mathbf{v}^{\text{tar}}$ is the velocity of the dynamic target, $d_i = \sqrt{(e_i^x)^2 + (e_i^y)^2}$ is the horizontal distance to the i-th obstacle, and $\mathbf{H}_i$ is the Hessian matrix for the i-th obstacle's repulsive field. A few properties are to be satisfied when designing $U_{\text{rep}_i}(\cdot) : (0, d] \to \mathbb{R}$:

- $U_{\text{rep}_i}(x)$ should be monotonically decreasing with respect to x;
- $||\nabla_{\mathbf{p}} U_{\text{rep}_i}(x)||$ should be monotonically decreasing with respect to x;
- $\lim_{x \to 0^+} ||\nabla_{\mathbf{p}} U_{\text{rep}_i}(x)|| = +\infty$ and $||\nabla_{\mathbf{p}} U_{\text{rep}_i}(d) = 0||$;
- $U_{\text{rep}_i}(x)$ should be second-order differentiable.

The first requirement is to ensure that the force is repulsive. The second property states that the influence of the obstacle weakens as the distance increases. The third property implies that the repulsion force approaches infinity at the surface of the obstacle, while it becomes zero when the obstacle is at the boundary of the sensing range. The final requirement guarantees the existence of the Hessian matrix. Since the repulsive APF is only applied through its derivative forms, it is reasonable to directly design $\mathbf{F}_{\text{rep}_i}$ as

$$\mathbf{F}_{\text{rep}_i} = \begin{cases} \lambda \left(\frac{1}{d_i^2} - \frac{1}{d^2} \right) \frac{\mathbf{e}_i}{||\mathbf{e}_i||} & d_i \leq d \\ \mathbf{0} & d_i > d \end{cases}, \tag{23}$$

where λ is a manually set factor that scales the intensity of repulsion.

Considering that the calculation of $\mathbf{H}_i$ necessitates significant computational resources and is time-consuming, we approximate $\mathbf{v}^{\text{tar}}$ by utilizing the time difference of $\mathbf{F}_{\text{rep}}$. If the radial component of $\mathbf{v}^{\text{tar}}$ points towards the target, we eliminate it, as in such cases, the obstacle poses no threat to the agent. To be more specific, we define a distinct target velocity in the radial direction using the following approach:

$$\mathbf{v}^{\text{rad}} = \begin{cases} \mathbf{v}^{\text{tar}} & \mathbf{h}^T \mathbf{v}^{\text{tar}} > 0 \\ \mathbf{0} & \mathbf{h}^T \mathbf{v}^{\text{tar}} \le 0 \end{cases} . \tag{24}$$

This technique prevents e^r from exceeding the bound d and therefore avoids overshooting. Ultimately, the horizontal states in the testing phase are modified as

$$\begin{aligned} s^{\text{rad}} &= [e^{\text{rad}}, (\tilde{v}/\bar{v})(\mathbf{v} - \mathbf{v}^{\text{rad}})^T \mathbf{h}, \text{sign}(e^{\text{rad}})e^z, \text{sign}(e^{\text{rad}})v^z, \tilde{\theta}, \dot{\tilde{\theta}}]^T, \\ s^{\text{tan}} &= [e^{\text{tan}}, (\tilde{v}/\bar{v})(\mathbf{v} - \mathbf{v}^{\text{tar}})^T T(\mathbf{h}), \text{sign}(e^{\text{tan}})e^z, \text{sign}(e^{\text{tan}})v^z, \tilde{\phi}, \dot{\tilde{\phi}}]^T. \end{aligned} \tag{25}$$

This design enables the seamless integration of RL motion control and APF obstacle avoidance, without requiring any modifications to the underlying structure of the pretrained motion controllers. The modification of states only occurs during the testing phase. What sets this strategy apart from traditional APF approaches is that, instead of directly applying forces to the QUAV, they are utilized to propel the target. This approach is equivalent to generating a sequence of waypoints that are carefully designed to avoid any potential obstacles.

5. Simulation Experiments

5.1. Training Setup

The QUAV's dynamics followed Crazyflie 2.0 [46]. All agents were trained using TD3, whose hyperparameters are summarized in Table 1. The time step was set as 0.02 s to ensure control accuracy, while the upper bounds of translational and angular acceleration were $c_t = 5 \, \text{m/s}^2$ and $c_r = 20 \, \text{rad/s}^2$, respectively.

Table 1. TD3 hyperparameter settings.

Symbols	Descriptions	Values
$\|\mathcal{D}\|$	Replay Buffer Size	1×10^4
N	Batch Size	512
γ	Discount Factor	0.99
τ	Target Update Rate	0.1
δ	Learning Rate	1×10^{-3}
$\tilde{\sigma}$	Noise Variance	0.2
c	Noise Clip	0.5
d	Policy Update Delay	2

An MLP with two layers, each consisting of 32 nodes, was utilized to train each controller. The optimizer used in this process was Adam. The targets were randomly sampled from a uniform distribution. Specifically, angular targets were chosen from the range of $(-\frac{\pi}{2}, \frac{\pi}{2})$, while translational targets were selected from the range of $[-5, 5]$. It is worth noting that the horizontal network was trained only after the angular and vertical networks had been obtained. The training stage of each controller comprised a total of 200 training episodes, with episode lengths set at 100 and 500 for angular and translational control respectively. This ensured that the agent had sufficient time to reach the target before an episode concluded. Throughout the training phase, no obstacles or external disturbances were taken into consideration.

5.2. Testing Setup

We designed three different controllers for all experiments. Controller 1 utilized the default setting, while controller 2 excluded the axial target velocity cancellation operation.

On the other hand, controller 3 solely employed APF to influence position errors, without affecting velocity errors.

The testing environment consisted of an open space with a floor and a ceiling that was 5 m high. Additionally, there were several cylindrical obstacles, each standing at a height of 5 m, strategically placed around the center of the space. The positions of these obstacles were determined by uniformly selecting the distances of their axes from the center within the range of $[0, 6]$. Furthermore, the radii of the cylinders were uniformly chosen from the range of $[0.9, 1.1]$. It was ensured that the minimum distance between any two obstacles was at least d_{min}. For each trial, three QUAVs were utilized, which acted as dynamic obstacles and generated APF repulsive forces similar to the fixed obstacles in the environment.

They were initialized at $[8\sin(\frac{2k\pi}{3}), 8\cos(\frac{2k\pi}{3})]$, and the corresponding targets were at $[-8\sin(\frac{2k\pi}{3}), -8\cos(\frac{2k\pi}{3})]$, where $k = 0, 1, 2$. As the agents moved, proximity sensors mounted on them could detect nearby obstacles. The maximum sensing range d was 2 m, the position error clipping bound d_c was 0.5 m, while the repulsive force scaling factor λ was set as 0.18. The manually set velocity upper bound $\bar{v}$ was set as 2 m/s, and through experiments, it was found that the natural velocity upper bound $\tilde{v} = 1.5$ m/s.

The agents faced the challenge of reaching the targets on the opposite side without any collisions. The success of navigation was determined by whether each agent entered the 0.1 m proximity of its corresponding target. We created four different environments with specific values for d_{min} and N: $(0.7, 15), (0.7, 12), (0.9, 12)$, and $(0.9, 9)$. In each environment, we conducted 1000 trials for every controller setting. Based on these trials, we calculated the success rates and the average time taken for successful trials.

To evaluate path smoothness, we calculated the curvatures of projection of the actual paths on the horizontal plane according to

$$\kappa = \frac{|\dot{x}\ddot{y} - \ddot{x}\dot{y}|}{(\dot{x}^2 + \dot{y}^2)^{\frac{3}{2}}}, \tag{26}$$

in which the time-derivatives of positions are approximated by the differences between two time steps. It should be emphasized that κ may be either positive or negative, depending on the turning direction. Path oscillations are reflected by the occurrence of sharp turns, which correspond to huge absolute curvatures. In other words, if a path contains multiple points with huge absolute curvatures, it is oscillatory. In this work, the curvature threshold for a sharp turn is defined as 1000.

5.3. Experiment Results

5.3.1. General Performance

Figure 7 is a vertical view of the agents' trajectories as they moved around in the four environments, where the circles represent cylindrical obstacles, the curves in orange, green, and blue illustrate the trajectories, and the stars at their ends are the targets. A smaller d_{min} indicates a more challenging scenario, since the minimum width of the passageway between two obstacles was smaller, while a bigger N means there were more obstacles around the origin. It can be seen that the hierarchical control framework we have proposed is capable of navigating the agents to the corresponding targets smoothly without collisions or detours, even in the densest environment.

To validate the effectiveness of the proposed modifications, we conducted 1000 trials for the three controllers in various environments. The success rates and average reaching times are presented in Table 2. In the environment with low density (0.9, 9), all controllers achieved a high success rate. Among them, controller 2 demonstrated the fastest performance, with an average reaching time of only 22.801 s. This can be attributed to controller 2's tendency to accelerate more dramatically when the repulsive force contributes to the attractive force. However, as the density increased, controller 2 experienced a more significant decrease in success rate compared to controller 1. This is because the higher acceleration increases the risk of overshooting, which will be discussed in the following section. Furthermore, controller 3, which did not consider any velocity errors, exhibited a severe decline in performance. It not only took more time to reach the targets but also struggled to handle dense environments. This was due to the delay between the assumed target and the actual target, modified by the repulsive APF forces. As a result, oscillatory behaviors were observed. In summary, the introduction of velocity errors effectively mitigated oscillations, while the cancellation of the target velocity radial component enhanced the agents' ability to address overshooting and the resulting oscillations. However, this came at the expense of slower agent movement.

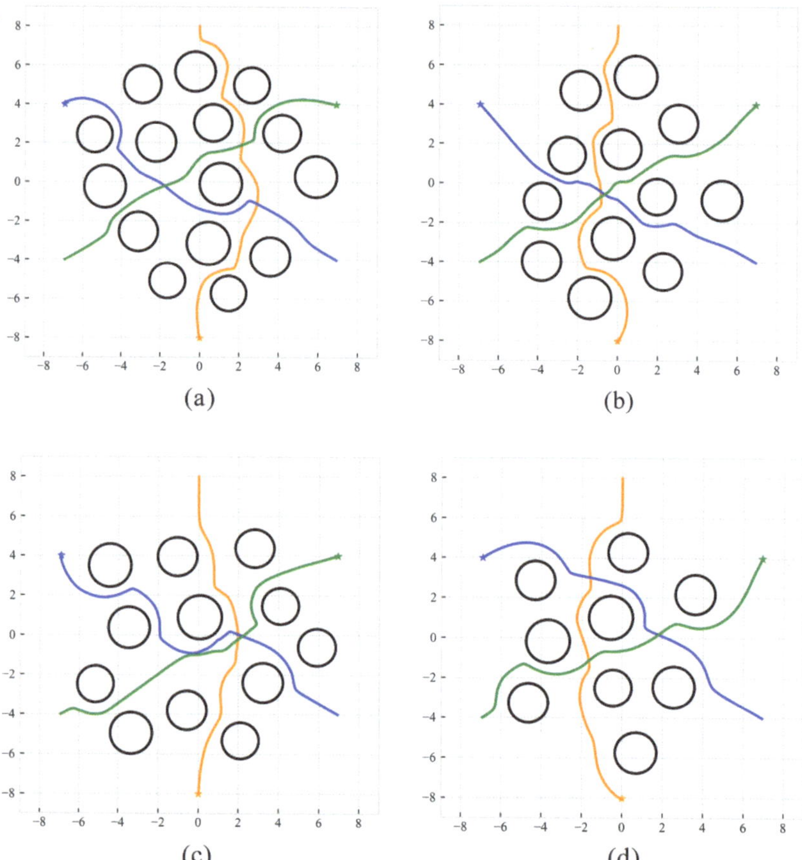

Figure 7. QUAV performance with the designed controllers in four environments, with the densest on the leftmost side and the sparsest on the rightmost side. (**a**) $d_{\min} = 0.7, N = 15$. (**b**) $d_{\min} = 0.7, N = 12$. (**c**) $d_{\min} = 0.9, N = 12$. (**d**) $d_{\min} = 0.9, N = 9$.

Table 2. Success rate (SCR) and average reaching time (ART) of different controllers.

Settings	Controller 1		Controller 2		Controller 3	
	SCR	ART (s)	SCR	ART (s)	SCR	ART (s)
(0.7, 15)	0.950	29.545	0.886	27.433	0.364	32.236
(0.7, 12)	0.968	27.194	0.945	25.223	0.675	29.012
(0.9, 12)	0.986	26.000	0.953	24.181	0.950	27.836
(0.9, 9)	0.989	24.529	0.975	22.801	0.968	25.984

The method is also applicable to an environment with more QUAVs, as shown in Figure 8. The environment setting is similar to that discussed in the testing setup. In general, the agents are able to reach their targets without collisions. For example, in Figure 8b, where three pairs of agents move head to head, they can still reach the targets safely.

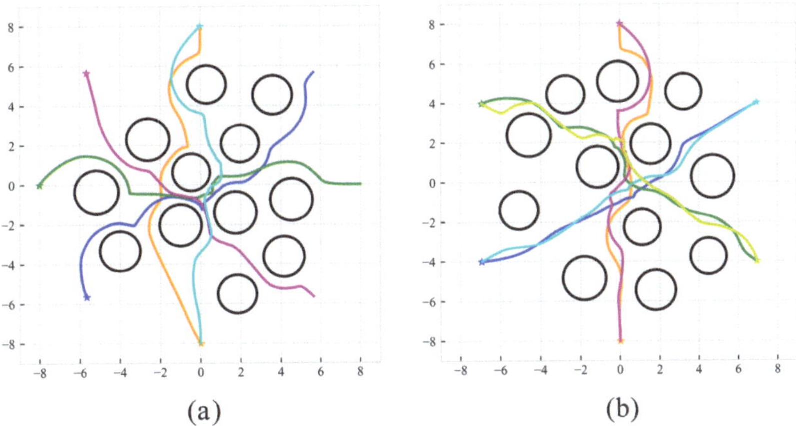

(a)   (b)

Figure 8. The method is applicable to cases with more QUAVs. The environment settings are $d_{\min} = 0.7, N = 12$. (**a**) The case with 5 agents. (**b**) The case with 6 agents.

5.3.2. Oscillation Reduction and Overshooting Prevention

Table 3 lists the average number of points with an absolute curvature greater than 1000 for the three controllers over all successful trials. It can be seen that controller 3 demonstrates the worst performance in terms of path smoothness, which testifies that target velocity information plays an essential role in reducing path oscillations. On the other hand, controllers 1 and 2 are close, with the path of the former being slightly more oscillatory. It should be noted that overshooting is not equivalent to global path smoothness, since it only occurs at certain areas, such as the triangular trap. Therefore, the oscillations caused by overshooting cannot easily be reflected by path curvature.

Table 3. The average number of sharp turns ($|\kappa| > 1000$) of different controllers.

Settings	Controller 1	Controller 2	Controller 3
(0.7, 15)	79.427	73.855	122.403
(0.7, 12)	54.788	51.731	90.938
(0.9, 12)	41.925	33.658	75.417
(0.9, 9)	30.434	25.136	53.372

To give quantitative analysis to the effectiveness of radial target velocity cancellation operation as defined by Equation (24), another experiment setting was used. Concretely, three cylindrical obstacles with a radius of 1.5 m were randomly placed on a ring whose inner and outer radii were 2.3 m and 2.5 m, respectively. The three QUAVs were initialized at $[0.6\sin(\frac{2k\pi}{3}), 0.6\cos(\frac{2k\pi}{3})]$, and the corresponding targets were at $[-8\sin(\frac{2k\pi}{3}), -8\cos(\frac{2k\pi}{3})]$, where $k = 0, 1, 2$. In this way, they were trapped by the encircled obstacles at first, and the task was to escape the triangular trap. Ideally, the maximum distance between any QUAV and the corresponding target position is 8.6 m, otherwise it is caused by overshooting. In this experiment, overshooting cases are defined as cases where at least one QUAV's distance to its target exceeded 10 m, as shown in Figure 9a. Furthermore, we define trapped cases where at least one QUAV's distance to the origin is within 4 m, as shown in Figure 9b. Table 4 shows the results. It can be seen that controller 1 encountered the fewest overshooting cases, which shows that the velocity cancellation operation helps preventing overshooting. Moreover, since this operation simplifies the obstacle environment by neglecting obstacles behind it, trapped cases are largely avoided as well.

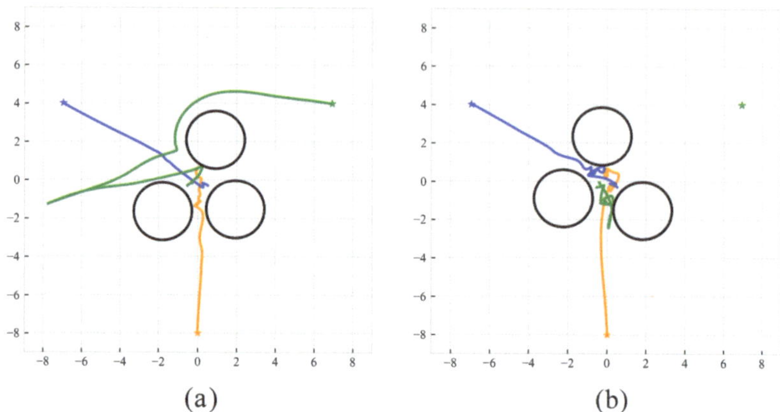

(a)   (b)

Figure 9. (**a**) Overshooting case. (**b**) Trapped case.

Table 4. The number of overshooting and trapped cases of different controllers.

	Controller 1	Controller 2	Controller 3
Overshooting Cases	23	145	481
Trapped Cases	24	199	641
SCR	0.969	0.732	0.237
ART (s)	13.735	13.958	18.566

An intuitive comparison of the three controllers can be found in Figure 10. In Figure 10b, the agent represented by the green trajectory experienced oscillation within the triangular area formed by obstacles 1, 2, and 3. The reason behind this oscillation was that the combined force exerted by obstacle 1 and obstacle 2 was directed towards the target and was not counteracted even though they did not pose a threat to the agent. Consequently, the force towards the goal became excessively strong, leading to the agent overshooting in this direction. Furthermore, disregarding velocity information entirely resulted in more frequent oscillations. The reason was that, in this case, the agent assumed the intermediate target to be fixed, while in reality the intermediate target was affected by the relative position between the agent and the obstacles. Oscillations occurred when the intermediate target updated before the agent reached the previous one. This can be observed in Figure 10c, where all agents' trajectories exhibit a zigzag pattern, especially when maneuvering between obstacles. This increases the risk of collision with nearby obstacles.

5.3.3. The Effect of a Clipping Bound

In the mentioned experiments, we utilized a position error clipping bound of $d_c = 0.5$ m. This choice was made based on previous experiments that indicated that it was the safest option. We varied the value of d_c for controller 1, ranging from 0.3 m to 1.0 m with an increment of 0.1 m. A total of 1000 trials were conducted with the environment parameters set at (0.7, 12). Additionally, we observed that, for d_c values of 0.3 m and 0.4 m, the agent's velocity was relatively low, resulting in a longer time to reach the target. To account for more trials that had the potential to reach the targets, we relaxed the time requirement for success. This adjustment allowed us to consider a broader range of trials. The results of these experiments are presented in Table 5. From the table, it can be observed that the average reaching time is inversely correlated with d_c. Furthermore, $d_c = 0.5$ yielded the highest success rate. This can be attributed to the fact that, when the agent moved too slowly, it lacked the flexibility to handle imminent collisions. Conversely, when the agent moved too quickly, the outputs of the low-level flight controllers were not sufficiently strong to decelerate the agent in time, thereby increasing the likelihood of collisions.

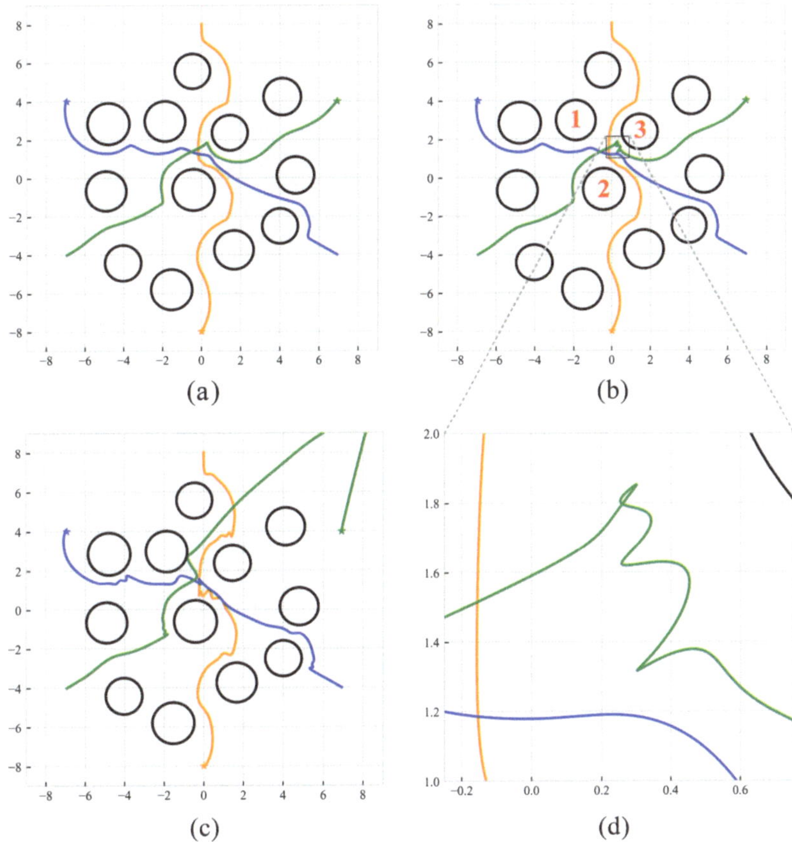

Figure 10. Performance comparison of different controllers in the same environment, where $d_{\min} = 0.7$ and $N = 12$. (**a**) Controller 1. All the designs mentioned previously were used. (**b**) Controller 2. Radial component of the target velocity was not canceled even when it contributed to the attractive force. (**c**) Controller 3. No velocity errors were introduced to the states. (**d**) A close-up view of controller 2's path.

Table 5. Success rate (SCR) and average reaching time (ART) of different controllers.

d_c	0.3 *	0.4 *	0.5	0.6	0.7	0.8	0.9	1.0
SCR	0.517	0.930	**0.968**	0.941	0.851	0.714	0.585	0.568
ART (s)	43.437	34.983	27.194	22.340	19.606	18.705	19.146	**17.710**

* The time requirement for success was relaxed from 3000 steps to 5000 steps.

5.3.4. Comparison to Existing Methods

This section compares the proposed method to two other existing methods in terms of success rate, average reaching time, average absolute curvature, and the number of sharp turns. The first method to be compared uses A* as the path planner and the SE(3) control law proposed in [45] as the motion controller. This method discretizes the environment into cubes with side length 0.2 m, which are also known as voxels. It should be noted that the environment information of this algorithm is known in advance instead of being obtained with sensors. The second method to be compared [47] is a hierarchical DRL method that uses a DQN network to generate waypoints based on the sensor information returned by 16 rangefinders. The waypoints then serve as the target points of the motion controller trained with PPO. The testing environment was a single-angle scenario, where the agent was initialized at [0, 4.5, 2.5], and the target position was [0, −4.5, 2.5]. Twelve cylindrical obstacles with a radius of 0.4 m were randomly placed around the origin. The positions of these obstacles were determined by uniformly selecting the distances of their axes from the center within the range of $[0, 3]$ and $d_{\min} = 0.6$. The results are listed in Table 6, which also includes the performance of controller 3 in this environment. We can see that the proposed method outperformed the other two algorithms in terms of success rate and average reaching time, with the path slightly more oscillatory. Although path oscillation is an inherent problem with APF, it can be reduced to a large extent with the help of velocity information. After addressing this issue, the controller can navigate the QUAV in a more safe and efficient manner.

Figure 11 illustrates the paths of the four methods in the same environment. It can be seen that the proposed method's path is smooth on a macro level and is far away from obstacles. The path of A* + SE(3) is mostly straight because the waypoints returned by A* are exactly on the centers of the voxels. Figure 11c shows a similar path, in that there are only 16 possible directions to choose from when updating the waypoints. The path of controller 3, on the other hand, is the most oscillatory one.

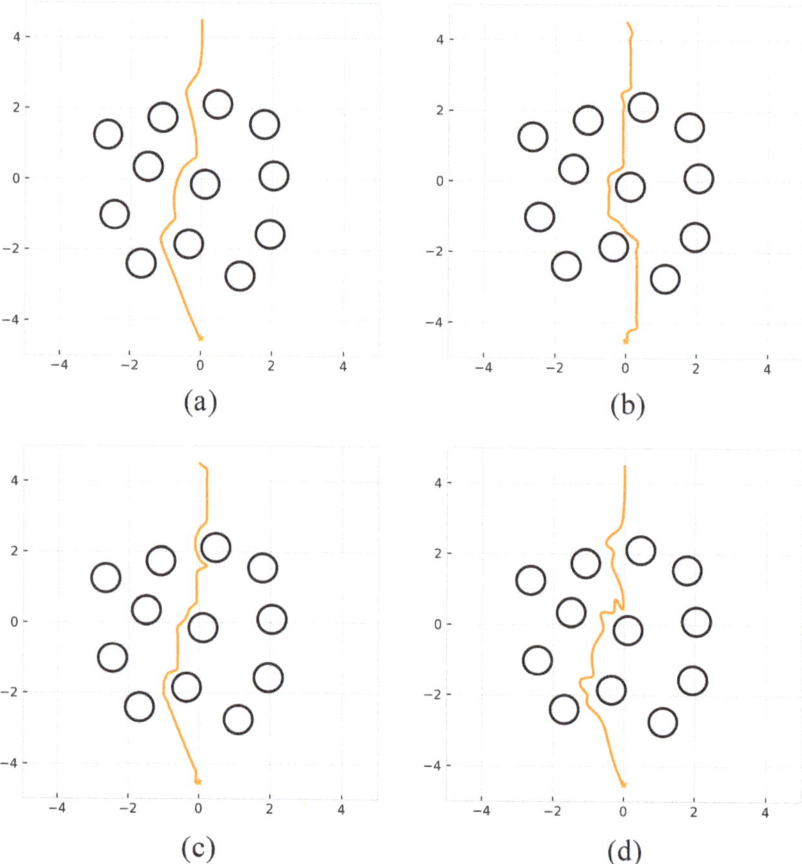

Figure 11. Performance comparison of the four methods in the same environment. (**a**) Proposed method. (**b**) A* + SE(3). (**c**) DQN + PPO. (**d**) Controller 3.

Table 6. Comparison of the proposed method to existing methods.

	APF + TD3	A* + SE(3)	DQN + PPO	Controller 3
SCR	**0.983**	0.969	0.885	0.761
ART (s)	**15.312**	19.458	36.694	18.249
$|\kappa|$	7.121	**4.794**	6.610	11.426
sharp turns	0.651	0.480	**0.431**	1.112

6. Discussion

6.1. Limitations

Although the designed controller can achieve high success rates with moderate reaching time even in a dense environment, there is still room for improvement. We considered cases as failures when the agents were close to the targets but still outside the 0.1 m neighborhood. This issue is referred to as the GNRON problem, which is discussed in Section 2. Table 7 provides the number of GNRON cases for different controllers, where the success requirement is relaxed to entering the 0.3 m neighborhood of targets. It is evident that the success rates could have been higher if we had employed a strategy that actively addressed this problem. By introducing a damping term in the repulsive forces when the agents are in close proximity to the targets, this problem can be effectively resolved. Additionally, there are other limitations to the proposed method, such as the absence of explicit collision detection, leading to occasional collisions, and the hyperparameter d_c, which needs to be manually adjusted.

Another limitation of the proposed method is that it fails to escape local minima created by U-shaped obstacles, which is shown in Figure 12a. Moreover, when the QUAV's moving direction is almost collinear with the joint force, the path is usually not smooth, as shown in Figure 12b. One possible solution is to combine the proposed method with the vortex potential field method discussed in [36,37], since in this way the agent will move smoothly around the obstacle instead of heading directly towards it and spending a long time adjusting the path.

Table 7. The number of GNRON cases for different controllers.

Environments	(0.7, 15)	(0.7, 12)	(0.9, 12)	(0.9, 9)
Controller 1	19	12	10	8
Controller 2	20	10	11	3
Controller 3	6	4	4	1

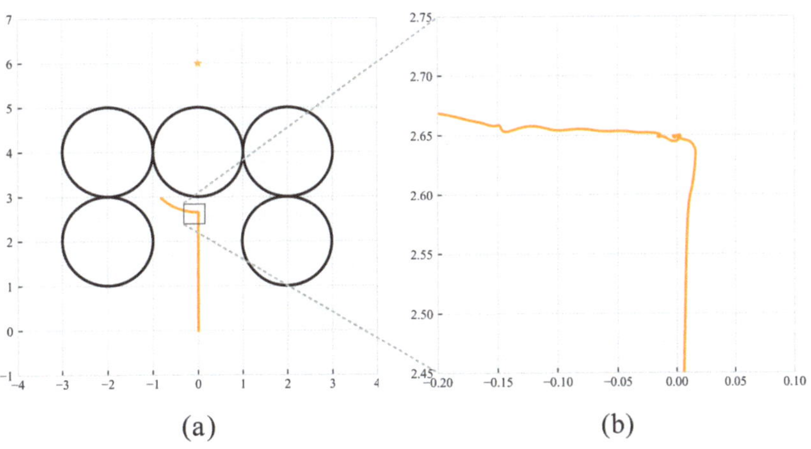

Figure 12. (**a**) The QUAV failed to escape the local minimum caused by a U-shaped obstacle. (**b**) When the approaching angle is close to zero, the path is not smooth.

6.2. Applicability to Dynamic Environments

The proposed method has the potential to be transferred to dynamic environments since collision avoidance behavior is only determined by the repulsive forces updated at every time step. In fact, the preceding experiments have verified this ability because all the QUAVs are dynamic, generating repulsive forces to other QUAVs. As shown in Figure 13, when they move close to each other, the QUAVs are able to take actions to avoid collisions. This behavior resembles the cooperative collision avoidance mentioned in [36]. One significant difference between agents and obstacles is that the former ones usually actively take action to avoid collisions while the latter ones' behavior may be unpredictable. Therefore, avoiding collisions between moving QUAVs and moving obstacles is more challenging. One possible solution is to combine the current framework with reciprocal velocity obstacles [48] based on the current velocity information of the obstacles.

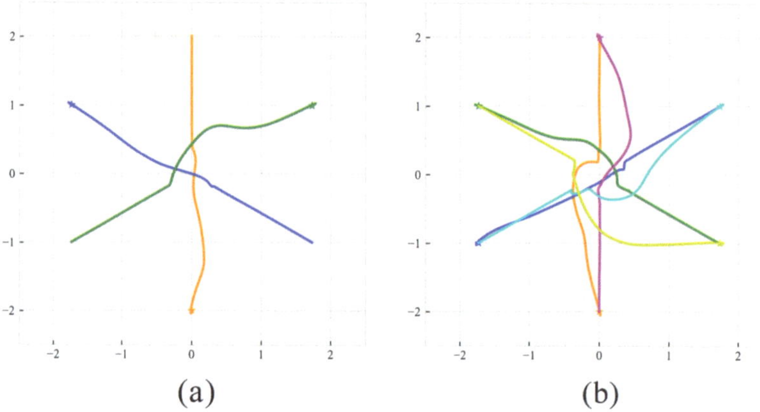

Figure 13. Cooperative collision avoidance. (**a**) The case with 3 agents. (**b**) The case with 6 agents.

6.3. Irregular Obstacles

In real applications, the shapes of obstacles may not be as simple as cylinders. Since this work assumes that the QUAV only detects nearby obstacles with sensors, it is impractical to generate a global potential field in advance. There are two simple ways to adapt this framework so as to make it applicable to environments with irregular obstacles. The first method, as shown in Figure 14a, is to replace the surface of the sensed part with smaller cylinders. In this way, the irregular obstacle's repulsive force equals the joint repulsive force of all cylindrical obstacles. The second method is to assume the part behind the sensed surface as solid and calculate its center of gravity. The distance to the obstacle is obtained as shown in Figure 14b. Some works deal with irregular obstacles in advanced ways. For example, Ge et al. [49] integrated the repulsive force of every part of the obstacle to calculate the ultimate repulsive force. This can model the obstacles more accurately, but it requires the QUAV to have prior global information of the environment. Guo et al. [50] used the closest point on the obstacle's surface to generate repulsive force, which is the same as our method when the obstacles are cylindrical or round. However, separate discussions are needed for obstacles with corners or edge points. This makes it difficult to achieve stable control using the Hessian matrix considering that its calculation will be even more complex under such circumstances.

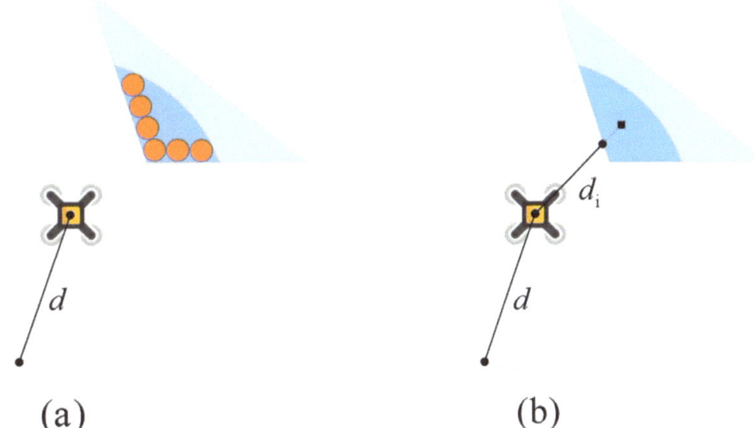

Figure 14. Two possible ways to model irregular obstacles, where the sensed parts are shown in dark blue. (**a**) The surface of the sensed part is replaced with smaller cylindrical obstacles, illustrated as orange circles. (**b**) Assuming the sensed part is a solid object.

7. Conclusions

In this study, we have proposed an efficient hierarchical controller that combines the use of an APF for path planning and obstacle avoidance with DRL motion controllers to generate rotor signals. Our approach differs from traditional APF methods in that the APF force is no longer directly applied to the QUAV but instead used to modify the position errors, effectively pushing the sensed target position. Additionally, we have introduced the calculation of the time-derivative of the APF force and incorporated a second-order term in the velocity error of the DRL controllers' inputs. This modification has successfully reduced oscillations that occur when an agent is surrounded by multiple obstacles. The oscillation technique not only ensures smoother paths but also significantly improves the robustness of the controller. Furthermore, we have demonstrated that canceling the projection of the target velocity on the attractive force when they are in the same direction is an effective method to prevent overshooting and enhance the success rate. Although this approach may slightly decrease the speed at which the target is reached, it is a reasonable compromise that prioritizes safety, especially in dense environments. We would like to emphasize the importance of the clipping factor d_c as a hyperparameter in our design. It is crucial to find an optimal value for d_c, as both excessively small and excessively large values can lead to significant performance degradation. Lastly, we discussed the limitations of the proposed method and provided some possible solutions. The control framework in this work can be applied to a real testing environment considering that it is based on real parameters of a QUAV. Potential challenges include the accumulated errors of the discrete system that generates outputs based on differences, a key point that needs to be addressed in future work.

Author Contributions: Conceptualization, H.H.; methodology, H.H.; software, Z.X.; validation, Z.X.; formal analysis, Z.X. and H.H.; investigation, Z.X., H.H., J.C. and M.L.; writing—original draft preparation, Z.X. and H.H.; writing—review and editing, J.C. and M.L.; visualization, Z.X.; supervision, J.C. and M.L.; project administration, J.C.; funding acquisition, J.C. All authors have read and agreed to the published version of the manuscript.

Funding: This research was funded in part by the NNSFC&CAAC under Grant U2233209 and in part by the Natural Science Foundation of Sichuan, China, under Grant 2023NSFSC0484.

Data Availability Statement: Data are contained within the article.

Conflicts of Interest: The authors declare no conflicts of interest.

References

1. Pugliese, L.D.P.; Guerriero, F.; Macrina, G. Using Drones for Parcels Delivery Process. *Procedia Manuf.* **2020**, *42*, 488–497 [CrossRef]
2. Shakhatreh, H.; Sawalmeh, A.H.; Al-Fuqaha, A.; Dou, Z.; Almaita, E.; Khalil, I.; Othman, N.S.; Khreishah, A.; Guizani, M. Unmanned Aerial Vehicles (UAVs): A survey on Civil Applications and Key Research Challenges. *IEEE Access* **2019**, *7*, 48572–48634. [CrossRef]
3. Huang, Y.; Thomson, S.J.; Hoffmann, W.C.; Lan, Y.; Fritz, B.K. Development and Prospect of Unmanned Aerial Vehicle Technologies for Agricultural Production Management. *Int. J. Agric. Biol. Eng.* **2012**, *6*, 1–10.
4. Muchiri, G.N.; Kimathi, S. A Review of Applications and Potential Applications of UAV. In Proceedings of the Sustainable Research and Innovation Conference (SRI), Pretoria, South Africa, 20–24 June 2022; pp. 280–283.
5. Valsan, A.; Parvathy, B.; GH, V.D.; Unnikrishnan, R.S.; Reddy, P.K.; Vivek, A. Unmanned Aerial Vehicle for Search and Rescue Mission. In Proceedings of the 2020 4th International Conference on Trends in Electronics and Informatics (ICOEI), Tirunelveli, India, 16–18 April 2020; pp. 684–687.
6. Silvagni, M.; Tonoli, A.; Zenerino, E.; Chiaberge, M. Multipurpose UAV for Search and Rescue Operations in Mountain Avalanche Events. *Geomat. Nat. Hazards Risk* **2017**, *8*, 18–33. [CrossRef]
7. Pinto, M.F.; Melo, A.G.; Marcato, A.L.; Urdiales, C. Case-based Reasoning Approach Applied to Surveillance System Using an Autonomous Unmanned Aerial Vehicle. In Proceedings of the 2017 IEEE 26th International Symposium on Industrial Electronics (ISIE), Edinburgh, UK, 19–21 June 2017; pp. 1324–1329.
8. Lv, M.; Wang, N. Distributed Control for Uncertain Multi-agent Systems with the Powers of Positive-odd Numbers: A Low-complexity Design Approach. *IEEE Trans. Autom. Control.* **2024**, *69*, 434–441 . [CrossRef]
9. Wang, Y.; Dong, L.; Sun, C. Cooperative Control for Multi-player Pursuit-evasion Games with Reinforcement Learning. *Neurocomputing* **2020**, *412*, 101–114. [CrossRef]
10. Dijkstra, E.W. A Note on Two Problems in Connexion with Graphs. *Numer. Math.* **1959** , *1*, 269–271. [CrossRef]
11. Kavraki, L.E.; Svestka, P.; Latombe, J.C.; Overmars, M.H. 1996. Probabilistic Roadmaps for Path Planning in High-dimensional Configuration Spaces. *IEEE Trans. Robot. Autom.* **1996**, *12*, 566–580. [CrossRef]
12. Elbanhawi, M.; Simic, M. Sampling-based Robot Motion Planning: A Review. *IEEE Access* **2014**, *2*, 56–77. [CrossRef]
13. Khatib, O. Real-time Obstacle Avoidance for Manipulators and Mobile Robots. *Int. J. Robot. Res.* **1986**, *5*, 90–98. [CrossRef]
14. Tang, L.; Dian, S.; Gu, G.; Zhou, K.; Wang, S.; Feng, X. A Novel Potential Field Method for Obstacle Avoidance and Path Planning of Mobile Robot. In Proceedings of the 2010 3rd International Conference on Computer Science and Information Technology (ICCSIT), Chengdu, China, 9–11 July 2010; pp. 633–637.
15. Han, H.; Cheng, J.; Xi, Z.; Lv, M. Symmetric Actor–critic Deep Reinforcement Learning for Cascade Quadrotor Flight Control. *Neurocomputing* **2023**, *559*, 126789. [CrossRef]
16. Han, H.; Xi, Z.; Cheng, J.; Lv, M. Obstacle Avoidance Based on Deep Reinforcement Learning and Artificial Potential Field. In Proceedings of the 9th International Conference on Control, Automation and Robotics (ICCAR), Beijing, China, 21–23 April 2023; pp. 215–220.
17. Han, H.; Xi, Z.; Lv, M.; Cheng, J. Acceleration of Formation Control Based on Hessian Matrix of Artificial Potential Field. In Proceedings of the 2023 42th Chinese Control Conference, Tianjin, China, 24–26 July 2023; pp. 5866–5871.
18. Han, H.; Cheng, J.; Xi, Z.; Yao, B. Cascade Flight Control of Quadrotors Based on Deep Reinforcement Learning. *IEEE Robot. Autom. Lett.* **2022**, *7*, 11134–11141. [CrossRef]
19. Hart, P.E.; Nilsson, N.J.; Raphael, B. A Formal Basis for the Heuristic Determination of Minimum Cost Paths. *IEEE Trans. Syst. Man, Cybern.* **1968**, *4*, 100–107. [CrossRef]
20. Stentz, A. Optimal and Efficient Path Planning for Partially-known Environments. In Proceedings of the 1994 IEEE International Conference on Robotics and Automation (ICRA), San Diego, CA, USA, 8–13 May 1994; pp. 3310–3317.
21. Farooq, M.U.; Ziyang, Z.; Ejaz, M. Quadrotor UAVs Flying Formation Reconfiguration with Collision Avoidance using Probabilistic Roadmap Algorithm. In Proceedings of the 2017 International Conference on Computer Systems, Electronics and Control, Dalian, China, 25–27 December 2017; pp. 866–870.
22. Ma'arif, A.; Rahmaniar, W.; Vera, M.A.M.; Nuryono, A.A.; Majdoubi, R.; Çakan, A. Artificial Potential Field Algorithm for Obstacle Avoidance in UAV Quadrotor for Dynamic Environment. In Proceedings of the 2021 IEEE International Conference on Communication, Networks and Satellite (COMNETSAT), Online, 17–18 July 2021; pp. 184–189.
23. Koren, Y.; Borenstein, J. Potential Field Methods and Their Inherent limitations for Mobile Robot Navigation. In Proceedings of the 1991 International Conference on Robotics and Automation, Sacramento, CA, USA, 9–11 April 1991; pp. 1398–1404.
24. Kim, J.O.; Khosla, P. Real-time Obstacle Avoidance Using Harmonic Potential Functions. In Proceedings of the 1991 International Conference on Robotics and Automation, Sacramento, CA, USA, 9–11 April 1991; pp. 790–796.
25. Park, M.G.; Jeon, J.H.; Lee, M.C. Obstacle Avoidance for Mobile Robots Using Artificial Potential Field Approach with Simulated Annealing. In Proceedings of the 2001 IEEE International Symposium on Industrial Electronics, Pusan, Republic of Korea, 12–16 June 2001; pp. 1530–1535.
26. Wang, D.; Li, C.; Guo, N.; Song, Y.; Gao, T.; Liu, G. Local Path Planning of Mobile Robot Based on Artificial Potential Field. In Proceedings of the 2020 39th Chinese Control Conference, Shenyang, China, 27–29 July 2020; pp. 3677–3682.

27. Lai, D.; Dai, J. Research on Multi-UAV Path Planning and Obstacle Avoidance Based on Improved Artificial Potential Field Method. In Proceedings of the 2020 3rd International Conference on Mechatronics, Robotics and Automation (ICMRA), Shanghai, China, 16–18 October 2022; pp. 84–88.
28. Doria, N.S.F.; Freire, E.O.; Basilio, J.C. An Algorithm Inspired by the Deterministic Annealing Approach to Avoid Local Minima in Artificial Potential Fields. In Proceedings of the 2013 16th International Conference on Advanced Robotics, Montevideo, Uruguay, 25–29 November 2013; pp. 1–6.
29. Ge, S.S.; Cui, Y.J. New Potential Functions for Mobile Robot Path Planning. *IEEE Trans. Robot. Autom.* **2000**, *16*, 615–620. [CrossRef]
30. Ren, J.; McIsaac, K.A.; Patel, R.V. Modified Newton's Method Applied to Potential Field-based Navigation for Mobile Robots. *IEEE Trans. Robot.* **2006**, *22*, 384–391.
31. Biswas, K.; Kar, I. On Reduction of Oscillations in Target Tracking by Artificial Potential Field Method. In Proceedings of the 2014 9th International Conference on Industrial and Information Systems (ICIIS), Gwalior, India, 15–17 December 2014; pp. 1–6.
32. Zhao, M.; Lv, X. Improved Manipulator Obstacle Avoidance Path Planning Based on Potential Field Method. *J. Robot.* **2020**, *2020*, 1701943. [CrossRef]
33. Zhang, W.; Xu, G.; Song, Y.; Wang, Y. An Obstacle Avoidance Strategy for Complex Obstacles Based on Artificial Potential Field Method. *J. Field Robot.* **2023**, *40*, 1231–1244. [CrossRef]
34. Zheng, S.; Luo, L.; Zhang, J. Non-oscillation Path Planning Based on Artificial Potential Field. In Proceedings of the IEEE International Conference on Control, Electronics and Computer Technology (ICCETC), Jilin, China, 28–30 April 2023; pp. 1225–1228.
35. Tran, H.N.; Shin, J.; Jee, K.; Moon, H. Oscillation Reduction for Artificial Potential Field Using Vector Projections for Robotic Manipulators. *J. Mech. Sci. Technol.* **2023**, *37*, 3273–3280. [CrossRef]
36. Martis, W.P.; Rao, S. Cooperative Collision Avoidance in Mobile Robots using Dynamic Vortex Potential Fields. In Proceedings of the International Conference on Automation, Robotics and Applications (ICARA), Abu Dhabi, United Arab Emirates, 10–12 February 2023; pp. 60–64.
37. Szczepanski, R. Safe Artificial Potential Field-Novel Local Path Planning Algorithm Maintaining Safe Distance from Obstacles. *IEEE Robot. Autom. Lett.* **2023**, *8*, 4823–4830. [CrossRef]
38. Meradi, D.; Benselama, Z.A.; Hedjar, R.; Gabour, N.E.H. Quaternion-based Nonlinear MPC for Quadrotor's Trajectory Tracking and Obstacles Avoidance. In Proceedings of the International Conference on Advanced Electrical Engineering (ICAEE), Constantine, Algeria, 29–31 October 2022; pp. 1–6.
39. Valencia, D.; Kim, D. Quadrotor Obstacle Detection and Avoidance System Using a Monocular Camera. In Proceedings of the Asia-Pacific Conference on Intelligent Robot Systems (ACIRS), Singapore, 21–23 July 2018; pp. 78–81.
40. Gageik, N.; Benz, P.; Montenegro, S. Obstacle Detection and Collision Avoidance for a UAV with Complementary Low-cost Sensors. *IEEE Access* **2015**, *3*, 599–609. [CrossRef]
41. Yao, Q.; Zheng, Z.; Qi, L.; Yuan, H.; Guo, X.; Zhao, M.; Liu, Z.; Yang, T. Path Planning Method with Improved Artificial Potential Field—A Reinforcement Learning Perspective. *IEEE Access* **2020**, *8*, 135513–135523. [CrossRef]
42. Xing, T.; Wang, X.; Ding, K.; Ni, K.; Zhou, Q. Improved Artificial Potential Field Algorithm Assisted by Multisource Data for AUV Path Planning. *Sensors* **2023**, *23*, 6680. [CrossRef]
43. Panerati, J.; Zheng, H.; Zhou, S.; Xu, J.; Prorok, A.; Schoellig, A.P. Learning to Fly—A Gym Environment with Pybullet Physics for Reinforcement Learning of Multi-agent Auadcopter Control. In Proceedings of the IEEE/RSJ International Conference on Intelligent Robots and Systems (IROS), Prague, Czech Republic, 28–30 September 2021; pp. 7512–7519.
44. Lv, M.; Ahn, C.K.; Zhang, B; Fu, A. Fixed-time Anti-saturation Cooperative Control for Networked Fixed-wing Unmanned Aerial Vehicles Considering Actuator Failures. *IEEE Trans. Aerosp. Electron. Syst.* **2023**, *59*, 8812–8825. [CrossRef]
45. Goodarzi, F.; Lee, D.; Lee, T. Geometric Nonlinear PID Control of a Quadrotor UAV on SE(3). In Proceedings of the 2013 European Control Conference (ECC), Zurich, Switherland, 17–19 July 2013; pp. 3845–3850.
46. Bitcraze. *Crazyflie 2.0.* Available online: https://www.bitcraze.io/products/old-products/crazyflie-2-0/ (accessed on 2 January 2024).
47. Xi, Z.; Han, H.; Zhang, Y.; Cheng, J. Autonomous Navigation of QUAVs Under 3D Environments Based on Hierarchical Reinforcement Learning. In Proceedings of the 2023 42nd Chinese Control Conference (CCC), Tianjin, China, 24–26 July 2023; pp. 4101–4106.
48. Van den Berg, J.; Lin, M.; Manocha, D. Reciprocal Velocity Obstacles for Real-time Multi-agent Navigation. In Proceedings of the 2008 IEEE International Conference on Robotics and Automation (ICRA), Pasadena, CA, USA, 19–23 May 2008; pp. 1928–1935.
49. Ge, S.S.; Liu, X.; Goh, C.H.; Xu, L. Formation Tracking Control of Multiagents in Constrained Space. *IEEE Trans. Control Syst. Technol.* **2015**, *24*, 992–1003. [CrossRef]
50. Guo, Y.; Chen, G.; Zhao, T. Learning-based Collision-free Coordination for a Team of Uncertain Quadrotor UAVs. *Aerosp. Sci. Technol.* **2021**, *119*, 107127. [CrossRef]

Disclaimer/Publisher's Note: The statements, opinions and data contained in all publications are solely those of the individual author(s) and contributor(s) and not of MDPI and/or the editor(s). MDPI and/or the editor(s) disclaim responsibility for any injury to people or property resulting from any ideas, methods, instructions or products referred to in the content.

 *drones*

Article

Iterative Trajectory Planning and Resource Allocation for UAV-Assisted Emergency Communication with User Dynamics

Zhilan Zhang [1], Yufeng Wang [2,*], Yizhe Luo [3], Hang Zhang [1], Xiaorong Zhang [1] and Wenrui Ding [2]

[1] School of Electronics and Information Engineering, Beihang University, Beijing 100083, China; zzl1223@buaa.edu.cn (Z.Z.); zhangh102@buaa.edu.cn (H.Z.); zhangxiaorong@buaa.edu.cn (X.Z.)
[2] Institute of Unmanned System, Beihang University, Beijing 100083, China; ding@buaa.edu.cn
[3] School of Computer and Artificial Intelligence, Zhengzhou University, Zhengzhou 450001, China; luoyizhe@zzu.edu.cn
* Correspondence: wyfeng@buaa.edu.cn

Abstract: The demand for air-to-ground communication has surged in recent years, underscoring the significance of unmanned aerial vehicles (UAVs) in enhancing mobile communication, particularly in emergency scenarios due to their deployment efficiency and flexibility. In situations such as emergency cases, UAVs can function as efficient temporary aerial base stations and enhance communication quality in instances where terrestrial base stations are incapacitated. Trajectory planning and resource allocation of UAVs continue to be vital techniques, while a relatively limited number of algorithms account for the dynamics of ground users. This paper focuses on emergency communication scenarios such as earthquakes, proposing an innovative path planning and resource allocation algorithm. The algorithm leverages a multi-stage subtask iteration approach, inspired by the block coordinate descent technique, to address the challenges presented in such critical environments. In this study, we establish an air-to-ground communication model, subsequently devising a strategy for user dynamics. This is followed by the introduction of a joint scheduling process for path and resource allocation, named ISATR (iterative scheduling algorithm of trajectory and resource). This process encompasses highly interdependent decision variables, such as location, bandwidth, and power resources. For mobile ground users, we employ the cellular automata (CA) method to forecast the evacuation trajectory. This algorithm successfully maintains data communication in the emergency-stricken area and enhances the communication quality through bandwidth division and power control which varies with time. The effectiveness of our algorithm is validated by evaluating the average throughput with different parameters in various simulation conditions and by using several heuristic methods as a contrast.

Keywords: unmanned aerial vehicles; resource allocation; trajectory planning; iterative scheduling; cellular automata

Citation: Zhang, Z.; Wang, Y.; Luo, Y.; Zhang, H.; Zhang, X.; Ding, W. Iterative Trajectory Planning and Resource Allocation for UAV-Assisted Emergency Communication with User Dynamics. *Drones* **2024**, *8*, 149. https://doi.org/10.3390/drones8040149

Academic Editor: Emmanouel T. Michailidis

Received: 13 March 2024
Revised: 7 April 2024
Accepted: 9 April 2024
Published: 11 April 2024

Copyright: © 2024 by the authors. Licensee MDPI, Basel, Switzerland. This article is an open access article distributed under the terms and conditions of the Creative Commons Attribution (CC BY) license (https://creativecommons.org/licenses/by/4.0/).

1. Introduction

UAV-assisted mobile communication takes the role of an efficient technology that uses unmanned aerial vehicles (UAVs) as communication nodes in wireless networks, with UAVs performing as aerial base stations, communication relays, and data connection stations. In research on B5G/6G communication, UAVs have already been widely applied [1]. They can provide enhanced coverage, capacity, and connectivity for applications in various communication scenarios, such as surveillance management [2], smart agriculture [3], and aerial delivery [4]. Drone-assisted IoT (the Internet of Things) systems are also called IoD (Internet of Drones) [5]. The authors in [6] have demonstrated the data collection ability of drones, which is also applied in emergency [7] or MEC cases [8]. More than a data transmitting node, a UAV can also serve as a cloud computation center with limited ground processing capability [9]. In other applications, UAVs provide sensing [10], target search [11], and healthcare supply service [12,13] based on their mobility.

In recent years, the integration of UAVs into emergency communication systems has caught significant attention due to their ability to overcome limitations in traditional ground base stations [14]. In emergency-stricken cases, disasters such as earthquakes, hurricanes, and wildfires can inflict severe damage on ground communication infrastructure, making them either disabled or inaccessible. Consequently, timely and reliable communication services in such contexts play a vital role in rescue and response. Both civil and military institutions can benefit from UAV-assistance communication. Emergency response agencies like FEMA (Federal Emergency Management Agency) would be interested in leveraging UAVs to enhance communication capabilities during disasters or emergencies, while telecommunication companies could benefit from UAV-assisted communication planning algorithms to improve their capacity for assisting in affected areas. During the 2021 Henan flood, China Mobile dispatched Yilong drones to temporarily restore communication in the disaster area. UAV-assisted mobile communication has several advantages over traditional terrestrial or satellite-based communication systems, such as flexibility, mobility, scalability, ease of deployment, and low cost, therefore they can match the requirements needed in emergency communication cases.

However, the field also encounters many challenges, especially in intricate problems. In UAV deployment, costs of moving and energy consumption need to be considered, which provide constraints on assistance quality optimization. In problems of trajectory planning or routing, the communication environment changes rapidly, resulting in imperfect channel state information. Coordination, security, and power management also matter in related problems, making the algorithm design rather complicated [9]. Generally speaking, research on UAV-assisted communication mainly involves two aspects: one is the establishment of an air-ground communication model, and the other is the design and optimization of drone scheduling algorithms.

1.1. Related Works

Differing from conventional ground-based communication, air-to-ground (A2G)) communication is subject to the influence of altitude differentials, thus presenting a more complex and dynamic model of the environment [15]. In recent years, many researchers have focused on the design of evaluating indicators, along with communication models in both general and certain environments [16] for air-to-ground channels. To assess the effectiveness and robustness of UAV-assisted networks, various metrics have been proposed for A2G networks. The authors in [17] considered energy efficiency, while [18] considered throughput. In [19], the average completion time of subtasks was applied for assessment. Other evaluating factors include the arrival rate, spectral efficiency, and channel capacity, while factors such as time delay, coverage [20], and outage probability also affect A2G networks.

Communication parameters of A2G channels differ in typical scenes involving different environments such as urban, dense urban, suburban, etc. In [21], researchers developed the A2G path loss model in the urban environment, while [22] focused on multilink channel model analysis at 2.4 and 5.9 GHz, both in low altitude circumstance. For aerial sensor networks, [23] introduced a realistic channel model leveraging cooperative UAVs in order to reach maximum spatial exploration efficiency. In [24], the statistical characteristics of the airframe shadowing loss were further analyzed. In [25], the authors gave a spatially and temporally correlated A2G channel model in cellular-connected UAV swarms, as well as a design for performance analysis. Ref. [26] considered atmospheric refractivity and precipitation, and they obtained path loss along the range and altitude. Moreover, [27] proposed a clustering method to analyze time-varying channels.

Blessed with auto-mobility and self-decision ability, UAVs can usually perform as aerial base stations in emergency cases when ground base stations are blocked. To execute communication assistance missions, researchers need to design trajectory planning and resource allocation algorithms for UAVs' scheduling. In the field of UAV trajectory planning, researchers have proposed various algorithms and methods to address path-planning problems in different environments. Heuristic-based methods employ heuristic algorithms such as genetic algorithms, simulated annealing, etc., to find optimal paths. Heuristic-based methods are often suitable for complex environments but may exhibit lower efficiency for large-scale problems. Ref. [28] introduced a 3-D path planning method improved from ant colony optimization, and Ref. [29] searched the UAV configuration space with a modified Mayfly algorithm. For collision avoidance, the slime mold algorithm (SMA) performs well with a design preventing it from local optimization points [30]. Inspired by a genetic algorithm, [31] proposed the ANSGA-III method with enhanced planning ability in complex environments. Graph theory-based methods model the environment as a graph and use classical graph algorithms to find the shortest or optimal paths, which also tend to perform well in simple environments. Ref. [32] discussed an approach of dynamic coloring for UAV planning in emergency cases, and [33] proposed a 3-D deployment method based on Dijkstra's algorithm, with UAV playing both as an aerial base station and relays. The idea of TSP (Traveling Salesman Problem) was applied in [18] with classified stressed regions, while [34] combined graph theory and convex optimization. Moreover, recent years have witnessed remarkable advancements in path planning facilitated by deep learning methodologies. Convolutional neural networks (CNNs) and recurrent neural networks (RNNs) have been employed to predict flight trajectories and have shown impressive performance in complex environments. Deep reinforcement learning (DRL) techniques have also been harnessed for this purpose, further contributing to enhanced performance. The trajectory planning method varies with the environment. Path planning for UAVs in windy environments was proposed in [35], with simulated moving targets for UAVs to pursue. Ref. [36] took advantage of the traditional collision avoidance method and DRL method, resulting in long trajectory planning with unknown obstacles. In [37], UAVs serve in a warehouse for stock inventory, updating real-time paths with image recording.

Resource allocation constitutes another critical aspect of ensuring UAV assistance. The efficiency of UAVs is often limited by their battery life, prompting research into spectrum resource allocation and efficient energy management [38], including task scheduling and dynamic recharging strategies. Given that the resource allocation problem is NP-hard, researchers strive for sub-optimal solutions using methods from convex optimization, machine learning, etc. Zhang et al., in [39], proposed a safe-DQN method to optimize UAV trajectories, considering constraints such as user equipment (UE) energy limits and obstacles in emergency scenarios. Furthermore, for target assignment in multi-object scenes, multi-agent reinforcement learning (MARL) has been demonstrated to be effective, as shown in [40].

In the context of an evacuation, the trajectory of moving ground users requires statistical data or simulation for accurate consideration [41]. Since the moving behavior of users varies with the environment, cellular automata can be employed to adjust their trajectories. Cellular automata model the map as a two-dimensional grid space and assign different values to different grids to represent users, obstacles, exits, etc. Therefore, the state transition of certain grids can be predicted according to the states of their neighboring grids. Ref. [42] utilized cellular automata in forest fire spread prediction, specified influencing factors to adjust cell state and cell transition rules, and gave a 3D visualization for the fire spread model. In [43], cellular automata and the ant colony algorithm were used to optimize the evacuation model, which is applicable in emergency scenarios.

1.2. Our Contributions

To address the challenges in the communication of emergency cases, we propose an iterative scheduling algorithm for trajectory and resource (ISATR) for path planning and limited resource allocation in a UAV-assisted emergency communication network. In addition to commonly discussed variables, we consider the dynamics of ground users and present a comprehensive approach using a quasi-convex method to optimize UAV path, power, and bandwidth allocation across different time slots. This approach spans dimensions of time, space, spectrum, and energy; therefore, it can provide a rather accurate and comprehensive plan. The main contributions of this paper are summarized as follows:

1. UAV-assisted communication model with the dynamic environment. For emergency cases, few researchers test their algorithm with dynamic users. We have established an A2G communication model with moving ground users, where the energy consumption and assistance communication quality are jointly optimized.
2. Dynamic bandwidth allocation. For resource allocation in UAV-assisted communication, few researchers focus on bandwidth, due to its high complexity. Our work tackles dynamic bandwidth allocation, providing an algorithm for real UAV planning.
3. Designed iterative algorithm. For the NP-hard optimization problem of UAV planning in emergency communication, we leverage the idea of subtask iterative algorithm and work out an effective iterative scheduling algorithm of trajectory and resource.
4. Simulation analysis. Experiments are implemented to evaluate the effectiveness of the proposed optimization algorithm, which achieves obvious performance compared with non-optimized and several other methods and can maintain the performance in different environments.

Benefiting from its high accuracy, ISATR can serve as a necessary baseline in case of emergency communication situations, applied for a pre-planning scheme derived before emergency strikes.

This paper is organized as follows:

*   The A2G communication model, user moving strategy, and mathematical optimization model are established in Section 2.
*   The iterative scheduling algorithm for trajectory and resource (ISATR) is elaborated on in Section 3.
*   In Section 4, the results and performance are discussed, with comparisons in different environments.
*   Section 5 concludes the paper.

2. Modeling of UAV-Assisted Communication

2.1. UAV Air-to-Ground Channel Model

In dense urban environments, communication is primarily supported by ground-based stations serving mobile users distributed across cellular networks. However, during certain emergencies such as earthquakes, urban communication infrastructure may fail. When the ground-based station in the earthquake-stricken area malfunctions, it results in the loss of signal coverage within the community. This disruption complicates communication for ground users, leading to panic and impeding emergency relief efforts. Therefore, unmanned aerial vehicles (UAVs) are introduced as temporary aerial base stations to provide communication services in emergency scenarios. In this section, we present the basic model of UAV-assisted A2G communication.

When the ground base station malfunctions, the UAV can be utilized as an aerial base station to serve urgent communication. Before establishing a UAV-assisted emergency communication model, it is necessary to discuss the air-to-ground channel. In A2G communication, there are LoS (line of sight) paths and NLoS (not line of sight) paths, in which the NLoS is obstructed by obvious obstacles. While A2G communication can enjoy LoS channel in common cases, NLoS channel occurs when obstacles exist. As Figure 1 demonstrates, urban architecture and natural landscape both perform as obstacles in UAV-assisted communication. The occurrence probability for LoS/NLoS channel differs according to the location of UAV and the density of obstacles.

It can be seen that the ratio of LoS and NLoS path varies with the height of UAV. Therefore, a probabilistic LoS/NLoS channel model is applied [21] with parameter p_{LoS} to denote the occurrence probability of LoS channel and p_{NLoS} for that in NLoS channel. In consideration of varied environmental factors, p_{LoS} and p_{NLoS} for a single A2G link can be calculated by Equations (1) and (2).

$$p_{LOS} = \alpha \left(\frac{180\theta}{\pi} - 15 \right)^{\gamma}, \tag{1}$$

$$p_{NLOS} = 1 - \alpha \left(\frac{180\theta}{\pi} - 15 \right)^{\gamma}, \tag{2}$$

in which θ refers to the elevation angle between each user equipment and UAV, reflecting the impact of UAV height and position on the link. Parameters α and γ are influenced by environmental characteristics at the same time.

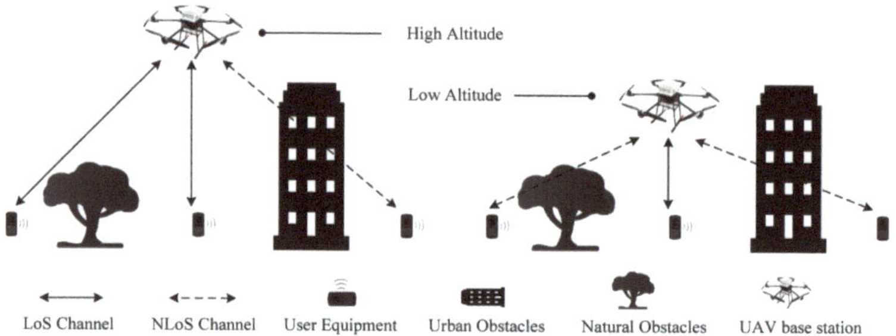

Figure 1. LoS/NLoS A2G channels in UAV-assisted communication. The model consists of a UAV aerial base station, ground user equipment, urban obstacles, and natural obstacles, which can influence LoS probability and then determine the A2G channel.

In the discussed earthquake-stricken cellular cells, mobile users evacuate from buildings and move toward certain exits. The ground base station breaks, interrupting communication. To plan a UAV for urgent communication, we need to model the environment of the cell, and then adjust the UAV's resources and location to maximize communication quality.

As Figure 2 displays, the path of a UAV can be planned according to ground users' evacuation trajectory in order to enhance communication quality and increase coverage.

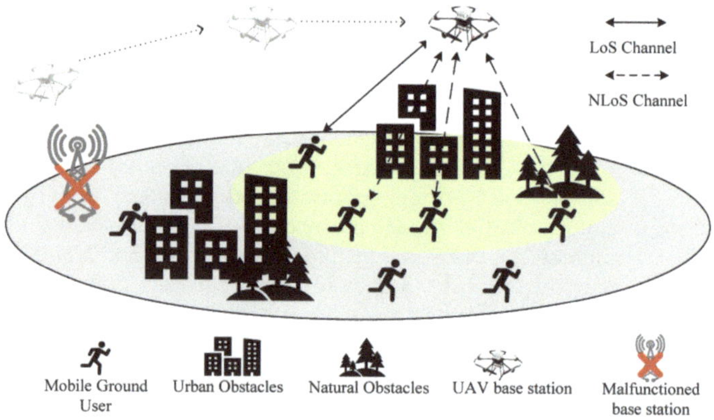

Figure 2. UAV-assisted emergency communication with malfunctioned ground base station and moving users. The model considers obstacles that impact both the trajectory of users and the signal transmission with the UAV base station, and our research is dedicated to the scheduling of UAV trajectory and resources to optimize the efficiency of communication during the evacuation.

The LoS path of UAV is obstructed by several obstacles, including compound buildings and greenery such as trees. Buildings are also regarded as obstacles for mobile users to evacuate from. Moreover, at each communication timeslot, transmission power and spectrum division of the UAV are determined.

As the proposed problem mainly focuses on the path planning and resource allocation algorithm of the UAV base station, a number of assumptions are introduced to simplify the model with no influence on algorithm construction. The UAV and valid moving ground users are constrained in the area of earthquake-stricken cells, which means one user turns invalid as soon as he/she runs out of the exit. When optimizing the trajectory, the height of UAV is fixed in each experiment. The FDMA (frequency division multiple access) technique is applied; thus, there is no interference between ground users and between ground users to drones.

To build up this UAV-assisted communication model, specifically defined parameters, formulas, and functions are introduced as follows. Briefly speaking, air-to-ground communication quality is used as an evaluation index and expressed in terms of system throughput.

Communication throughput depends on the arrival rate of bi-directional links. Therefore, transmission power of uplink and downlink channels in A2G communication can be calculated, followed by the corresponding SINR (Signal to Interference plus Noise Ratio). P_{ug} denotes received power at ground users in downlink communication, with P_{gu} defined for received power at UAV in uplink mode. Taking the link parameter into consideration, $P(u, g_j)$ describes transmission power between link j, in which u refers to UAV and g_j refers to j^{th} ground user. The probabilistic LoS/NLoS space power propagation model is clarified in (3) and (4), which defines transmission power P, received gain G, euclidean distance d, and environmental loss parameter k_0. P_{LoS} and P_{NLoS} refer to the occurrence possibility of LoS and NLoS channel, while ϕ_{LoS} and ϕ_{NLoS} refer to the corresponding shadow parameters.

$$P_{ug}(u, g_j) = \frac{P_u G(d_j)}{(k_0 d_{ij})^n (p_{LoS}\phi_{LoS} + p_{NLoS}\phi_{NLOS})}. \tag{3}$$

$$P_{gu}(u, g_j) = \frac{P_g G_0}{(k_0 d_{ij})^n (p_{LoS}\phi_{LoS} + p_{NLoS}\phi_{NLOS})}. \tag{4}$$

$$SINR_{ug} = \frac{P_{ug}(u_i, e_j)}{\sigma^2}, SINR_{gu} = \frac{P_{gu}(u_i, e_j)}{\sigma^2}. \tag{5}$$

Finally, the throughput is calculated through Shannon's law as follows, as the considered communication occurs in the channel with AWGN (Additive White Gaussian Noise).

$$C = B \times log_2(1 + SINR). \tag{6}$$

2.2. User Trajectory Prediction Model

Before conducting optimization for UAV path planning and resource allocation, it is essential to define the ground users' moving trajectories as the initial input data for optimizing UAV strategies, as we consider user dynamics. In this context, we introduce the cellular automata (CA) method as a means to simulate and predict the users' moving trajectories.

Cellular automata (CA) is a discrete grid-based dynamic model that encompasses discrete representations of time, space, and state variables. Notably, it exhibits a localized spatial interaction and temporal causality, enabling it to simulate the spatiotemporal evolution process of intricate systems. CA methods have been found with extensive applications, including fire spread simulations and other domains. Due to its capability to achieve a balance between accuracy and efficiency, CA is also well-suited for simulating evacuation scenarios.

The CA model is kind of a multi-dimensional dynamic programming method to some degree, as shown in Figure 3. Grids represent states at each location and are influenced by neighbor grids based on certain transition probability matrices. With the idea of the CA model, the method for predicting users' trajectories is then derived. To simplify the evacuation model, we suppose all the mobile users in the earthquake-stricken compound have already left their residential buildings and, therefore, are initially located within so-called valid areas. Once a user successfully escapes from the evacuation exit, his or her location becomes invalid and is no longer considered in the calculation of the overall communication throughput.

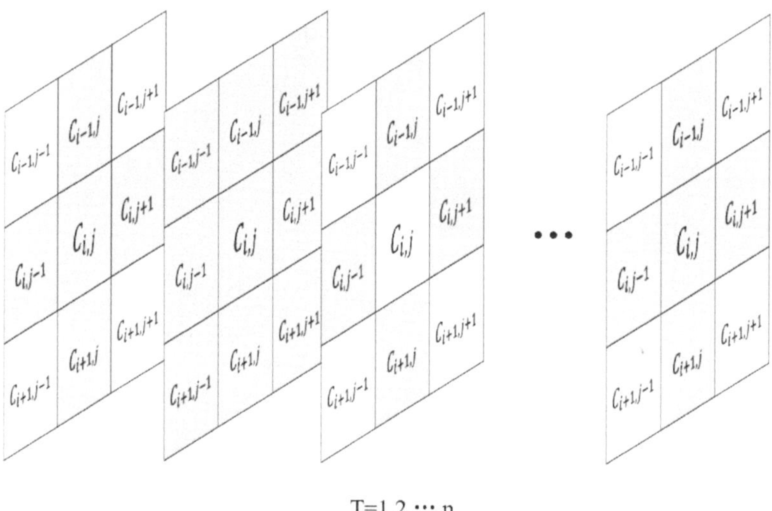

T=1,2,···,n

Figure 3. Cellular automata schemes. States of grids change according to states of neighbors; thus, they can predict data along time sequence with an initial input. CA model is applied to generate the evacuation trajectory data of users.

Considering that the moving ground users are residents of the neighborhood, it is reasonable to assume that they possess knowledge of the map, including the distribution of obstacles and the location of exits. Accordingly, it can be supposed that the moving trajectory of each ground user follows the shortest path from their current location to the evacuation exit.

Suppose there are overall K users placed randomly in the valid area of the cellular zone, with only one exit, while residential buildings play as obstacles in the area. The environment is modeled as a 2-dimensional grid map, and the moving direction of the ground user can be defined in a discrete direction set D = {east, south, west, north, northeast, southeast, northwest, southwest}, mapping to numbers 0 to 7.

Mobile users select their moving direction from the direction set D according to the current location, and adjust moving speed regarding current crowd density, namely the influence of neighbor grids. The moving speed affected by the current density of adjacent users is defined in the following equation:

$$v = v_0/\rho \ (m/s),\tag{7}$$

in which v_0 refers to the typical moving speed of humans with no obstacles and neighbors. ρ changes with the current number of users in a neighbor range, being an integer no less than 1 m/s. Therefore the state of any grid can be initialized and then transformed into a next state step by step, until reaching convergence.

By this means, after gridding the current map, the evacuation trajectories of users with different initial distributions can be obtained through the CA method, and obtain the data of user positions at each time point. The gridded map is illustrated in Figure 4.

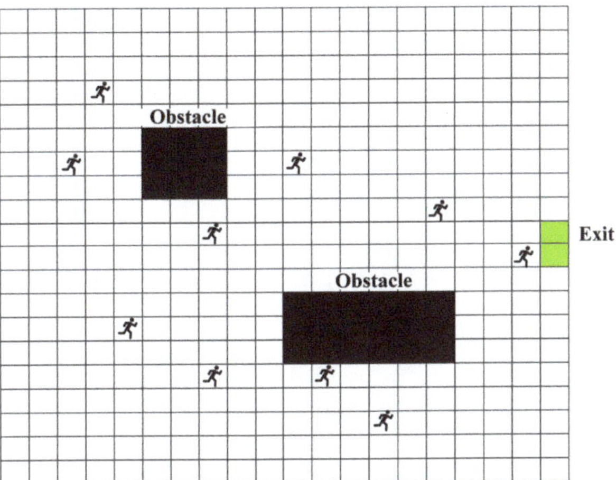

Figure 4. Grid map of valid evacuation area. The area includes obstacles, one exit, and 10 ground users in the simulation. Each grid equals a square area of 1 m × 1 m in actual simulation.

2.3. Optimization Mathematical Model

In the UAV-assisted emergency communication discussed above, we propose the air-to-ground communication model and simulation of the user evacuation trajectory. The optimization mathematical functions are defined in this section, followed by complexity analysis.

The main task is to recover the interrupted service and enhance communication quality; thus, we set the total throughput as the objective function in optimization, as shown in Equation (8). In the planning model with K users and N time slots, user location, bandwidth division, and transmission power are considered decision variables. Both uplink and downlink communication happens, regarding the UAV as the aerial base station.

$$R_{i,n} = \frac{1}{2} \cdot B_n \cdot [log_2(1 + SINR_{ug}) + log_2(1 + SINR_{gu})]. \tag{8}$$

Subscript i represents the current moving user and subscript n represents the current time slot. Moreover, design for constraint functions is also necessary. In the above discussions, it can be seen that the objective function consists of the bi-directional arriving rate $R_{i,n}$ at each resource block. On one side, as UAVs have real physical characteristics, the energy, velocity, and power are limited. On the other, the communication resource is also limited, involving maximum bandwidth.

Therefore, the constraint variables are divided into 3 sets, which are $U = [v_n, a_n]$, $B = [B_n]$, and $P = [E_n^C, E_n^F, P_n]$, referring to variables of UAV's location and velocity, variables of bandwidth resource, and variables of energy and power consumption. E_n^C represents communication energy cost at time slot n, and E_n^F refers to flight consumption. To ensure data communication, the lower bound of R_i is also specified. The total optimization functions are shown in Equation (9),

$$\max_{UBP} \sum_n^N \sum_i^K R_{i,n}$$

$$s.t. \begin{cases} 0 \leq P_u \leq P_{max} \\ \sum_i^C E_{i,n} + E_n^F \leq E_{max} \\ \sum B_{i,n} \leq B_i \\ v_n \leq v_{max} \\ R_{min} \leq R_i \end{cases} \tag{9}$$

The complexity of the discussed problem is proved to be NP-hard (non-deterministic polynomial-time hard) in this section, which can reveal the significance and effectiveness of our algorithm.

NP means a problem can not be resolved in polynomial time, and all the NP problems can reduced to the NP-hard problem. Reduction is a reversible process, which means that the solvers of both problems can be transformed into one. Therefore, any certain problem can be proved NP-hard if it can be reduced to other typical NP-hard problems.

According to the list of proven NPH problems, the bounded knapsack problem (BKP) is a NPH model with a basic optimization model:

$$\max \sum_{i=1}^{n} v_i x_i$$

$$s.t. \quad \left\{ \sum_{i=1}^{n} w_i x_i \leq W, \quad x_i \in 0, 1, 2, \ldots, c \right. \tag{10}$$

x_i refers to decision sequence, while w_i stands for weight cost and v_i represents value. Similarly, in our problem, the decision sequence for UAV aims to maximize total throughput with certain v_i factors, and weight constraints are controlled in the communication energy aspect. Thus, our problem can be reduced to a bounded knapsack problem, which indicates the NPH characteristic of our model. Therefore, the joint optimization problem discussed can only achieve an approximate solution, necessitating highly accurate algorithms.

3. Iterative Scheduling Algorithm for Throughput Optimization

To address the multi-objective optimization in the evacuation scenario, two primary stages require deliberation, after the establishment of the communication model outlined in the preceding section. The initial stage involves predicting unknown environmental information, specifically the trajectories of moving ground users. Subsequently, the second stage revolves around formulating an optimization algorithm for UAV to enhance total communication efficiency based on these predictions.

3.1. Algorithm Architecture of ISATR

In the non-linear optimization field, the coordinate descent method differs from the gradient descent method, as it searches for optimal values along all coordinates. The block coordinate descent method adds the problem division stage to the traditional coordinate descent method, which means it performs coordinate descent on several designed sub-problems.

$$\mathbf{x}_i^{(k)} = \arg\min_{\mathbf{x_i}} \ f(\mathbf{x}_1^{(k)}, \ldots, \mathbf{x}_{i-1}^{(k)}, x_i, \mathbf{x}_{i+1}^{(k-1)}, \ldots, \mathbf{x}_n^{(k-1)}). \tag{11}$$

As shown in (11) and Algorithm 1, the block coordinate descent algorithm groups all the variables into several blocks including $x_1, x_2, \ldots, x_n$, and then in each iteration, only the variables in one block are optimized, while the variables in the other blocks remain unchanged. By updating the variables in different blocks alternately, the objective function is finally reduced.

To tackle the throughput optimization problem with high complexity, we designed the ISATR (iterative scheduling algorithm of trajectory and resource) for this solution, inspired by the idea of BCD. The variables are divided into three categories, which are location, transmission power, and bandwidth allocation of UAV at all considered time slots.

The energy cost of flight consumption is calculated in Equation (12), followed by communication cost derived in Equation (13). E^C represents communication energy cost, with normalized emission power P_0 of unit space distance. E^F refers to flight consumption. Other relative variables and abbreviations are listed in Table 1.

$$E^C = \sum_{n}^{N} \sum_{i} P_0 \Delta T. \tag{12}$$

$$E^F = \sum_n^N mv_n^2. \tag{13}$$

Algorithm 1: Block coordinate descent algorithm

Data: Initial variables in n designed blocks $X = \{x_1^0, x_2^0, \ldots, x_n^0\}$
Result: Optimal $\{x_1, x_2, \ldots, x_n\}$
1 $X \leftarrow X0;$
2 **for** $k = 1, 2, \ldots$ **do**
3   **for** $i = 1, 2, \ldots$ **do**
4          $x_i^k \leftarrow x_i^k$, update x_i^k with other blocks fixed;
5   **end**
6 **end**
7 **if** stopping criteria satisfied **then**
8   return $\{x_1^k, x_2^k, \ldots, x_n^k\};$
9 **end**

Table 1. Abbreviation definitions.

Variable	Definition
m	Mass of UAV
u_n	UAV location at nth time slot
a_n	Flight direction of UAV at nth time slot
v_n	Flight velocity of UAV at nth time slot
v_{max}	Maximum flight speed of UAV
$e_{i,n}$	ith user's location at nth time slot
$P_{u,n}$	Transmission power of UAV at nth time slot
P_{max}	Upper bound of transmission power
$B_{i,n}$	Bandwidth allocated to ith user at nth time slot
B_i	Total bandwidth for A2G communication at time slot i
$E_{i,n}^C$	UAV communication power consumption with ith user at nth time slot
E_n^F	UAV flight energy consumption in nth time slot
E_{max}	Upper bound of total energy consumption
R_i	Throughput for communication with user i
R	Total throughput of A2G communication

The simulation process terminates when the last moving ground user has left the evacuation exit. Details of the iterative algorithm are clarified in Equation (14), followed by its pseudo-code Algorithm 2, with complexity also discussed.

$$\max_{UBP} \sum_n^N \sum_i^K R_{i,n}$$

$$s.t. \begin{cases} 0 \leq P_u \leq P_{max} \\ \sum_i^C E_{i,n} + E_n^F \leq E_{max} \\ \sum B_{i,n} \leq B_i \\ v_n \leq v_{max} \\ R_{min} \leq R_i \end{cases} \tag{14}$$

The complexity of ISATR is also derived. For the block of UAV trajectory planning, the time complexity is O(NK), where N represents the number of time steps and K denotes the number of users. In each time step, the algorithm iterates over each user to calculate the distance between the UAV and each user. This involves a loop nested within the time step loop, resulting in a time complexity proportional to the product of N and K. Moreover, constant-time calculations are performed to determine the signal-to-interference-plus-noise

---

Algorithm 2: ISATR (Iterative Scheduling Algorithm of Trajectory and Resource)

Data: Initial variables in designed blocks $X = \{x_U^0, x_B^0, x_P^0\}$
Result: Optimal sets $\{U, B, P\}$

1   $x_U^0 \leftarrow \{u_n\}, u_0 = [0, L/2], u_n = u_{n-1} + [v_0 \cdot cosa_0, v_0 \cdot sina_0]$;
2   $x_B \leftarrow \{B_i\}, B_i = B/K$;
3   $x_P \leftarrow \{P_0\}$;
4   **for** $t = 1, 2, \ldots$ **do**
5   **for** $k = 1, 2, \ldots, K$ **do**
6   **for** $i = 1, 2, \ldots, N$ **do**
7         $x_B \leftarrow x_B^{t-1}, x_P \leftarrow x_P^{t-1}$;
8   update $x_U^t(i, k)$ with other blocks fixed;
9         $x_U \leftarrow x_U^{t-1}, x_P \leftarrow x_P^{t-1}$;
10   update $x_B^t(i, k)$ with other blocks fixed;
11        $x_U \leftarrow x_U^{t-1}, x_B \leftarrow x_B^{t-1}$;
12   update $x_P^t(i, k)$ with other blocks fixed;
13   **end**
14   **end**
15   **end**
16   **if** stopping criteria satisfied **then**
17   return $\{x_U^T, x_B^T, x_P^T\}$;
18   **end**

---

ratio (SINR) and to update the total throughput. Therefore, the time complexity of this block is O(NK). Similar to the UAV positions optimization, the time complexity of the other two blocks is also O(NK).

Therefore, the complexity of our method can be derived as O(NK), in which K represents user number, and N represents time slots, which can be regarded as $O(n^2)$.

3.2. UAV Trajectory Subtask Optimization

The first sub-optimization problem block focuses on UAV path planning, with full knowledge of external information acquired in the previous iteration round, including resource allocation and environmental states. The sum of R_i stands for the objective function, namely the total throughput of the A2G communication links.

$$
\begin{aligned}
Objevtive &= \sum R_i \\
&= \sum B * log_2(1 + SINR) \\
&= \sum B * log_2(1 + \frac{P_{ug}(u_i, e_j)}{\sigma^2}).
\end{aligned}
\tag{15}
$$

As the UAV location changes, the parameters of the A2G communication channel model and the distance between UAV base station and ground users also change, which can be determined by UAV position and known environmental information. Environmental parameters also depend on the UAV's location, thus influencing the SINR value of communication links. Therefore, when fixing the other two sub-optimization blocks, the objective function can be regarded as a function of the UAV's position vector.

To design the location optimization block, constraints of acceleration and energy limitation are considered, which are performed as unequal constraints.

$$\max_{u_n, v_n} \sum_i^K R_i$$

$$s.t. \begin{cases} \sum_n^N \sum_i E_{i,n}^C + \sum_n E_n^F \le E_{max} \\ 0 \le u_n \le [x_{max}, y_{max}] \\ -\pi \le a_n \le \pi \\ v_n \le v_{max} \end{cases} \tag{16}$$

Decision variables u_n and v_n represent sets of the two-dimensional location and speed of UAV base station at all time slots, while K and N represent the total number of moving ground users and communication time slots. ΔT is the interval of each time slot and P_0 counts for typical transmission power of UAV; therefore, $\sum_n^N \sum_i E_{i,n}^C$ represents energy consumption of communication, while $\sum_n^N E_n^F$ is flight energy consumption. u_n stands for location of UAV at time slot n, and a_n indicates the direction of UAV.

3.3. Transmission Power Subtask Optimization

When the location and bandwidth sub-optimization blocks are fixed, the sup-optimization problem of UAV transmission power consumption can be built by the same format. As transmission power multiplying P_{ug} term, it can influence the objective function:

$$\begin{aligned} Objevtive &= \sum R_i \\ &= \sum B * log_2(1 + \frac{P_{ug}(u_i, e_j)}{\sigma^2}) \\ &= \sum B * log_2(1 + \frac{P_u G(d_j)}{(k_0 d_{ij})^n (p_{LoS}\phi_{LoS} + p_{NLoS}\phi_{NLOS})} \cdot \frac{1}{\sigma^2}). \end{aligned} \tag{17}$$

At every time slot, the objective function R_i corresponding to ground user i can be calculated based on transmission power, effective communication bandwidth distributed to user i, and current UAV location. In the power optimization block, the initial bandwidth allocation and UAV moving strategy and fixed as follows, which are simply compliant with velocity and bandwidth constraints:

$$\begin{aligned} B_i &= B/K. \\ u_n &= u_{n-1} + [v_0 \cdot cosa_0, v_0 \cdot sina_0]. \end{aligned} \tag{18}$$

When other environmental conditions are static, communication efficiency can grow with transmission power. Thus, to optimize UAV transmission power $P_{u,n}$ at time slot n, necessary constraints need to be specified. Relative constraints focus on the upper range of transmission power and total energy consumption, and the sub-optimization problem is shown as follows:

$$\max_{P_{u,n}} \sum_n^N \sum_i^K R_{i,n}$$

$$s.t. \begin{cases} 0 \le P_{u,n} \le P_{max} \\ \sum_n^N P_{u,n}\Delta T + E_n^F \le E_{max} \end{cases} \tag{19}$$

Single decision variable $P_{u,n}$ denotes the transmission power of UAV at time slot n, while the upper bound is set as P_{max}. Energy constraints are consistent with that in the location optimization block discussed above, limiting transmission power from selecting the maximum at all times.

3.4. Communication Bandwidth Allocation Subtask Optimization

The third sub-optimization block pertains to the allocation of bandwidth, which encompasses a decision variable dimension of significant magnitude.

$$\begin{aligned} Objevtive \;\; &= \sum R_i \\ &= \sum B_i * log_2(1 + SINR). \end{aligned} \tag{20}$$

To ensure optimal communication efficiency for all users, it is imperative to guarantee that the allocated bandwidth does not fall below the minimal threshold required for the successful transmission of communication data, resulting in constraint on R_i. An equal constraint is also given in this sub-block, as the sum of allocated bandwidths needs to be no more than B, but with no waste.

$$\max_{B_i} \sum_n^N \sum_i^K R_{i,n}$$
$$s.t. \quad \begin{cases} R_{min} \leq R_i \\ \sum_i^K B_i = B. \end{cases} \tag{21}$$

Decision variable B_i refers to bandwidth resource allocated to user i at a certain time slot. After these three sub-optimization tasks iterate and eventually converge, a multi-stage planning scheme can be proposed as follows, with sets of decision variables including U, B, and P, regarding UAV location, bandwidth allocation, and power control strategy, respectively. After all, our designed multi-subtask algorithm is established, with three sub-block problem consisting the total optimization function.

$$\max_{UBP} \sum_n^N \sum_i^K R_{i,n}$$
$$s.t. \quad \begin{cases} 0 \leq P_u \leq P_{max} \\ \sum_i^C E_{i,n} + E_n^F \leq E_{max} \\ \sum B_{i,n} \leq B_i \\ v_n \leq v_{max} \\ R_{min} \leq R_i \end{cases} \tag{22}$$

4. Numerical Results and Analysis

In this section, we validate the efficiency and robustness of the proposed optimization method under different UAV heights and environments. The optimized planned trajectory and time sequence of communication resources are also demonstrated, under our assumptions introduced in the model establishment, and trajectory prediction of moving ground users. Figure 5 demonstrates path planning for several scenarios.

To train UAV planning, the trajectory of moving ground users needs specification at first. With the CA model mentioned above, we have obtained the predicted data of ground users' location along the time sequence successfully.

For a simple demonstration and test, the number of ground users is set as 10, and the range of earthquake-stricken area is rectangular with length = 100 m and width = 50 m, with several obstacles scattered in, resembling residential buildings. The typical speed of a moving user in adequate wide space is 5 m/s, with evacuation time discretized into a time series with an interval of 1 s. All the residents have already left architecture at the first second, with random locations in the valid areas. As time passes, all users move in the shortest path directed to escape. In this simulation, the evacuation is completed in 80 s, thus we obtain the predicted trajectory of moving ground users. The software interface of the simulation is shown in Figure 6.

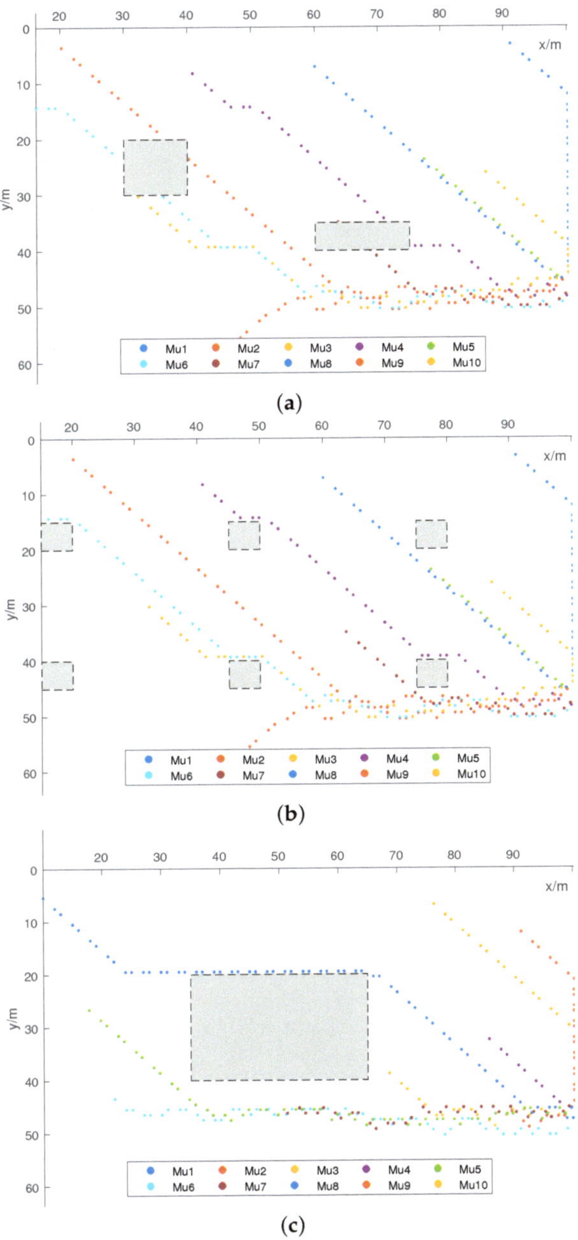

Figure 5. Trajectory of ground users predicted by CA model. The moving trajectories of considered users are depicted using scatter plots in different colors, showing that they effectively avoided obstacles and found the shortest possible paths to the exit, in different tested scenarios. (**a**) Predicted trajectory of moving users with 2 obstacles of different sizes. (**b**) The predicted trajectory of moving users with 6 scattered obstacles. (**c**) Predicted trajectory of moving users with 1 centered obstacle.

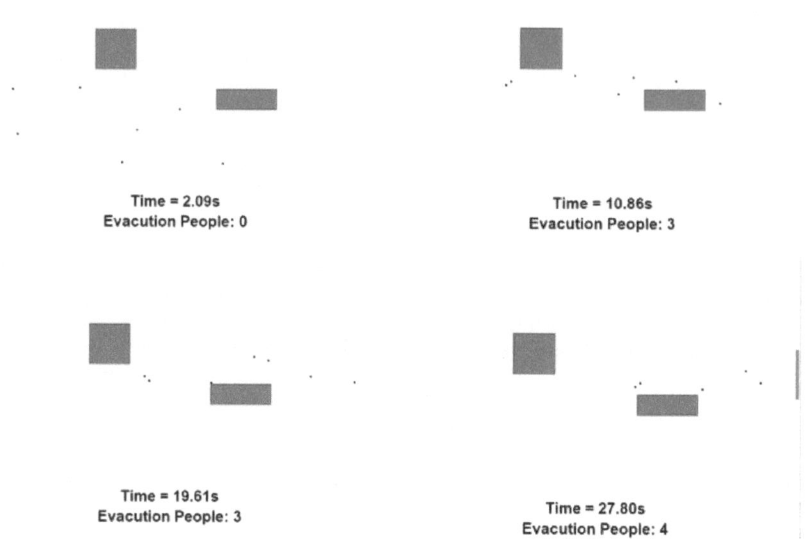

Figure 6. Software Illustration of ground users' moving trajectory simulation. As time passes, the number of evacuation people updates and provides a real-time demonstration of user evacuation movement based on CA simulation.

As the simulated results demonstrate, the trajectory of users can be predicated and the data can be applied to train ISATR later. We have generated different user trajectories to evaluate the adaptability of our proposed method, with scenarios varied in user distribution, obstacles, and exits.

4.1. Evaluation of UAV Trajectory planning

Based on the environmental information and user trajectories determined above, trajectory planning for the UAV base station has been achieved through the iterative optimization algorithm for trajectory sub-block problem, with parameters in the other two sub-block problems fixed.

As shown in Figure 7, the proposed trajectory planning method achieves the fastest convergence. Compared to several path planning methods including A* and genetic algorithm (GA), our optimized path planning strategy performs best in the discussed environment.

Visualization of the planned UAV trajectory is given in Figure 8, with trajectories of 10 moving users also displayed.

In each sub-graph, we release the aerial base station from different initial points and test the algorithm with different user distributions, finding that it maintains stability and moves synchronously with the trajectories of users toward the exit. In Figure 8a,b, there are two different-sized obstacle buildings in the environment, with the exit located at the center-right of the map. The UAVs are released from different initial positions, demonstrating their adaptability to different initial release positions. In Figure 8c,d, the exit is set at the bottom-right corner of the map. The scenarios with six dispersed obstacles and one central obstacle in the map are tested, thus validating the effectiveness of the algorithm in different scenarios.

Meanwhile, the number of inflection points on the convergence curve matches the number of users specified in the simulation, lending reasonable support to the optimization's correctness. Therefore, the effectiveness and robustness of the proposed trajectory planning algorithm are validated.

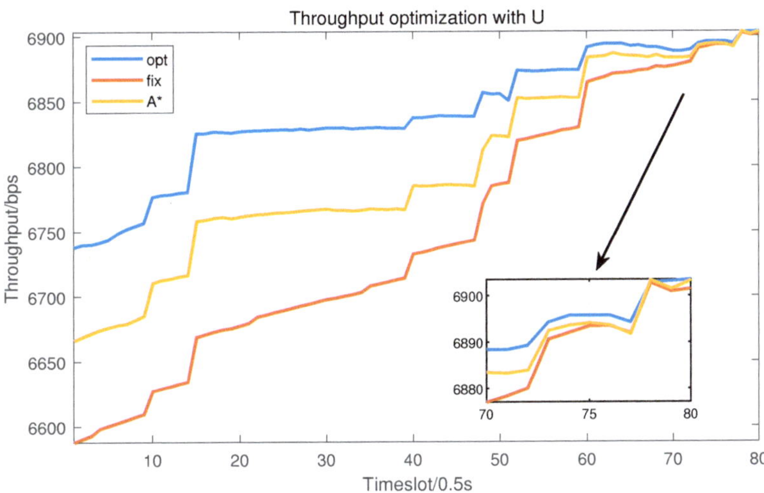

Figure 7. Throughput optimized with UAV trajectory in different methods. It can be found that our ISATR opt method outperforms the traditional A* algorithm and the algorithm with fixed UAV trajectory.

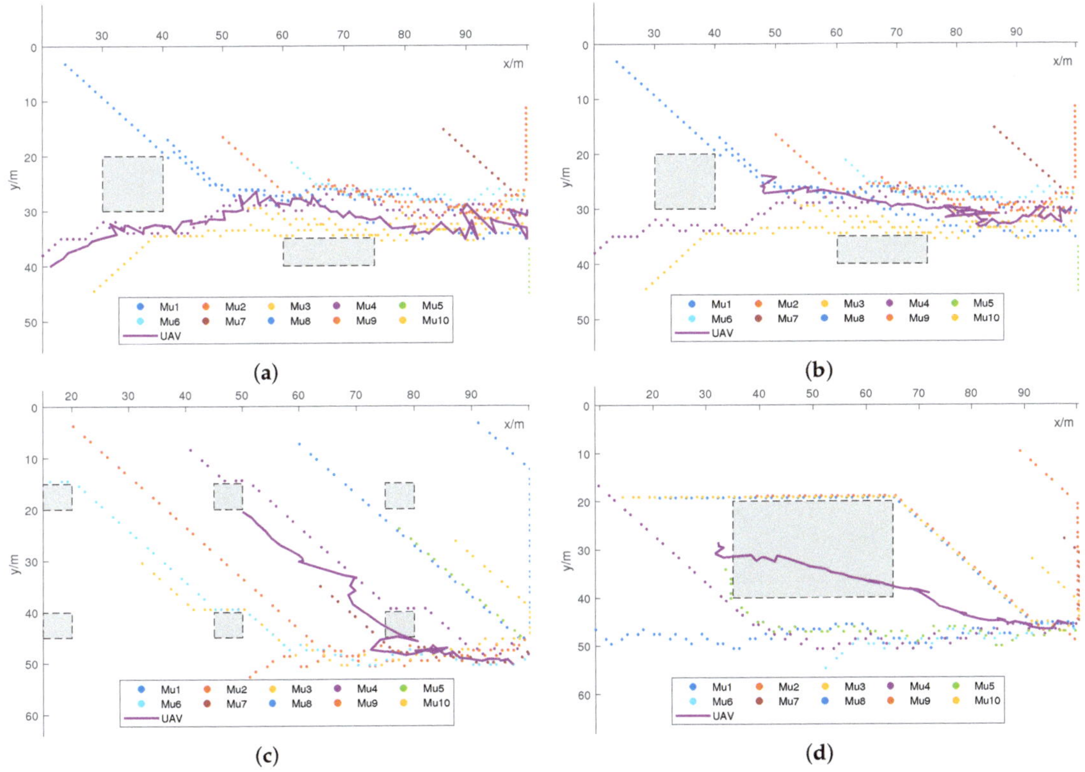

Figure 8. Trajectory planning illustration. Different initial positions for UAV trajectory optimization via ISATR are tested to verify the effectiveness and robustness of our algorithm. It can be seen that the trajectory of UAV is always consistent with users. (**a**) ISATR opted path 1. (**b**) ISATR opted path 2. (**c**) ISATR opted path 3. (**d**) ISATR opted path 4.

4.2. Evaluation of Resource Allocation

Besides path planning of the UAV base station, communication resource allocation also matters. Transmission power and bandwidth allocation for all ground users at different time slots are derived via our designed optimization algorithm.

It can be seen in Figures 9 and 10 that as time passes, the bandwidth allocation varies as the power of the UAV increases because the number of users in the valid area decreases. It is worth noting that there are 10 significant changes in P of the UAV base station, which is equal to the number of moving ground users, making it consistent with the logic of algorithm optimization.

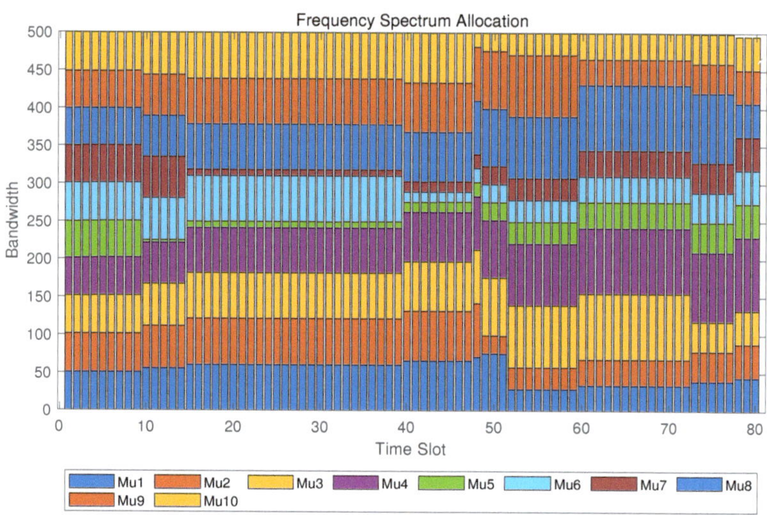

Figure 9. Bandwidth allocation illustration. Bandwidth allocation scheduling is influenced by the positions of UAV and users, as well as the power of UAV, also adhering to the minimum throughput constraint.

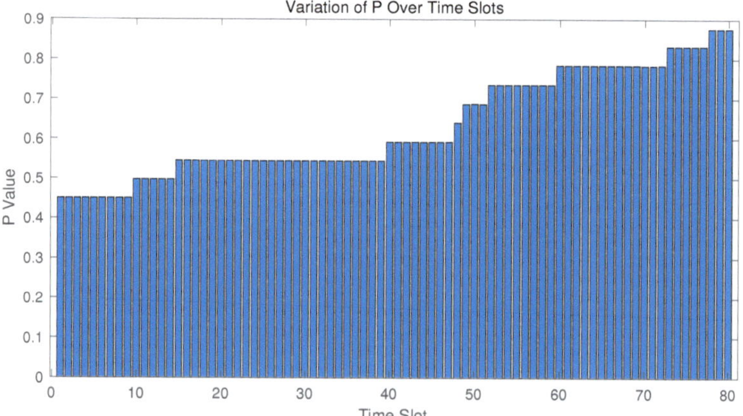

Figure 10. Power control illustration. UAV transmission power control is influenced by the positions of UAV and users, as well as the current bandwidth allocation, also adhering to the minimum throughput constraint.

4.3. Evaluation of Trajectory and Resource Joint Optimization

With UAV location, bandwidth allocation, and transmission power all optimized, total throughput in the communication system has an obvious increase, as illustrated in Figure 11.

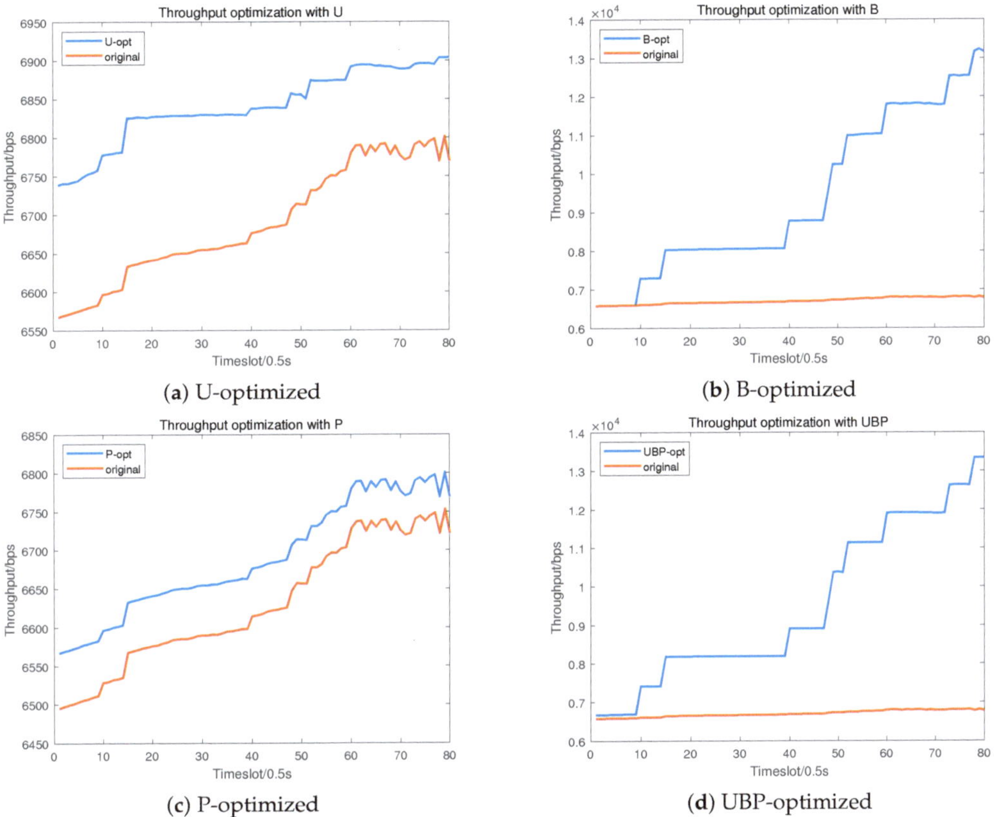

(a) U-optimized

(b) B-optimized

(c) P-optimized

(d) UBP-optimized

Figure 11. Throughput optimized with different decision variables. (**a**) Throughput with U (UAV location) optimized. (**b**) Throughput with B (bandwidth allocation) optimized. (**c**) Throughput with P (power of UAV) optimized. (**d**) Throughput with U, B, and P optimized.

The optimization of the three components, U, B, and P, can be seen to have all contributed to the improvement in throughput, with the combined optimization demonstrating better performance compared to the contrast algorithms. Among them, the optimization of B presents a step-like pattern, attributed to the consideration of the issue where ground users leaving the valid area are not involved in allocation in this algorithm. It can be observed that the number of steps is consistent with the total number of users.

Moreover, we select the average total throughput observed over the time sequence of simulation as the evaluation metric and compare our method with several alternative strategies, experimenting with various sets of environmental information. At different heights, the UAV adjusts the routing plan and resource allocation strategy according to the optimization functions, maintaining a stable optimization effect, as shown in Figure 12.

In Figure 13, the throughput optimization achieved by various methods is depicted, highlighting the superior performance of our ISATR method compared to the others. The compared methods include A*, GA, and fixed algorithm, in which "fixed" refers to the non-optimized case.

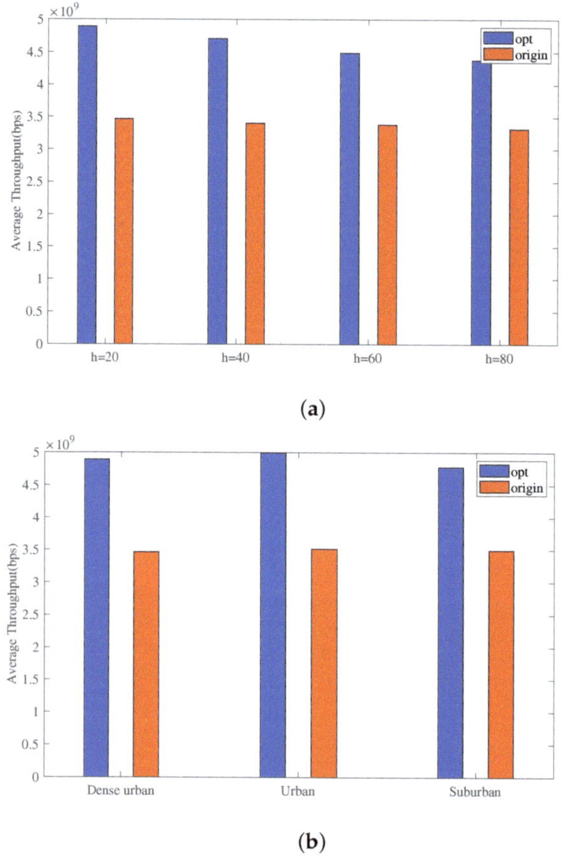

(a)

(b)

Figure 12. Average total throughput illustration in typical environments. (**a**) Throughput optimized at different heights. (**b**) Throughput optimized in different environments. Compared to the non-optimized value, the ISATR method has an increase of 42%.

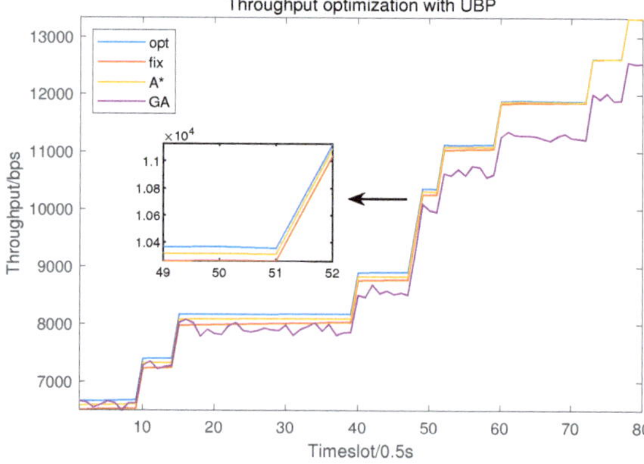

Figure 13. Throughput optimization with different methods. Throughput with location, bandwidth allocation, and power control optimized via different algorithms, including the proposed opt method (ISATR), GA method, A* method, and fixed case. ISATR outperforms the others.

5. Discussion

In the above sections, we have introduced an iterative algorithm ISATR to tackle the joint optimization problem of UAV-assisted communication, involving trajectory planning, power control, and bandwidth allocation. This proposed algorithm is dedicated to providing the pre-planning scheme of a UAV as a temporary base station in an emergency scenario. It can serve as a necessary baseline in case of emergencies, benefiting from its high accuracy.

Therefore, it is necessary to combine ISATR with other real-time algorithms, such as deep reinforcement learning (DRL), for practical implications. The main difference between ISATR and DRL is that their application scenarios are complementary. The former is applied to pre-planning which requires high accuracy, while the latter is applied to dynamic planning that requires real-time response. DRL-based UAV planning algorithm will be studied in future work to provide a dynamic response.

Physical limitations of UAVs are also essential in UAV-assisted systems. One notable drawback lies in the current weight and complexity of UAV systems, which can pose challenges in rapid deployment, particularly in emergency scenarios where swift action is imperative. To address these limitations, future research could focus on advancing lightweight UAV designs and streamlined deployment mechanisms. Integration of advanced materials and miniaturized components could significantly reduce the weight and size of UAVs, facilitating quick and agile deployment even in constrained environments.

6. Conclusions

This paper discusses a UAV-assisted communication scenario in an earthquake-stricken cellular cell. As the ground base station is devastated and blocked, a UAV is dispatched as a temporary aerial base station. An ISATR (iterative scheduling algorithm of trajectory and resource) is constructed to solve optimization questions to enhance the UAV's communication efficiency. A trajectory prediction model is derived via cellular automata and provides location data of ground users in evacuation for the UAV's decision. Path planning and resource allocation including bandwidth distribution and transmission power control are involved in the decision space, and the total throughput of A2G channels is considered as the objective function in optimization. With our designed multi-stage subtask iteration optimization algorithm, the total throughput is enhanced. Compared to the traditional optimization method GA and path planning method A*, our method has an advantage in higher optimization performance. Finally, we have an enhancement of approximately 40% total throughput tested in several typical environments, compared to non-optimized cases, which indicates that the proposed method can serve as an effective algorithm for pre-planning emergency UAV scheduling tasks.

Author Contributions: Conceptualization, Z.Z. and Y.L.; methodology, Z.Z., Y.W., and Y.L.; software and validation, Z.Z. and X.Z.; formal analysis, H.Z.; investigation, Z.Z. and W.D.; resources, Y.W.; data curation, Y.L.; writing—original draft preparation, Z.Z. and X.Z.; writing—review and editing, Y.L. and Y.W.; visualization, Z.Z. and H.Z.; supervision, W.D.; project administration, W.D.; funding acquisition, W.D. All authors have read and agreed to the published version of the manuscript.

Funding: This research was funded by National Natural Science Foundation of China (U20B2042).

Data Availability Statement: The data used to support this study have not been made available.

Conflicts of Interest: The authors declare no conflicts of interest.

References

1.   Zeng, Y.; Zhang, R.; Lim, T.J. Wireless communications with unmanned aerial vehicles: Opportunities and challenges. *IEEE Commun. Mag.* **2016** , *54*, 36–42. [CrossRef]
2.   Kim, H.; Ben-Othman, J. A Collision-Free Surveillance System Using Smart UAVs in Multi Domain IoT. *IEEE Commun. Lett.* **2018**, *22*, 2587–2590. [CrossRef]

3. Hasan, K.M.; Suhaili, W.S.; Shah Newaz, S.H.; Ahsan, M.S. Development of an Aircraft Type Portable Autonomous Drone for Agricultural Applications. In Proceedings of the 2020 International Conference on Computer Science and Its Application in Agriculture (ICOSICA), Bogor, Indonesia, 16–17 September 2020; pp. 1–5. [CrossRef]

4. Alsawy, A.; Hicks, A.; Moss, D.; Mckeever, S. An Image Processing Based Classifier to Support Safe Dropping for Delivery-by-Drone. In Proceedings of the 2022 IEEE 5th International Conference on Image Processing Applications and Systems (IPAS), Genova, Italy, 5–7 December 2022; Volume 5, pp. 1–5. [CrossRef]

5. Alrayes, F.S.; Alzahrani, J.S.; Alissa, K.A.; Alharbi, A.; Alshahrani, H.; Elfaki, M.A.; Yafoz, A.; Mohamed, A.; Hilal, A.M. Dwarf Mongoose Optimization-Based Secure Clustering with Routing Technique in Internet of Drones. *Drones* **2022**, *6*, 247. [CrossRef]

6. Samir, M.; Sharafeddine, S.; Assi, C.M.; Nguyen, T.M.; Ghrayeb, A. UAV Trajectory Planning for Data Collection from Time-Constrained IoT Devices. *IEEE Trans. Wirel. Commun.* **2020**, *19*, 34–46. [CrossRef]

7. Huang, Z.; Chen, C.; Pan, M. Multiobjective UAV Path Planning for Emergency Information Collection and Transmission. *IEEE Internet Things J.* **2020**, *7*, 6993–7009. [CrossRef]

8. Xu, J.; Ota, K.; Dong, M. Big Data on the Fly: UAV-Mounted Mobile Edge Computing for Disaster Management. *IEEE Trans. Netw. Sci. Eng.* **2020**, *7*, 2620–2630. [CrossRef]

9. Jeong, S.; Simeone, O.; Kang, J. Mobile Edge Computing via a UAV-Mounted Cloudlet: Optimization of Bit Allocation and Path Planning. *IEEE Trans. Veh. Technol.* **2018**, *67*, 2049–2063. [CrossRef]

10. Asad, M.; Aidaros, O.A.; Beg, R.; Dhahri, M.A.; Neyadi, S.A.; Hussein, M. Development of autonomous drone for gas sensing application. In Proceedings of the 2017 International Conference on Electrical and Computing Technologies and Applications (ICECTA), Ras Al Khaimah, United Arab Emirates, 21–23 November 2017; pp. 1–6. [CrossRef]

11. Wu, C.; Ju, B.; Wu, Y.; Lin, X.; Xiong, N.; Xu, G.; Li, H.; Liang, X. UAV Autonomous Target Search Based on Deep Reinforcement Learning in Complex Disaster Scene. *IEEE Access* **2019**, *7*, 117227–117245. [CrossRef]

12. Abeygunawaradana, P.; Gamage, N.; De Alwis, L.; Ashan, S.; Nilanka, C.; Godamune, P. E-Medic—Autonomous Drone for Healthcare System. In Proceedings of the 2021 International Conference on Computing, Communication, and Intelligent Systems (ICCCIS), Greater Noida, India, 19–20 February 2021; pp. 994–999. [CrossRef]

13. Bitar, A.; Jamal, A.; Sultan, H.; Alkandari, N.; El-Abd, M. Medical Drones System for Amusement Parks. In Proceedings of the 2017 IEEE/ACS 14th International Conference on Computer Systems and Applications (AICCSA), Hammamet, Tunisia, 30 October–3 November 2017; pp. 19–20. ISSN: 2161-5330 . [CrossRef]

14. Besada, J.A.; Bernardos, A.M.; Bergesio, L.; Vaquero, D.; Campaña, I.; Casar, J.R. Drones-as-a-service: A management architecture to provide mission planning, resource brokerage and operation support for fleets of drones. In Proceedings of the 2019 IEEE International Conference on Pervasive Computing and Communications Workshops (PerCom Workshops), Kyoto, Japan, 11–15 March 2019; pp. 931–936. [CrossRef]

15. Yan, C.; Fu, L.; Zhang, J.; Wang, J. A Comprehensive Survey on UAV Communication Channel Modeling. *IEEE Access* **2019**, *7*, 107769–107792. [CrossRef]

16. Khawaja, W.; Guvenc, I.; Matolak, D.W.; Fiebig, U.C.; Schneckenburger, N. A Survey of Air-to-Ground Propagation Channel Modeling for Unmanned Aerial Vehicles. *IEEE Commun. Surv. Tutorials* **2019**, *21*, 2361–2391. [CrossRef]

17. Alsamhi, S.H.; Shvetsov, A.V.; Shvetsova, S.V.; Hawbani, A.; Guizani, M.; Alhartomi, M.A.; Ma, O. Blockchain-Empowered Security and Energy Efficiency of Drone Swarm Consensus for Environment Exploration. *IEEE Trans. Green Commun. Netw.* **2023**, *7*, 328–338. [CrossRef]

18. Srivastava, K.; Pandey, P.C.; Sharma, J.K. An Approach for Route Optimization in Applications of Precision Agriculture Using UAVs. *Drones* **2020**, *4*, 58. [CrossRef]

19. Luan, Q.; Cui, H.; Zhang, L.; Lv, Z. A Hierarchical Hybrid Subtask Scheduling Algorithm in UAV-Assisted MEC Emergency Network. *IEEE Internet Things J.* **2022**, *9*, 12737–12753. [CrossRef]

20. Cabreira, T.M.; Brisolara, L.B.; Ferreira, P.R., Jr. Survey on Coverage Path Planning with Unmanned Aerial Vehicles. *Drones* **2019**, *3*, 4. [CrossRef]

21. Al-Hourani, A.; Kandeepan, S.; Jamalipour, A. Modeling air-to-ground path loss for low altitude platforms in urban environments. In Proceedings of the 2014 IEEE Global Communications Conference, Austin, TX, USA, 8–12 December 2014; pp. 2898–2904. ISSN: 1930-529X. [CrossRef]

22. Lyu, Y.; Wang, W.; Sun, Y.; Yue, H.; Chai, J. Low-Altitude UAV Air-to-Ground Multilink Channel Modeling and Analysis at 2.4 and 5.9 GHz. *IEEE Antennas Wirel. Propag. Lett.* **2023**, *22*, 2135–2139. [CrossRef]

23. Goddemeier, N.; Daniel, K.; Wietfeld, C. Role-Based Connectivity Management with Realistic Air-to-Ground Channels for Cooperative UAVs. *IEEE J. Sel. Areas Commun.* **2012**, *30*, 951–963. [CrossRef]

24. Ge, C.; Zhai, D.; Jiang, Y.; Zhang, R.; Yang, X.; Li, B.; Tang, X. Pathloss and Airframe Shadowing Loss of Air-to-Ground UAV Channel in the Airport Area at UHF- and L-Band. *IEEE Trans. Veh. Technol.* **2023**, *72*, 8094–8098. [CrossRef]

25. Li, H.; Ding, L.; Wang, Y.; Wang, Z. Air-to-Ground Channel Modeling and Performance Analysis for Cellular-Connected UAV Swarm. *IEEE Commun. Lett.* **2023**, *27*, 2172–2176. [CrossRef]

26. Wang, S.; Lim, T.H.; Choo, H. Path Loss Analysis Considering Atmospheric Refractivity and Precipitation for Air-to-Ground Radar. *IEEE Antennas Wirel. Propag. Lett.* **2021**, *20*, 1968–1972. [CrossRef]

27. Cui, Z.; Guan, K.; Oestges, C.; Briso-Rodríguez, C.; Ai, B.; Zhong, Z. Cluster-Based Characterization and Modeling for UAV Air-to-Ground Time-Varying Channels. *IEEE Trans. Veh. Technol.* **2022**, *71*, 6872–6883. [CrossRef]

28. Wan, Y.; Zhong, Y.; Ma, A.; Zhang, L. An Accurate UAV 3-D Path Planning Method for Disaster Emergency Response Based on an Improved Multiobjective Swarm Intelligence Algorithm. *IEEE Trans. Cybern.* **2023**, *53*, 2658–2671. [CrossRef]
29. Wang, X.; Pan, J.; Yang, Q.; Kong, L.; Snášel, V.; Chu, S. Modified Mayfly Algorithm for UAV Path Planning. *Drones* **2022**, *6*, 134. [CrossRef]
30. Zheng, L.; Tian, Y.; Wang, H.; Hong, C.; Li, B. Path Planning of Autonomous Mobile Robots Based on an Improved Slime Mould Algorithm. *Drones* **2023**, *7*, 257. [CrossRef]
31. Shen, Y.; Zhu, Y.; Kang, H.; Sun, X.; Chen, Q.; Wang, D. UAV Path Planning Based on Multi-Stage Constraint Optimization. *Drones* **2021**, *5*, 144. [CrossRef]
32. Wang, B.; Sun, Y.; Sun, Z.; Nguyen, L.D.; Duong, T.Q. UAV-Assisted Emergency Communications in Social IoT: A Dynamic Hypergraph Coloring Approach. *IEEE Internet Things J.* **2020**, *7*, 7663–7677. [CrossRef]
33. Prasad, N.L.; Ramkumar, B. 3-D Deployment and Trajectory Planning for Relay Based UAV Assisted Cooperative Communication for Emergency Scenarios Using Dijkstra's Algorithm. *IEEE Trans. Veh. Technol.* **2023**, *72*, 5049–5063. [CrossRef]
34. Zhang, S.; Zeng, Y.; Zhang, R. Cellular-Enabled UAV Communication: A Connectivity-Constrained Trajectory Optimization Perspective. *IEEE Trans. Commun.* **2019**, *67*, 2580–2604. [CrossRef]
35. Jayaweera, H.; Hanoun, S. Path Planning of Unmanned Aerial Vehicles (UAVs) in Windy Environments. *Drones* **2022**, *6*, 101. [CrossRef]
36. Zhang, S.; Li, Y.; Ye, F.; Geng, X.; Zhou, Z.; Shi, T. A Hybrid Human-in-the-Loop Deep Reinforcement Learning Method for UAV Motion Planning for Long Trajectories with Unpredictable Obstacles. *Drones* **2023**, *7*, 311. [CrossRef]
37. Gubán, M.; Udvaros, J. A Path Planning Model with a Genetic Algorithm for Stock Inventory Using a Swarm of Drones. *Drones* **2022**, *6*, 364. [CrossRef]
38. Zhang, L.; Ma, X.; Zhuang, Z.; Xu, H.; Sharma, V.; Han, Z. Q-Learning Aided Intelligent Routing With Maximum Utility in Cognitive UAV Swarm for Emergency Communications. *IEEE Trans. Veh. Technol.* **2023**, *72*, 3707–3723. [CrossRef]
39. Zhang, T.; Lei, J.; Liu, Y.; Feng, C.; Nallanathan, A. Trajectory Optimization for UAV Emergency Communication With Limited User Equipment Energy: A Safe-DQN Approach. *IEEE Trans. Green Commun. Netw.* **2021**, *5*, 1236–1247. [CrossRef]
40. Qie, H.; Shi, D.; Shen, T.; Xu, X.; Li, Y.; Wang, L. Joint Optimization of Multi-UAV Target Assignment and Path Planning Based on Multi-Agent Reinforcement Learning. *IEEE Access* **2019**, *7*, 146264–146272. [CrossRef]
41. Wang, F.; Xu, X.; Chen, M.; Nzige, J.; Chong, F. Simulation Research on Fire Evacuation of Large Public Buildings Based on Building Information Modeling. *Complex Syst. Model. Simul.* **2021**, *1*, 122–130. [CrossRef]
42. Hou, Z.; Sun, Y.; Cai, M. 3D Visualization of Forest Fire Spread Model Based on Cellular Automata. In Proceedings of the 2023 8th International Conference on Computer and Communication Systems (ICCCS), Guangzhou, China, 21–23 April 2023; pp. 880–883. [CrossRef]
43. Ye, Z.; Yin, Y.; Zong, X.; Wang, M. An Optimization Model for Evacuation Based on Cellular Automata and Ant Colony Algorithm. In Proceedings of the 2014 Seventh International Symposium on Computational Intelligence and Design, Hangzhou, China, 13–14 December 2014; Volume 1, pp. 7–10. [CrossRef]

Disclaimer/Publisher's Note: The statements, opinions and data contained in all publications are solely those of the individual author(s) and contributor(s) and not of MDPI and/or the editor(s). MDPI and/or the editor(s) disclaim responsibility for any injury to people or property resulting from any ideas, methods, instructions or products referred to in the content.

Article

HHPSO: A Heuristic Hybrid Particle Swarm Optimization Path Planner for Quadcopters

Jiabin Lou, Rong Ding * and Wenjun Wu

State Key Laboratory of Software Development Environment, School of Artificial Intelligence, Beihang University (BUAA), Beijing 100191, China; loujiabin@buaa.edu.cn (J.L.); wwj09315@buaa.edu.cn (W.W.)
* Correspondence: dingr@buaa.edu.cn

Abstract: Path planning for quadcopters has been proven to be one kind of NP-hard problem with huge search space and tiny feasible solution range. Metaheuristic algorithms are widely used in such types of problems for their flexibility and effectiveness. Nevertheless, most of them cannot meet the needs in terms of efficiency and suffer from the limitations of premature convergence and local minima. This paper proposes a novel algorithm named Heuristic Hybrid Particle Swarm Optimization (HHPSO) to address the path planning problem. On the heuristic side, we use the control points of cubic b-splines as variables instead of waypoints and establish some heuristic rules during algorithm initialization to generate higher-quality particles. On the hybrid side, we introduce an iteration-varying penalty term to shrink the search range gradually, a Cauchy mutation operator to improve the exploration ability, and an injection operator to prevent population homogenization. Numerical simulations, physical model-based simulations, and a real-world experiment demonstrate the proposed algorithm's superiority, effectiveness and robustness.

Keywords: particle swarm optimization; path planning; motion capture; unmanned aerial vehicles (UAVs); aerial systems; applications

Citation: Lou, J.; Ding, R.; Wu, W. HHPSO: A Heuristic Hybrid Particle Swarm Optimization Path Planner for Quadcopters. *Drones* **2024**, *8*, 221. https://doi.org/10.3390/drones8060221

Academic Editors: Jihong Zhu, Heng Shi, Zheng Chen and Minchi Kuang

Received: 19 April 2024
Revised: 22 May 2024
Accepted: 22 May 2024
Published: 28 May 2024

Copyright: © 2024 by the authors. Licensee MDPI, Basel, Switzerland. This article is an open access article distributed under the terms and conditions of the Creative Commons Attribution (CC BY) license (https://creativecommons.org/licenses/by/4.0/).

1. Introduction

Path planning is the cornerstone of unmanned aerial systems, enabling Unmanned Aerial Vehicles (UAVs) to handle complex scenarios [1]. For any given scenario, there are generally three elements: task, individual, and environment, which together comprise the path planning optimization problem. The planned path should be optimal within specific criteria associated with the task. For example, in time-sensitive tasks such as air delivery and military transport, the core principle is to minimize the distance between the drone access locations, thereby reducing the time and fuel cost [2]. However, for search and rescue, patrol, and other search-related tasks, the performance metric is usually to maximize the area coverage within a fixed amount of time [3]. Meanwhile, the planned path should be able to safely address environmental threats and smoothly respond to individual maneuver properties. In general, solving this optimization problem is not a complicated task. However, as a task becomes more urgent, the environment becomes more complex, or the UAV becomes less manoeuvrable, it remains a challenge to calculate feasible optimal paths to avoid all threats within an acceptable time [4].

Researchers have proposed a series of approaches to solve the UAV path planning problem. Previously, the path planning problem was generally equivalent to the shortest path problem, and deterministic search algorithms were widely adopted, for instance, the Dijkstra algorithm [5], the Voronoi diagram method [6] and the A* algorithm [7,8]. However, as researchers began to consider the specific demands of different scenarios, the optimal path has become associated with the average altitude, fuel consumption, environmental threats and so on, making it clear that the problem has become increasingly complex. We know that the path planning problem is an NP-hard problem, which is difficult to solve with deterministic algorithms when the scale of the problem becomes large. Therefore, researchers have slowly shifted from using deterministic to using non-deterministic algorithms.

Metaheuristics are one kind of non-deterministic algorithm that can provide a sufficiently good solution to an optimization problem with limited computation capacity. They are nature-inspired, population-based, and generation-iterated, facilitating powerful global searching and rapid convergence capability. In recent years, metaheuristic algorithms have been increasingly favored for UAV path planning because of their ability to explore low-dimensional manifolds in high-dimensional space [9]. For example, ref. [10] proposed an initial population enhancement method in a Genetic Algorithm(GA), which speeds up the convergence process. Ref. [11] proposed a spherical vector-based Particle Swarm Optimization (PSO) to solve the problem within complicated environments subjected to multiple threats. In addition, other meta-heuristic algorithms such as Differential Evolution (DE) [12–14], Ant Colony Optimization (ACO) [15,16], and Wolf Pack Search (WPS) [17] have been extensively studied in recent years.

Among these algorithms, PSO is much simpler to implement while maintaining excellent efficiency, effectiveness and scalability, and thus has been successfully applied in many drone fields. Ref. [18] presented a comprehensively improved Particle Swarm Optimization (CIPSO) algorithm with the chaos-based logistic map initialization and mutation strategy to solve this problem in war scenarios. Ref. [19] proposed a multiobjective particle swarm optimization algorithm with multimode collaboration based on reinforcement learning (MCMOPSO-RL) algorithm to find the optimal path and handle threats simultaneously. Ref. [20] proposed the SHOPSO algorithm, which combines the Selfish Swarm Optimizer (SHO) and the PSO, to accomplish a given combat mission at a meager cost. Nevertheless, the scenarios used in these studies are relatively simple: the number of threats was small, and the terrain was small-scale or flat, making these algorithms impractical for complex real scenes. The main reason is that PSO, as a general optimizer, does not analyze the built-in physics for specific problems. Although it inherently provides approaches for automatically abstracting features from iterations, the iterations require sufficient time to play a part. Therefore, with the limited execution time, PSO always suffers from premature convergence limitations, hindering its promotion in complex, high-dimensional, and noisy bounded scenarios.

Compared to other scenarios, the drone scenarios often require path planning algorithms that can respond quickly and ensure safety [21]. On the one hand, the drone tasks are often urgent, resulting in limited planning time for the algorithm. On the other hand, a drone scenario has many threats and complex terrain, which makes the algorithm easily to fall into local optimum. For the former problem, we established heuristic rules during particle initialization to prevent invalid searching and inspire the powerful search efficiency of the algorithm. For the latter problem, we hybridized a Cauchy mutation operator, an injection operator and a penalty function to enhance the exploration capabilities of the algorithm. The new algorithm, called Heuristic Hybrid Particle Swarm Optimization (HHPSO), strikes a good balance between exploration and exploitation and significantly improves the convergence, robustness and constraint-handling ability. Numerical simulations, Unreal Engine 4 (UE4) [22] simulations and a real-drone experiment confirmed the results.

The remainder of this paper is organized as follows: Section 2 proposes the problem scenarios and optimization model. The heuristic rules and hybrid operators are introduced in Section 3. The experimental results are provided in Section 4. Section 5 concludes this article.

2. Problem Statement

The quadcopters path planning problem can be treated as a multi-objective constrained optimization problem. In this section, scenario representation and the optimization model are discussed.

2.1. Scenario Representation

In a drone scenario, two elements must be considered for the path planning task. One the terrain, which imposes physical constraints on drones. The other is threats, which constitute dangers that drones may encounter in their missions.

2.1.1. Terrain

The terrain of the drone scenario is totally open, implying that we can discretize the broad planning space into a surface. On this basis, the terrain can be generated using several perlin noises layers [23] with diverse frequencies and amplitudes. This terrain ensures that the height z_{ij} is unique and continuous over the entire plane (x_i, y_j). To save computing resources, we sample the entire plane at specific intervals to obtain a point cloud map of the terrain, as shown in Figure 1a. However, such a discretized representation of terrain cannot constrain all points on the plane, so we used triangle interpolation among every three nearest points, thus confirming the unique mapping of (x_i, y_j), denoted as $z_{ij} = Map(x_i, y_j)$, as shown in Figure 1b.

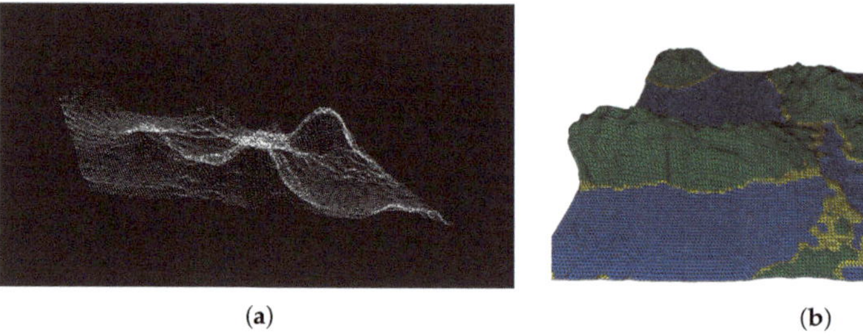

(a)   (b)

Figure 1. Terrain representation. (**a**) Terrain representation with a point cloud. (**b**) Terrain representation with triangle interpolation.

2.1.2. Threats

In general, quadcopters should remain concealed and secure when performing tasks in a drone scenario. We assume that the enemy will use radar and missiles to detect and attack drones. In addition, there are some No-fly Zones (NFZs) in the scene that drones cannot approach.

The probability of radar and missile affecting flight safety can be calculated by (1) and (2) [24].

$$P_R = \begin{cases} \dfrac{1}{1+\zeta_2\left(d^4/RCS\right)^{\zeta_1}} & \text{if } d \leq R_R \\ 0 & \text{otherwise} \end{cases} \tag{1}$$

$$P_M = \begin{cases} R_M^4/\left(R_M^4 + d^4\right), & \text{if } d \leq R_M \\ 0, & \text{otherwise.} \end{cases} \tag{2}$$

where R_R and R_M denote the maximum influence distance of the radar and missile, respectively, d is the distance between the drone's position and the missile and radar deployment center, and ζ_1 and ζ_2 depend on the radar used. RCS denotes the radar cross-section, which can be calculated according to the drone's position and velocity [25].

For the NFZs, we need to determine whether the waypoints are within their range, so we defined the following equation:

$$\text{InNFZs}(w_i) = \begin{cases} 1, & \text{if } w_i \text{ falls in any NFZs} \\ 0, & \text{otherwise} \end{cases} \tag{3}$$

where $w_i(i = 1, 2, ..., n)$ are waypoints.

2.2. Optimization Model

Mathematically, the path planning problem can be modeled as a Multi-Objective Constraint Satisfaction Problem, which comprises three components, i.e., variables, objectives, and constraints.

2.2.1. Variables

Typically, a path planning problem uses waypoints as planning variables. Suppose we start from start point $S : (x_S, y_S, z_S)^T$, and go to target point $T : (x_T, y_T, z_T)^T$. The path planning problem can be depicted as finding a series of waypoints $W = \{w_1, w_2, ..., w_{n-1}\}$ through which the UAV can reach its destination successfully. However, the outputs obtained in the three-dimensional configuration space cannot guarantee differential flat control. For example, these paths may contain sharp turns that challenge the kinematics and dynamics of the drone.

Several methods have been proposed to generate a smoothing path from the control points $P = \{p_0, p_1, ..., p_m\}$. In some previous research, the Dubins curve was used to smooth the path [26]. A Dubins curve uses a series of arcs and straight line segments to form the motion path of the drone, as shown in the Figure 2a. This method is unsuitable for parameterization, because it may generate many arcs without curvature continuity. Another method used in recent research is the Rauch-Tung-Striebel (RTS) smoother, which consists of two stages, Kalman forward filtering and RTS backward smoothing [27], as depicted in Figure 2b. At a higher computational cost, RTS smoother achieves relatively low tracking errors between generated paths and control points. However, it is unnecessary in the proposed algorithm because the UAV does not require flying directly through the control points. A Bezier curve [28] and b-spline curve [29] are two of the most well-known path smoothing methods, as shown in Figure 2c,d. Of these two curves, the latter evolves from the former and inherits all the advantages, including geometrical invariance, convexity-preserving, and affine invariance. Compared to the Bezier curve, the b-spline curve overcomes the disadvantage that moving one control point affects the entire curve and does not increase the degree of the polynomial no matter how many control points are added [30].

The b-spline function used in this paper can be defined as:

$$w_i = \sum_{j=0}^{m} p_j B_{j,k}\left(\frac{i}{n+1}\right) \tag{4}$$

where $w_i (i = 1, 2, ..., n)$ are waypoints, $p_j (j = 0, 1, ..., m)$ denote control points, $B_{j,k}()$ are the k-order normalized b-spline basic functions defined by the de Boor–Cox recursion formula as follows:

$$\begin{cases} B_{j,0}(t) = \begin{cases} 1, & \text{if } u_j \leq t \leq u_{j+1} \\ 0, & \text{otherwise} \end{cases} \\ B_{j,k}(t) = \frac{t-u_j}{u_{j+k}-u_j} B_{j,k-1}(t) + \frac{u_{j+k+1}-t}{u_{j+k+1}-u_{j+1}} B_{j+1,k-1}(t) \\ \text{define } 0/0 = 0 \end{cases} \tag{5}$$

where $t \in [0, 1]$ and $U = \{u_0, u_1, \cdots, u_{k+m}\}$ is a non-decreasing sequence of parameters called the knot vector.

B-spline enjoys C^{k-1} continuous property and the derivative of a b-spline is still a b-spline curve with order $k - 1$. Therefore, we choose the cubic b-spline curve with C^2 continuity (this guarantees that the quadcopters will not be commanded to change their propeller speed sharply) to convert the control points P into waypoints W. Concretely, we fixed the p_0 at the starting point S, the p_m at the target point T and defined all remaining intervening control points as decision variables:

$$\xi = \{p_1, p_2, ..., p_{m-1}\} \tag{6}$$

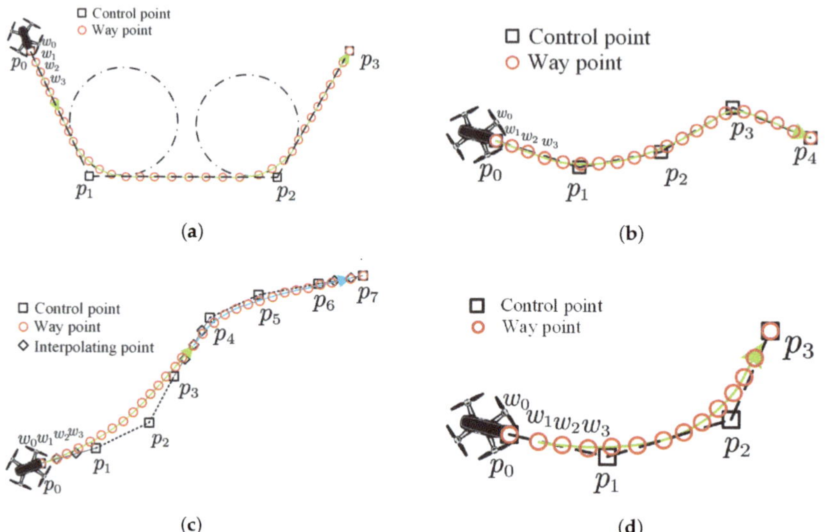

Figure 2. Smooth methods. (**a**) Tangent circle curve. (**b**) RTS Smoother. (**c**) Six-order Bezier Curve. (**d**) Cubic b-spline Curve

2.2.2. Objectives

In order to obtain a short, safe and smooth trajectory, five objectives, i.e., length cost, flight altitude, radar detection, missile attack and turning angle are considered. Generally, these objectives are somewhat contradictory so we treat them into a weighted function (7).

$$F = \omega_1 f_1 + \omega_2 f_2 + \omega_3 f_3 + \omega_4 f_4 + \omega_5 f_5 \tag{7}$$

where ω_1, ω_2, ω_3, ω_4, ω_5 are weights that sum to 1, and f_1, f_2, f_3, f_4, f_5 denote different objectives.

(1) Length Cost

Traditionally, the goal of a planner is to find the shortest path conforming to constraints, and the normalized approximate length of a path is defined as (8).

$$f_1 = \frac{\sum_{i=1}^{n} \sqrt{\|w_i - w_{i-1}\|}}{\sqrt{\|w_n - w_0\|}} \tag{8}$$

where $w_i = [x_i, y_i, z_i]$ denotes the coordinate of the ith waypoint.

(2) Flight Altitude

A lower altitude is desired for the sake of using ground effect to avoid radars and saving fuel. The mean flight altitude of a path is denoted by (9).

$$f_2 = \sum_{i=1}^{n-1} \text{FA}_i \quad \text{with}$$

$$\text{FA}_i = \begin{cases} 0, & \text{if } z_i \leq Map(x_i, y_i) \\ (z_i - Map(x_i, y_i))/n, & \text{otherwise} \end{cases} \tag{9}$$

where $Map(x_i, y_i)$ represent the terrain height at (x_i, y_i).

(3) Radar Detection

The probability of the quadcopters being detected by radars can be calculated as follows:

$$f_3 = \sum_{i=1}^{n-1} \sum_{j=1}^{R} P_{Rij} \tag{10}$$

where R denotes the number of radars in the scenario, P_R can be obtained according to (1).

(4) Missile Attack

The probability of the quadcopters being attacked by missiles is expressed as follows:

$$f_4 = \sum_{i=1}^{n-1} \sum_{j=1}^{M} P_{Mij} \tag{11}$$

where M denotes the number of missiles in the scenario, P_M is calculated according to (2).

(5) Smoothness

This is designed to evaluate the smoothness of the planned path. Values closer to 0 indicate smoother paths, and 0 means a straight path [31].

$$f_5 = \frac{1}{n-1} \sum_{i=1}^{n-1} \theta_i \quad \text{with}$$
$$\theta_i = \arccos\left(\frac{(x_i - x_{i-1}, y_i - y_{i-1}) \cdot (x_{i+1} - x_i, y_{i+1} - y_i)^T}{\| (x_i - x_{i-1}, y_i - y_{i-1}) \| \cdot \| (x_{i+1} - x_i, y_{i+1} - y_i) \|} \right) \tag{12}$$

To ensure the single-valuedness and continuity of the arccos function, that is, to ensure each value corresponds to a unique θ_i, we choose to restrict the function's range to $[0, \pi]$.

2.2.3. Constraints

It's well accepted that quadcopters should meet the following constraints to ensure safe and stable flight.

(1) Climbing/Gliding Angle

Since the maneuverability of quadcopters, the slope s_i should be restricted in the range of maximum climbing angle α_i and minimum gliding angle β_i [20]. This forms the constraint functions g_1 and g_2:

$$g_1 = \max(s_i - \alpha_i) \leq 0 \tag{13}$$
$$g_2 = \max(\beta_i - s_i) \leq 0 \tag{14}$$

For i in $1, \cdots, n-1$, where

$$\alpha_i = -1.5377 \times 10^{-10} z_i^2 - 2.6997 \times 10^{-5} z_i + 0.4211 \tag{15}$$

$$\beta_i = 2.5063 \times 10^{-9} z_i^2 - 6.3014 \times 10^{-6} z_i - 0.3257 \tag{16}$$

$$s_i = \frac{z_{i+1} - z_i}{\sqrt{(x_{i+1} - x_i)^2 + (y_{i+1} - y_i)^2}} \tag{17}$$

(2) Turning angle constraint

The turning angle θ_i at waypoint w_i can be calculated according to (12). Due to the maneuverability constraints of the quadcopters, the turning angle should not be greater than its upper bound, which can be written as

$$g_3 = \max(\theta_i - \theta_i^{max}) \leq 0 \tag{18}$$

(3) Minimum flight altitude

For safety reasons, quadcopters should be at a certain level with the ground, as described in (19).

$$g_3 = H_{\text{safe}} - \min(z_i - Map(x_i, y_i)) \leq 0 \tag{19}$$

where H_{safe} denotes the minimal safe flight height.

(4) Forbidden flying area

According to mission requirements, the quadcopters have to keep away from NFZs. We describe this as a hard constraint as follows:

$$h_1 = \sum_{i=1}^{n-1} \text{InNFZs}(w_i) = 0 \tag{20}$$

3. Approach

3.1. Standard Particle Swarm Optimization

The standard PSO algorithm was developed by Kennedy and Eberhart in 1995 based on social and cognitive behavior [32], and is widely used in engineering. It solves problems by generating candidate solutions (particles) and moving those particles through a search space based on their positions and velocities, as seen in (21) and (22).

$$V_k = wV_k + c_1 \cdot r_1 \cdot (pb_k - \xi_k) + c_2 \cdot r_2 \cdot (gb - \xi_k) \tag{21}$$

$$\xi_k = \xi_k + V_k \tag{22}$$

where the Equation (21) updates a new velocity for the k-th particle according to its previous velocity V_k, its current position ξ_k, its best historical position pb_k and the current global best position gb. And w is the inertia weight that determines the particle to maintain its original trend, r_1 and r_2 denote two random numbers, c_1 and c_2 are learning factors. The Equation (22) updates each particle's position based on its updated velocity from the former.

To further improve the algorithm's efficiency for solving path planning problems, we introduce the heuristic rules to guide the search and hybrid some operators to speed up the convergence.

3.2. Heuristic Rules

When a drone move from the start position to its goal, there are strong constraints inside a path for respecting the kinematic and the dynamic limits. Embedding these limits into initialization can provide informative priors, i.e., strong physics constraints and inductive biases to guide the search.

Thanks to the convexity-preserving property, constraining the derivatives of the control points is sufficient for constraining the entire b-spline [33]. Therefore we set up heuristic rules for control points.

3.2.1. Rotated Coordinate System

Searching points within a 3-D Cartesian coordinate system has been widely used in current studies. But it is usually inefficient since the heading direction is rarely consistent with axes, so the inherent sequential relationships between waypoints or control points are underutilized. In this paper, we use the rotated coordinate system O_R-$X_R Y_R Z$ to initialize the candidate solutions, as shown in Figure 3, where X_R is the direction from S to T. The transformation between the two coordinates can be obtained by the rotating matrix:

$$\begin{pmatrix} x \\ y \\ z \end{pmatrix} = \begin{pmatrix} x^R \\ y^R \\ z^R \end{pmatrix} \begin{pmatrix} \cos\theta & \sin\theta & 0 \\ -\sin\theta & \cos\theta & 0 \\ 0 & 0 & 1 \end{pmatrix} + \begin{pmatrix} x_S \\ y_S \\ 0 \end{pmatrix} \tag{23}$$

where θ is the angle from the X axis to the X_R axis, x_S and y_S denote the x coordinate and y coordinate of the start point S, and the superscript R represents the coordinates in the rotating coordinate system.

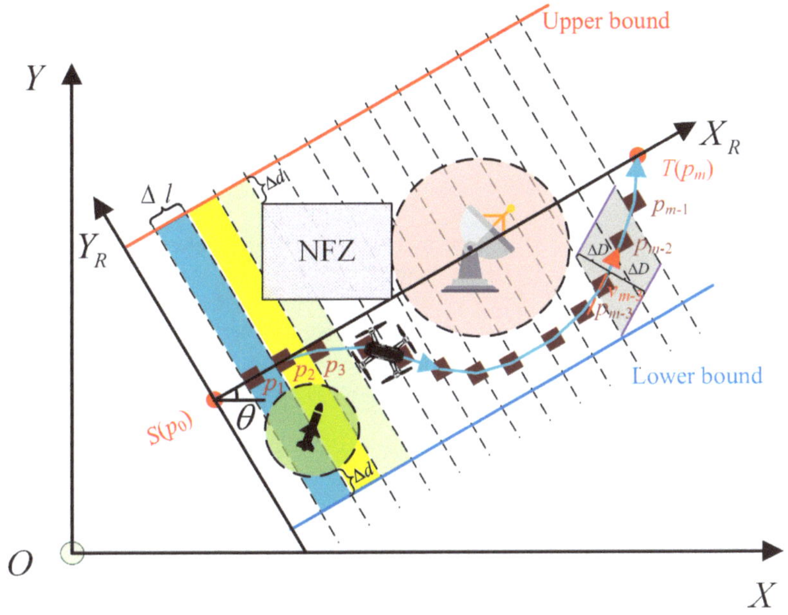

Figure 3. Division of UAV mission space.

In O_R-$X_R Y_R Z$, we assume that the control points are monotonically increasing along the X_R axis, which means the quadcopters cannot move backwards. Actually, some other scholars [12,34,35] have also utilized this benefit in the phase of path initialization and they have proved that the solution set is almost never lost.

3.2.2. Physical Plausibility

Physically plausible paths should ensure the quadcopters keep motor speed changing smoothly, implying that the distance moved in each time interval should not differ much. According to this rule, we divided the $\overline{ST}$ by $(n + 1)$ Δl segments and defined each equilateral point as the expected position of the control points. Considering that the distribution of each point is different and correlated, we limited a control point's x^R value between the expected position of its previous and next point in the initialization phase.

Similarly, the y^R coordinate can be constrained. As shown in Figure 3, we define two boundaries parallel to the velocity that constrain the velocity direction variation trend with the metric ΔD. The value of ΔD is determined based on the maximum flight speed of the drone. In this paper, We set ΔD equal to Δl, so that the components of the path along the x and y axes are of the same order of magnitude at initialization. Therefore, we can roughly determine the x^R and y^R range of p_j according the position of p_{j-1} (e.g., p_{m-2} is locked in the gray area). This restrict is modeled as a Markov Chains, which require a considerable computational cost to be handled, and also in order not to lose the possible solution, it's only used in the initialization phase.

In addition, we defined a mission space. On the X_R-axis, the x^R coordinate is confined between 0 and $|\overline{ST}|$. And on the Y_R-axis, two lines in Figure 3 determine the upper and lower bound, which are obtained by extending outward a constant distance Δd of the points from the nearest safe areas around $\overline{ST}$. Under the boundary, the y^R coordinate is restricted in $[y_{min}, y_{max}]$, which are calculated as (24) and (25), respectively:

$$y_{\min} = \min\left\{\min_i\{y^*_{\text{threat},i} - R_i\}, 0\right\} - \Delta d \tag{24}$$

$$y_{\max} = \max\left\{\max_i\{y^*_{\text{threat},i} + R_i\}, 0\right\} + \Delta d \tag{25}$$

where R_i is the radius or the circumradius of the ith threat, $y^*_{\text{threat},i}$ is the vertical coordinate of ith threat in the rotated coordinate frame O_R.

3.2.3. Initialization

In practice, we introduce three criteria to the initialization process.

- x^R coordinate

In the rotated coordinate system we established before, the x^R value is monotonically increasing. In addition, we assume x_j^R obeying the normal distribution with mean $(j\Delta l)$ and standard deviation $(\Delta l/3)$ so that its value will fall in the range we expected with 99.7% probability according to the pauta criterion [36].

- y^R coordinate

For y coordinate, it's highly dependent on the state of the previous point. So we initialize the y_i^R in the range of $\left(-\Delta D + y'_j, \Delta D + y'_j\right)$ under uniform distribution, where y'_j reflects the state of the previous point and is calculated by (26).

$$y_j' = \begin{cases} 0 & , j < 1 \\ \frac{y^R_{j-1} - y^R_{j-2}}{x^R_{j-1} - x^R_{j-2}}\left(x_j^R - x^R_{j-1}\right) + y^R_{j-1} & , \text{ otherwise} \end{cases} \tag{26}$$

- z coordinate

Heuristically, the drone's path follows the ups and downs of the terrain. So we initialize the z-value obeying the normal distribution with mean z'_i and standard deviation Δh, where Δh is set roughly equal to $\Delta l/3$ in this paper according to the maneuverability of quadcopters.

$$z_j' = z_{j-1} + Map(x_j, y_j) - Map(x_{j-1}, y_{j-1}) \tag{27}$$

3.3. Hybrid Operators

3.3.1. Penalty Function

We use the penalty function as the constraint-handling method to evaluate the particles better. This approach defines the particle's fitness function as the sum of the objective function and the penalty term due to constraint violation:

$$fit(\xi_k) = F(\xi_k) + \varphi\left(\xi_k, r^{(\alpha)}\right) \tag{28}$$

where $\varphi\left(\xi_k, r^{(\alpha)}\right)$ indicates the penalty term of the kth particle in α^{th} generation and it can be calculated as follows:

$$\varphi\left(\xi_k, r^{(\alpha)}\right) = r^{(\alpha)}\left(\sum_{i=1}^{4}(\text{Max}[0, g_i(\xi_k)])^2 + h_1(\xi_k)\right) \tag{29}$$

where $r^{(\alpha)}$ is the penalty factor, it changes with iteration to ensure that infeasible individuals suffer from more selection pressure at the later stage of iteration:

$$0 < r^{(1)} < r^{(2)} < \cdots < r^{(\alpha-1)} < \lim_{\alpha \to \infty} r^{(\alpha)} = +\infty \tag{30}$$

In this paper, the penalty coefficient is designed as a quadratic function of α.

$$r_\alpha = 10 \times \alpha^2 / I_{\max}^2 \qquad (31)$$

where $I_{\max}$ denotes the maximum iteration algebra.

3.3.2. Cauchy Mutation

Like other variants of PSO, the heuristic PSO faces the problem of premature convergence; that is, its particles converge to a local optimum in some scenarios. Intuitively, a stochastic mechanism might help the premature particles escape from the current trapped local optimum, thus avoiding premature convergence. Genetic algorithms have a similar mechanism called **mutation** to help the individual escape from local optima [37]. Suppose we apply the mutation operator to the premature particles, and then use the fitness function to evaluate the results. A good mutation means that the mutated particle has a better fitness value than the original one. Following this criterion, we design a mutation operator for PSO.

Since the Cauchy distribution has a small peak value at the origin but a long distribution at both ends, it can generate a larger disturbance near the individual to jump out of the local optimum, thus we choose Cauchy distribution to generate trail variable. The operator process is shown below:

Firstly, we sorted all particles according to their fitness and took out the inferior half of the particles as the target of the mutation operator.

Secondly, for each selected particle, the trail variable is generated from its current position in the following way:

$$\xi_k' = \xi_k + C(0, \gamma) \qquad (32)$$

where ξ_k denote the position of the kth particle and $C(0, \gamma)$ represents a Cauchy random vector of the same dimension as ξ_k with a location parameter 0 and a scale parameter γ.

Finally, replace the origin particle with the trail variable if the mutated particle has better fitness value.

3.3.3. Injection

Besides, there is an approach commonly used in the meta-heuristic algorithm to increase the randomness of the population, namely **injection** [38]. Due to the role of heuristic rules, the initial particles of our algorithm are of high quality. Therefore, if the inferior particle can be initialized with a higher fitness value; it is beneficial for the population to escape from the current trapped local optimum. Similar to the rules of mutation, the injection operator operates on the sorted particles at the following scale:

$$\lambda^{(\alpha+1)} = \zeta \lambda^{(\alpha)} \qquad (33)$$

where $\lambda^{(\alpha)}$ indicates the number of injected particles in α-th generation, ζ is the decay factor.

Since the substantial irregularity of the injected particles, the new particle swarm is sorted to ensure the effectiveness of the injection, and the λ_α particles at the bottom are eliminated.

3.4. Algorithm Presentation

To summarize, we use the standard PSO as the prototype, initialize particles with heuristic rules, hybrid Penalty function operator, Cauchy mutation operator, and Injection operator to improve the search capability and propose the Heuristic Hybrid Particle Swarm Algorithm (HHPSO). The pseudo-code of HHPSO is given in Algorithm 1. It's worth noting that the fitness $fit(\xi)$ in (28) converts the control points $[p_0, \xi, p_m]$ to waypoints using a cubic b-spline curve.

Algorithm 1: HHPSO

---

1 **Initialize:** N particles with velocity 0 following the heuristic rules in Section 3.2;
2 Assign the best historical position pb of each particle to its position ξ;
3 **for** $\alpha = 1$ *to* $I_{\max}$ **do**
4      $gb \leftarrow \mathrm{argmin}[fit(\xi_k, r^\alpha)]$, for $k = 1$ to N;
5   **for** $k = 1$ *to* N **do**   ▷ standard PSO
6 | Update ξ_k and V_k using Equations (21) and (22);
7   **end**
8   **for** $k = 1$ *to* $N/2$ **do**   ▷ Mutation
9 | Generate trail vector ξ'_k according to Equation (32);
10 | Replace ξ_k with ξ'_k and reset its velocity to 0 if $fit(\xi'_k) < fit(\xi)$;
11   **end**
12   Generate λ_α particles and inject them into the swarm;   ▷ Injection
13   Delete λ_α inferior particles according to their fitness;
14   Calculate $\lambda_{\alpha+1}$ and $r_{\alpha+1}$ according to the Equations (31) and (33) ;
15 **end**
16 **return** gb

---

4. Experimental Results

To verify the effectiveness of the proposed algorithm, numerical simulations, physical model-based simulations and a real-drone experiment are designed. In the numerical simulation, we deploy the HHPSO to handle various scenarios with increasing obstacles and different terrains and set up several respective algorithms to draw comparisons. However, sim-to-real translation has been known to be a long-standing problem in robotics. But it was difficult for us to set up an accurate test site with the terrain and various constrained areas. As a compromise, we built similar scenarios on UE4 as physical model-based simulations. And for the real-drone experiment, we choose NOKOV motion caputrue system as the global location system and use Bitcraze Crazyflie 2.1 nano-quadrotors [39] as the flying platform for its characteristics of small volume (9cm rotor-to-rotor), lightweight (34 g), and suitable for indoor flying. All experiments were conducted on a desktop computer featuring an Intel Core i9-9900K CPU, 32 GB of DDR4 RAM, a 1 TB NVMe SSD, and an NVIDIA GeForce RTX 3070Ti GPU.

4.1. Numerical Simulation

In this paper, the proposed algorithm was run in four scenarios with different terrains and increasing obstacles, and some other recently proposed metaheuristic planners, i.e., GA [10], CIPSO [18], CIPDE [13], JADE [12], mWPS [17], were selected as the compared algorithms, the hyperparameters of these algorithms are shown in the Supplementary Materials (Tables S1–S6). Besides, the heuristic-PSO and the hybrid-PSO are also put as comparative algorithms to further discuss the influence of the two operators proposed in this paper.

For comparisons, all algorithms are shared with the same basic parameters: the population size of 30, the maximum iteration algebra of 25 and the waypoints number of 35. In addition, the main control parameters of the mentioned algorithms are shown in Table 1, the definitions of these parameters can be found from their original papers. The heuristic-PSO and the hybrid-PSO share the same control parameters as HHPSO. Since the high security requirements of the scenarios, we have set the weights ($[\omega_1 = 0.2,\ \omega_2 = 0.1,\ \omega_3 = 0.3,\ \omega_4 = 0.3,\ \omega_5 = 0.1]$) among the five objective functions.

Table 1. Main Control Parameters of Used Algorithms.

Algorithm	GA			CIPSO				JADE		
parameter	α	β	fr	w	c	μ	a	u_{CR}	u_F	
value	0.5	0.5	10	[0.4, 0.9]	[0.5, 3.5]	4	2	0.5	0.5	

Algorithm	CIPDE			mWPS		HHPSO				
parameter	μ_F	μ_{CR}	c	Eliminated-Qty	Safari-Wolves-Qty	w	c_1	c_2	γ	ζ
value	0.7	0.5	0.1	5	5	1	1.5	1.5	2	0.9

In order to test the performance of algorithms under different threat density and terrain, we designed four scenarios. Scenarios 1–3 are on the same flat terrain with increasing threats, and scenario 4 is extremely challenging with rugged terrain and crowded threats. The threats are randomly generated with the numbers 1, 4, 10, and 20 for each type, the radar and missile influence area is a sphere of a radius of 10, the RCS is set to -23.8 [40], and the no-fly zone randomly covers an area of 10–30.

All algorithms are implemented in scenarios 1–4, and the numerical experiment results are shown in Figure 4. To better analyze the fitness of the path that satisfies the constraint, which is critical in the drone scenario, we truncate the portion with a fitness value greater than 5.

From the 2D view of planning results, all the algorithms can complete the task in simple scenarios (Scenarios 1 and 2). But in the complex scenarios (scenarios 3 and 4), our algorithm can always find a relatively good location to avoid the threat and reach the destination, while the rest paths frequently enter the radar areas, missile areas and NFZs. More clearly seen from the fitness figure, as the scenario gets more complicated, the performance gaps between the proposed algorithm and others become more and more apparent. As the penalty coefficient r^α increases, all algorithm except HHPSO can't handle constraints well. For example, in scenario 3, there are two algorithms that complete the task, i.e., heuristic-PSO and HHPSO, while the path generated by heuristic-PSO is clearly stuck in a local optimum. But in scenario 4, only HHPSO is left. Further, we look at constraint functions. In all scenarios, HHPSO starts with a low constraint value and ends up satisfying the constraints well. This is mainly because the method fully considers the problem-dependent heuristics during initialization and the introduction of penalty term results in higher selection pressure along with iteration. And the heuristic-PSO, which also benefits from the heuristic initialization, has the same property of low starting constraint value. Moreover, it shows that the mixture of Cauchy mutation and injection operators makes our method less prone to falling into local optima and prematurity. Without these operators, the algorithms easily fall into local minimum, like the heuristic-PSO. But without the heuristical initialization, the algorithm doesn't even converge; the hybrid-PSO provides a good example. And other algorithms perform fine in simple scenarios but miserably in complex ones. Therefore, it can be inferred that the heuristic rules and the hybrid operators are of great help to planners, especially in complex scenarios.

To compare these algorithms more rigorously, we ran each algorithm 100 times in each scenario and adopted several metrics to measure the algorithm performance, i.e., Successful Rate (SR), Average Fitness (AF), Average Constraints value(AC) and Average Time (AT). Here we define the path that satisfies most conditions (at least 3), and the constraint value does not exceed 0.1 as a successful plan. The results are recorded in Table 2. Among all algorithms, HHPSO achieves the best fitness with a success rate of over 90% in all scenarios. The worst performer is CIPSO, probably because its original paper environment has no dense threats and thus weaker constraint control. Some other planners like GA, JADE, CIPDE and mWPS may get a tolerable AF in simple scenarios but perform poorly in scenarios 3 and 4. In addition, the two variants of HHPSO, i.e., heuristic-PSO and hybrid-PSO, performed reasonably well. Heuristic-PSO shows strong constraint handling ability and requires less running time than HHPSO. But its average fitness value is weaker than HHPSO, meaning it often falls into the local minimum. Hybrid-PSO also requires less running time than HHPSO, but due to its weaker constraint handling capability, it does not perform as well as the other two.

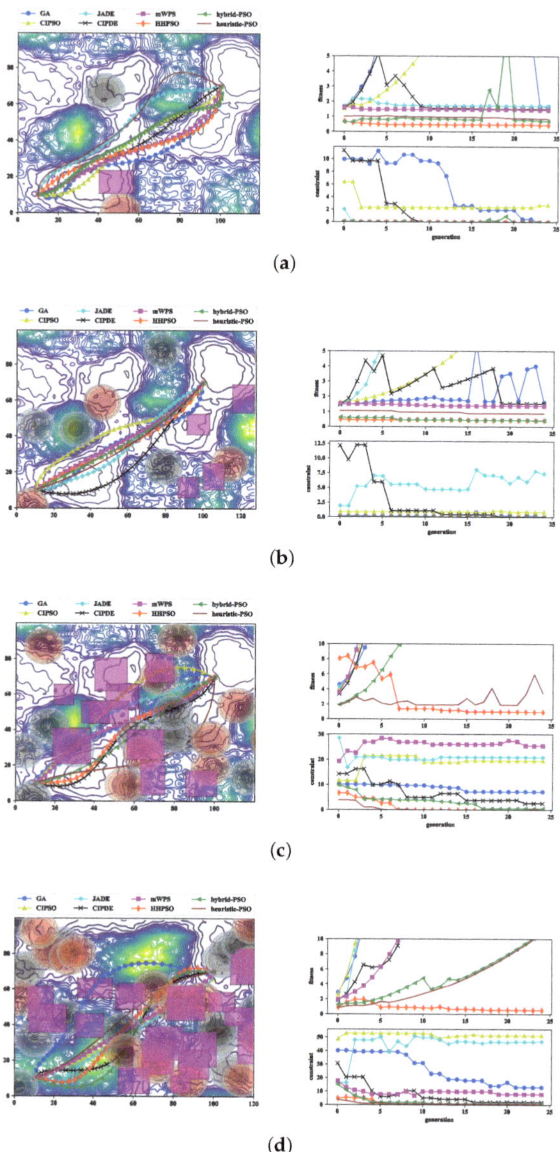

Figure 4. The comparative results among different algorithms. (**a**) Scenario 1. (**b**) Scenario 2. (**c**) Scenario 3. (**d**) Scenario 4.

From the above discussions, it can be seen that the proposed planner is more effective and efficient than the compared algorithms. In terms of effectiveness, HHPSO achieves the highest *SR* and the best *AF* in all scenarios, which is crucial in drone scenarios. And in terms of efficiency, HHPSO runs in less than 0.4 s and can be implemented for urgent tasks.

Table 2. Statistical Results for Different Algorithms.

		GA	CIPSO	JADE	CIPDE	mWPS	HHPSO	Heuristic-PSO	Hybrid-PSO
Scenario 1	$SR(\%)$	87.00 ± 2.61	82.00 ± 2.46	96.00 ± 3.12	92.00 ± 2.76	95.00 ± 2.85	$\mathbf{100.00 \pm 0.00}$	99.00 ± 2.97	97.00 ± 2.91
	AF	1.73 ± 0.03	18.20 ± 0.91	1.75 ± 0.04	1.28 ± 0.03	1.45 ± 0.04	$\mathbf{0.65 \pm 0.03}$	0.85 ± 0.02	0.97 ± 0.02
	AC	0.07 ± 0.00	1.78 ± 0.04	0.09 ± 0.00	0.08 ± 0.00	0.05 ± 0.00	$\mathbf{0.00 \pm 0.00}$	0.00 ± 0.00	0.07 ± 0.00
	$AT(s)$	0.39 ± 0.01	0.54 ± 0.01	0.62 ± 0.02	0.60 ± 0.02	0.89 ± 0.02	0.23 ± 0.01	$\mathbf{0.19 \pm 0.02}$	0.21 ± 0.01
Scenario 2	$SR(\%)$	86.00 ± 2.58	81.00 ± 2.43	75.00 ± 2.25	84.00 ± 2.52	89.00 ± 2.67	$\mathbf{98.00 \pm 0.13}$	96.00 ± 2.88	89.00 ± 2.67
	AF	1.76 ± 0.04	14.85 ± 0.74	18.26 ± 0.91	1.83 ± 0.04	1.55 ± 0.04	$\mathbf{0.49 \pm 0.01}$	0.51 ± 0.01	0.85 ± 0.02
	AC	0.13 ± 0.00	1.41 ± 0.03	1.75 ± 0.04	0.12 ± 0.00	0.09 ± 0.00	$\mathbf{0.00 \pm 0.00}$	0.01 ± 0.00	0.05 ± 0.00
	$AT(s)$	0.33 ± 0.01	0.45 ± 0.01	0.45 ± 0.01	0.47 ± 0.01	0.77 ± 0.02	0.22 ± 0.00	$\mathbf{0.15 \pm 0.05}$	0.19 ± 0.00
Scenario 3	$SR(\%)$	26.00 ± 0.78	12.00 ± 0.36	20.00 ± 0.60	48.00 ± 1.44	27.00 ± 0.81	$\mathbf{94.00 \pm 1.35}$	68.00 ± 2.04	25.00 ± 0.75
	AF	79.27 ± 2.38	208.20 ± 4.16	228.90 ± 5.73	45.50 ± 1.37	295.50 ± 8.87	$\mathbf{1.16 \pm 0.02}$	3.75 ± 0.11	10.85 ± 0.33
	AC	7.73 ± 0.2	20.54 ± 0.62	22.75 ± 0.68	4.48 ± 0.13	29.45 ± 0.88	$\mathbf{0.02 \pm 0.00}$	0.32 ± 0.01	0.97 ± 0.03
	$AT(s)$	0.41 ± 0.01	0.58 ± 0.02	0.58 ± 0.02	0.61 ± 0.02	0.92 ± 0.03	0.34 ± 0.00	$\mathbf{0.27 \pm 0.05}$	0.32 ± 0.00
Scenario 4	$SR(\%)$	15.00 ± 0.45	6.00 ± 0.18	11.00 ± 0.33	25.00 ± 0.75	2.00 ± 0.06	$\mathbf{92.00 \pm 2.32}$	52.00 ± 1.56	15.00 ± 0.45
	AF	15.41 ± 0.46	516.30 ± 15.49	472.80 ± 14.18	22.27 ± 0.67	81.80 ± 2.45	$\mathbf{1.35 \pm 0.31}$	10.50 ± 0.32	13.50 ± 0.41
	AC	13.20 ± 0.40	51.50 ± 1.55	47.20 ± 1.42	2.18 ± 0.07	8.05 ± 0.24	$\mathbf{0.05 \pm 0.01}$	0.92 ± 0.03	1.27 ± 0.04
	$AT(s)$	0.44 ± 0.01	0.68 ± 0.02	0.68 ± 0.02	0.73 ± 0.02	1.23 ± 0.04	0.38 ± 0.00	$\mathbf{0.27 \pm 0.03}$	0.29 ± 0.00

4.2. Physical-Based Simulation

We built a suitable terrain in the realistic physics engine Unreal Engine 4 (UE4) and deployed AirSim [41] as the dynamics model of the quadcopters. The terrain is shown in Figure 5a and the drone model used in the AirSim is the Ar Drone, which is shown in Figure 5b.

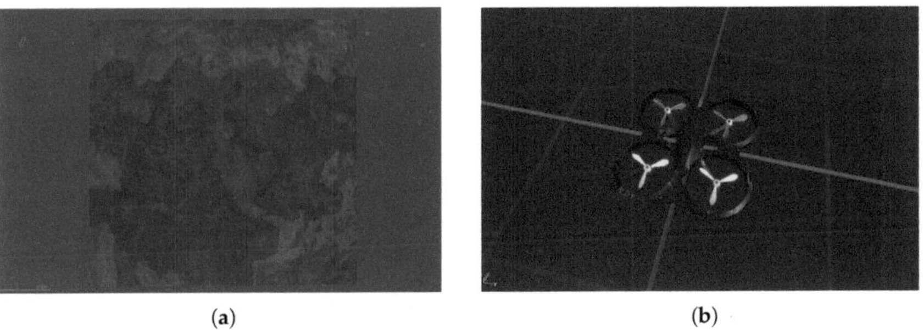

(a)   (b)

Figure 5. Basic components of the physical simulation. (**a**) The terrain in UE4. (**b**) Basic components of the physical simulation.

The terrain is augmented with different threats to generate two experimental scenarios; scenario 1 has [10, 10, 10] radars, missiles and NFZs and scenario 2 with [30, 30, 30]. The radar and missile detection radius is 20 for scenario 1 and 10 for scenario 2. Those threats, together with the interpolation point clouds for terrain scanning and the target point are used as the inputs of HHPSO to generate waypoints. The results are shown in Figure 6, the demonstration videos are available at https://youtu.be/9-ZV-A0M3b4 and https://youtu.be/VLxXbAZJzQQ, (accessed on 7 January 2024). It turns out that drones can follow planned paths to complete missions, and our algorithms can address diverse scenarios, such as larger threat zones and numerous and fragmented threats.

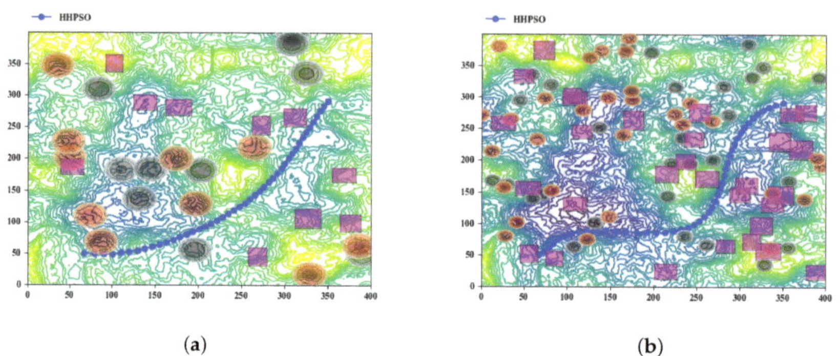

(a)   (b)

Figure 6. The planned paths of HHPSO in UE4. (**a**) Sparse environment. (**b**) Dense environment.

4.3. Real-Drone Experiment

Although UE4 simulates physical conditions, it is still just a simulation. We have carried out am in-door experiment to verify the validity of the proposed algorithm. Concretely, we use a 3×1 m test site indoors to map to the simulation system at a ratio of 1 to 10. For such a small venue, we chose Bitcraze CrazyFlie 2.1 [39], the world's smallest quadcopter, as the executor of the algorithm. As shown in Figure 7, we used some boxes to simulate the NFZs and set the safe height as 5 m (0.5 m in the real world).

Figure 7. Real-world experimental environment

To verify the real-time performance of the algorithm, we set up a series of checkpoints for the quadcopters to pass through (point A to D in Figure 8). The mission of the planner is to plan and execute a path to the next checkpoint when the quadcopters approaches a checkpoint. Although the quadcopters can move in any direction, we force the drone to keep its heading consistent with the direction of moving during the experiment. Besides, we add the last three points of the previous plan to the next plan, and calculate the turning angle according to (12). In this way, the quadcopters can do continuous planning without hovering over an intermediate point.

The path recorded from a motion capture system is shown in Figure 8, and the video is available at https://youtu.be/Fis1Fm25z04, (accessed on 7 January 2024). We can see that the first path of the quadcopters from point A to B is not consistent with the subsequent planning, this is because that they arrive at point A from different directions, resulting in different initial states. Moreover, due to the limited space of the physical room, we introduced a 10x zoom, resulting in some sharp turns in the flight of the quadcopters. But the actual planning result (Figure 8) is very smooth. This experiment demonstrates that our algorithm is efficient enough to be deployed for real-time path planning of quadcopters.

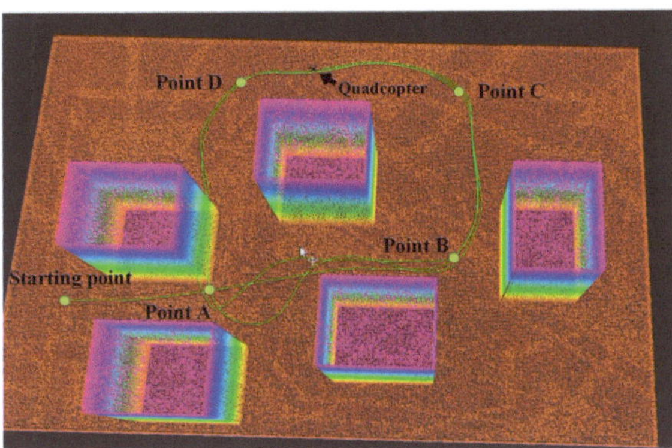

Figure 8. Trajectory of the Crazyflie

5. Conclusions and Discussions

Path planning plays a crucial role in autonomous unmanned systems. This paper presents an effective and efficient PSO-based path planning algorithm that allows the quadcopters to complete navigation tasks in complex scenarios. Concretely, we set up a series of heuristic rules during population initialization to generate high-quality particles to avoid invalid searches. But the heuristic-PSO is easy to fall into the local optimum, so we hybrid the penalty function, Cauchy mutation operator and Injection operator further to improve the global search ability of the algorithm. The proposed algorithm is named HHPSO; comparative numerical simulations of four scenarios with increasing obstacles show that HHPSO outperforms other state-of-the-art meta-heuristic algorithms. Furthermore, the physical-based simulations in UE4 show that our method can be successfully deployed in simulation models to perform complex missions on the battlefield. Finally, a real-world experiment demonstrates that the proposed method is efficient and can be used for continuous real-time path planning for quadcopters. Although the performance of HHPSO is remarkable, it has not yet been able to handle confrontational scenarios, which means enemy aircraft will also be deployed to scout and defend. We will focus on such type of scenario in the future and we have built a simulation environment in UE4. Finally, we hope this work could facilitate the applications of intelligent algorithms in path planning.

Supplementary Materials: The following supporting information can be downloaded at: https://www.mdpi.com/article/10.3390/drones8060221/s1. Table S1: Binary encoding of all algorithms; Table S2: 11-bits genes for Initialization; Table S3: 13-bits genes for sorting and selection; Table S4: 12-bits genes for exploitation and exploration; Table S5: 13-bits genes for Ending Criterion; Table S6: 15-bits genes for other operators.

Author Contributions: Conceptualization, J.L. and R.D.; methodology, J.L.; software, J.L.; validation, J.L., R.D. and W.W.; formal analysis, J.L.; investigation, R.D.; resources, J.L.; data curation, R.D.; writing—original draft preparation, J.L.; writing—review and editing, R.D.; visualization, R.D.; supervision, W.W.; project administration, W.W.; funding acquisition, W.W. All authors have read and agreed to the published version of the manuscript.

Funding: This research was funded by the National Key R&D Program of China (No. 2022ZD0116401) and the State Key Laboratory of Software Development Environment (Funding No. SKLSDE-2023ZX-20).

Data Availability Statement: The datasets generated during the current study are not publicly available due to confidential agreement but are available from the corresponding author on reasonable request.

Conflicts of Interest: The authors declare no conflicts of interest.

References

1. Huang, H.; Savkin, A.V.; Ni, W. Decentralized Navigation of a UAV Team for Collaborative Covert Eavesdropping on a Group of Mobile Ground Nodes. *IEEE Trans. Autom. Sci. Eng.* **2022**, *19*, 3932–3941. [CrossRef]
2. Brunner, G.; Szebedy, B.; Tanner, S.; Wattenhofer, R. The Urban Last Mile Problem: Autonomous Drone Delivery to Your Balcony. In Proceedings of the 2019 International Conference on Unmanned Aircraft Systems (ICUAS), Atlanta, GA, USA, 11–14 June 2019; pp. 1005–1012.
3. Dissanayaka, D.; Wanasinghe, T.R.; Silva, O.D.; Jayasiri, A.; Mann, G.K.I. Review of Navigation Methods for UAV-Based Parcel Delivery. *IEEE Trans. Autom. Sci. Eng.* **2022**, *21*, 1068–1082. [CrossRef]
4. He, W.; Qi, X.; Liu, L. A novel hybrid particle swarm optimization for multi-UAV cooperate path planning. *Appl. Intell.* **2021**, *51*, 7350–7364. [CrossRef]
5. Julius Fusic, S.; Ramkumar, P.; Hariharan, K. Path planning of robot using modified dijkstra Algorithm. In Proceedings of the 2018 National Power Engineering Conference (NPEC), Madurai, India, 9–10 March 2018; pp. 1–5.
6. Pehlivanoglu, Y.V. A new vibrational genetic algorithm enhanced with a Voronoi diagram for path planning of autonomous UAV. *Aerosp. Sci. Technol.* **2012**, *16*, 47–55. [CrossRef]
7. Meng, B. UAV Path Planning Based on Bidirectional Sparse A* Search Algorithm. In Proceedings of the 2010 International Conference on Intelligent Computation Technology and Automation, Changsha, China, 11–12 May 2010; Volume 3, pp. 1106–1109.
8. Li, J.; Deng, G.; Luo, C.; Lin, Q.; Yan, Q.; Ming, Z. A Hybrid Path Planning Method in Unmanned Air/Ground Vehicle (UAV/UGV) Cooperative Systems. *IEEE Trans. Veh. Technol.* **2016**, *65*, 9585–9596. [CrossRef]
9. Hangxuan, H.; Haibin, D. A multi-strategy pigeon-inspired optimization approach to active disturbance rejection control parameters tuning for vertical take-off and landing fixed-wing UAV. *Chin. J. Aeronaut.* **2022**, *35*, 19–30.
10. Pehlivanoglu, Y.V.; Pehlivanoglu, P. An enhanced genetic algorithm for path planning of autonomous UAV in target coverage problems. *Appl. Soft Comput.* **2021**, *112*, 107796. [CrossRef]
11. Phung, M.D.; Ha, Q.P. Safety-enhanced UAV path planning with spherical vector-based particle swarm optimization. *Appl. Soft Comput.* **2021**, *107*, 107376. [CrossRef]
12. Yang, P.; Tang, K.; Lozano, J.A.; Cao, X. Path Planning for Single Unmanned Aerial Vehicle by Separately Evolving Waypoints. *IEEE Trans. Robot.* **2015**, *31*, 1130–1146. [CrossRef]
13. Pan, J.S.; Liu, N.; Chu, S.C. A Hybrid Differential Evolution Algorithm and Its Application in Unmanned Combat Aerial Vehicle Path Planning. *IEEE Access* **2020**, *8*, 17691–17712. [CrossRef]
14. Yu, X.; Li, C.; Zhou, J.F. A constrained differential evolution algorithm to solve UAV path planning in disaster scenarios. *Knowl.-Based Syst.* **2020**, *204*, 106209. [CrossRef]
15. Li, D.; Wang, L.; Cai, J.; Ma, K.; Tan, T. Research on Terminal Distance Index-Based Multi-Step Ant Colony Optimization for Mobile Robot Path Planning. *IEEE Trans. Autom. Sci. Eng.* **2022**, *20*, 2321–2337. [CrossRef]
16. Yu, X.; Chen, W.N.; Gu, T.; Yuan, H.; Zhang, H.; Zhang, J. ACO-A*: Ant Colony Optimization Plus A* for 3-D Traveling in Environments With Dense Obstacles. *IEEE Trans. Evol. Comput.* **2019**, *23*, 617–631. [CrossRef]
17. YongBo, C.; YueSong, M.; JianQiao, Y.; XiaoLong, S.; Nuo, X. Three-dimensional unmanned aerial vehicle path planning using modified wolf pack search algorithm. *Neurocomputing* **2017**, *266*, 445–457. [CrossRef]
18. Shao, S.; Peng, Y.; He, C.; Du, Y. Efficient path planning for UAV formation via comprehensively improved particle swarm optimization. *ISA Trans.* **2020**, *97*, 415–430. [CrossRef] [PubMed]
19. Zhang, X.; Xia, S.; Li, X.; Zhang, T. Multi-objective particle swarm optimization with multi-mode collaboration based on reinforcement learning for path planning of unmanned air vehicles. *Knowl.-Based Syst.* **2022**, *250*, 109075. [CrossRef]
20. Zhao, R.; Wang, Y.; Xiao, G.; Liu, C.; Hu, P.; Li, H. A method of path planning for unmanned aerial vehicle based on the hybrid of selfish herd optimizer and particle swarm optimizer. *Appl. Intell.* **2022**, *52*, 16775–16798. [CrossRef]
21. Shin, J.J.; Bang, H. UAV path planning under dynamic threats using an improved PSO algorithm. *Int. J. Aerosp. Eng.* **2020**, *2020*. [CrossRef]
22. Sanders, A. *An Introduction to Unreal Engine 4*; CRC Press: Boca Raton, FL, USA, 2016.
23. Perlin, K. An Image Synthesizer. *SIGGRAPH Comput. Graph.* **1985**, *19*, 287–296. [CrossRef]
24. Besada-Portas, E.; de la Torre, L.; Jesus, M.; de Andrés-Toro, B. Evolutionary trajectory planner for multiple UAVs in realistic scenarios. *IEEE Trans. Robot.* **2010**, *26*, 619–634. [CrossRef]
25. Patel, J.S.; Fioranelli, F.; Anderson, D. Review of radar classification and RCS characterisation techniques for small UAVs or drones. *IET Radar Sonar Navig.* **2018**, *2*, 911–919. [CrossRef]
26. Anderson, E.; Beard, R.; McLain, T. Real-time dynamic trajectory smoothing for unmanned air vehicles. *IEEE Trans. Control. Syst. Technol.* **2005**, *13*, 471–477. [CrossRef]
27. Wu, X.; Bai, W.; Xie, Y.; Sun, X.; Deng, C.; Cui, H. A hybrid algorithm of particle swarm optimization, metropolis criterion and RTS smoother for path planning of UAVs. *Appl. Soft Comput.* **2018**, *73*, 735–747. [CrossRef]
28. Elhoseny, M.; Tharwat, A.; Hassanien, A.E. Bezier curve based path planning in a dynamic field using modified genetic algorithm. *J. Comput. Sci.* **2018**, *25*, 339–350. [CrossRef]
29. Zhou, X.; Zhu, J.; Zhou, H.; Xu, C.; Gao, F. EGO-Swarm: A Fully Autonomous and Decentralized Quadrotor Swarm System in Cluttered Environments. In Proceedings of the 2021 IEEE International Conference on Robotics and Automation (ICRA), Xian, China, 30 May–5 June 2021; pp. 4101–4107.

30. Qu, C.; Gai, W.; Zhang, J.; Zhong, M. A novel hybrid grey wolf optimizer algorithm for unmanned aerial vehicle (UAV) path planning. *Knowl.-Based Syst.* **2020**, *194*, 105530. [CrossRef]
31. Xue, Y.; Sun, J.Q. Solving the path planning problem in mobile robotics with the multi-objective evolutionary algorithm. *Appl. Sci.* **2018**, *8*. [CrossRef]
32. Kennedy, J.; Eberhart, R. Particle swarm optimization. In Proceedings of the ICNN'95—International Conference on Neural Networks, Perth, WA, Australia, 27 November–1 December 1995; Volume 4, pp. 1942–1948.
33. Zhou, X.; Wang, Z.; Ye, H.; Xu, C.; Gao, F. EGO-Planner: An ESDF-Free Gradient-Based Local Planner for Quadrotors. *IEEE Robot. Autom. Lett.* **2021**, *6*, 478–485. [CrossRef]
34. Zhang, X.; Duan, H. An improved constrained differential evolution algorithm for unmanned aerial vehicle global route planning. *Appl. Soft Comput. J.* **2015**, *26*, 270–284. [CrossRef]
35. Wang, J.; Chi, W.; Li, C.; Wang, C.; Meng, M.Q.H. Neural RRT*: Learning-Based Optimal Path Planning. *IEEE Trans. Autom. Sci. Eng.* **2020**, *17*, 1748–1758. [CrossRef]
36. Zheng, L.; Zhang, P.; Tan, J.; Chen, M. The UAV Path Planning Method Based on Lidar. In *Intelligent Robotics and Applications*; Yu, H., Liu, J., Liu, L., Ju, Z., Liu, Y., Zhou, D., Eds.; Springer: Cham, Switzerland, 2019; pp. 303–314.
37. Tao, X.; Guo, W.; Li, Q.; Ren, C.; Liu, R. Multiple scale self-adaptive cooperation mutation strategy-based particle swarm optimization. *Appl. Soft Comput.* **2020**, *89*, 106124. [CrossRef]
38. Salhi, S.; Petch, R.J. A GA Based Heuristic for the Vehicle Routing Problem with Multiple Trips. *J. Math. Model. Algorithms* **2007**, *6*, 591–613. [CrossRef]
39. Giernacki, W.; Skwierczyński, M.; Witwicki, W.; Wroński, P.; Kozierski, P. Crazyflie 2.0 quadrotor as a platform for research and education in robotics and control engineering. In Proceedings of the 2017 22nd International Conference on Methods and Models in Automation and Robotics (MMAR), Międzyzdroje, Poland, 28–31 August 2017; pp. 37–42.
40. Guay, R.; Drolet, G.; Bray, J.R. Measurement and modelling of the dynamic radar cross-section of an unmanned aerial vehicle. *IET Radar Sonar Navig.* **2017**, *11*, 1155–1160. [CrossRef]
41. Shah, S.; Dey, D.; Lovett, C.; Kapoor, A. AirSim: High-Fidelity Visual and Physical Simulation for Autonomous Vehicles. In *Field and Service Robotics*; Hutter, M., Siegwart, R., Eds.; Springer: Cham, Switzerland, 2018; pp. 621–635.

Disclaimer/Publisher's Note: The statements, opinions and data contained in all publications are solely those of the individual author(s) and contributor(s) and not of MDPI and/or the editor(s). MDPI and/or the editor(s) disclaim responsibility for any injury to people or property resulting from any ideas, methods, instructions or products referred to in the content.

Article

Research on Unmanned Aerial Vehicle (UAV) Visual Landing Guidance and Positioning Algorithms

Xiaoxiong Liu *, Wanhan Xue , Xinlong Xu, Minkun Zhao and Bin Qin

School of Automation, Northwestern Polytechnical University, Xi'an 710129, China;
xuewanhan@mail.nwpu.edu.cn (W.X.); xuxinlong@mail.nwpu.edu.cn (X.X.);
minkun_zhao@mail.nwpu.edu.cn (M.Z.); binq3638@mail.nwpu.edu.cn (B.Q.)
* Correspondence: nwpulxx@outlook.com

Abstract: Considering the weak resistance to interference and generalization ability of traditional UAV visual landing navigation algorithms, this paper proposes a deep-learning-based approach for airport runway line detection and fusion of visual information with IMU for localization. Firstly, a coarse positioning algorithm based on YOLOX is designed for airport runway localization. To meet the requirements of model accuracy and inference speed for the landing guidance system, regression loss functions, probability prediction loss functions, activation functions, and feature extraction networks are designed. Secondly, a deep-learning-based runway line detection algorithm including feature extraction, classification prediction and segmentation networks is designed. To create an effective detection network, we propose efficient loss function and network evaluation methods Finally, a visual/inertial navigation system is established based on constant deformation for visual localization. The relative positioning results are fused and optimized with Kalman filter algorithms. Simulation and flight experiments demonstrate that the proposed algorithm exhibits significant advantages in terms of localization accuracy, real-time performance, and generalization ability, and can provide accurate positioning information during UAV landing processes.

Keywords: computer vision; deep neural networks; autonomous landing; combined navigation

Citation: Liu, X.; Xue, W.; Xu, X.; Zhao, M.; Qin, B. Research on Unmanned Aerial Vehicle (UAV) Visual Landing Guidance and Positioning Algorithms. *Drones* **2024**, *8*, 257. https://doi.org/10.3390/drones8060257

Academic Editors: Jihong Zhu, Heng Shi, Zheng Chen and Minchi Kuang

Received: 25 April 2024
Revised: 31 May 2024
Accepted: 7 June 2024
Published: 12 June 2024

Copyright: © 2024 by the authors. Licensee MDPI, Basel, Switzerland. This article is an open access article distributed under the terms and conditions of the Creative Commons Attribution (CC BY) license (https://creativecommons.org/licenses/by/4.0/).

1. Introduction

Landing is a critical phase in unmanned aerial vehicle (UAV) flight. Currently, there are three main navigation methods used during UAV landings: instrument landing system (ILS), microwave landing system (MLS), and Global Positioning System (GPS). However, these navigation methods heavily rely on external equipment. Furthermore, they have drawbacks such as expensive equipment, poor maneuverability, difficulty in installation, susceptibility to signal interference, and vulnerability to deception. Therefore, the development of a fully autonomous, reliable, and stable autonomous landing navigation system has become an urgent problem.

With the continuous development of visual perception and navigation technologies, the application of visual navigation in the autonomous landing process of unmanned aerial vehicles (UAVs) has gained widespread attention. Visual navigation offers several advantages: (1) It does not require establishing an information link with the outside world, making it a completely autonomous navigation system that is immune to interference. (2) There is no need to set up expensive communications equipment on the ground, which is less costly. (3) It requires minimal prior information about the landing airport, allowing UAVs to land in relatively unfamiliar or temporary airfields.

UAV visual guidance landing image processing is particularly important, the current research methods on image processing can be divided into traditional methods to detect the runway line, traditional methods to detect cooperative signs and deep learning methods to detect the runway. Traditional methods are faster and easier to deploy but are more sensitive to the environment; traditional methods to detect cooperative signs are more accurate but require certain labeling conditions; the deep learning methods used in this paper is environmentally robust and require fewer external conditions. These three methods have their own advantages and disadvantages and are used in different application scenarios

The detection of runway lines using traditional methods generally involves five steps: image preprocessing, feature extraction, feature selection, runway line fitting, and runway line detection and classification. In [1], the authors employ multi-sensor image fusion to obtain image data of the runway. They use support vector machine (SVM) for runway recognition and extract runway edges and ground lines using edge detection and the Hough transform. Finally, they obtain the aircraft's attitude data. In [2], the authors extract the horizon and runway edges using the Hough transform. They estimate the aircraft's pose separately using the horizon and runway edges and track the runway using template matching. In [3], the authors detect runway lines using the Canny edge detector and Hough transform.

The detection of cooperative markers using traditional methods is similar to the process of detecting runway lines. However, adjustments need to be made based on the characteristics and requirements of the cooperative markers. Generally, it involves five steps: image preprocessing, feature extraction, feature selection, marker detection and localization, and marker classification and recognition. In [4], a monocular single-frame vision measurement algorithm for autonomous landing of unmanned helicopters is derived. By detecting square cooperative targets, the position and attitude of the unmanned helicopter are calculated, totaling six parameters. In [5], the landing guidance process for unmanned aerial vehicles (UAVs) is divided into two stages. In the initial stage of landing, runway corner features are used as guidance markers, and in the later stage of landing, Apriltag labels are recognized to guide the UAV's landing. In [6], a fixed-wing UAV autonomous landing method based on binocular vision is proposed.

Using deep learning for runway line detection involves several steps, including data preparation and preprocessing, building the deep learning model, model training, test evaluation, model optimization, and model inference. Deep learning methods enable end-to-end training using a large amount of image data, allowing for the learning of more complex feature representations and improving the performance and robustness of runway line detection. Additionally, deep learning methods can adapt to different lighting conditions, complex backgrounds, and exhibit good generalization capability. In [7], a deep learning approach is used for UAV detection and tracking. By performing triangulation and filtering calculations on the detected objects in a binocular vision system, the spatial position of the UAV is estimated. In the positioning stage, a Kalman filter is used to smooth the spatial trajectory, approximating the area where the target is likely to appear in the current frame. This improves the accuracy of estimation while reducing the difficulty of tracking. In [8], an onboard-YOLO algorithm suitable for lightweight and efficient usage on UAV onboard systems is proposed. It utilizes separable convolutions instead of conventional convolutional kernels, effectively improving the detection speed.

However, single visual navigation alone may not satisfy the requirements for precision and reliability in autonomous UAV landing. It is necessary to complement the limitations of visual navigation by leveraging measurement information from other sensors. Traditional methods of combined navigation include GPS/INS(combined navigation), INS/visual combined navigation, GPS/INS/visual combined navigation, and so on. As the number of sensors increases, the accuracy and robustness of the navigation system continuously improve, enhancing the performance of autonomous UAV landing. In [9], a visual–inertial navigation fusion algorithm is proposed, where position and attitude alignment are achieved using Kalman filtering. The position alignment estimates velocity errors and accelerometer biases, while the attitude alignment estimates attitude errors and gyroscope drift. The estimated alignment errors and the attitude information output by the visual navigation system are used to correct the inertial navigation attitude. In [10], YOLOv3 is used to detect the runway region of interest (ROI), and an RDLines algorithm is employed to extract the left and right runway lines from the ROI. A visual/inertial combined navigation model is then designed within the framework of square-root unscented Kalman filtering.

In a visual navigation system, accurate detection of the runway and runway lines is crucial for the system's performance [11]. Traditional methods for runway line detection, such as those based on the Hough transform and LSD line detection, offer good real-time performance. However, their generalization to different scenarios is poor and they heavily rely on manually designed features and parameters. Although researchers have proposed more comprehensive feature description methods (e.g., SIFT, ORB), their robustness and accuracy still cannot fully meet the demands of practical applications. Therefore, traditional line detection methods are not suitable for use in visual navigation systems. With the advancement of parallel computing capabilities brought about by hardware such as GPUs, deep neural networks have greatly improved detection accuracy and robustness. This has led to a new stage in object detection and the development of a series of deeper, faster training, and more accurate deep neural networks. Utilizing deep learning for runway line detection is a promising choice. Furthermore, considering the limitations of single visual navigation, integrating IMU (inertial measurement unit) information and visual localization results can be a good solution. This combination can provide complementary information and enhance the overall performance of the navigation system.

This paper proposes a deep-learning-based UAV localization method to address the navigation problem in autonomous UAV landing. The simulation and experimental results demonstrate that the proposed algorithm exhibits good robustness, accuracy, and real-time performance. These findings suggest that the algorithm can be used effectively for autonomous UAV landing.

The main contributions of this paper are as follows:

(1) A runway line detection and visual positioning system during visual guidance landing is constructed. The system is divided into four parts: runway ROI selection, runway line detection, visual positioning and combined navigation, thereby providing an end-to-end navigation solution for UAV visual guidance landing.

(2) In view of the requirements of navigation accuracy and real-time performance in this application scenario, the image processing end algorithm is optimized and designed, including optimizing the loss function, optimizing the feature extraction network and feature fusion network, adding an attention mechanism, and optimizing the network structure.

(3) In order to further improve the visual positioning accuracy, the Kalman filter algorithm is used to fuse the IMU information and the visual positioning information. The simulation results show that the combined navigation algorithm can effectively improve the positioning accuracy.

The rest of this paper is organized as follows: The Section 2 of the paper presents the framework of the visual-guided landing localization algorithm. The Section 3 focuses on the runway ROI selection algorithm in the visual landing detection, while the Section 4 explores the runway line detection algorithm for airport runways within the visual landing detection. The Section 5 discusses the visual localization and combined navigation algorithm. Finally, the Section 6 describes the deployment of the algorithm on edge computing devices and presents the results of simulation experiments.

2. Vision-Guided Landing Positioning Algorithm Framework

This paper focuses on visual-guided landing for small fixed-wing unmanned aerial vehicles (UAVs). The main research objectives are image processing algorithms and visual localization algorithms for UAV landing. The specific research content includes runway ROI selection networks, runway line detection networks, UAV position estimation algorithms, and combined navigation algorithms. The effectiveness of these algorithms is then validated in a constructed simulation system. The overall algorithm framework is illustrated in Figure 1.

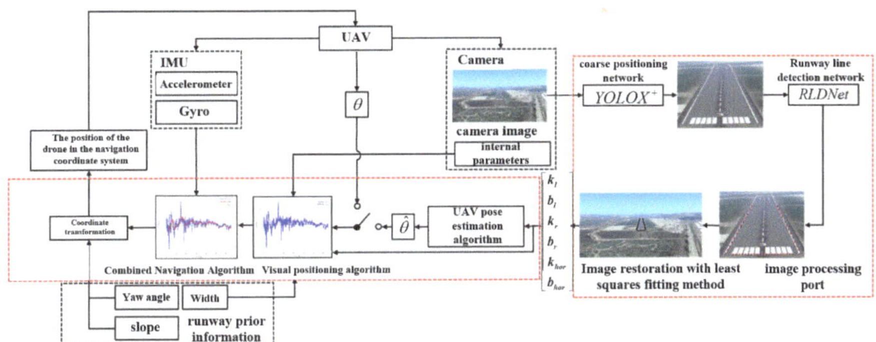

Figure 1. Overall flowchart for visually guided landing.

The algorithm system designed in this paper consists of two main components: the image processing side and the pose estimation combined navigation side. In the image processing side, the images captured by the camera are processed. The YOLOX runway ROI selection network is used to perform rough positioning of the runway lines, which helps exclude interfering objects and ensures that the runway is evenly distributed in the image. The detected bounding boxes are then input into the runway line detection network (RLDNet) for line detection. In this stage, instead of using image segmentation techniques, a specific row (or column) classification method is employed, reducing computational complexity and improving real-time inference. The line detection outputs information such as the slope and intercept of the runway lines. The pose estimation and combined navigation side mainly consist of two algorithms: visual localization and combined navigation. The visual localization algorithm utilizes prior information about the runway, camera intrinsic parameters, UAV attitude information, and information obtained from the detected runway lines to estimate the UAV's position. The visual localization algorithm in this study utilizes a vision-based localization algorithm based on the concept of homogeneous transformation. The position information derived from visual localization is fused with the position information obtained from the IMU using Kalman filtering, resulting in accurate and reliable localization results.

The algorithm designed in this study is applicable to UAV visual landing guidance scenarios. However, most currently available datasets that include airport runways are composed of aerial images, such as the runway images in NWPU-RESISC45, which are all remote sensing aerial images. These datasets cannot meet the design requirements of the algorithm proposed in this study. Therefore, this study combines runway images from the landing perspective in the virtual simulation system Vega Prime, real airport runway images, affine-transformed NWPU-RESISC45 runway images, and simplified runway images. After manual processing and annotation, these images form the dataset used in this study, which includes a total of 2500 landing perspective runway images. Additionally, for the purpose of simulating and validating the visual positioning algorithm and combined navigation algorithm, the dataset collected from the virtual simulation system Vega Prime includes UAV's real poses and IMU data corresponding to the images. The four types of airport runway images included in this dataset are shown in Figure 2.

Furthermore, in order to save annotation time, this study initially annotates the left and right runway lines, as well as the starting runway line. Then, by calculating the position of the rough runway localization box based on the annotated pixel coordinates of the runway line endpoints, the airport runway rough localization dataset is automatically generated. After cropping the original image using the rough localization box, the resolution is adjusted to generate the airport runway line dataset. This allows for the simultaneous generation of datasets for both the airport runway rough localization and the airport runway line detection tasks through a single annotation process.

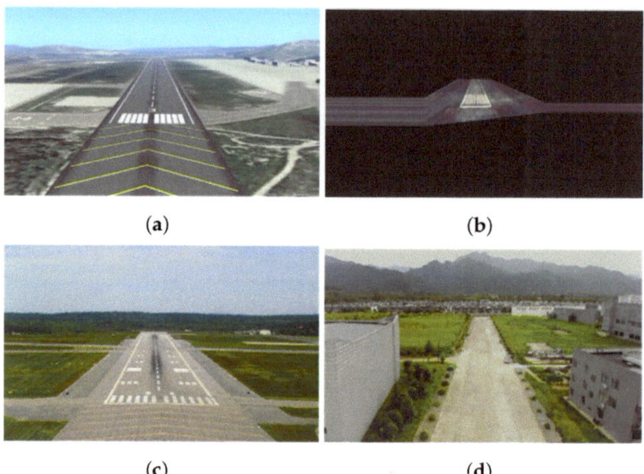

(a)   (b)

(c)   (d)

Figure 2. Construction of the visual landing guidance datasets. (**a**) Runway in Vega Prime; (**b**) Remote sensing image; (**c**) Real runway; (**d**) Airstrip runway.

In order to effectively utilize the dataset and validate the algorithm's performance, this study divided the constructed runway rough localization dataset and runway line detection dataset into training and testing sets using a ratio of 4:1. The training set is utilized to train the runway rough localization network and runway line detection network. The testing set is then used to evaluate the prediction performance of the rough localization network and runway line detection network, as well as to calculate their performance metrics. This division helps with the optimization and adjustment of the networks.

3. Airport Runway Rough Localization Algorithm

As a general-purpose object detection framework, YOLOX has high detection accuracy and speed for most object detection applications. However, due to the higher requirements for image processing accuracy and real-time performance in the scenario of this study, further optimization and design of YOLOX are needed.

3.1. Design of Probability Prediction Loss Function

The probability prediction loss of YOLOX, L_{obj}, is calculated using the binary cross-entropy loss [12]. For a given sample, the binary cross-entropy loss is computed as Equation (1).

$$l(y_i, \hat{y}_i) = \sum_{i=0}^{C} -(y_i \log(\hat{y}_i) + y_i \log(1 - \hat{y}_i)) \tag{1}$$

where y_i represents the ground truth and $\hat{y}_i$ represents the predicted value. For all samples, the binary cross-entropy loss function value is the average of the loss function values for all positive and negative samples. The calculation method is as Equation (2).

$$BCELoss = \frac{1}{N} \sum_{i=1}^{N} l(y_i, \hat{y}_i) \tag{2}$$

where N represents the total number of positive and negative samples.

Focal Loss is a solution for addressing the issue of sample imbalance [13]. Its calculation method is as Equation (3).

$$Focal\ Loss = \begin{cases} -(1 - \hat{p})^{\gamma} \log(\hat{p}) & if\ y_i = 1 \\ -\hat{p}^{\gamma} \log(1 - \hat{p}) & if\ y_i = 0 \end{cases} \tag{3}$$

Let

$$p_t = \begin{cases} \hat{p} & \text{if } y_i = 1 \\ 1 - \hat{p} & \text{else} \end{cases} \tag{4}$$

The expression for Focal Loss can be uniformly represented as Equation (5).

$$Focal\ Loss = -(1 - p_t)^{\gamma} \log(p_t) \tag{5}$$

where p_t reflects the degree of proximity between the predicted value and the ground truth. The larger the value of p_t, the closer the predicted value is to the ground truth, indicating a more accurate classification. Where $\gamma > 0$ is an adjustable factor. Similarly, the expression for the binary cross-entropy loss function can be uniformly represented as Equation (6).

$$L_{ce} = -\log(p_t) \tag{6}$$

Compared to the binary cross-entropy loss function, Focal Loss does not modify the loss function value for inaccurately classified samples, while reducing the weight of the loss function value for accurately classified samples. This ultimately increases the weight of inaccurately classified samples in the overall loss function.

The calculation method of the Focal Loss loss function used in the training process is as Equation (7).

$$Loss = -\alpha(1 - p_t)^{\gamma} \log(p_t) \tag{7}$$

That is, in the traditional Focal Loss, a coefficient is introduced, and $\alpha = 0.25$, $\gamma = 2$. At this point, the model accuracy will slightly improve. In all subsequent experiments in this study, the binary cross-entropy loss is replaced with Focal Loss by default.

3.2. Design of Regression Loss Function

YOLOX calculates the position regression loss for predicting bounding boxes and ground truth boxes using the IoU loss. When the predicted box and the ground truth box do not intersect, the IoU loss function cannot reflect the distance between the predicted box and the ground truth box. In this case, the loss function is non-differentiable, making it unable to optimize the scenario where the two boxes do not intersect. Therefore, this paper replaces the calculation method of the position regression loss with EIoU. The EIoU loss, which reduces the contribution of a large number of anchor frames with less overlap area with the target frame to the predictor frame regression, makes the regression of the predictor frame more focused on high-quality anchor frames. The EIoU is calculated as Equation (8).

$$L_{EIoU} = L_{IoU} + L_{dis} + L_{asp} = 1 - IoU + \frac{\rho^2(c, c^{gt})}{d^2} + \frac{\rho^2(w, w^{gt})}{C_w^2} + \frac{\rho^2(h, h^{gt})}{C_h^2} \tag{8}$$

where $\frac{\rho^2(c, c^{gt})}{d^2}$ denotes the centroid loss, $\frac{\rho^2(w, w^{gt})}{C_w^2}$ is the width loss, $\frac{\rho^2(h, h^{gt})}{C_h^2}$ is the height loss, and C_w and C_h are the widths and heights of the smallest outer bounding box containing the prediction box and the target box.

3.3. Design of Feature Extraction Network

The feature extraction network in YOLOX is a multi-branch residual structure called CSPDarknet53. Since the algorithm in this paper needs to be deployed on edge devices, in order to further compress the model parameter size while improving model accuracy, the feature extraction network of YOLOX is replaced with EfficientRe [14], GhostNet [15], MobileNetV3-Large, and MobileNetV3-Smallin [16] in separate experiments. The performance of different feature extraction networks is tested, and the experimental results are shown in Table 1.

Table 1. The ablation experiments on different feature extraction networks.

Feature Extraction Network	Evaluation Metrics					
	$AP_{0.75}$	F1	*Recall* (%)	Precision	FLOPS (G)	Param (M)
EfficientRep	93.42	0.86	90.08	82.09	50.229	17.05
GhostNet	94.95	0.86	90.6	82.22	20.309	9.06
MobileNetV3-Small	93.82	0.85	89.11	81.98	**18.473**	**6.849**
MobileNetV3-Large	**95.01**	**0.87**	**91.37**	**82.35**	31.703	25.974

From the experimental results, it can be seen that for the dataset used in this paper, introducing EfficientRep does not significantly improve the model's performance. On the contrary, the introduction of EfficientRep leads to a significant increase in model size and computational complexity. When using MobileNetV3-Large as the feature extraction network, the model's performance is significantly improved; however, the trade-off is a substantial increase in both model size and computational complexity. As a comparison, this paper uses MobileNetV3-Small as the feature extraction network, which significantly reduces the model size and computational complexity. Although $AP_{0.75}$ is slightly higher than EfficientRep, it is much lower than MobileNetV3-Large. When using GhostNet as the feature extraction network, there is a significant improvement in $AP_{0.75}$, recall, and precision. In terms of parameter size, the model is comparable to using CSPDarknet53 as the feature extraction network, but the computational complexity is substantially reduced. Given the requirements of the application scenario in this paper regarding model performance and computational complexity, from this point forward, the experiments in this paper default to using GhostNet as the feature extraction network.

In addition, the feature extraction network in YOLOX includes the SiLU activation function. As the network deepens, models that use the SiLU activation function tend to experience a noticeable decrease in classification accuracy. In this paper, on the basis of the YOLOX network structure, the SiLU activation function is replaced with the Mish activation function. With the deepening of the network, the Mish activation function can still maintain a higher classification accuracy. Equation (9) is the expression of the Mish activation function [17].

$$Mish = x \times \tanh(\ln(1 + e^x)) \tag{9}$$

3.4. Feature Fusion Networks and Channel Attention Mechanisms

In this paper, the ordinary convolutions in the YOLOX feature fusion network are replaced with group shuffle convolution (GSConv). GSConv reduces the model's parameter count while preserving the connections between channels in the feature layers, ensuring that the model's accuracy is not compromised [18,19]. After the feature layers go through ordinary convolutions, GSConv applies depth-wise separable convolutions, and then, concatenates the feature layers before the depth-wise separable convolutions in the channel direction. Finally, the shuffle structure is used to fuse the feature layers from both ordinary convolutions and depth-wise separable convolutions. Additionally, if GSConv is used throughout the entire model, the model will become deeper and may have an impact on real-time performance. Therefore, in this paper, only the ordinary convolutions in the YOLOX feature fusion network are replaced with GSConv, specifically, replacing the BottleNeck in CSPLayer with GSBottleNeck.

Attention mechanism can allocate computational resources to more important tasks without increasing computational complexity significantly, especially when resources are limited.

The efficient channel attention (ECA) mechanism builds upon the SE channel attention mechanism [20] by replacing the fully connected layer with a (1×1) convolutional layer. This allows for learning the weight information between channels without reducing the channel dimension, and it also helps reduce the number of parameters [21]. The ECA mechanism first applies global average pooling to the input feature layer to obtain $(1 \times 1 \times C)$-dimensional feature maps. Then, through $(1 \times 1 \times k)$ convolutional operations, it learns the importance of different channels.

The size of the convolutional kernel affects the receptive field, and larger convolutional kernels are needed for feature layers with a larger number of channels. Therefore, the kernel size can be dynamically adjusted using a function. The calculation method for the convolutional kernel is as Equation (10).

$$k = \psi(C) = \left| \frac{\log_2(C)}{2} + \frac{1}{2} \right|_{odd} \tag{10}$$

In this context, k represents the number of channels in the convolutional kernel, C represents the number of channels in the input convolutional layer, and $||_{odd}$ indicates that the size of the convolutional kernel must be an odd number.

In this paper, we added the channel attention mechanism ECA in the middle of YOLOX's feature extraction network and PAFPN.

4. Airport Runway Line Detection Algorithm

4.1. Detection Principle

In this paper, the idea behind designing the runway line detection algorithm is to select the correct location of the left and right runway lines in a predefined row anchor box and the start location of the runway line in a predefined column anchor box using global features. Therefore, the first step is to partition the input image into row anchor boxes and column anchor boxes. Then, each row and column anchor box is further divided into grid cells. In this way, runway line detection can be defined as selecting specific cells within the predefined row/column anchor boxes to represent the positions of the left and right runway lines and the starting runway line.

Assume the maximum number of runways is C, the number of row anchor boxes is h, and the number of grid cells in each row/column anchor box is w, and let X denote the global features of the image. Let f^{ij} represent the classifier for the runway line positions on the ith row/column anchor box of the jth runway. Then, the prediction of the runway line can be expressed as Equation (11).

$$P_{i,j,:} = f^{ij}(X) \tag{11}$$

where $i \in [1, C]$, $j \in [1, h]$, $P_{i,j,:}$ is an S−dimensional vector that represents the probability of the Nth grid of the Mth runway line; F denotes the global features of the image, and it is a $(w + 1)$−dimensional vector. It represents the probability of the $(w + 1)$th grid cells for the ith runway line. For each grid in every row/column anchor box, the network predicts the probability of the corresponding grid. Thus, the grid with the highest probability represents the predicted position of the runway line. If no runway line is predicted on a particular row/column anchor box, then the probability of the last grid in that anchor box is set to 1.

4.2. Network Structure

The network architecture consists of three parts: feature extraction, classification prediction, and segmentation. The feature extraction part is responsible for extracting the features of the runway lines from the image. The classification prediction part is used to classify these features, while the segmentation part helps to fuse multi-scale features, improving the detection accuracy. To improve the network's inference speed, the segmentation part is only used during training and not utilized during the inference prediction stage [22]. The network structure is illustrated in Figure 3.

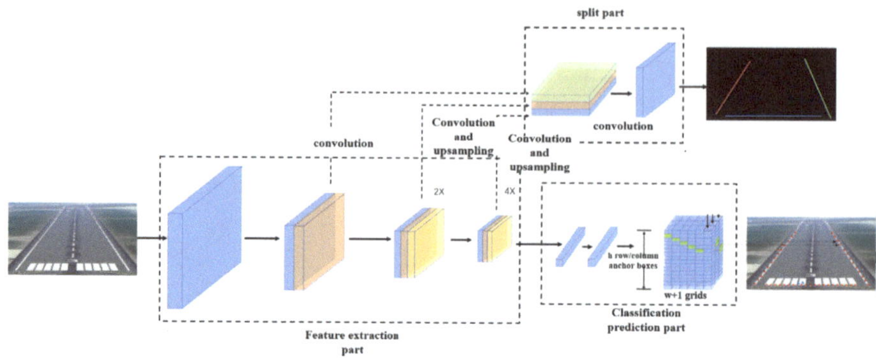

Figure 3. Diagram of runway line detection network structure.

The role of the feature extraction part is to extract the features of the runway lines and provide them to the classification prediction part. Common feature extraction networks, such as ResNet, VGG, MobileNet, ShuffleNet, have been proven to exhibit strong feature extraction capabilities for classification tasks. In this algorithm, ResNet is used as the feature extraction network. ResNet is a type of residual network that addresses the problem of increased loss with increasing network depth [23]. Considering the need for extracting a relatively limited set of features and the requirement for real-time processing on board computers, the algorithm utilizes the lightest variant of ResNet, which is ResNet18 (18 represents the number of layers that require parameter updating through training).

In the classification prediction, the last feature layer of ResNet18 is initially downsampled by the convolutional operation, reducing the number of channels. Then, the resulting feature layer is flattened into a column, resulting in a dimension of $(1 \times 1 \times 1800)$. Next, the feature layer dimension is transformed to $(1 \times 1 \times 13,635)$ using a fully connected layer and the ReLU activation function. Finally, the dimension of the fully connected layer is reshaped to $((w+1) \times h \times 3)$ using the reshape operation. In this context, $w+1$ represents the number of grid cells for each row/column anchor box, h represents the number of row/column anchor boxes, and 3 corresponds to the total number of runway lines. And the condition in Equation (12) needs to be satisfied.

$$h \times (w+1) \times 3 = 13635 \tag{12}$$

Performing softmax on each row/column anchor box for the three runway lines can compute the grid with the highest probability within each anchor box. This is used as the predicted position of the track line and is utilized for calculating classification loss, structural loss, and association loss.

In the segmentation network, the last three feature layers of ResNet18 are first subjected to convolution and upsampling operations. These three feature layers are then concatenated along the channel dimension. Subsequently, convolution is applied to reduce the number of channels in the feature layer to four, these are used for calculating the segmentation loss.

4.3. Loss Function

The classification loss during the network training process can be represented as Equation (13).

$$L_{cls} = \sum_{i=1}^{C} \sum_{j=1}^{h} L_{CE}\left(P_{i,j,:}, T_{i,j,:}\right) \tag{13}$$

Here, L_{CE} represents the cross-entropy loss, and $T_{i,j,:}$ represents the ground truth of the track line position on the jth row/column anchor box of the ith track line.

In addition to the classification loss, several other loss functions are used in the algorithm based on the structural prior information of the runway lines. These loss functions are utilized to represent the position relationships of the runway lines, allowing the neural network to learn the structural information of the runway lines. Since each track line must be continuous, the predicted points of the runway lines in adjacent row/column anchor boxes should be as close as possible. Therefore, the continuity of the predicted runway lines can be achieved by constraining the distribution of the classification vectors on adjacent row anchors. The loss function can be represented as Equation (14).

$$L_{sim} = \sum_{i=1}^{C} \sum_{j=1}^{h-1} \|P_{i,j,:} - P_{i,j+1,:}\|_1 \tag{14}$$

Here, $P_{i,j,:}$ represents the predicted track line position on the jth row/column anchor box of the ith track line, $P_{i,j+1,:}$ represents the ground truth of the track line position on the $(j+1)$th row/column anchor box of the ith track line. In the loss function, the distance between the predicted track line positions and the ground truth is minimized through an L_1-norm constraint.

Additionally, based on the prior information that each track line is a straight line, the predicted track points can be constrained using second-order differences. The formula for the second-order difference can be represented as Equation (15).

$$L_{shp} = \sum_{i=1}^{C} \sum_{j=1}^{h-2} \|((Loc_{i,j} - Loc_{i,j+1}) - (Loc_{i,j+1} - Loc_{i,j+2})\|_2 \tag{15}$$

Here, $Loc_{i,j}$ represents the predicted point on the jth row/column anchor box of the jth track line, and its calculation method is given as Equation (16).

$$Loc_{i,j} = \sum_{k=1}^{w} k \cdot Prob_{i,j,k} \tag{16}$$

Here, $Prob_{i,j,k}$ represents the probability of the ith track line in the kth grid of the jth row/column anchor box, and its calculation method is given as Equation (17).

$$Loc_{i,j,:} = softmax(P_{i,j,1:w}) \tag{17}$$

Based on the above, the overall structural loss of the network can be represented as Equation (18).

$$L_{str} = L_{sim} + \lambda L_{shp} \tag{18}$$

In addition to the classification loss and structural loss, this paper incorporates an auxiliary segmentation task that utilizes multi-scale features for local feature modeling during the training process. The auxiliary segmentation loss is calculated using the cross-entropy function. To improve the performance of the algorithm, this segmentation task is removed during the testing phase.

Real runway lines are parallel to each other, but due to perspective, the left and right runway lines in the image become closer as they move upward. Based on this prior condition, this paper designs the association loss for the runway lines. The design logic is as follows: if the left and right runway lines above the image are farther apart compared to the left and right runway lines below, a loss is generated; otherwise, no loss is generated. The calculation process of the association loss is shown in Algorithm 1, where ε_T represents the tolerable error threshold, which is defined in terms of vertical grids.

Algorithm 1 Correlation Loss Calculation Process

Input: The predicted track line point on the $j - th$ row anchor box of the $i - th$ track line:
$Loc_{i,j}, i \in [0,1], j \in [0, h-1]$.
Output: The value of the association loss :L_{reduce}

1: **for** $j = 0$ to $h - 1$ **do**
2:      $D_j = Loc_{1,j} - Loc_{0,j}$
3:      $\Delta D_j = D_j - D_{j-1}, j \in [1, h-1]$
4:      $M_j = 0.5 \times (|\Delta D_j| - \Delta D_j) - \varepsilon_T, j \in [1, h-1]$
5:      $N_j = 0.5 \times (|M_j| + M_j), j \in [1, h-1]$
6:      $L_{reduce} = \sum\limits_{j=1}^{h-1} \|N_j\|_1, j \in [1, h-1]$;
7: **end for**

In summary, the overall loss of the algorithm can be represented as Equation (19).

$$L_{total} = \alpha L_{cls} + \beta L_{str} + \gamma L_{seg} + \theta L_{reduce} \tag{19}$$

Here, L_{cls} represents the classification loss, L_{str} represents the structural loss, L_{seg} represents the segmentation loss, and L_{reduce} represents the association loss. α, β, γ, and θ represent the weights assigned to the classification loss, structural loss, segmentation loss, and association loss, respectively.

4.4. Evaluation

In order to ensure the stability of distance calculation and accurately reflect the differences between the ground truth and predicted points, this paper first uses the predicted points on the runway lines for least squares fitting to obtain the slope of the fitted line (k). This further allows us to calculate the distance threshold (ε) between the predicted and ground truth points. The calculation method is as Equation (20).

$$\varepsilon = \frac{RealDistance}{cos(arctan(k))} \tag{20}$$

where RealDistance represents the pixel distance between the predicted points and the ground truth points in the horizontal or vertical direction. Considering the actual angle between the left/right runway lines and the x-axis of the pixel coordinate system is close to 90 degrees, and the angle with the y-axis is smaller; the starting track line is close to 90 degrees with the y-axis, and the angle with the x-axis is smaller. Therefore, when calculating the slope (k) of the line, the left/right runway lines adopt the line equation $'x = ky + b'$, and the starting track line adopts the line equation $'y = kx + b'$. This means that the angle with the y-axis of the pixel coordinate system is used when calculating the threshold for the left/right runway lines, and the angle with the x-axis is used when calculating the threshold for the starting track line.

Since the runway lines predicted by the neural network are obtained by fitting the grid points on the predicted track line using the least squares method, the least squares method reduces the impact of prediction errors on track line predictions to a certain extent. Therefore, to evaluate the accuracy of the track line predictions, it is necessary to quantitatively calculate the similarity between the predicted and ground truth runway lines. The evaluation metrics include accuracy, miss rate, and over-detention rate. Accuracy represents the similarity in slope between the predicted and ground truth runway lines, and its calculation method is as Equation (21).

$$acc = \frac{\sum\limits_{j=1}^{w} \sum\limits_{i=1}^{h} \Lambda_{ij}}{w \times h} \tag{21}$$

$$\Lambda_{i,j} = \begin{cases} 1 & |\arctan(k_{real,i,j}) - \arctan(k_{pred,i,j})| < \varepsilon \\ 0 & else \end{cases} \qquad (22)$$

Additionally, the miss rate represents the proportion of the dataset that has ground truth runway lines but no corresponding predicted results. The over-detection rate represents the proportion of the dataset that has predicted runway lines but no corresponding ground truth.

5. Algorithms for Visual Positioning and Combined Navigation

5.1. Algorithm for Visual Positioning

The coordinate systems involved in performing position solving in this paper include the navigation coordinate system, runway coordinate system, airframe coordinate system, camera coordinate system, phase plane coordinate system, and pixel coordinate system, etc., which are defined as Figure 4:

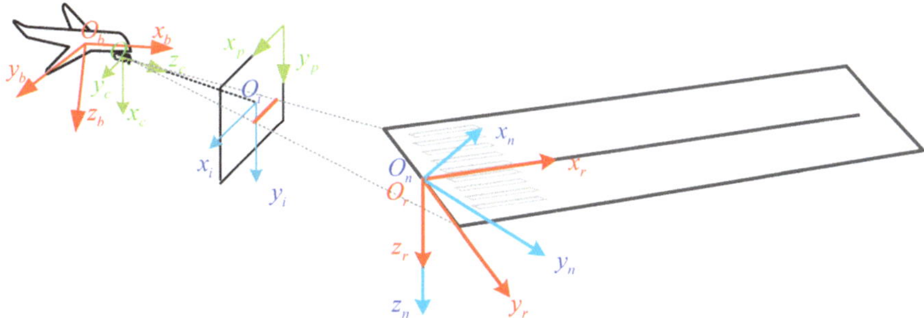

Figure 4. The coordinate systems involved in position algorithm.

The navigation coordinate system $(O_n x_n y_n z_n)$ is defined as the northeast ground coordinate system. The runway coordinate system $(O_r x_r y_r z_r)$ is the base coordinate system for the position solving in the visual guidance landing process of the UAV in this paper. The fuselage coordinate system $(O_b x_b y_b z_b)$ has the origin O_b at the center of the UAV IMU. The camera coordinate system $(O_c x_c y_c z_c)$ is solidly connected to the camera. The image plane coordinate system $(O_i x_i y_i z_i)$ has the image plane located in front of the camera at z = f (f is the focal length). The pixel coordinate system $(O_p x_p y_p z_p)$ is used to describe the original image information.

The conversion relationship from runway coordinate system to pixel coordinate system is as Equation (23) [24].

$$\begin{aligned}
\vec{v}_p &= \frac{1}{Z_c} K\vec{v}_c = \frac{1}{Z_c} K\mathbf{C}_\gamma(\vec{v}_b - \vec{v}_1) \\
&= \frac{1}{Z_c} K\mathbf{C}_\gamma\left[\mathbf{C}_n^b(\vec{v}_n - \vec{v}_0) - \vec{v}_1\right] \\
&= \frac{1}{Z_c} K\mathbf{C}_\gamma\left[\mathbf{C}_n^b(\mathbf{C}_r^n\vec{v}_r - \vec{v}_0) - \vec{v}_1\right] \\
&= \frac{1}{Z_c} K\mathbf{C}_\gamma\mathbf{C}_n^b\mathbf{C}_r^n\left[\mathbf{I}_{3\times3} \mid -\left(\mathbf{C}_n^b\mathbf{C}_r^n\right)^{-1}\left(\mathbf{C}_n^b\vec{v}_0 + \vec{v}_1\right)\right]\begin{bmatrix}\vec{v}_r \\ 1\end{bmatrix} \\
&= \frac{1}{Z_c} K\mathbf{R}_{cr}\left[\mathbf{I}_{3\times3} \mid -\vec{t}_{cr}\right]\begin{bmatrix}\vec{v}_r \\ 1\end{bmatrix} = \frac{1}{Z_c}\mathbf{P}\begin{bmatrix}\vec{v}_r \\ 1\end{bmatrix}
\end{aligned} \qquad (23)$$

where $P = KR_{cr}[I_{3\times3}\mid -\vec{t}_{cr}]$ represents the projection matrix from the runway coordinate system to the pixel coordinate system; $R_{cr} = C_\gamma C_n^b C_r^n = C_\gamma C_r^b$ represents the rotation matrix from the runway coordinate system to the camera coordinate system; and

$\vec{t}_{cr} = \left(C_n^b C_r^n\right)^{-1}\left(C_n^b \vec{v}_0\right)$ represents the coordinates of the camera coordinate system origin in the runway coordinate system. $\vec{v}_r$ indicates the coordinates of a point in the runway coordinate system; $\overrightarrow{v_p}$ denotes the coordinates of the corresponding point in the pixel coordinate system; $\overrightarrow{v_{0r}}$ denotes the coordinates of the origin of the airframe coordinate system in the runway coordinate system; $\overrightarrow{v_0}$ denotes the coordinates of the origin of the fuselage coordinate system in the navigation coordinate system; $\overrightarrow{v_1}$ denotes the coordinates of the camera optical center in the body coordinate system; K is an internal parameter of the camera that can be obtained by calibrating; Z_c is the z-axis coordinate of the point in the camera coordinate system; γ is the camera mounting angle; $\mathbf{C}_\gamma$ is the rotation matrix of the camera mounting angles; $\mathbf{C}_r^b$ represents the rotation matrix from the runway system to the machine system; and $\mathbf{C}_r^n$ represents the rotation matrix from the runway coordinate system to the navigation coordinate system. Since the role of the runway coordinate system relative to the world coordinate system is very small and can be neglected, $\mathbf{C}_r^n$ can be represented as Equation (24).

$$\mathbf{C}_r^n = \left(\mathbf{C}_n^r\right)^T = \left(\mathbf{C}_{\phi_r}\mathbf{C}_{\theta_r}\mathbf{C}_{\psi_r}\right)^T \approx \left(\mathbf{C}_{\theta_r}\mathbf{C}_{\psi_r}\right)^T \tag{24}$$

If the equation of a line in the pixel coordinate system is $y = kx + b$, then any point, $\vec{v}_p = \begin{bmatrix} x_p & y_p & 1 \end{bmatrix}^T$ on that line must satisfy Equation (25).

$$\begin{bmatrix} k & -1 & b \end{bmatrix}\vec{v}_p = \begin{bmatrix} k & -1 & b \end{bmatrix}\begin{bmatrix} x_p \\ y_p \\ 1 \end{bmatrix} = 0 \tag{25}$$

By multiplying both sides of Equation (23) by $\begin{bmatrix} k & -1 & b \end{bmatrix}$, we can obtain

$$\begin{bmatrix} k & -1 & b \end{bmatrix}K\mathbf{R}_{cr}\begin{bmatrix} \mathbf{I}_{3\times3} | & -\vec{t}_{cr} \end{bmatrix}\begin{bmatrix} \vec{v}_r \\ 1 \end{bmatrix} = 0 \tag{26}$$

So $\vec{A} = \begin{bmatrix} a_1 & a_2 & a_3 \end{bmatrix}K\mathbf{R}_{cr} = \begin{bmatrix} a_1 & a_2 & a_3 \end{bmatrix}$, and

$$\begin{bmatrix} a_1 & a_2 & a_3 \end{bmatrix}\begin{bmatrix} 1 & 0 & 0 & -x_{cr} \\ 0 & 1 & 0 & -y_{cr} \\ 0 & 0 & 1 & -z_{cr} \end{bmatrix}\begin{bmatrix} x_r \\ y_r \\ z_r \\ 1 \end{bmatrix} \tag{27}$$

Expanding the above equation, we can obtain

$$a_1 x_r + a_2 y_r + a_3 z_r = a_1 x_{cr} + a_2 y_{cr} + a_3 z_{cr} \tag{28}$$

Assuming the width of the runway is denoted as W_r, then in the runway coordinate system, the coordinates of any point on the left runway line can be represented as Equation (29):

$$\begin{bmatrix} x_r & -\frac{W_r}{2} & 0 \end{bmatrix}^T \tag{29}$$

where x_r is an arbitrary variable. By substituting Equation (29) into Equation (28), we can obtain

$$a_1 x_r - a_2 \frac{W_r}{2} = a_1 x_{cr} + a_2 y_{cr} + a_3 z_{cr} \tag{30}$$

Since Equation (30) holds true for any variation in x_r, we can conclude that $a_1 = 0$, which leads to

$$-\frac{W_r}{2} = y_{cr} + \frac{a_3}{a_2}z_{cr} \tag{31}$$

so

$$\begin{bmatrix} 1 & \frac{a_3^l}{a_2^l} \end{bmatrix} \begin{bmatrix} y_{cr} \\ z_{cr} \end{bmatrix} = -\frac{W_r}{2} \tag{32}$$

Similarly, for the right runway, we have

$$\begin{bmatrix} 1 & \frac{a_3^r}{a_2^r} \end{bmatrix} \begin{bmatrix} y_{cr} \\ z_{cr} \end{bmatrix} = \frac{W_r}{2} \tag{33}$$

By combining Equation (32) and Equation (33), we have

$$\begin{bmatrix} 1 & \frac{a_3^l}{a_2^l} \\ 1 & \frac{a_3^r}{a_2^r} \end{bmatrix} \begin{bmatrix} y_{cr} \\ z_{cr} \end{bmatrix} = \begin{bmatrix} -\frac{W_r}{2} \\ \frac{W_r}{2} \end{bmatrix} \tag{34}$$

For the starting runway line, it must satisfy the following equation in the runway coordinate system:

$$\vec{v}_r = \begin{bmatrix} 0 & x_r & 0 \end{bmatrix}^T \tag{35}$$

where x_r is an arbitrary variable. By substituting Equation (35) into Equation (28), we can obtain

$$a_2^s x_r = a_1^s x_{cr} + a_2^s y_{cr} + a_3^s z_{cr} \tag{36}$$

To ensure that Equation (36) holds true for any variation in x_r, we have $a_2^s = 0$. Therefore, we can conclude that

$$x_{cr} = -\frac{a_3^s}{a_2^s} z_{cr} \tag{37}$$

For the variables a_2 and a_3 in Equation (33) and Equation (37), they can be calculated from $\vec{A}$, as Equation (38).

$$\begin{cases} \vec{A}_l = \begin{bmatrix} k_l & -1 & b_l \end{bmatrix} KR_{cr} = \begin{bmatrix} a_1^l & a_2^l & a_3^l \end{bmatrix} \\ \vec{A}_r = \begin{bmatrix} k_r & -1 & b_r \end{bmatrix} KR_{cr} = \begin{bmatrix} a_1^r & a_2^r & a_3^r \end{bmatrix} \\ \vec{A}_s = \begin{bmatrix} k_s & -1 & b_s \end{bmatrix} KR_{cr} = \begin{bmatrix} a_1^s & a_2^s & a_3^s \end{bmatrix} \end{cases} \tag{38}$$

5.2. Algorithm for Combined Navigation

The visual/inertial fusion navigation system designed in this paper consists of an IMU (gyroscope, accelerometer) and a visual localization system [25]. Initially, the paper utilizes the visual localization system to obtain the UAV's position in the runway coordinate system. The Kalman filter utilizes the difference between the visual localization system's output position and the current position calculated by the combined navigation algorithm as the measurement information for position error [26].

$$\begin{cases} \dot{q}_t = \frac{1}{2} q_t \otimes w_t \\ \dot{v}_t = C_{b,t}^n (a_t) \\ \dot{p}_t = v_t \\ \dot{\varepsilon}_r = -\frac{1}{\tau_g} \varepsilon_r + w_{\varepsilon r} \\ \dot{\nabla}_r = -\frac{1}{\tau_a} \nabla_r + w_{\nabla r} \end{cases} \tag{39}$$

where

$$w_t = w_m - \varepsilon_g \tag{40}$$

$$a_t = a_m - \nabla_a \tag{41}$$

In the equation, ω_m represents the gyroscope measurement value, and a_m represents the accelerometer measurement value. Since the above kinematic state equation contains noise terms that cannot be directly eliminated in practical measurements, it is necessary to estimate these noise terms through the Kalman filter [27]. According to the inertial system error model, after eliminating the noise terms, the IMU error and its derivative should be constant. By using the above two sets of equations to calculate the error for each state variable, we can obtain the error state equation for the state variable $[\theta, v, p, \varepsilon, \nabla_r]^T$:

$$\begin{cases} \delta\dot{\theta} = -[\omega_m - \varepsilon_r]_\times \delta\theta - \varepsilon_r - w_\varepsilon \\ \delta\dot{v} = -C_b^n[a_m - \nabla_r]_\times \delta\theta - C_b^n\nabla_r - C_b^n w_\nabla \\ \delta\dot{p} = \delta v \\ \dot{\varepsilon}_r = -\frac{1}{\tau_g}\varepsilon_r + w_{\varepsilon r} \\ \dot{\nabla}_r = -\frac{1}{\tau_a}\nabla_r + w_{\nabla_r} \end{cases} \tag{42}$$

By taking partial derivatives of each state variable, we can obtain the state-space equations for the following prediction model:

$$\dot{X}(t) = F(t)X(t) + G(t)w(t) \tag{43}$$

where

$$\dot{X}(t) = [\delta\dot{\theta}, \delta\dot{v}, \delta\dot{p}, \dot{\varepsilon}_r, \dot{\nabla}_r]^T \tag{44}$$

$$X(t) = [\delta\theta, \delta v, \delta p, \varepsilon_r, \nabla_r]^T \tag{45}$$

The continuous-time state transition matrix is as Equation (46).

$$F(t) = \begin{bmatrix} -[\omega_m - \varepsilon_r]_\times & 0_{3\times3} & 0_{3\times3} & -I_{3\times3} & 0_{3\times3} \\ -C_b^n[a_m - \nabla_r]_\times & 0_{3\times3} & 0_{3\times3} & 0_{3\times3} & -C_b^n \\ 0_{3\times3} & I_{3\times3} & 0_{3\times3} & 0_{3\times3} & 0_{3\times3} \\ 0_{3\times3} & 0_{3\times3} & 0_{3\times3} & -\frac{1}{\tau_g}I_{3\times3} & 0_{3\times3} \\ 0_{3\times3} & 0_{3\times3} & 0_{3\times3} & 0_{3\times3} & -\frac{1}{\tau_a}I_{3\times3} \end{bmatrix} \tag{46}$$

The continuous-time noise matrix is as Equation (47).

$$G(t) = \begin{bmatrix} -I_{3\times3} & 0_{3\times3} & 0_{3\times3} & 0_{3\times3} \\ 0_{3\times3} & -C_b^n & 0_{3\times3} & 0_{3\times3} \\ 0_{3\times3} & 0_{3\times3} & 0_{3\times3} & 0_{3\times3} \\ 0_{3\times3} & 0_{3\times3} & I_{3\times3} & 0_{3\times3} \\ 0_{3\times3} & 0_{3\times3} & 0_{3\times3} & I_{3\times3} \end{bmatrix} \tag{47}$$

The noise term is given by Equation (48).

$$w(t) = [w_\varepsilon, w_\nabla, w_{\varepsilon r}, w_{\nabla r}]^T \tag{48}$$

Since the position output by the visual localization system is in a discrete form, the measurement equation in the discrete form can be represented as Equation (49).

$$Z_{k+1} = H_{k+1}X_{k+1} + v_{k+1} \tag{49}$$

where

$$Z_{k+1} = \begin{bmatrix} X_n - \hat{p}_x \\ Y_n - \hat{p}_y \\ Z_n - \hat{p}_z \end{bmatrix} \tag{50}$$

$[X_n, Y_n, Z_n]^T$ represents the three-dimensional position coordinates of the UAV in the runway coordinate system, calculated by the visual localization system.

The measurement noise v_{k+1} can be represented as Equation (51).

$$v_{k+1} = n_p \tag{51}$$

And n_p satisfies Equation (52).

$$n_p \sim N(0, \sigma_{n_p}^2) \tag{52}$$

The measurement equation can be represented as Equation (53).

$$\begin{bmatrix} X_n - \hat{p}_x \\ Y_n - \hat{p}_y \\ Z_n - \hat{p}_z \end{bmatrix} = H_p X_{k+1} + n_p \tag{53}$$

where

$$H_p = \begin{bmatrix} \mathbf{0}_{3\times6} & I_{3\times3} & \mathbf{0}_{3\times6} \end{bmatrix} \tag{54}$$

In summary, the structure of the visual/inertial navigation system developed in this paper is shown in Figure 5.

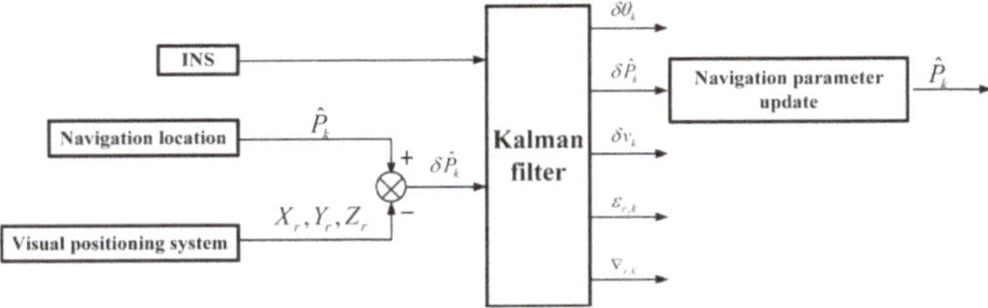

Figure 5. Vision/inertial combined navigation structure diagram.

6. Simulation Results of Detection and Localization Algorithms

6.1. Runway ROI Selection Network Training and Testing

This paper tests the optimized network, and the experimental results are shown in Figure 6. The runway ROI selection network can give the pixel location, category, and probability of the runway, which lays the foundation for the subsequent segmentation of the image based on the pixel location and detection of the runway line.

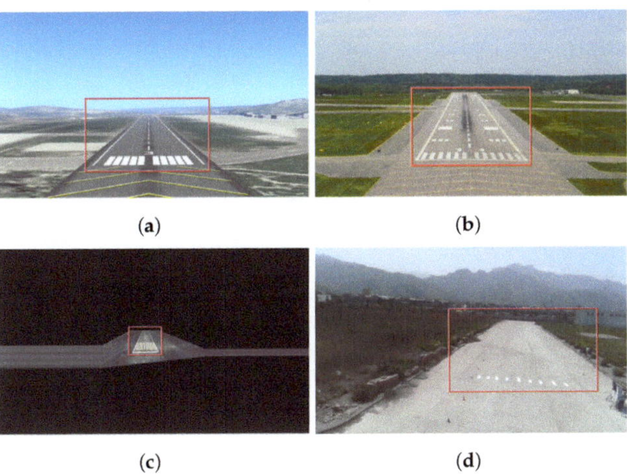

(a)   (b)

(c)   (d)

Figure 6. Prediction results of the runway ROI selection network. (**a**) Airport runway line detection results in virtual environment; (**b**) Real airport runway line detection results; (**c**) Airport runway line detection results after transformation; (**d**) Simple airport runway line detection results.

This paper also tested the performance of the improved model on a constructed dataset and plotted the P-R curves before and after optimization. The experimental results are shown in Figure 7, where YOLOX represents the model before optimization, and YOLOX$^+$ is used to represent the optimized model for ease of description. From the test results, it can be observed that the P-R curve of the optimized model has a larger enclosed area with respect to the horizontal and vertical axes. This confirms the effectiveness of the proposed optimization measures in improving the model's performance.

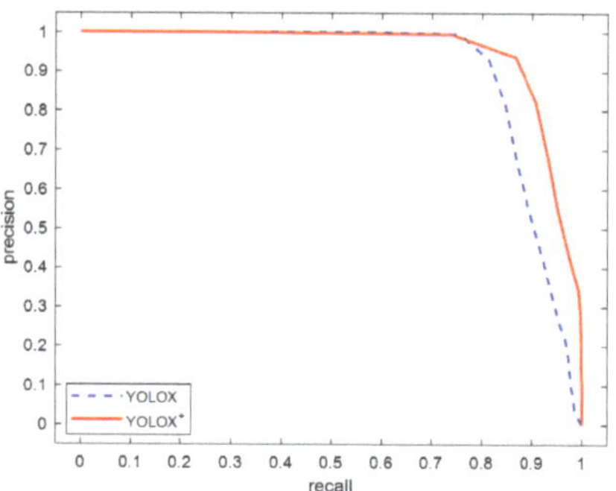

Figure 7. The P-R curves before and after the improvement of YOLOX.

6.2. Runway Line Detection Network Training and Testing

In this paper, the input image resolution for training the runway line detection network is set to 480×480. The Adam optimization strategy is chosen for model optimization due to its ease of implementation, efficiency, and low memory requirements. Additionally, a cosine decay strategy is employed to update the learning rate, with an initial learning rate set to $4 \times e^{-4}$. The batch size is set as 32, and the training process is conducted for 100 epochs. The weights assigned to the classification loss (α), structural loss (β), segmentation loss (γ), and correlation loss (θ) are set to 1, 1, 1, and 0.6, respectively. The row/column anchor box grid is configured with 50 cells.

The trained network is tested using a partitioned test dataset, and the predicted results of the network are fitted using the least squares method. The experimental results are illustrated in Figure 8. From the test results, it can be observed that the runway line detection network accurately predicts the positions of the three runway lines.

Figure 8. Prediction results of the runway line detection network. (**a**) Airport runway line detection results in virtual environment; (**b**) Real airport runway line detection results; (**c**) Airport runway line detection results after transformation; (**d**) Simple airport runway line detection results.

6.3. Visual Localization Simulation Results

In this paper, a test dataset was used to validate the localization algorithm. The test dataset includes the runway width, relative position between the UAV and the runway, and relative attitude angles. The UAV's flight trajectory in the dataset is illustrated in Figure 9, where x and y represent the UAV's position in the runway coordinate system, and height represents the UAV's altitude above the ground.

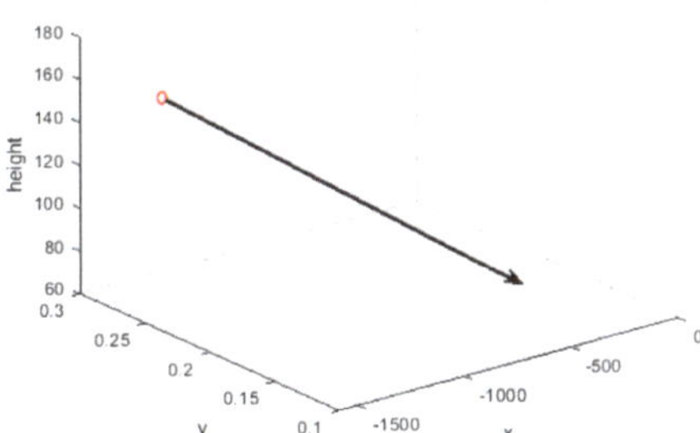

Figure 9. Schematic diagram of the flight path of the drone.

The visual localization results are shown in Figure 10. From the experimental results, it can be observed that the visual localization results follow a similar trend to the ground truth. However, there is some deviation and significant fluctuations between the visual localization results and the ground truth in the initial stage. This deviation is mainly due to the UAV being farther from the runway, resulting in a smaller representation of the runway in the image. Under the same detection accuracy, this leads to larger pixel errors. As the UAV comes closer to the runway, the deviation between the visual localization results and the ground truth gradually reduces.

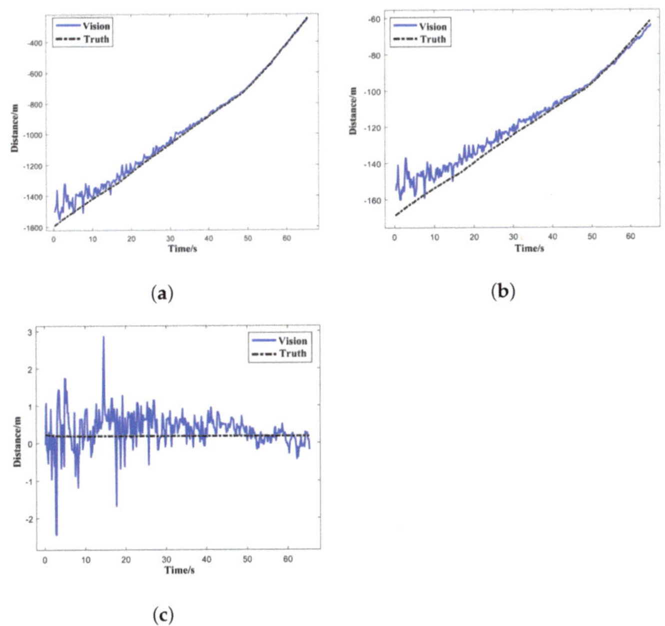

Figure 10. Visual positioning simulation results. (**a**) x-direction visual positioning results; (**b**) z-direction visual positioning results; (**c**) y-direction visual positioning results.

To quantitatively describe the error characteristics of visual localization, this paper calculated the MAE (mean absolute error) and RMSE (root-mean-square error) of the visual localization in three directions. The calculation results are shown in Table 2. Due to the larger pixel errors as the UAV is farther from the runway, the localization errors in all three directions are larger and more fluctuating in the initial stage, with some outliers present. As the UAV approaches the runway, the localization errors in all three directions gradually reduce, and the number of outliers decreases. Additionally, from the error curves, it can be observed that when approaching the runway, the localization error in the x-direction is approximately 4 m. Since the runway has a certain margin in the x-direction, this localization error meets the landing requirements for the UAV on the runway. The localization error in the y-direction is about 0.3 m, which can satisfy the accuracy requirement for UAV landing. However, the localization error in the z-direction is around 2.5 m, which indicates the need for further fusion with other sensor data to improve the positioning accuracy in the z-direction.

Table 2. Visual positioning error characteristics.

Axis	Evaluation Metrics	
	MAE (m)	RMSE (m)
x-axis	23.7810	40.375
y-axis	0.3548	0.5030
z-axis	4.3948	6.2123

6.4. Simulation Results of Combined Navigation Algorithm

During the simulation process, this paper set the time update frequency and measurement update frequency of the Kalman filter and the adaptive fading Kalman filter to 100 Hz and 10 Hz, respectively. Additionally, the initial state covariance matrix P_0, process noise covariance matrix Q_k, and measurement noise covariance matrix R_k for the traditional Kalman filter and the adaptive fading Kalman filter were set as Equation (55).

$$
\begin{cases}
P_0 = diag\{(0.1°)^2, (0.1°)^2, (0.1°)^2, \\
\quad (0.1 \text{ m/s})^2, (0.1 \text{ m/s})^2, (0.1 \text{ m/s})^2, \\
\quad (1 \text{ m})^2, (1 \text{ m})^2, (1 \text{ m})^2, \\
\quad (0.01)^2 \cdot I_{3\times3}, (0.01)^2 \cdot I_{3\times3}\}_{15\times15} \\
Q_k = diag\{(0.01 \text{ rad/s})^2 \cdot I_{3\times3}, (0.01 \text{ m/s}^2)^2 \cdot I_{3\times3}, 0_{9\times9})\}_{3\times15} \\
R_k = diag\{(40 \text{ m})^2, (5 \text{ m})^2, (10 \text{ m})^2\}_{3\times3}
\end{cases}
\tag{55}
$$

This paper compared the localization results of visual localization, traditional Kalman filter-based combined navigation, and adaptive fading Kalman filter-based combined navigation [28,29]. The experimental results are shown in Figure 11. In the initial stage, where the UAV is far from the runway and the runway occupies a small portion of the camera's field of view, the pixel errors on the image processing side are larger. As a result, the visual localization results in the initial stage exhibit significant fluctuations and outliers. The traditional Kalman filter adjusts the weights of the state prediction and innovation in the state prediction process based on the state noise covariance matrix and measurement noise covariance matrix. This helps to reduce the number of outliers to some extent. However, the filtering performance of the traditional Kalman filter heavily relies on the accuracy of the measurement noise covariance matrix. If the statistical properties of the measurement noise are not well understood, the improvement in filtering accuracy may be limited. In the adaptive fading Kalman filter, an adaptive factor is introduced, which reduces the dependence on the accuracy of the measurement noise covariance matrix [30]. This enables better removal of outliers in the visual localization results, resulting in improved filtering performance.

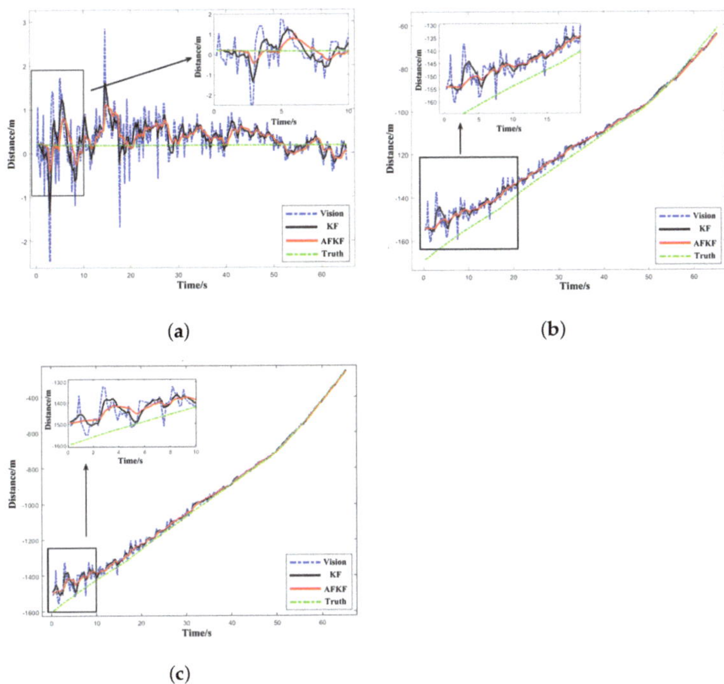

Figure 11. Comparison results of the true value, visual localization, combined navigation based on traditional Kalman filter and combined navigation based on adaptive fading Kalman filter. (**a**) Positioning result in y-direction; (**b**) Positioning result in z-direction; (**c**) Positioning result in x-direction.

In order to quantitatively describe the error characteristics of the two filtering algorithms, this paper calculates the MAE and RMSE of the two filtering algorithms in three directions, and the experimental results are shown in Table 3. From the experimental results, it can be seen that the MAE and RMSE of the traditional Kalman filter and the adaptive fading Kalman filter are smaller than the visual positioning results, and the accuracy has been improved. Since the adaptive factor is added to the adaptive fading Kalman filter, the covariance matrix of the measurement can be dynamically adjusted, thereby reducing the proportion of the measurement in the prediction and estimation, so the filtering effect is better, and its MAE and RMSE are better than those of the traditional Kalman filter. The Mann filter is small, and the positioning accuracy is further improved. The experimental results show the effectiveness of the combined navigation algorithm designed in this paper.

Table 3. Error characteristics of combined navigation algorithm.

Filtering Algorithm	Axis	Evaluation Metrics	
		MAE (m)	RMSE (m)
KF	x	22.7146	36.6027
	y	0.2819	0.3673
	z	4.3260	5.7779
AFKF	x	22.7146	36.6027
	y	0.2819	0.3673
	z	4.3260	5.7779

6.5. Flight Test in Real Scenario

6.5.1. Model Compression and Acceleration

To accelerate the model training process, this paper utilized a high-performance NVIDIA GeForce RTX3090 during training. However, the trained model needs to be deployed on an onboard computer with limited computational power. Direct deployment may result in insufficient inference speed and latency, which may not meet the real-time requirements for UAV landing navigation. Furthermore, if models built using different deep learning frameworks are directly deployed, conflicts may arise, preventing them from running simultaneously on the onboard computer. Therefore, it is necessary to use a unified framework to refactor these models. In this paper, when deploying the runway coarse positioning network and the runway line detection network on the onboard computer, ONNX and TensorRT were used to compress and accelerate both models [31]. The program execution flow is illustrated in Figure 12. The speed of the algorithm on an NVIDIA Jetson Xavier NX is shown in Table 4. The results show that the real-time requirements of the algorithm can be achieved after model compression and acceleration.

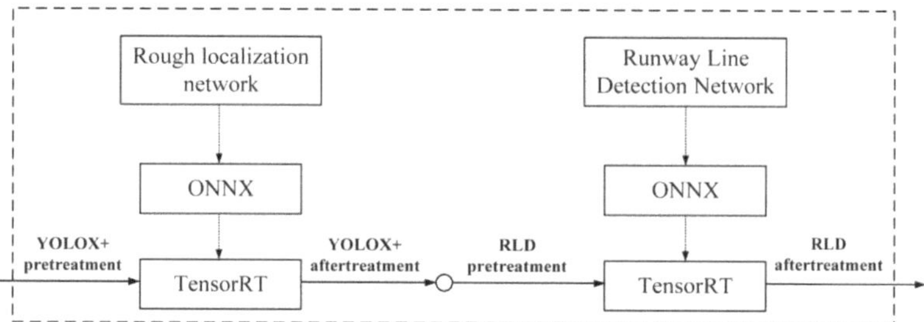

Figure 12. Model compression and acceleration process.

Table 4. Comparison of runway line detection and localization algorithm update frequency.

	ONNX and TensorRT	Update Frequency (Hz)
Detection algorithms	×	44
	√	56
Detection and localization algorithms	×	40
	√	50

6.5.2. Introduction of Flight Test Equipment

The fixed-wing UAV platform built during the actual flight test is shown in Figure 13a, with a wingspan of 1.8 m and a maximum flight height of 120 m. The equipment carried by the UAV mainly includes cameras, onboard computers, flight controllers, and power supplies. Equipment such as the control and power supply are installed inside the fuselage. The runway coarse positioning network, runway line detection network, and visual positioning algorithm are deployed on the onboard computer for image processing and visual navigation data calculation, and the visual navigation results are transmitted to the flight control terminal through USART. After the control rate in the flight controller solves the control command, it controls the steering gear in the form of PWM, and finally, controls the UAV to land smoothly.

The onboard computer selected in this paper is an NVIDIA Jetson Xavier NX. Due to its powerful computing power and small size, it can be widely used in drones, small robots, and security systems. Considering the image quality, volume, weight and other factors, the camera selected in this paper is a Logitech C1000e in the flight test, and the onboard computer and camera are shown in Figure 13b. In addition, the flight control system can ensure the stability and controllability of the UAV flight. The flight control used in this paper is self-developed flight control, an onboard IMU, barometer, magnetometer, and other common sensors. The physical picture is shown in Figure 13c.

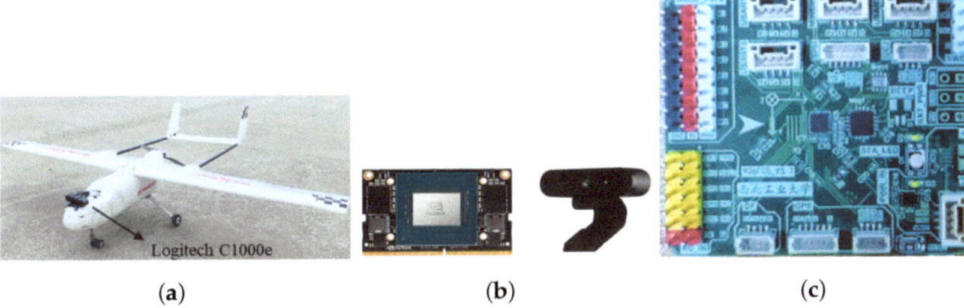

(a)   (b)   (c)

Figure 13. Real flight test platform. (**a**) UAV platform; (**b**) Onboard computer and camera; (**c**) Physical diagram of flight control system.

6.5.3. Flight Test Results

In the work, the landing process of the UAV on a simplified runway, treating the left and right edges of the runway as the left and right runway lines are simulated, respectively. Additionally, a starting runway line was added to the simplified runway. The runway line detection results are shown in Figure 14. From the experimental results, it can be observed that the designed model accurately detects all three runway lines, demonstrating good robustness and generalization.

Figure 14. Results of real runway line detection.

In the flight experiment, the visual localization results and the corresponding localization error curves in the x-, y-, and z-directions are shown in Figure 15. Due to various factors such as the short length of the simplified runway, surrounding building interference, fast UAV flight speed, and significant UAV maneuverability, the runway appears in the camera's field of view for a short time. As a result, there is limited effective localization data, and the errors are larger compared to the simulation results. However, overall, the localization trends in all three directions align with the ground truth.

From the error curves, it can be observed that the localization error in the x-direction gradually decreases as the UAV approaches the runway, reaching around 2 m near the runway. Since the runway has some margin in the x-direction, this accuracy level is sufficient for UAV landing requirements. In the y-direction, the localization error is around 2.7 m, which deviates significantly from the simulation results. The reason could be the curved edges of the left and right runway lines and possible image distortion due to the rolling shutter effect of the camera, or there may be some deviation in the ground truth setting. In the z-direction, the localization error is approximately 1.5 m, which cannot be applied to autonomous navigation during the UAV landing process. Therefore, it is necessary to further incorporate data such as IMU and laser sensor data to improve the localization accuracy.

In addition, In the work, the error characteristics of the localization results in the three directions is also calculated, as shown in Table 5. The accuracy of the errors in the y-axis and z-axis is low, and the root-mean-square error in the x-axis reaches 22.1081 m. However, given that the accuracy in the x-direction is not very demanding during the flight landing process, the visual localization algorithm proposed in this paper is effective. Furthermore, in the flight experiment, only visual localization is validated. If the adaptive fading extended Kalman filter algorithm mentioned earlier is further used to fuse the visual localization results and IMU data, it can theoretically improve the localization accuracy.

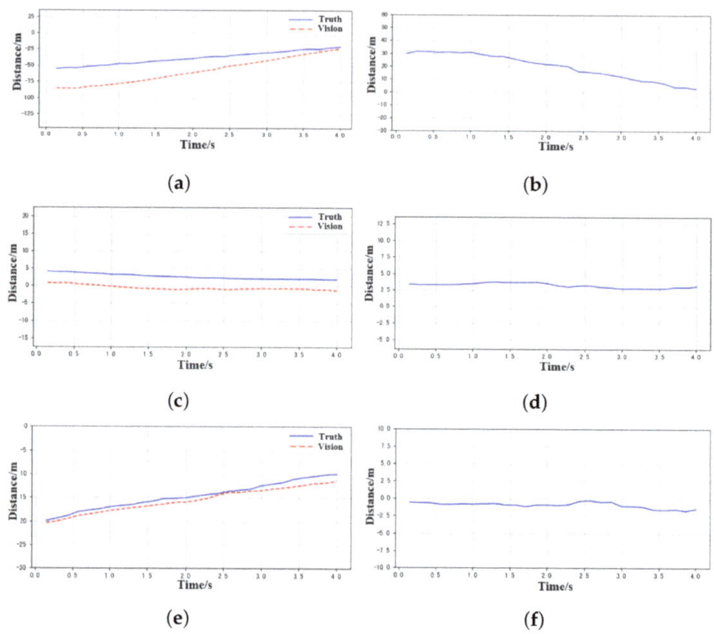

Figure 15. Experiment error curves in real scenario. (**a**) Visual positioning error in x-direction; (**b**) Visual positioning error in x-direction; (**c**) Visual positioning error in y-direction; (**d**) Visual positioning error in y-direction; (**e**) Visual positioning error in z-direction; (**f**) Visual positioning error in z-direction.

Table 5. Error characteristics of visual localization algorithm.

Axis	Evaluation Metrics	
	MAE (m)	**RMSE (m)**
x-axis	19.8979	22.1081
y-axis	3.2175	3.2331
z-axis	0.9074	0.9831

7. Conclusions

In recent years, UAVs have been used in various fields on a large scale. Landing as an important stage of flight, realizing unmanned autonomous landing is of great significance for UAV intelligence. This paper takes the visual navigation algorithm for autonomous UAV landing as the research purpose, constructs an end-to-end visual guidance landing navigation system, and optimizes the detection algorithm at the image processing end, and fuses the IMU information and visual localization data at the localization end by combining the navigation algorithms, in response to the requirements of this paper's application scenarios for accuracy and real-time. The innovations of this paper are as follows:

(1) To meet the requirements of UAV visual-guided landing, a deep-learning-based system for runway ROI detection, runway line detection, visual localization, and combined navigation is constructed.

(2) The paper optimizes the runway ROI detection algorithm and runway line detection algorithm to meet the navigation accuracy and real-time performance requirements in the application scenario.

(3) To further improve visual localization accuracy, the paper utilizes the Kalman filtering algorithm to fuse IMU information and visual localization results.

Simulation and experimental results demonstrate the significant advantages of the proposed algorithms in terms of detection accuracy, real-time performance, and generalization ability. The paper provides a reliable solution for the visual navigation problem in UAV landing.

Author Contributions: Conceptualization, X.L. and W.X.; methodology, W.X.; software, W.X.; validation, W.X. and X.X.; formal analysis, W.X.; investigation, W.X.; resources, X.L.; data curation, X.L., W.X., B.Q., and X.X.; writing—original draft preparation, W.X.; writing—review and editing, W.X., X.X., and B.Q.; visualization, W.X., X.X., and M.Z.; supervision, X.L.; project administration, X.L.; funding acquisition, X.L. All authors have read and agreed to the published version of the manuscript.

Funding: This research was funded by the National Natural Science Foundation of China, grant number No.62073266, and the Aeronautical Science Foundation of China, grant number No.201905053003.

Institutional Review Board Statement: Not applicable.

Informed Consent Statement: Not applicable.

Data Availability Statement: Data are contained within the article.

Acknowledgments: Gratitude is extended to the Shaanxi Province Key Laboratory of Flight Control and Simulation Technology.

Conflicts of Interest: The authors declare that they have no known competing financial interests or personal relationships that could have appeared to influence the work reported in this paper.

References

1. Jiao, S.; Ding, H.; Zhong, Y.; Yao, X.; Zheng, J.J. A UAV Target Tracking and Control Algorithm Based on SiamRPN. *J. Syst. Simul.* **2023**, *35*, 1372–1380.
2. Xu, W.; Li, P.; Han, B. An attitude estimation method for MAV based on the detection of vanishing point. In Proceedings of the 2010 8th World Congress on Intelligent Control and Automation, Jinan, China, 7–9 July 2010; pp. 6158–6162.
3. Chong-ming, W.; Xiao-dan, W.; Jin, G.; Wen, Q. Image Matching Method Based on the Modified Hough Transform and the Line Characteristics. In Proceedings of the 2010 Chinese Conference on Pattern Recognition (CCPR), Beijing, China, 22–24 October 2008.
4. Feng, G.; Zhang, D.; Wu, W. Pose estimation of moving object based-on dual quaternion from monocular camera. *Geomat. Inf. Sci. Wuhan Univ.* **2010**, *35*, 1147–1150.
5. Zhenghong, X.; Qiang, Z.; Xinping, Z. Research on photoelectric surveillance warning system and design scheme for airport surface. *China Saf. Sci. J.* **2020**, *30*, 136.
6. Wei-Dong, Z.; Xiao-Cheng, L.; Peng, H. Progress and challenges of overwater unmanned systems. *Acta Autom. Sin.* **2020**, *46*, 847–857.
7. Tao, L.; Hong, T.; Chao, X. Drone identification and location tracking based on YOLOv3. *Chin. J. Eng.* **2020**, *42*, 463–468.
8. Liu, J.; Wang, W.; He, Q.; Kong, X.; Ye, B.; Wang, S. Autonomous patrol technology and system of leapfrogcharging UAV (II): Automatic charging control based on machine vision. *J. Electr. Power Sci. Technol.* **2022**, *36*, 182–188.
9. Yin, H.; Zhang, X.; Zhang, X.; Liu, X. Interference analysis to aerial flight caused by UHV lines using airborne GPS. *Geomat. Inf. Sci. Wuhan Univ.* **2009**, *34*, 774–777.
10. Zhang, L.; Wang, Y.J.; Sun, H.H.; Yao, Z.J.; Wu, P. Adaptive scale object tracking with kernelized correlation filters. *Guangxue Jingmi Gongcheng Optics Precis. Eng.* **2016**, *24*, 449–459.
11. Liu, X.; Li, C.; Xu, X.; Yang, N.; Qin, B. Implicit Neural Mapping for a Data Closed-Loop Unmanned Aerial Vehicle Pose-Estimation Algorithm in a Vision-Only Landing System. *Drones* **2023**, *7*, 529. [CrossRef]
12. Ge, Z.; Liu, S.; Wang, F.; Li, Z.; Sun, J. Yolox: Exceeding yolo series in 2021. *arXiv* **2021**, arXiv:2107.08430.
13. Lin, T.Y.; Goyal, P.; Girshick, R.; He, K.; Dollár, P. Focal loss for dense object detection. In Proceedings of the IEEE International Conference on Computer Vision, Venice, Italy, 22–29 October 2017; pp. 2980–2988.
14. Weng, K.; Chu, X.; Xu, X.; Huang, J.; Wei, X. EfficientRep: An efficient RepVGG-style convnets with hardware-aware neural network design. *arXiv* **2023**, arXiv:2302.00386.
15. Han, K.; Wang, Y.; Tian, Q.; Guo, J.; Xu, C.; Xu, C. Ghostnet: More features from cheap operations. In Proceedings of the IEEE/CVF Conference on Computer Vision and Pattern Recognition, Seattle, WA, USA, 14–19 June 2020; pp. 1580–1589.

16. Howard, A.; Sandler, M.; Chu, G.; Chen, L.C.; Chen, B.; Tan, M.; Wang, W.; Zhu, Y.; Pang, R.; Vasudevan, V.; et al. Searching for mobilenetv3. In Proceedings of the IEEE/CVF International Conference on Computer Vision, Seoul, Republic of Korea, 27–28 October 2019; pp. 1314–1324.
17. Misra, D. Mish: A self regularized non-monotonic activation function. *arXiv* **2019**, arXiv:1908.08681.
18. Liu, S.; Huang, D.; Wang, Y. Learning spatial fusion for single-shot object detection. *arXiv* **2019**, arXiv:1911.09516.
19. Li, H.; Li, J.; Wei, H.; Liu, Z.; Zhan, Z.; Ren, Q. Slim-neck by GSConv: A better design paradigm of detector architectures for autonomous vehicles. *arXiv* **2022**, arXiv:2206.02424.
20. Hu, J.; Shen, L.; Sun, G. Squeeze-and-excitation networks. In Proceedings of the IEEE Conference on Computer Vision and Pattern Recognition, Salt Lake City, UT, USA, 18–22 June 2018; pp. 7132–7141.
21. Wang, Q.; Wu, B.; Zhu, P.; Li, P.; Zuo, W.; Hu, Q. ECA-Net: Efficient channel attention for deep convolutional neural networks. In Proceedings of the IEEE/CVF Conference on Computer Vision and Pattern Recognition, Seattle, WA, USA, 13–19 June 2020; pp. 11534–11542.
22. Qin, Z.; Wang, H.; Li, X. Ultra fast structure-aware deep lane detection. In Proceedings of the Computer Vision–ECCV 2020: 16th European Conference, Glasgow, UK, 23–28 August 2020; Part XXIV 16. Springer: Berlin/Heidelberg, Germany, 2020; pp. 276–291.
23. He, K.; Zhang, X.; Ren, S.; Sun, J. Deep residual learning for image recognition. In Proceedings of the IEEE Conference on Computer Vision and Pattern Recognition, Las Vegas, NV, USA, 27–30 June 2016; pp. 770–778.
24. Jing-hong, L.; Bo, J.; Gang, L.; Qian-fei, Z. Geometric correction of oblique images for array CCD aerial cameras. *Chin. J. Liq. Cryst. Displays* **2015**, *30*, 505–531.
25. Liu, J.; Zhao, Z.; Hu, N.; Huang, G.; Gong, X.; Yang, S. Summary and prospect of indoor high-precision positioning technology. *Geomat. Inf. Sci. Wuhan Univ.* **2022**, *47*, 997–1008.
26. Jing, B.; Jian-ye, L.; Xin, Y. Study of fuzzy adaptive kalman filtering technique. *Inf. Control* **2002**, *31*, 193–197.
27. Chen, X.; Zhou, J.; Li, J.; Guo, L. Data processing of wind profiler radar based on nonlinear filtering. *Nanjing Xinxi Gongcheng Daxue Xuebao* **2013**, *5*, 533.
28. Llerena Caña, J.P.; García Herrero, J.; Molina López, J.M. Error Reduction in Vision-Based Multirotor Landing System. *Sensors* **2022**, *22*, 3625. [CrossRef]
29. Wubben, J.; Fabra, F.; Calafate, C.T.; Krzeszowski, T.; Marquez-Barja, J.M.; Cano, J.C.; Manzoni, P. Accurate landing of unmanned aerial vehicles using ground pattern recognition. *Electronics* **2019**, *8*, 1532. [CrossRef]
30. Gao, W.; Yang, Y.; Cui, X.; Zhang, S. Application of adaptive Kalman filtering algorithm in IMU/GPS combined navigation system. *Geo-Spat. Inf. Sci.* **2007**, *10*, 22–26. [CrossRef]
31. Yang, X.; Jiawei, W.; Jianxue, L.; Jun, L. A dynamic routing algorithm based on deep reinforcement learning. *Inf. Commun. Technol. Policy* **2020**, *46*, 48.

Disclaimer/Publisher's Note: The statements, opinions and data contained in all publications are solely those of the individual author(s) and contributor(s) and not of MDPI and/or the editor(s). MDPI and/or the editor(s) disclaim responsibility for any injury to people or property resulting from any ideas, methods, instructions or products referred to in the content.

Article

A Heuristic Routing Algorithm for Heterogeneous UAVs in Time-Constrained MEC Systems

Long Chen [1,2], **Guangrui Liu** [1,2], **Xia Zhu** [1,2,*] **and Xin Li** [1,2]

[1] School of Computer Science and Engineering, Southeast University, Nanjing 211189, China; chen_long@seu.edu.cn (L.C.); liu_guangrui@seu.edu.cn (G.L.); lixinseu@163.com (X.L.)

[2] Key Laboratory of New Generation Artificial Intelligence Technology and Its Interdisciplinary Applications (Southeast University), Ministry of Education, Nanjing 211189, China

* Correspondence: zhuxia@seu.edu.cn

Abstract: The rapid proliferation of Internet of Things (IoT) ground devices (GDs) has created an unprecedented demand for computing resources and real-time data-processing capabilities. Integrating unmanned aerial vehicles (UAVs) into Mobile Edge Computing (MEC) emerges as a promising solution to bring computation and storage closer to the data sources. However, UAV heterogeneity and the time window constraints for task execution pose a significant challenge. This paper addresses the multiple heterogeneity UAV routing problem in MEC environments, modeling it as a multi-traveling salesman problem (MTSP) with soft time constraints. We propose a two-stage heuristic algorithm, heterogeneous multiple UAV routing (HMUR). The approach first identifies task areas (TAs) and optimal hovering positions for the UAVs and defines an effective fitness measurement to handle UAV heterogeneity. A novel scoring function further refines the path determination, prioritizing real-time task compliance to enhance Quality of Service (QoS). The simulation results demonstrate that our proposed HMUR method surpasses the existing baseline algorithms on multiple metrics, validating its effectiveness in optimizing resource scheduling in MEC environments.

Keywords: UAV routing; Mobile Edge Computing; heterogeneous multiple UAVs; IoT ground devices; heuristic

Citation: Chen, L.; Liu, G.; Zhu, X.; Li, X. A Heuristic Routing Algorithm for Heterogeneous UAVs in Time-Constrained MEC Systems. *Drones* **2024**, *8*, 379. https://doi.org/10.3390/drones8080379

Academic Editor: Carlos Tavares Calafate

Received: 21 June 2024
Revised: 31 July 2024
Accepted: 3 August 2024
Published: 6 August 2024

Copyright: © 2024 by the authors. Licensee MDPI, Basel, Switzerland. This article is an open access article distributed under the terms and conditions of the Creative Commons Attribution (CC BY) license (https://creativecommons.org/licenses/by/4.0/).

1. Introduction

Mobile Edge Computing (MEC) provides high-speed processing and large-scale distributed computing capabilities for many typical computing-intensive applications, such as automatic navigation, augmented reality, and remote control aircraft [1]. The implementation of MEC relies on dispersed large-scale IoT devices and their wireless communication. The number of global IoT devices has increased rapidly in recent years [2]. The increase in a large number of IoT devices has brought new challenges to mobile communication and service quality. At the same time, due to the limited computing power of IoT devices, additional computing resources are needed to support the effectiveness of services. In this context, UAVs have aroused significant research interest in academia due to their high maneuverability. As a result, UAV-assisted MEC has emerged as a prominent area of study [3]. Meanwhile, the short-range communication characteristics of the UAV means that it cannot transmit data over limited resources over long distances. This requires the dispatch of UAVs to IoT-device-intensive areas through task allocation, path planning, and trajectory optimization to assist in completing tasks.

This paper considers assigning multiple UAVs to multiple task clusters for auxiliary calculations, and effectively formulating UAV flight routes through a heuristic strategy. Taking into account the different flight capabilities of the UAV and the constraints of the task time window, the goal of maximizing the QoS is achieved. In this paper, we consider both the number of tasks resolved by users and the data throughput in relay communications as elements of the Quality of Service (QoS) considerations. To our best knowledge, in the

MTSP, the vast majority of UAVs studied are isomorphic [2]. However, there are many UAV manufacturers, and there is no fixed unified standard for the UAVs produced, so the heterogeneous UAV model used in this article has different flight capabilities, i.e., maximum flight coverage [4], to adapt to different UAVs produced by different manufacturers, so that various types of UAVs can be applied to our solution. Meanwhile, compared to previous studies on static environmental systems, we consider the real-time constraints of tasks generated by GDs, which means that the UAVs can only serve tasks during the open time window of the task. Real-time tasks are widely present in real environments, such as a traffic intersection having a large amount of traffic flow during specific time periods and a company's security system needing to monitor a certain area at specific times every day. We model the UAV routing problem with the above characteristics as a heterogeneous multi-traveling salesman routing problem with soft time constraints. Based on the NP-hard characteristics of the traveling salesman problem (TSP) and the MTSP, the heterogeneous multiple traveling salesman routing problem (HMTSRP) is also NP-hard. For UAV-assisted MEC, as shown in Figure 1, the UAV departs from the airport, is deployed to a TA, that is an area where a large number of GDs are concentrated, interacts with the GD, and offloads tasks to the base station.

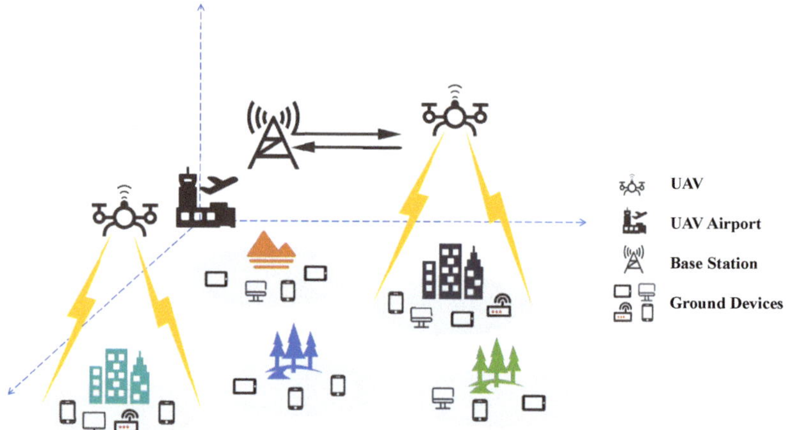

Figure 1. UAV-assisted MEC schematic diagram.

Providing high QoS for GDs in this model is a challenging issue: (a) A large number of dispersed GDs may make it difficult to determine the TAs, further leading to difficulties in selecting the optimal hovering position. (b) The heterogeneity of UAVs results in different flight capabilities, can lead to different results when different UAVs perform the same task combination, and significantly increase the solution space of the allocation strategy, then making the allocation of UAVs a difficult problem. (c) The time window constraint of a task defines the time it can be served, making it necessary to consider the time factor in UAV path planning. Only by scheduling the UAV to the task location within the time window can effective service be provided.

For the challenges mentioned above, we propose a heuristic UAV scheduling algorithm, referred to as HMUR, for the problem under study. The main contributions are as follows:

- The task allocation and path planning problem of heterogeneous UAVs in MEC systems is modeled as a heterogeneous MTSP with soft time constraints, which is an NP-hard problem. This article proposes a solution called HMUR, which applies a two-stage heuristic algorithm to obtain an approximate optimal solution for improving the QoS.

- Using the k-means clustering method to cluster the geographic coordinates of a large number of GDs solves the problem of too many IoT devices.
- A method for calculating the fitness between heterogeneous UAVs with different flight capabilities and different task assignments is proposed to determine UAV allocation under resource constraints.
- The time window constraint of the task is considered, and a heuristic method is used to formulate travel paths within the responsibility range of each UAV to achieve a higher QoS.

The rest of this paper is organized as follows. Section 2 reviews the existing literature on the problem under study. Section 3 proposes a modeling method for the system, and the problem is described and formulated. Section 4 introduces the HMUR algorithm proposed in this paper. Section 5 shows the experimental results, and Section 6 provides the conclusion.

2. Related Work

In a MEC environment assisted by UAVs, many studies focus on path planning, trajectory optimization, task offloading strategies, and other aspects of UAVs. In this section, we discuss the achievements of existing research and analyze the focus that can be further studied.

The integration of energy-efficient routing protocols is a critical issue in prolonging the operating time of UAVs [5]. The research focuses primarily on the following areas: Energy-Aware Routing Protocol (AODV [6]): This protocol extends the overall network lifetime by prioritizing UAVs with a higher remaining energy to transmit data. Location-Aware Routing Protocol (GPSR [7]): This protocol uses a greedy algorithm to select the node closest to the destination, thereby reducing data transmission distance and energy consumption. Cluster-Based Routing Protocol [8]: This protocol extends the lifetime of the UAV network by periodically rotating the cluster heads to balance energy consumption. Hybrid Routing Protocol (ZRP [9]): This protocol combines reactive and proactive routing mechanisms and dynamically selects the optimal routing strategy based on the distance between nodes and their energy status. Security is another critical concern in UAV-assisted MEC systems, where the integrity and confidentiality of data transmission must be maintained. Ensuring secure communication involves implementing robust encryption methods and secure communication protocols. Using advanced cryptographic techniques and secure transmission channels [10,11], the risk of data breaches and unauthorized access can be minimized.

Regarding the routing issue of a single UAV, Dariush Ebrahimi et al. [12] proposed a reinforcement learning method to enable UAVs to autonomously locate and formulate trajectories within the task area; this work improved the positioning accuracy of multiple objects while considering time and path length to reduce UAV energy consumption. Liang Zhang et al. [13] proposed an energy-efficient trajectory-optimization scheme for UAVs based on reinforcement learning; they jointly applied reinforcement learning methods to solve formulaic optimization problems, considering the average data rate, total energy consumption, and coverage fairness of IoT terminals to optimize the trajectory design of UAVs. Marceau Coupechoux et al. [14] proposed an algorithm based on the Hamilton Jacobi equation to solve the single UAV-trajectory-optimization problem. This work finds the optimal trajectory to minimize costs while considering velocity and service traffic.

Many studies on routing and trajectory optimization for multiple UAVs have received attention. Kai Wang et al. [15] proposed an iterative algorithm and made the first attempt to solve the collaborative path planning problem of multiple unmanned aerial vehicles. This work successfully transforms the problem into an integer linear programming problem by creating a new directed acyclic graph of the UAV state transition and proposes an iterative algorithm with a constant approximation ratio to solve this problem. Abhishek Bera et al. [16] modeled the multi-UAV path planning problem as a capacitated single-depot vehicle-routing problem (CSDVRP) and proposed a routing-adjustment algorithm to optimize the trajectory of the UAVs; this algorithm considers the trade-off between different activation modes of IoT devices and UAV travel time. Yejun He et al. [17] investigated a 3D multi-UAV trajectory optimization based on GDs selecting the target UAV for task computing; this algorithm theoretically derives and proves an optimal selection and uninstallation strategy for IoT devices. Hongyue Wu et al. [18] proposed an improved tabu search algorithm under the background of UAV-assisted edge computing, focusing on path planning, while effectively optimizing the number of UAVs to speed up the unloading of computing tasks. Xiaohan Qiu et al. [19] proposed an integrated host and content-centric routing mechanism that takes advantage of both mechanisms to address the issues of multi-UAV formation and integrated management. Hao Song et al. [20] modeled the UAV network using the Poisson clustering process, divided the UAV network into multiple clusters, and designed an enhanced flood routing protocol based on random network coding and clustering to achieve efficient routing management of the UAV cluster network. Anna Gaydamaka et al. [21] proposed a dynamic topology organization and maintenance method for UAV swarms and, based on this, designed advanced functions for dynamic cluster merging and separation, making it suitable for practical applications. Xutong Yang et al. [22] proposed an adaptive routing scheme based on the Bodies mobile model to complete the task of efficient unmanned aerial vehicle (UAV) networks. They developed a biologically inspired Bodies-based Social Force Model (BSFM) and designed an adaptive UAV routing scheme to improve the communication performance of the UAV network while maintaining the topology. Mohammed Gharib et al. [23] proposed a method based on linear programming modeling to address the problem of a self-organizing unmanned-aerial-vehicle-routing protocol to find the shortest path. Pengcheng Zhao et al. [24] proposed a blockchain and deep Q-network (DQN) co-evolutionary routing (BDCoER) scheme for UAV networks, which completes the path co-evolution of the entire network based on each UAV training its own DQN model.

Many works on collision avoidance in UAV routing also have achieved accomplishments. Yu-Hsin Hsu et al. [25] proposed a reinforcement learning method to help UAVs learn collision avoidance without prior knowledge of other UAV paths, while using optimization theory to find the shortest backward path for each UAV to ensure that the UAV collects data from all relevant IoT devices. Jinyu Fu et al. [26] proposed a multi-layer projection clustering algorithm for multiple UAVs, developed a straight-line flight judgment to reduce the computational complexity of obstacle avoidance, and proposed an improved adaptive window probability roadmap algorithm to plan obstacle avoidance paths.

Table 1 shows a comparison between some related work and the work of this paper. The comparison of some related works with certain features of this paper, including the methods, the number of drones, and whether the drones are heterogeneous, reveals that many studies on multiple UAVs consider UAVs that are isomorphic, with identical battery capacity, signal transmission power, CPU operating speed, and so on. To better fit the actual situation and enrich the diversity of choices, we introduce the heterogeneity of UAVs in this study and then consider routing decisions for multiple UAVs. At the same time, we consider the real-time characteristics of the task based on the static MEC environment, to be closer to the real situation of UAV-assisted MEC.

Table 1. Summary of related work.

Related Works	Method	Number of UAVs	Heterogeneity of UAVs
[12,13]	Reinforcement Learning	one	-
[14]	Heuristic Algorithm	one	-
[15–17,19–21]	Heuristic Algorithm	Multiple	Isomorphic
[18,22]	Intelligent Algorithm	Multiple	Isomorphic
[23]	Linear Programming	Multiple	Isomorphic
[24]	Deep Learning	Multiple	Isomorphic
This paper	Heuristic Algorithm	Multiple	Heterogeneous

3. System Model

In our proposed algorithm, each GD possesses an auxiliary calculation task. These tasks should be offloaded to the high-computing-ability base station (BS) by choosing proper UAVs as the relay devices, so as to optimize the QoS performance metrics (i.e., the number of tasks to be solved and the amount of offloaded workload). Several essential factors are included in the process, such as the number of tasks, the amount of offloaded workload, energy consumption, and the number of dispatched UAVs. We simulated this problem by establishing the following models: (a) scenario model, (b) communication model, (c) task-assisted model, and (d) energy consumption model.

The notations to be used are listed in Notations section.

3.1. Scenario Model

As illustrated in Figure 2, we abstracted the entire environment into two parts: the air part and the ground part; GDs are fixed in the ground part and clustered as TAs, generating tasks waiting for UAV-assisted computation; $D = \{d_1, d_2, ..., d_M\}$ denotes a collection of M GDs. Each GD is identified by a unique number, and a location is provided in the system. In this system, to simplify the computation, we assumed that each GD only generates one task within the same time interval, i.e., $A = \{a_1, a_2, ..., a_M\}$; task set A corresponds one-to-one to GD set D. Each task $a_i \in A$ is restricted by a real-time window $\delta_i = [\delta_i^{start}, \delta_i^{end}]$ during which the task can only be performed. L_i represents the workload of task a_i.

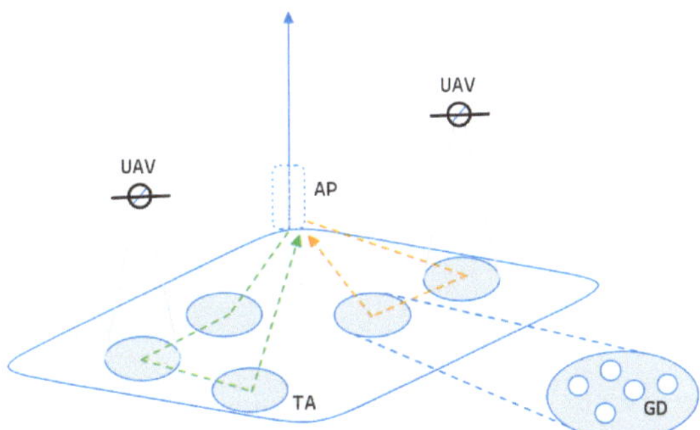

Figure 2. System model.

Due to the large number of GDs, we need to identify TAs containing several GDs in the environment as a prerequisite for dispatching UAVs, i.e., $B = \{b_1, b_2, ..., b_N\}$ denotes N TAs. Due to our simplified model, tasks correspond one-to-one to the GDs, and we can consider that each TA contains several tasks, i.e., $B_n = \{a_1^n, a_2^n, ..., a_s^n\}$ indicates that TA b_n contains a total of s tasks. For each TA, we determine a unique hover position to facilitate UAV positioning services. In addition, the BS and airport (AP) are also fixed on the ground, representing high-computing-power service base stations and UAV aprons, respectively.

In the air part, $U = \{u_1, u_2, ..., u_K\}$ denotes K UAVs moving in the air. In this paper, task allocation and path planning are performed on them.

3.2. Communication Model

Based on many previous studies, such as [2,3,16,27,28], we used the air-to-ground (ATG) [2] LoS channel probability model to simulate the communication between the UAV and the GD. Let $(x_k^U(t), y_k^U(t), h_k^U(t))$ and $(x_i^D, y_i^D, 0)$ denote the coordinates of UAV u_k and GD d_i at time t; the Euclidean distance between them can be calculated as

$$d_{ik}^{DU}(t) = \sqrt{(x_k^U(t) - x_i^D)^2 + (y_k^U(t) - y_i^D)^2 + (h_k^U(t))^2} \tag{1}$$

The average path loss between UAV u_k and GD d_i can be expressed as a probability averaged ATG-LoS model [2,16] into

$$\overline{PL_{ik}} = Pr_{ik}^{LoS} PL_{ik}^{LoS} + Pr_{ik}^{NLoS} PL_{ik}^{NLoS} \tag{2}$$

where PL_{ik}^{LoS} and PL_{ik}^{NLoS} are the pass loss for the LoS and Non-Line of Sight (NLoS) link and Pr_{ik}^{LoS} and Pr_{ik}^{NLoS} denote the probabilities of LoS communication and NLoS communication, respectively. PL_{ik}^{LoS} and PL_{ik}^{NLoS} are given as

$$PL_{ik}^{LoS} = 20 \log_{10} \frac{4\pi f_c d_{ik}^{DU}(t)}{c} + \eta_{LoS} \tag{3}$$

$$PL_{ik}^{NLoS} = 20 \log_{10} \frac{4\pi f_c d_{ik}^{DU}(t)}{c} + \eta_{NLoS} \tag{4}$$

where f_c is the carrier frequency, c is the velocity of light, and η_{LoS} and η_{NLoS} are the average additional loss for LoS and NLoS, respectively. Pr_{ik}^{LoS} and Pr_{ik}^{NLoS} are given as

$$Pr_{ik}^{LoS} = (1 + \mathbf{X}^{[-\mathbf{Y}(\theta_{ik} - \mathbf{X})]})^{-1} \tag{5}$$

$$Pr_{ik}^{NLoS} = 1 - Pr_{ik}^{LoS} \tag{6}$$

where $\mathbf{X}$ and $\mathbf{Y}$ are the environmental constants, depending on the environmental condition in which the system is located. θ_{ik} is the elevation angle between GD d_i and UAV u_k. It is expressed in radians, as

$$\theta_{ik} = \frac{180}{\pi} \arcsin \frac{h_k^U(t)}{d_{ik}} \tag{7}$$

Therefore, the data transfer rate (in bps) between GD d_i and UAV u_k is

$$R_{ik}^{DU} = B_0 \log_2(1 + \frac{P_i^D / \overline{PL_{ik}}}{\sigma^2}) \tag{8}$$

According to the orthogonal frequency-division multiple access (OFDMA) communication [2], B^{bw} is the total channel bandwidth, $B_0^{bw} = B^{bw}/k_0$, k_0 is the number of GDs serviced simultaneously, P_i^D is the transmission power of GD d_i, and σ^2 is the noise power.

Similarly, the transmission rate R_k^{UB} of the offloaded data from the UAV to the BS is calculated by

$$R_k^{UB} = B_0^{bw} \log_2(1 + \frac{P_k^U / \overline{PL_k}}{\sigma^2}) \tag{9}$$

where P_k^U is the transmission power of UAV u_k.

3.3. Task-Assisted Model

This work considers using UAVs as mobile relay points, which means that tasks can be offloaded from UAV relays to the BS for computation. As in many previous studies [2,29,30], because its size is much smaller than the data size that must be offloaded, we ignored the delay in sending the data results from the BS back to the UAV and from the UAV back to the GD.

This work divides the working time of the entire system into several parts of a sufficiently small constant δ^t of equal size and considers that the UAV is stationary within a time slot. Due to our communication being divided into two parts, GD to UAV and UAV to BS, we divided a time slot into the two same parts, i.e., UAV u_k assists task a_i in offloading data, and we have $\delta_{ik}^{DU} + \delta_k^{UB} = \delta^t$, where δ_{ik}^{DU} is the time that GD d_i offloads data to UAV u_k. δ_k^{UB} is the time that UAV u_k offloads data to BS. To ensure the complete processing of a portion of data within one time slot, i.e., the amount of data processed in both parts is the same, the time relationship between the two parts should meet

$$\delta_{ik}^{DU} R_{ik}^{DU}(t) = \delta_{ik}^{UB} R_k^{UB}(t) \tag{10}$$

Based on the above, there are

$$\delta_{ik}^{DU} = \delta^t \cdot \frac{R_k^{UB}(t)}{R_k^{UB}(t) + R_{ik}^{DU}(t)} \tag{11}$$

$$\delta_{ik}^{UB} = \delta^t \cdot \frac{R_{ik}^{DU}(t)}{R_k^{UB}(t) + R_{ik}^{DU}(t)} \tag{12}$$

At this point, the amount of data processed within one time slot, i.e., single time slot data processing speed, is represented as

$$\beta_{ik} = \delta^t \cdot \frac{R_{ik}^{DU}(t) R_k^{UB}(t)}{R_{ik}^{DU}(t) + R_k^{UB}(t)} \tag{13}$$

The service duration of UAV u_k for GD d_i is $t_{ik} = L_i / \beta_{ik}$, where L_i is the amount of data that needs to be offloaded by d_i. At this point, the total duration of the UAV u_k hovering over TA b_j is

$$t_{kj}^{hover} = \max_{d_i \in b_j} t_{ik} \tag{14}$$

The total flight time of UAV u_k from TA b_p to TA b_q is $t_{pq}^k = v_k \times D_{pq}$, where D_{pq} is the Euclidean distance between TAs b_p and b_q.

3.4. Energy Consumption Model

Due to the fact that the communication energy consumption of the UAVs is much lower than that of motion (including traveling and hovering), we ignore the communication energy consumption of the UAVs. Therefore, the energy consumption of UAV u_k executing tasks along the path r_k is expressed as

$$E_k^{total} = E_k^{trav} + E_k^{hover} \tag{15}$$

Based on the alternate fixed rotary wing UAV energy consumption calculation method [2,31] used, when UAV travels at velocity v, the unit time flight energy consumption is

$$\begin{aligned} P(||v||) = P_0 &\left(1 + \frac{3||v||^2}{U_{tip}^2}\right) \\ &+ P_i \left(\sqrt{1 + \frac{||v||^4}{4v_0^4}} - \frac{||v||^2}{2v_0^2}\right)^{\frac{1}{2}} \\ &+ \frac{1}{2} d_0 \rho s A ||v||^3 \end{aligned} \tag{16}$$

where P_0, P_i, v_0, U_{tip}^2, d_0, ρ, s, and A are the aerodynamic parameters of the UAV. Specific definitions and settings can be found in Table 2. Then, the total flight energy consumption of UAV u_k is measured by

$$E_k = \delta^t \sum_{n=1}^{N} P(||v||) \tag{17}$$

Table 2. Parameter setting.

Parameter	Value	Definition
f_c	2 GHz [16]	Carrier frequency
c	3×10^8 m/s	Velocity of light
η_{LoS}	1 [2]	Average additional loss in LoS
η_{NLoS}	20 [2]	Average additional loss in NLoS
$\mathbf{X}$	10.39 [16]	Environmental parameter
$\mathbf{Y}$	0.05 [16]	Environmental parameter
σ^2	-174 dBm/Hz [28]	Gaussian noise
P_0	158.76 w [3]	Blade profile power in hovering status
P_i	88.63 w [3]	Induced power in hovering status
U_{tip}	120 m/s [3]	Tip speed of the rotor blade
v_0	4.03 [3]	Mean rotor-induced velocity in hover
d_0	0.3 [3]	Fuselage drag ratio
ρ	1.225 km/m^3 [3]	Air density
s	0.05 [3]	Rotor solidity
A	0.503 m^2 [3]	Rotor disc area

3.5. Problem Formulation

To optimize the QoS with the restriction of energy consumption and the number of dispatched UAVs, a soft time constraint model named HMTSRP is proposed as follows. We define

$$Y_{total} = \sum_{i}^{A} x_i \tag{18}$$

$$L_{total} = \sum_{i}^{A} x_i L_i \tag{19}$$

Among these, Y_{total} represents the total number of serviced tasks, L_{total} represents the total amount of offloaded data workload for serviced tasks, and L_i represents the amount of data offloaded by task a_i. Boolean variable x_i indicates whether task a_i is served by the UAV:

$$x_i = \begin{cases} 0, & \text{if } a_i \text{ is not be served,} \\ 1, & \text{if } a_i \text{ is served.} \end{cases} \tag{20}$$

Therefore, we formulate a minimization problem as

$$P1 : \max Y_{total} \tag{21a}$$

$$P2 : \max L_{total} \tag{21b}$$

$$s.t. \quad C1 : E_k^{total} \leq E_k^{max}, \forall u_k \in U \tag{21c}$$

$$C2 : q_k[0] = q_{AP}, \forall u_k \in U \tag{21d}$$

$$C3 : q_k[n] = q_{AP}, \forall u_k \in U \tag{21e}$$

$$C4 : \sum_{j=1, j \neq j'}^{||B||} \phi(b_{j'}, b_j) = 1 \tag{21f}$$

$$C5 : \sum_{j'=1, j' \neq j}^{||B||} \phi(b_{j'}, b_j) = 1 \tag{21g}$$

(21a) and (21b) are the optimization objectives, maximizing the total number of service tasks and data offloading, subject to the following: (21c) represents the battery capacity constraint for the UAVs; (21d) and (21e) represent the starting and ending points of the UAVs, which must be the AP; among these, $q_k[n]$ is the position of UAV u_k at time n and q_{AP} is the location of the AP; (21f) and (21g) represent that, for all TAs, there only exist one departure and one arrival for one UAV, respectively. The definition of Boolean variable $\phi(b_{j'}, b_j)$ is as follows:

$$\phi(b_{j'}, b_j) = \begin{cases} 1, & \text{if UAV selects path from } b_{j'} \text{ to } b_j, \\ 0, & \text{otherwise.} \end{cases} \tag{22}$$

4. Problem Analysis and Solution Approach

In order to effectively allocate tasks to BSs for computation by relaying among multiple UAVs with different capacities, a three-step algorithm named HMUR is proposed in this paper. HMUR consists of three steps: (a) TA determination: establishing TAs for related GDs and determining the optimal hover positions of UAVs; (b) UAV allocation: assigning TAs to suitable UAVs; (c) subpath determination: determining the final task execution path for each UAV.

4.1. HMUR

Algorithm 1 shows the framework of HMUR. Throughout the entire process, HMUR first applies the k-means clustering algorithm to cluster the geographic coordinates of the GD and uses its clustering center as the optimal hover location. Secondly, HMUR executes Algorithm 2 for the UAV allocation to perform tasks. Finally, HMUR executes Algorithm 3 to determine the final subpath for each UAV.

In this process, we consider using energy consumption as a constraint for task partitioning in Algorithm 2 to ensure that at least one route for the UAV to be able to complete the designated task. Meanwhile, we use the QoS as an evaluation metric in Algorithm 3 to determine the final subpath, to optimize the service QoS as much as possible within the allowable range of energy consumption.

Algorithm 1 HMUR algorithm.

Input: D, A, and U;
Output: Final subpath set R;
  1: k-means clustering algorithm generates B from D and determines the optimal hover position
  2: Algorithm 2 determines UAV allocation for B
  3: Algorithm 3 determines the final subpath set R
  4: **return** Final subpath set R

Algorithm 2 UAV allocation.

Input: B, U;
Output: Allocation set K for UAV;
  1: Generate graph G from B
  2: Generate minimum spanning tree T_{MST} from G
  3: Generate set of odd degree vertices O from T_{MST}
  4: Find minimum weight matching M from O
  5: Merge T and M to generate Eulerian circuit H_{EULAR}
  6: Generate Hamiltonian circuit R_C from H_{EULAR}
  7: $num = 0$
  8: **while** $num < len(R_C)$ **do**
  9:    $b_{start} = R_C[num]$
 10:    **for** $u_k \in U$ **do**
 11:      Simulate travel from b_{start}, and generate B_k
 12:      Record f_k
 13:    **end for**
 14:    Choose u_{best} with f_{best} for n TAs that it covered
 15:    Generate B_{best}, and allocate u_{best} for B_{best}
 16:    Add B_{best} to K
 17:    Remove u_{best} from U
 18:    $num = num + n$
 19: **end while**
 20: **return** Allocation set K

Algorithm 3 Subpath determination.

Input: U, K;
Output: Final subpath set R;
1: **for** $B_i \in K$ **do**
2:    **for all** $b_j \in B_i$ **do**
3:       $visit[b_j] = 0$
4:    **end for**
5:    $start = AP$
6:    **while** $B_i \neq \varnothing$ **do**
7:       **for all** $b_j \in B_i$ **do**
8:          **if** $visit[b_j] == 0$ **then**
9:             Calculate S_{b_j}
10:          **end if**
11:       **end for**
12:       Choose TA b_h with the highest S_{b_h} as the target
13:       $visit[b_h] = 1$
14:       $start = b_h$
15:       Add b_h to queue r_i
16:       Delete b_h from B_i
17:    **end while**
18:    **if** Energy consumption of r_i exceeds the maximum energy consumption of u_i **then**
19:       $r_i =$ predicted path generated by Algorithm 2
20:    **end if**
21:    Add r_i to set R
22: **end for**
23: **return** R

4.2. TA Determination

Since there are numerous GDs in the MEC environment, it is not realistic to assign one UAV to each GD for auxiliary computation. Thus, GDs with close geographical coordinates are aggregated into TAs, which are served by UAVs within certain time periods. Under constant environmental factors, the data transmission rate is mainly related to the distance $d_{ik}^{DU}(t)$ between UAV u_k and GD d_i. Obviously, the smaller $d_{ik}^{DU}(t)$ is, the higher the transmission rate. Therefore, in TA b_j, the optimal hover location is the position with the smallest distance from all GDs, and we formulate this problem as follows:

$$P2 : \min \sum_{d_i \in b_j} d_{ik}^{DU}(t) \tag{23}$$

The k-means clustering algorithm [16,32] is adopted here to establish clusters (i.e., TAs) and determine the cluster centers (i.e., the optimal hover positions of UAVs). The assignment of a GD to a nearby TA is iteratively updated by k-means until convergence. Then, the point with the minimum sum of distances from all the other GDs within one TA is considered as the optimal hover position.

4.3. UAV Allocation

After obtaining the TAs, each UAV will be allocated to several TAs. Let $K = \{B_1, B_2, ..., B_K\}$ be the set of allocations of K UAVs, of which each element B_k represents a mapping between u_k and the TA set B_k. Then, this paper takes the battery capacity carried by the UAV as a constraint and considers maximizing resource utilization for u_k to execute tasks in B_k. To ensure the continuity of UAV routes, a partitioning method is proposed by adding breakpoints to a single TSP path to partition TAs. Based on the Christofides approximation algorithm [16], an advanced fitness function f_k is introduced to efficiently evaluate the allocation results in TAs of different heterogeneous UAVs.

Algorithm 2 employs the Christofides algorithm to obtain an approximate solution for the traveling salesman problem (TSP). This approach utilizes a minimum spanning tree, perfect matching, an Eulerian circuit, and a Hamiltonian circuit from graph theory to provide a suboptimal solution. As mentioned in previous research [16], the performance ratio of the Christofides algorithm does not exceed 1.5 compared to the optimal solution, and it is a constructive solution that can be solved within polynomial time.

As shown in Algorithm 2, after creating a complete TSP path R_c by lines 1 to 6, some breakpoints are inserted to truncate the path into several predicted paths: when UAV u_k travels along R_C (denoting a TA sequence) and consumes energy during flight and mission execution; when a UAV finds that its energy consumption cannot reach the next TA and then returns to the AP, it will return to the AP and represent the TAs' path that has performed for the task as r_k. If u_k travels along the route r_k, the fitness can be designed as

$$f_k = \alpha \times \frac{E_k^{hover}}{E_k^{total}} + (1 - \alpha) \times \frac{E_k^{total}}{E_k^{max}} \tag{24}$$

where E_k^{hover} represents the energy consumption of u_k during task execution, E_k^{total} represents the total energy consumption of this travel path, and E_k^{max} represents the maximum battery capacity of u_k. So, the effective energy consumption rate $\rho_k^E = E_k^{hover} / E_k^{total}$. Due to the heterogeneity among UAVs, different UAVs executing tasks from the same route may insert virtual points into different positions (i.e., returning to the AP from different TA endpoints), resulting in different E_k^{trav}.

An example of Algorithm 2 is illustrated in Figures 3–5. Assume that travel consumption is an integer, and temporarily assume that the hover consumption of UAV at each concerned TA is 10, i.e., the energy consumption occurring when a UAV resolves tasks at a specific TA. The UAV executes tasks in the order of $(AP - b_1(TA_1) - b_2(TA_2) - b_3(TA_3))$, and the energy consumption rate of each situation is $\rho_1^E = 10/(10 + 10 + 10) = 33.33\%$, $\rho_2^E = 20/(10 + 10 + 12 + 10 + 20) = 32.26\%$, and $\rho_3^E = 30/(10 + 10 + 12 + 10 + 5 + 10 + 20) = 38.96\%$, respectively. To some extent, the higher ρ_k^E, the more tasks can be solved with the same energy consumption, further indicating that we can obtain a better solution by organizing the dispatch of heterogeneous UAVs. In addition, the second item of (24) with a gravity control factor α is used to control the resource utilization of UAVs, aiming to maximize the utilization of the total battery resource carried by the UAV.

4.4. Subpath Determination

After obtaining the allocation set for different UAVs, HMUR determines the final task execution path for each UAV. In this section, we take into account the real-time window constraint of tasks.

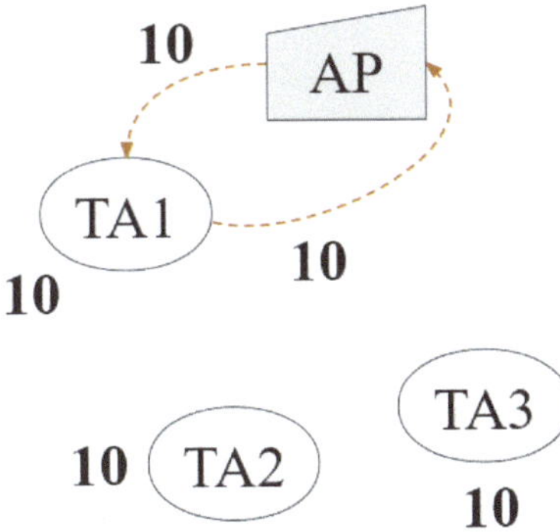

Figure 3. Route of u_1.

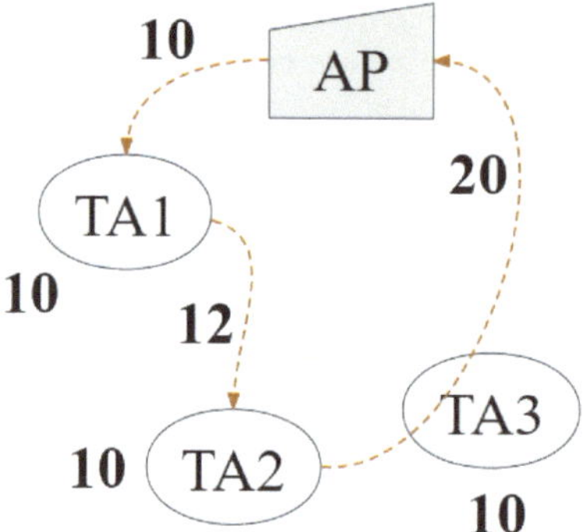

Figure 4. Route of u_2.

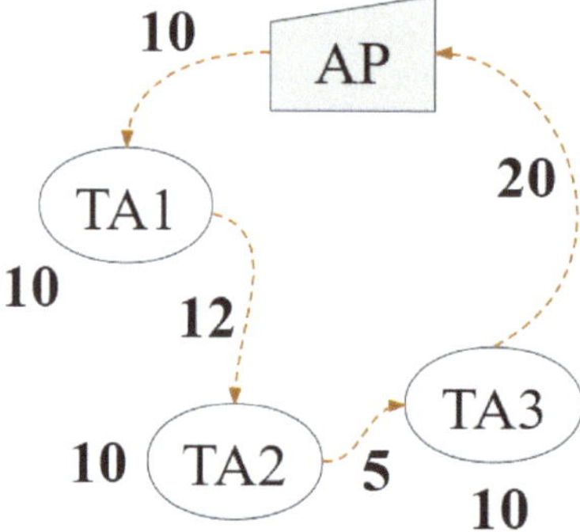

Figure 5. Route of u_3.

We process UAV allocation options K and rearrange the routing path within the energy consumption limitation. According to the predicted results of Algorithm 2, it can be seen that, among all the tasks the UAV is responsible for, each UAV has at least one route that can complete all tasks. Specifically, in the selection of the allocation of B_k, the UAV u_k starts from the AP, predicts the arrival time of the task points in B_k that have not been traversed, and calculates the score function for the selected TA. The score function is formulated as follows:

$$S_{b_j} = Y_{b_j}^B - \log_2 E_{b_{j'},b_j}^{trav} + \log_{10} L_{b_j}^B \tag{25}$$

where $Y_{b_j}^B$ is the number of tasks that open the window upon arrival, $E_{b_{j'},b_j}^{trav}$ is the energy consumption for flying from the previous point to b_j, and $L_{b_j}^B$ is total workload of data expected to be offloaded in b_j. The UAV selects the TA with the highest S_{b_j} and travels to TA b_j to complete as many tasks as possible during the window period. The optimization in this section focuses on adapting to the real-time window factor of the task, sacrificing some energy consumption to achieve greater QoS, and to some extent, neglecting energy consumption control. Therefore, during the adjustment process, there may be situations where the total energy consumption exceeds the maximum energy consumption of the UAV. In response to this situation, we abandon the path rearrangement and use the predicted path generated by Algorithm 2 as the final routing path.

This paragraph analyzes the time complexity of the algorithm to ensure the real-time applicability of the entire system. As previously mentioned, under the environment of n TAs and n UAVs, the Christofides algorithm can solve problems within polynomial time with a complexity of $O(n^2 logn)$. Additionally, lines 8-19 of Algorithm 2 iterate through each path divided by the Christofides algorithm and traverse m UAVs in each path, with a time complexity strictly less than $O(mn)$. Moreover, after initialization, Algorithm 3 enters a double loop, which simulates the value of the UAVs reaching different TAs on each subpath, with a time complexity of $O(m) \times O(n) = O(mn)$. Consequently, HMUR can solve the problem within polynomial time.

5. Performance Evaluation

5.1. Dataset

To evaluate HMUR performance, we used real data from IoT users and BS deployments in Melbourne CBD provided by the Australian Communications and Media Authority [33]. The dataset includes a total of 816 GDs and 125 BSs available for the UAVs to offload data, In Figure 6, black dots represent geographically fixed GDs and red dots (i.e., the center of each red circle) represent the service base stations. In the clustering results of Figure 7, we retain only the abstract representation of GDs and BSs, depicting GDs in different partitions with light-colored circles of various colors and marking the clustering centers of the clustering results with red crosses. The overall distribution of the data is between geographic coordinates of latitude (-37.822, -37.808) and longitude (144.950, 144.975). Figure 6 provides a realistic map image of the scene, and Figure 7 shows the location of IoT devices and the results after clustering into TAs (the number of TAs is 50).

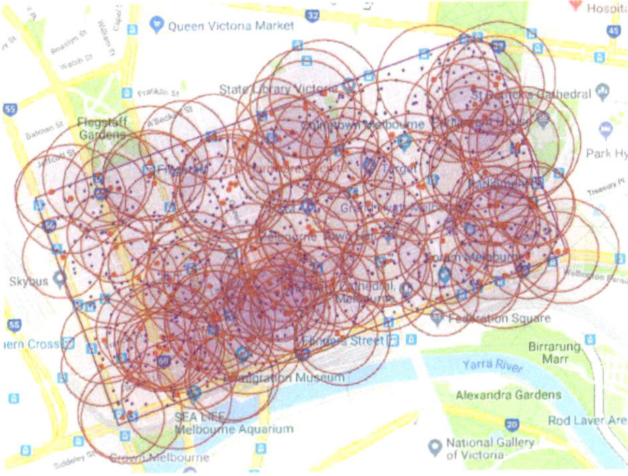

Figure 6. The real map.

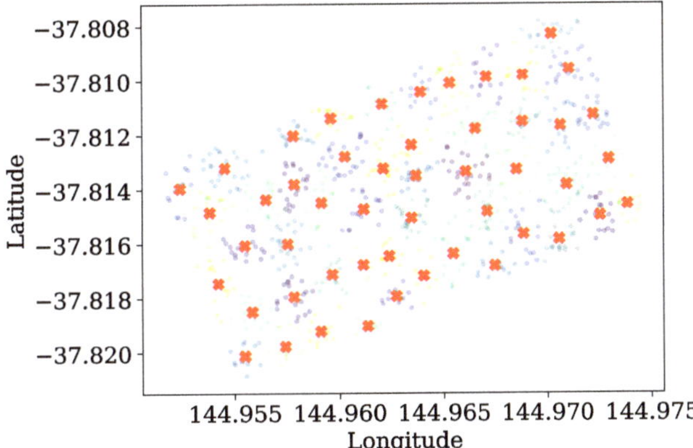

Figure 7. Cluster result.

5.2. Benchmarks

Due to conducting experimental simulations in urban environments, set $\mathbf{X} = 10.39$ $\mathbf{Y} = 0.05$ [16], $\eta_{LoS} = 1$, and $\eta_{NLoS} = 20$ [2]. Other parameters related to aerodynamics are shown in Table 2. To better validate the effectiveness of UAV assistance, we only retained one available BS $(144.966686, -37.815549)$ in the environment and set the AP at $(144.962344, -37.815303)$.

The settings and definitions of the instance factors are shown in Table 3. Among them three essential factors are used as independent variables to analyze the performance of the algorithm, which are: (a) the number of TAs, (b) the average real-time window size of the task, and (c) the average workload generated by the GDs. Four indicators were adopted to evaluate the compared algorithms, namely (a) the total number of solved tasks, (b) the amount of offloaded workload, (c) the quantity of dispatched UAVs, and (d) the number of tasks solved by the UAVs per minute.

Table 3. Factor setting.

Parameter	Value or Range of Value	Definition
n_{GD}	816	The number of GDs
t_{avg}^{A}	$[500, 1500]$ s	Average time window of task
n_{TA}	$[20, 60]$	The number of TAs
L_{avg}^{D}	$[1, 6]$ MB	GDs' average workload
E_{avg}^{U}	400 kJ	UAVs' average battery capacity
E^{U}	700 kJ, 700 kJ, 600 kJ, 500 kJ, 400 kJ, 300 kJ, 200 kJ, 100 kJ, 100 kJ	Heterogeneous UAVs' battery capacity
α	0.5	Weight of fitness
B^{bw}	5 GHz	Bandwidth

5.3. Performance Comparison

The architecture of this paper was written in Python 3.7, and the algorithm ran on the Ubuntu 22.04.1 LTS operating system, under an NVIDIA RTX 4090 cluster environment. To test the performance of the proposed algorithm HMUR, it was compared with the best-existing algorithms for the considered problem: CEDAN by Bera et al. [16] and the GA-based method (GAM) of Wang et al. [34]. In this paper, we set the energy payload of the UAVs in CEDAN and GAM to the average energy payload of heterogeneous UAVs in HMUR, denoted as E_{avg}^{U}. Furthermore, to evaluate the effectiveness of the subpath determination strategy (Algorithm 3), the proposed algorithm without this strategy (HMUWS) was also compared. Furthermore, in the diagram, as the task TA on the horizontal axis changes, we maintained the total number of GDs retained. Furthermore, we decreased the number of GDs included in each TA as the number of TAs increased.

The total number of solved tasks: Figure 8 shows the total number of solved tasks when varying the number of TAs. As the number of TAs increases, the number of tasks assisted by HMUR and that by the baseline algorithms both show an increasing trend, but HMUR is 73.86% higher than CEDAN and 80.22% higher than GAM, while 36.43% higher than HMUW3 (when the number of TAs is 30). Meanwhile, Figure 9 shows the numerical performance of the number of auxiliary tasks solved when the average duration of the task window changes. It reveals that the longer the duration of the task window, the more tasks the UAVs assist in solving. However, when the time window size exceeds a threshold (about 1000 s), the number of solved tasks obtained by the other three algorithms no longer increases, except that of HMUR. This is probably due to HMUR considering the soft time constraints; in each selection, HMUR tends to choose TAs with more tasks with open time windows.

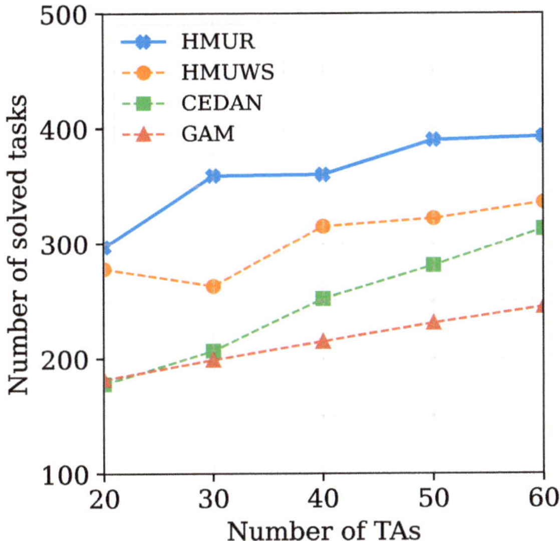

Figure 8. Total number of solved tasks (varying the number of TAs).

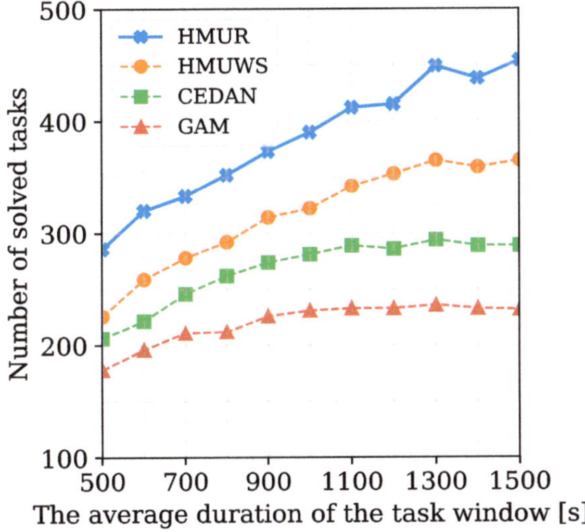

Figure 9. Total number of solved tasks (varying the average duration of the task window).

The total amount of offloaded workload: It can be seen from Figure 10 that the amount of offloaded workload generated by HMUR is higher than those of the other algorithms. With an increase in the number of solved tasks, the amount of offloaded workload gradually increases for all four algorithms. In Figure 11, we vary the average duration of the task windows to obtain the value of the workload offloading. HMUR selects routes that can achieve a higher QoS by calculating the TA's score function, thereby solving more tasks and offloading more data.

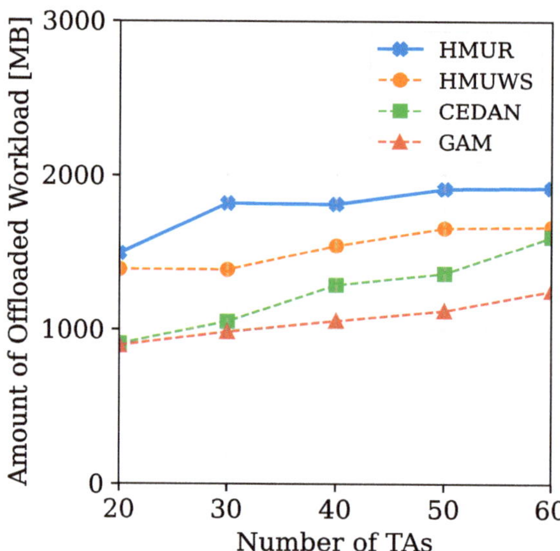

Figure 10. Total amount of offloaded workload (varying the number of TAs).

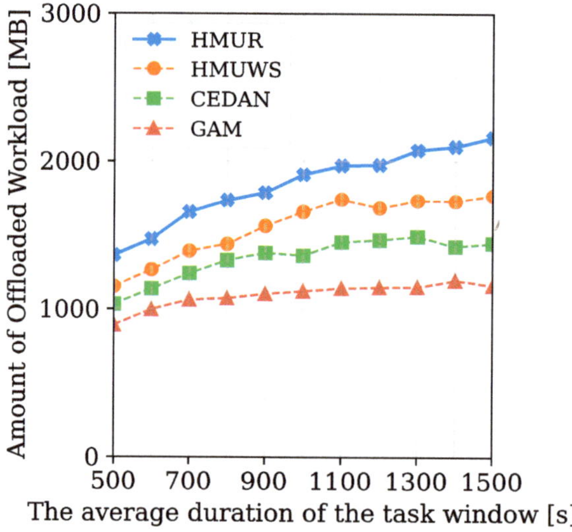

Figure 11. Total amount of offloaded workload (varying the average duration of the task window).

The quantity of dispatched UAVs: Figures 12–15 shows the difference in the number of UAVs dispatched by HMUR and CEDAN when the average amount of data generated by the GDs varies. The results show that the average number of UAVs dispatched by HMUR is 25.95% (the average task window is 500 s in Figure 12), 25.19% (the average task window is 1000 s in Figure 13), 41.63% (the average task window is 1500 s in Figure 14), and 24.62% (all tasks are open in Figure 15) less than that of CEDAN. This is because HMUR considers the compatibility between heterogeneous UAVs and paths, making more rational use of the UAV resources. Moreover, since HMUR is developed by adding Algorithm 3 to the base of HMUWS, and since Algorithm 3 does not involve changes in the number of unmanned aerial vehicles, HMUWS is not considered in this context.

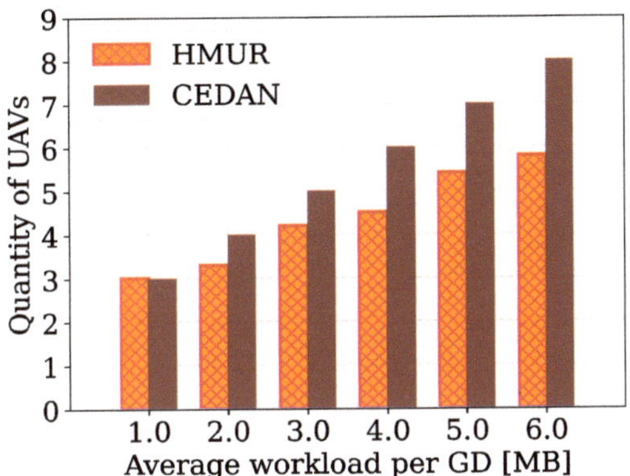

Figure 12. The quantity of dispatched UAVs (average duration of task window = 500 s).

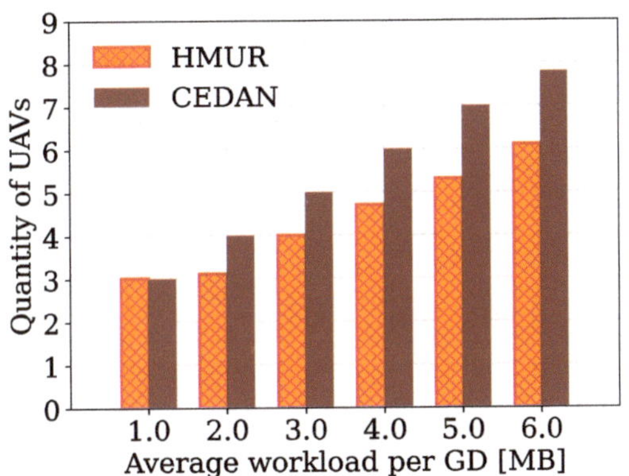

Figure 13. The quantity of dispatched UAVs (average duration of task window = 1000 s).

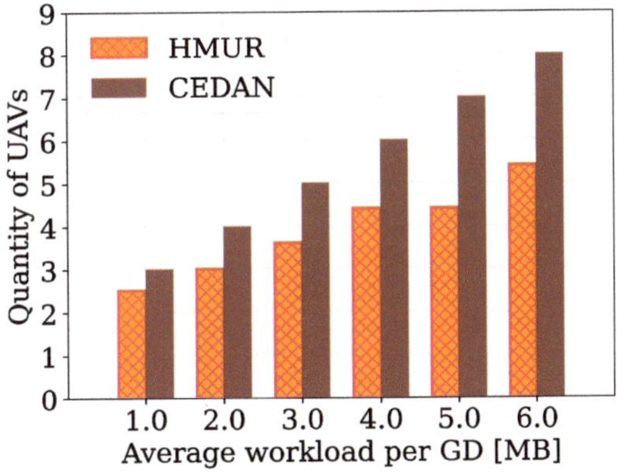

Figure 14. The quantity of dispatched UAVs (average duration of task window = 1500 s).

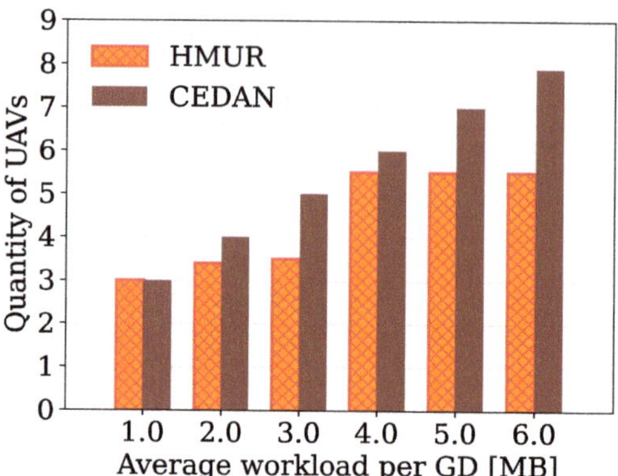

Figure 15. The quantity of dispatched UAVs (no time window constraint).

Number of tasks solved per minute: Figure 16 shows the time utilization of HMUR and the baseline algorithm, which we represent as the average number of tasks solved per minute. It can be seen that HMUR sacrifices some time to obtain a larger number of tasks, resulting in a decrease in time utilization. However, overall, the performance of HMUR is still acceptable.

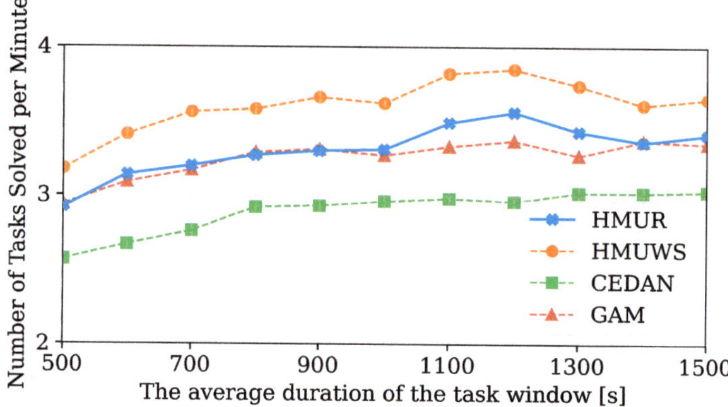

Figure 16. Number of tasks solved per minute.

In the above experiment, we tested the impact of system scale (number of TAs), task urgency (average duration of the task window), and task load (average workload generated by GD) on the system. The experimental results show that the number of auxiliary tasks and offloaded data increase with increasing system scale and average task window duration, while the quantity of dispatched UAVs mainly increases with the growth of the average workload generated by the GD. HMUR outperforms the other algorithms because of the following reasons: The clustering results of k-means exactly meet the requirements of the system model, that is the sum of distances between the optimal hover position and all GDs is the shortest. At the same time, in Algorithm 2, the fitness evaluates the effective energy consumption of the UAVs in executing tasks and maximizes the utilization of battery energy within the allowable range of energy consumption. Finally, in Algorithm 3, the score function considering the timeliness constraints helps the UAVs find the TA with the highest number of executable tasks at each time point, thus improving the QoS.

6. Conclusions and Future Work

In this paper, a multi-UAV routing algorithm named HMUR is proposed for the UAV-assisted MEC problem. Instead of isomorphic UAVs as commonly considered in the literature, we also took into account the heterogeneity of UAVs and the real-time characteristics of the task. An effective fitness measurement is defined to match the UAVs and different routing paths under certain energy constraints. A score function is designed to determine the final route with the highest QoS. An extensive comparison of the proposed algorithm is performed against the best existing approaches. The experimental results show that the proposed method is superior to the best existing algorithms on multiple metrics. We demonstrated the effectiveness of the algorithm through experiments divided into four parts, which collectively evaluated the performance of the service based on the number of tasks completed, the volume of data offloaded, the efficiency of task resolution, and the number of UAVs used. The volume of data offloaded is correlated with the number of tasks completed. Through the HMUR algorithm, heterogeneous UAVs are appropriately allocated in the UAV allocation segment and tasks are reorganized in the subpath determination segment to ensure that more tasks are within their window period when the UAV reaches the task area. Furthermore, since the fitness function considers the ratios of E_k^{hover} to E_k^{total} (reflecting the effective energy consumption of UAVs) and E_k^{total} to E_k^{max} (reflecting UAV battery utilization), it ensures that the energy of each UAV is efficiently used, thereby indirectly reducing the number of UAV deployments.

Future research avenues involve the consideration of the improvement of the proposed framework. For example, more UAV attributes such as output power, battery weight, bandwidth, and flight speed can be introduced as a UAV-type selection mechanism. It also seems worthwhile to apply the proposed algorithm with some heuristic offloading solutions for further improvement. Meanwhile, future work could consider replacing FDMA with communication methods such as TDMA and LoRaWAN to achieve new effects. At the same time, we will explore the integration of machine learning techniques and reinforcement learning for routing to adaptively allocate tasks based on historical data, improving the efficiency and responsiveness of our system. Additionally, we aim to develop a user-friendly tool for easy implementation and deployment of our algorithm in various MEC environments, enhancing user adoption and practicality.

Author Contributions: Conceptualization, L.C. and G.L.; methodology, L.C. and G.L.; software, G.L.; validation, G.L. and X.L.; formal analysis, G.L.; investigation, G.L.; resources, L.C. and X.L.; writing—original draft preparation, L.C. and G.L.; writing—review and editing, L.C., G.L. and X.Z.; supervision, X.Z.; project administration, G.L.; funding acquisition, L.C. and X.Z. All authors have read and agreed to the published version of the manuscript.

Funding: This work is supported by the National Key Research and Development Program of China (No. 2022YFB3305500), the National Natural Science Foundation of China (Nos. 62273089, 62102080), the Natural Science Foundation of Jiangsu Province (No. BK20210204), and the Fundamental Research Funds for the Central Universities (No. 2242022R10017).

Conflicts of Interest: The authors declare no conflicts of interest.

Notations

D	GD set, $D = \{d_1, d_2, ..., d_M\}$
A	Task set, $A = \{a_1, a_2, ..., a_M\}$
B	TA set, $B = \{b_1, b_2, ..., b_N\}$
B_n	Task set contained in TA b_n, $B_n = \{a_1^n, a_2^n, ..., a_s^n\}$
U	UAV set, $U = \{u_1, u_2, ..., u_K\}$
K	Allocation set, $K = \{B_1, B_2, ..., B_K\}$; B_k contains several TAs and represents the allocation of u_k to these TAs
R	Subpath set, $R = \{r_1, r_2, ..., r_K\}$; r_k and u_k correspond one-to-one
$\overline{PL_{ik}}$	Average path loss between d_i and u_k
PL_{ik}^{LoS}	Pass loss for LoS link between d_i and u_k
PL_{ik}^{NLoS}	Pass loss for NLoS link between d_i and u_k
Pr_{ik}^{LoS}	Probability of LoS link between d_i and u_k

Pr_{ik}^{NLoS}	Probability of NLoS link between d_i and u_k
$R_{ik}^{DU}(t)$	Data transfer rate between d_i and u_k at time t
$R_k^{UB}(t)$	Data transfer rate between u_k and the BS at time t
δ^t	Time slot
δ_{ik}^{DU}	Time for offloading data from d_i to u_k in a time slot
δ_k^{UB}	Time for offloading data from u_k to the BS in a time slot
β_{km}	Data processing rate between u_k and d_i
δ_k	Time for offloading data from u_k to the BS in a time slot
L_i	The workload of task a_i
t_{ik}	Time required for u_k to process a_i
t_{kj}^{hover}	Time for u_k to hover at b_j
E_k^{total}	Energy consumption for u_k throughout the entire process
E_k^{max}	Battery capacity of u_k
E_k^{trav}	Energy consumption during u_k movement
E_k^{hover}	Energy consumption during u_k hover
$E_{b_{j'},b_j}^{trav}$	Energy consumption for flying from the previous point to b_j

References

1. Long, T.; Ma, Y.; Xia, Y.; Xiao, X.; Peng, Q.; Zhao, J. A Mobility-Aware and Fault-Tolerant Service Offloading Method in Mobile Edge Computing. In Proceedings of the 2022 IEEE International Conference on Web Services (ICWS), Barcelona, Spain, 10–16 July 2022 ; pp. 67–72. [CrossRef]
2. Liao, Z.; Ma, Y.; Huang, J.; Wang, J.; Wang, J. HOTSPOT: A UAV-Assisted Dynamic Mobility-Aware Offloading for Mobile-Edge Computing in 3-D Space. *IEEE Internet Things J.* **2021**, *8*, 10940–10952. [CrossRef]
3. Zhang, T.; Xu, Y.; Loo, J.; Yang, D.; Xiao, L. Joint Computation and Communication Design for UAV-Assisted Mobile Edge Computing in IoT. *IEEE Trans. Ind. Inform.* **2020**, *16*, 5505–5516. [CrossRef]
4. Yan, H.; Bao, W.; Zhu, X.; Wang, J.; Liu, L. Data Offloading Enabled by Heterogeneous UAVs for IoT Applications Under Uncertain Environments. *IEEE Internet Things J.* **2023**, *10*, 3928–3943. [CrossRef]
5. Lu, Y.; Wen, W.; Igorevich, K.K.; Ren, P.; Zhang, H.; Duan, Y.; Zhu, H.; Zhang, P. Uav ad hoc network routing algorithms in space–air–ground integrated networks: Challenges and directions. *Drones* **2023**, *7*, 448. [CrossRef]
6. Perkins, C.; Belding-Royer, E.; Das, S. RFC3561: *Ad hoc On-Demand Distance Vector (Aodv) Routing*; ACM: New York, NY, USA , 2003.
7. Karp, B.; Kung, H.-T. Gpsr: Greedy perimeter stateless routing for wireless networks. In Proceedings of the of the 6th Annual International Conference on Mobile Computing and Networking, Boston, MA, USA, 6–11 August 2000; pp. 243–254.
8. Heinzelman, W.R.; Chandrakasan, A.; Balakrishnan, H. Energy-efficient communication protocol for wireless microsensor networks. In Proceedings of the 33rd Annual Hawaii International Conference on System Sciences, Maui, HI, USA, 7 January 2000 ; p. 10.
9. Haas, Z.J.; Pearlman, M.R. The performance of query control schemes for the zone routing protocol. *ACM SIGCOMM Comput. Commun. Rev.* **1998**, *28*, 167–177. [CrossRef]
10. Krichen, M.; Adoni, W.Y.H.; Mihoub, A.; Alzahrani, M.Y.; Nahhal, T. Security challenges for drone communications: Possible threats, attacks and countermeasures. In Proceedings of the 2022 2nd International Conference of Smart Systems and Emerging Technologies (SMARTTECH), Riyadh, Saudi Arabia, 9–11 May 2022; pp. 184–189.
11. Ko, Y.; Kim, J.; Duguma, D.G.; Astillo, P.V.; You, I.; Pau, G. Drone secure communication protocol for future sensitive applications in military zone. *Sensors* **2021**, *21*, 2057. [CrossRef]
12. Ebrahimi, D.; Sharafeddine, S.; Ho, P.-H.; Assi, C. Autonomous UAV Trajectory for Localizing Ground Objects: A Reinforcement Learning Approach. *IEEE Trans. Mob. Comput.* **2021**, *20*, 1312–1324. [CrossRef]
13. Zhang, L.; Celik, A.; Dang, S.; Shihada, B. Energy-Efficient Trajectory Optimization for UAV-Assisted IoT Networks. *IEEE Trans. Mob. Comput.* **2022**, *21*, 4323–4337. [CrossRef]
14. Coupechoux, M.; Darbon, J.; Kélif, J.-M.; Sigelle, M. Optimal Trajectories of a UAV Base Station Using Hamilton-Jacobi Equations. *IEEE Trans. Mob. Comput.* **2023**, *22*, 4837–4849. [CrossRef]
15. Wang, K.; Zhang, X.; Duan, L.; Tie, J. Multi-UAV Cooperative Trajectory for Servicing Dynamic Demands and Charging Battery. *IEEE Trans. Mob. Comput.* **2023**, *22*, 1599–1614. [CrossRef]
16. Bera, A.; Misra, S.; Chatterjee, C.; Mao, S. CEDAN: Cost-Effective Data Aggregation for UAV-Enabled IoT Networks. *IEEE Trans. Mob. Comput.* **2023**, *22*, 5053–5063. [CrossRef]
17. He, Y.; Gan, Y.; Cui, H.; Guizani, M. Fairness-Based 3-D Multi-UAV Trajectory Optimization in Multi-UAV-Assisted MEC System. *IEEE Internet Things J.* **2023**, *10*, 11383–11395. [CrossRef]
18. Wu, H.; Wu, M.; Peng, W.; Chen, S.; Feng, Z. ITS: Improved Tabu Search Algorithm for Path Planning in UAV-Assisted Edge Computing Systems. In Proceedings of the 2023 IEEE International Conference on Web Services (ICWS), Chicago, IL, USA, 2–8 July 2023; pp. 340–349. [CrossRef]
19. Qiu, X.; Zhang, S.; Wang, Z.; Luo, H. Integrated Host- and Content-Centric Routing for Efficient and Scalable Networking of UAV Swarm. *IEEE Trans. Mob. Comput.* **2023**, *23*, 2927–2942. [CrossRef]

20. Song, H.; Liu, L.; Shang, B.; Pudlewski, S.; Bentley, E.S. Enhanced Flooding-Based Routing Protocol for Swarm UAV Networks: Random Network Coding Meets Clustering. In Proceedings of the IEEE INFOCOM 2021—IEEE Conference on Computer Communications, Vancouver, BC, Canada, 10–13 May 2021; pp. 1–10. [CrossRef]
21. Gaydamaka, A.; Samuylov, A.; Moltchanov, D.; Ashraf, M.; Tan, B.; Koucheryavy, Y. Dynamic Topology Organization and Maintenance Algorithms for Autonomous UAV Swarms. *IEEE Trans. Mob. Comput.* **2024**, *23*, 4423–4439. [CrossRef]
22. Yang, X.; Wang, L.; Xu, L.; Zhang, Y.; Fei, A. Boids Swarm-based UAV Networking and Adaptive Routing Schemes for Emergency Communication. In Proceedings of the IEEE INFOCOM 2023—IEEE Conference on Computer Communications Workshops (INFOCOM WKSHPS), Hoboken, NJ, USA, 20 May 2023; pp. 1–6. [CrossRef]
23. Gharib, M.; Afghah, F.; Bentley, E. OPAR: Optimized Predictive and Adaptive Routing for Cooperative UAV Networks. In Proceedings of the IEEE INFOCOM 2021—IEEE Conference on Computer Communications Workshops (INFOCOM WKSHPS), Vancouver, BC, Canada, 10–13 May 2021; pp. 1–6. [CrossRef]
24. Zhao, P.; Lu, Y.; Wei, Y.; Leng, S. Blockchain and DQN Enabled Co-Evolutionary Routing Scheme in UAV Networks. In Proceedings of the IEEE INFOCOM 2023—IEEE Conference on Computer Communications Workshops (INFOCOM WKSHPS), Hoboken, NJ, USA, 20 May 2023; pp. 1–6. [CrossRef]
25. Hsu, Y.-H.; Gau, R.-H. Reinforcement Learning-Based Collision Avoidance and Optimal Trajectory Planning in UAV Communication Networks. *IEEE Trans. Mob. Comput.* **2022**, *21*, 306–320. [CrossRef]
26. Fu, J.; Sun, G.; Liu, J.; Yao, W.; Wu, L. On Hierarchical Multi-UAV Dubins Traveling Salesman Problem Paths in a Complex Obstacle Environment. *IEEE Trans. Cybern.* **2024**, *54*, 123–135. [CrossRef]
27. Qureshi, H.N.; Imran, A. On the Tradeoffs Between Coverage Radius, Altitude, and Beamwidth for Practical UAV Deployments. *IEEE Trans. Aerosp. Electron. Syst.* **2019**, *55*, 2805–2821. [CrossRef]
28. Li, J.; Zhao, H.; Wang, H.; Gu, F.; Wei, J.; Yin, H.; Ren, B. Joint Optimization on Trajectory, Altitude, Velocity, and Link Scheduling for Minimum Mission Time in UAV-Aided Data Collection. *IEEE Internet Things J.* **2020**, *7*, 1464–1475. [CrossRef]
29. Zhou, F.; Wu, Y.; Hu, R.Q.; Qian, Y. Computation Rate Maximization in UAV-Enabled Wireless-Powered Mobile-Edge Computing Systems. *IEEE J. Sel. Areas Commun.* **2018**, *36*, 1927–1941. [CrossRef]
30. Cao, X.; Xu, J.; Zhang, R. Mobile Edge Computing for Cellular-Connected UAV: Computation Offloading and Trajectory Optimization. In Proceedings of the 2018 IEEE 19th International Workshop on Signal Processing Advances in Wireless Communications (SPAWC), Kalamata, Greece, 25–28 June 2018; pp. 1–5. [CrossRef]
31. Zeng, Y.; Xu, J.; Zhang, R. Energy Minimization for Wireless Communication With Rotary-Wing UAV. *IEEE Trans. Wirel. Commun.* **2019**, *18*, 2329–2345. [CrossRef]
32. Bera, A.; Misra, S.; Chatterjee, C. QoE Analysis in Cache-Enabled Multi-UAV Networks. *IEEE Trans. Veh. Technol.* **2020**, *69*, 6680–6687. [CrossRef]
33. Lai, P.; He, Q.; Abdelrazek, M.; Chen, F.; Hosking, J.; Grundy, J.; Yang, Y. Optimal Edge User Allocation in Edge Computing with Variable Sized Vector Bin Packing. In Proceedings of the 16th International Conference on Service-Oriented Computing (ICSOC2018), Hangzhou, China, 12–15 November 2018; pp. 230–245.
34. Wang, S.; Jiang, Z.; Bao, X. Autonomous Trajectory Planning Method for Multi-UAV Collaborative Search. In Proceedings of the 2021 5th International Conference on Automation, Control and Robots (ICACR), Nanning, China, 25–27 September 2021; pp. 84–88. [CrossRef]

Disclaimer/Publisher's Note: The statements, opinions and data contained in all publications are solely those of the individual author(s) and contributor(s) and not of MDPI and/or the editor(s). MDPI and/or the editor(s) disclaim responsibility for any injury to people or property resulting from any ideas, methods, instructions or products referred to in the content.

Article

Unmanned Aerial Vehicle Obstacle Avoidance Based Custom Elliptic Domain

Yong Liao *, Yuxin Wu, Shichang Zhao and Dan Zhang

College of Air Traffic Management, Civil Aviation Flight University of China, Guanghan 618307, China; wyx10561@163.com (Y.W.); 17792325792@163.com (S.Z.); zdmail3108@163.com (D.Z.)
* Correspondence: liaoyong@cafuc.edu.cn; Tel.: +86-18283816015

Abstract: The velocity obstacles (VO) method is widely employed in real-time obstacle avoidance research for UAVs due to its succinct mathematical foundation and rapid, dynamic planning abilities. Traditionally, VO assumes a circle protection domain with a fixed radius, leading to issues such as excessive conservatism of obstacle avoidance areas, longer detour paths, and unnecessary avoidance angles. To overcome these challenges, this paper firstly reviews the fundamentals and pre-existing defects of the VO methodology. Next, we explore a scenario involving UAVs in head-on conflicts and introduce an elliptic velocity obstacle method tailored to the UAV's current flight state. This method connects the protection domain size directly to the UAV's flight state, transitioning from the conventional circle domain to a more efficient elliptic domain. Additionally, to manage the computational demands of Minkowski sums and velocity obstacle cones, an approximation algorithm for discretizing elliptic boundary points is introduced. A strategy to mitigate unilateral velocity oscillation had is developed. Comparative validation simulations in MATLAB R2022a confirm that, based on the experimental results for the first 10 s, the apex angle of the velocity obstacle cone for the elliptical domain is, on average, reduced by 0.1733 radians compared to the circular domain per unit simulation time interval, saving an airspace area of 13,292 square meters and reducing the detour distance by 14.92 m throughout the obstacle avoidance process, facilitating navigation in crowded situations and improving airspace utilization.

Keywords: unmanned aerial vehicle; velocity obstacles; elliptic domains; collision resolution; geometric optimization algorithm

Citation: Liao, Y.; Wu, Y.; Zhao, S.; Zhang, D. Unmanned Aerial Vehicle Obstacle Avoidance Based Custom Elliptic Domain. *Drones* **2024**, *8*, 397. https://doi.org/10.3390/drones8080397

Academic Editor: Abdessattar Abdelkefi

Received: 8 May 2024
Revised: 31 July 2024
Accepted: 13 August 2024
Published: 15 August 2024

Copyright: © 2024 by the authors. Licensee MDPI, Basel, Switzerland. This article is an open access article distributed under the terms and conditions of the Creative Commons Attribution (CC BY) license (https://creativecommons.org/licenses/by/4.0/).

1. Introduction

1.1. Related Prior Work

As unmanned aerial vehicles (UAVs) are widely used in various fields such as road traffic planning, military reconnaissance, agricultural production, and logistics distribution, the demand for low-altitude UAVs in segregated or integrated airspace will gradually increase in the foreseeable future. The operational safety of UAVs in low-altitude airspace and urban inter-traffic connections will be particularly important. A prominent issue in this regard is the collision avoidance problem of UAVs. There have been significant research achievements in addressing the collision avoidance problems of UAVs (or unmanned surface vessels, robots) in different research fields. The solutions can be categorized into global optimization control methods and local real-time avoidance methods.

The global optimal control method is based on the idea of mathematical optimization, and the required UAV navigation environment and flight state information is more complete, which can usually be planned to obtain a collision-free global navigation route under the specified constraints. Fu [1] utilized the additional flight distance of unmanned aerial vehicles as the maneuvering cost function. Initially, they computed the feasible solutions for initial collision avoidance using stochastic parallel gradient descent (SPGD) and then employed sequential quadratic programming (SQP) to determine the optimal collision avoidance heading. Sarim [2] proposed a combined solution based on A* mixed integer linear programming for the initial path planning of multiple unmanned aerial vehicles with individual task requirements and dynamic constraints. Sunberg [3] converted the multi-UAV conflict resolution problem into an approximate dynamic programming problem to solve. In addition, heuristic algorithms such as particle swarm optimization (PSO) and the genetic algorithm (GA) have considerable potential applications due to their efficient search capabilities in complex scenarios. Phung [4] effectively explored the configuration space of unmanned aerial vehicles by establishing corresponding relationships between particle positions and the UAV's speed, turning angle, and climb/descent angles. They utilize PSO to find the optimal path that minimizes the cost function. Pehlivanoglu [5] integrated GA, Voronoi diagrams, and clustering methods, proposing an initial population enhancement approach to accelerate the convergence process, thereby obtaining feasible optimal paths in a short time. The starting point of the above global optimization control methods is based on the entire process of aircraft conflict, aiming to satisfy the optimization goal of minimizing a certain cost payment. However, they suffer from poor real-time performance and limited operability. These methods are difficult to apply when the complete global aircraft flight data are unknown and the conflict positions are uncertain. Moreover, the challenge of solving large-scale problems with slow convergence speeds persists.

On the other hand, local real-time path planning methods do not optimize the entire flight path of the aircraft but rather focus on conflict detection and avoidance. The main methods include artificial potential field (APF) and velocity obstacle (VO). APF has the advantage of a short response time and small problem-solving scale but often fails to produce a flight trajectory directed towards the target point. By setting virtual targets, this issue can be addressed, allowing for the correct navigation of the drone to avoid obstacles and reach the target point [6,7]. Pan [8] also introduced a rotational potential field to help the drone escape local minima and oscillation phenomena, facilitating the navigation of drone formations and clusters. The VO method, originally proposed by Paolo Fiorini [9], is a prominent obstacle avoidance strategy that converts the positional collision potential in the motion space of an agent into the velocity vector space [10]. This method delineated the entire set of velocities that could lead to a collision with an obstacle within a finite time as the 'velocity obstacle space' (VO space). By assigning the UAV a new linear velocity vector outside this defined set, the method ensured that the UAV can circumnavigate the obstacle within a permissible time frame, thereby eliminating collision risks. The method has excellent geometric intuition and does not require a complex modeling process. The 2D VO theoretical model simplified the UAV as a circle domain; this geometric central symmetry significantly eases the calculation of the 'VO space'. However, this circle assumption uniformly scaled the conflict risk in all flight directions, potentially rendering conflict resolution schemes as overly conservative [10]. This could lead to operational failures in scenarios with high conflict numbers or dense UAV flights. For the same obstacle avoidance problem, there were also differences in whether or not to collide under circle protection domains with different radius scales, and if so, in the direction of the chosen resolution.

Current research on UAV obstacle avoidance using the VO method typically employs an empirical or artificially assigned fixed-radius circle protection domain [11–15]. Commonly, this radius is set by default as the UAV's detection range; when another UAV enters this range, a potential collision is presumed to have occurred. This radius of this protection domain was not determined by the specific obstacle avoidance capabilities of the UAV, which compromises its scientific validity. Such an approach unnecessarily reduces the available free space—the area not occupied by the UAV's flight path—potentially leading to suboptimal utilization of the navigable airspace.

From the perspective of minimizing the spatial size of the protection domain while ensuring sufficient safe space for UAV operation, substituting the circle domain with an elliptical domain is a practical option. This change not only maintains the smooth continuity of the boundary of the protection domain but also reflects the physical form of the intelligent agents more accurately. Circles tend to overestimate the necessary protection space for bodies that do not exhibit radial symmetry, while ellipses can more closely conform to the actual boundaries of the intelligent bodies. This adaptation was particularly beneficial in specific obstacle scenarios, such as navigating near walls, where circle domains may unnecessarily increase the detour distance [16,17]. Moreover, the geometric properties of elliptical domains offer broader applicability—circles are merely special cases of ellipses. Most agents, including ships and humanoid robots, do naturally conform to a elliptic boundary. Furthermore, the application of elliptical domains in the fields of pedestrian locomotion [18] and biomechanics [19] has demonstrated that ellipses provide a more accurate approximation of human movement.

In obstacle avoidance research utilizing the VO method, many researchers are aware of the drawbacks of circular domains but still compromise because of the simplicity of the computation. Still, several scholars had adopted an elliptical boundary domain. Lee [20] tackled the local obstacle avoidance problem for elliptical robots by approximating both the robots and obstacles with a minimum area boundary ellipse, implementing VO in two stages: acquiring a new linear velocity and correcting the angular velocity. Wang noted the substantial difference between longitudinal and transverse velocities in ground vehicle operations and designed an elliptical lattice boundary based on the vehicle's travel direction, velocity, minimum safe distance, and lane width to model the motion of connected automated vehicle (CAV) clusters [21]. Furthermore, various studies [22–24] on unmanned surface vehicles (USVs) hved adopted elliptic domains. Liu and Bucknall [25] also suggested a circular shape for slow-moving obstacles and elliptical shapes for fast-moving obstacles. Additionally, elliptical boundary agents have been applied to research on obstacle avoidance using potential field theory [26] and limit cycle theory [27].

In contrast, the development of elliptic domains has not been exploited in the field of UAV obstacle avoidance. Similar to the conclusions in the literature [21,25], UAVs also characteristically exhibit a significantly higher longitudinal velocity compared to their transverse velocity. Current VO obstacle avoidance studies typically only establish a circle protection domain whose size substantially exceeds that of the individual UAV, leading to notable issues of spatial redundancy and unscientifically set initial values for the radius. We investigated the obstacle avoidance flight process of UAVs in head-on conflict scenarios and innovatively proposed a customized elliptical domain structure related to the separation distance and flight performance of the conflicting UAV pairs. We conducted a qualitative comparison of the velocity obstacles for circular and elliptical domains, validating the advantages of the elliptical domain in simulations that achieve the same safety distance requirements for obstacle avoidance.

1.2. Organization

The remainder of this paper is structured as follows: Section 2 reviews the basic principles and related defects of the VO method. In light of these defects, we articulate the main contributions of our work. In Section 3, we explore the feasibility of elliptical domains and propose designing an elliptic domain size linked to the UAVs' flight state (velocity direction, velocity magnitude, angular velocity, and spatial distance), resulting in a custom elliptic domain. Section 4 introduces the elliptical velocity obstacles (EVO) method tailored for the boundaries of elliptic domains, establishing the elliptical absolute velocity obstacles (*EAVO*) as the core of our obstacle avoidance strategy. We also develop a velocity recovery rule to prevent one-sided velocity oscillations in Section 4. In Section 5, we validate our approach using MATLAB simulation experiments. Our results demonstrate that the *EAVO's* resolution strategy addresses the over-conservative velocity and space wastage issues prevalent in circle domain VO applications, offering a smaller spatial footprint, a broader range of potential velocity directions, and robust resolution performance, particularly in flight scenarios with narrow separation distances. Finally, the results are discussed and summarized, and studies to improve the method are proposed in Section 6. The research process and architecture of this study are shown in Figure 1.

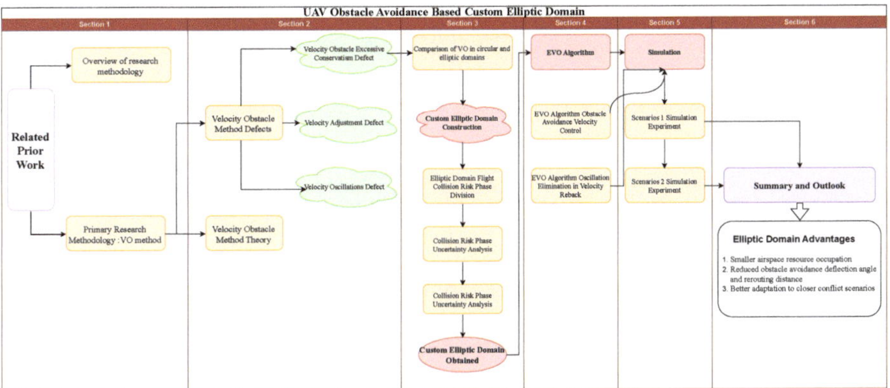

Figure 1. Schematic of research process and architecture.

2. Review of Velocity Obstacle Method

2.1. Velocity Obstacle Method Theory

2.1.1. Minkowski Sum

The Minkowski sum serves as the mathematical foundation of VO theory and is integral to forming the velocity obstacle space. Essentially, a computational geometry operation, the Minkowski sum in Euclidean spaces is calculated by adding each vector from two non-empty sets, 'A' and 'B'. Typically, this operation is defined as follows:

$$A \oplus B = \{a + b \mid a \in A, b \in B\} \tag{1}$$

where $\oplus$ is Minkowski sum notation.

Equation (1) delineates that for each vector in set 'A' and each vector in set 'B' a new vector in the Minkowski sum can be formed by their addition. This computation is straightforward for convex polygons. Non-convex challenges are often transformed into convex ones for resolution, lending the method broad applicability. In two dimensions, if 'A' and 'B' are polygons, their Minkowski sum results in a new polygon that encapsulates all possible combinations of their relative positions.

The Minkowski sum also provides a geometrically intuitive representation within the method. It is visualized as the area swept by set 'A' as it traces the perimeter of set 'B' combined with the area of set 'B' itself. If the elements of sets 'A' and 'B' within an algebraic system adhere to the properties of an Abelian group, the Minkowski sum also complies with the commutative law ($A \oplus B = B \oplus A$), reflecting the interchangeability of the summands. It is evident that the Minkowski sum is commutative within a two-dimensional real vector space. To clearly differentiate the research subjects in this paper, we define $A \oplus B$ to denote the Minkowski sum imposed by set 'A' on set 'B'. This symbolization helps define the construction of the velocity obstacle space in the region where set 'B' is affected. For two methods on the computation of convex polygonal Minkowski sums, see Appendix A.1.

2.1.2. Velocity Obstacle Cone Construction

The velocity obstacle method was developed with a focus on circle robots and obstacles whose instantaneous states are measured or known. This approach allows for the simplification that disregards the computational discrepancies caused by the rotation of objects and their varying orientations [9]. For ease of description, this paper not only uses A and B to denote the two UAVs but also to represents the circle or elliptical protection domains in which A and B are located. The collision cone $C_{A,B}$ is defined in the literature [9] as the set of relative velocities that could lead to collisions between A and B. Superimposing

$C_{A,B}$ onto the velocity of B defines the avoidance space in terms of B's absolute velocity. This can be denoted as follows:

$$\begin{cases} C_{A,B} = \{v_{A,B} | \lambda_{A,B} \cap \hat{B} \neq \varnothing\} \\ VO = C_{A,B} \oplus v_B \end{cases} \tag{2}$$

where $v_{A,B}$ is the velocity of A with respect to B, and $\lambda_{A,B}$ is the line on which vector $v_{A,B}$ lies. A 'collapses' to a point $\hat{A}$, and B 'expands' to $\hat{B}$, a circle region of twice the original radius. We prove this concisely in Appendix A.2. Equation (2) represents the region where relative velocity intersects with the velocity obstacle, forming a velocity obstacle cone. By superimposing the speed of Drone B, we can obtain the absolute velocity obstacle cone for Drone A.

The original definition of the velocity obstacle method involves adjusting the velocity of UAV-A outside the VO to avoid collisions. By integrating this classical VO definition and incorporating a time slice τ, the VO and EVO models are summarized and redefined into three expressions in this paper:

$$\begin{cases} \widehat{B_C} = A \oplus B \\ VO^{\tau}_{A|B} = \left\{v_{A,B} | \exists t \in [0, \tau] :: tv_{A,B} \cap \widehat{B_C} \neq \varnothing\right\} \quad (circle\ domain) \\ AVO^{\tau}_{A|B} = VO^{\tau}_{A|B} \oplus v_B \end{cases} \tag{3}$$

$$\begin{cases} \widehat{B_E} = A \oplus B \\ EVO^{\tau}_{A|B} = \left\{v_{A,B} | \exists t \in [0, \tau] :: tv_{A,B} \cap \widehat{B_E} \neq \varnothing\right\} \quad (elliptic\ domain) \\ EAVO^{\tau}_{A|B} = EVO^{\tau}_{A|B} \oplus v_B \end{cases} \tag{4}$$

For clarity of presentation, we made $VO^{\tau}_{A|B}, EVO^{\tau}_{A|B}$ represent $C_{A,B}$ in the original definition of the velocity obstacle method. At each time interval τ, the velocity obstacle cone for the drone can be obtained at each detection moment during the obstacle avoidance process, allowing for discrete conflict monitoring based on the calculated results.

The three variables above on the left side of Equations (3) and (4) are named the velocity obstacle set, the relative velocity obstacle cone, and the absolute velocity obstacle cone, respectively. The variable 't' represents time. According to the definitions used in this paper, the VO for circle domains with equal radii at any location is illustrated in Figure 2. The elliptic domain Minkowski sums $\widehat{B_E}$ cannot be generated directly from the superposition of the geometric radius; thus, we require specially designed algorithms for implementation.

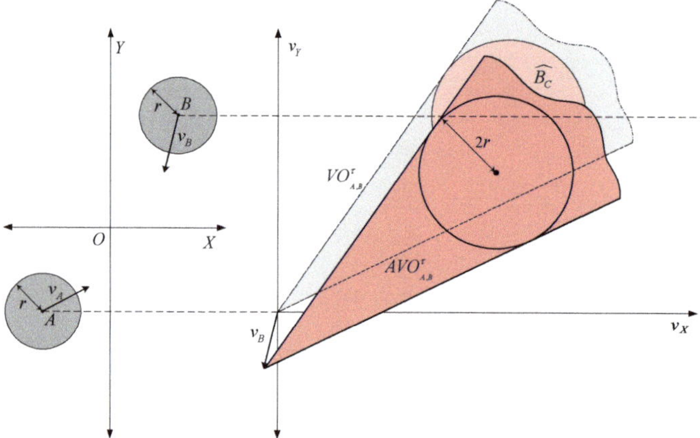

Figure 2. Schematic of AVO between equal-radius circle domains.

2.2. Velocity Obstacle Method Defects

2.2.1. Velocity Obstacle Excessive Conservatism Defect

The previous research related to the VO of the assumption of circle and elliptic domains has been briefly described in Section 1. Despite the computational benefits, we know the use of circles for multi-agent navigation results in many challenges. In many cases, a circle overestimates the actual profile of the robots that it represents [10]. In scenarios where the geometric appearance of agents closely resembles an ellipse, using an elliptical protection domain is a logical choice. The adoption of geometrical bodies that more closely resemble the appearance of the study subjects simplifies the obstacle avoidance modeling process. Essentially, this idea tends to ensure that the simulation results of abstract geometric bodies are more aligned with the actual movement of intelligent agents, eliminating redundant and unnecessary protection domains. From this perspective, assuming a circular radar range as the protective domain for UAVs is undeniably overly conservative. For a pair of UAVs in head-on conflict, if there is no interference, collisions would invariably occur in the direction of the velocity. Therefore, it is sensible to provide ample protection space in the direction of the velocity and reduce the protection space in unnecessary directions.

2.2.2. Velocity Adjustment Defect

The velocity obstacle method assumes that UAVs adjust their velocity states instantaneously upon detecting conflict risks, requiring highly sensitive sensors and efficient control algorithms for devising obstacle avoidance strategies [11]. If the length of time for velocity adjustment is considered, the VO space will be transformed into an area where the relative velocity obstacle cone sweeps through this length of time, necessitating greater velocity changes and occupying more space, which is disadvantageous for navigation in tight spaces. Moreover, the assumption that velocity changes instantaneously is not applicable when obstacle velocities are nonlinear, as the direction and magnitude of obstacle speeds can change at any moment, complicating the obstacle space dynamics. The discrete-time nonlinear velocity obstacle method (NLVO) [28–31] offers an effective solution for these challenges.

2.2.3. Velocity Oscillation Defects

Another notable problem is the oscillation of velocity selection. This issue arises when two conflicting entities simultaneously choose velocities for the next moment that are deemed collision-free. New velocities can lead to misjudgments about collision risks, prompting a return to original velocities. This cycle repeats, with each subsequent velocity adjustment reintroducing the potential for collisions. This phenomenon typically occurs due to a default preference for a higher-priority velocity direction v_{prefer}, often the initial velocity. As soon as a collision is detected or disappears, the velocity is approached toward v_{prefer} immediately at the next moment.

To address this common problem in VO applications, Van Den Berg [32] proposed the reciprocal velocity obstacle (RVO) method, where the original VO is shifted by a specified time–displacement distance by translating it along $(v_A + v_B)/2$. By referring to the newly defined obstacle cone, one can effectively judge and mitigate collisions and oscillations among autonomous agents. This approach implies that both conflicting parties equally share the responsibility for collision avoidance. The RVO method has been extensively utilized in research on obstacle avoidance and safe navigation for dynamic agents [33–35]. Nonetheless, challenges persist due to disagreements and desynchronization over navigation preferences [36], potentially leading to ineffective collaboration and a reciprocal dance of avoidance [37]. To address this issue, Jamie [38] proposed the hybrid reciprocal velocity obstacle (HRVO) method, which modifies the passive avoidance approach in RVO by basing the future trajectory of robots on more than just a simple estimation of the current velocity [39].

2.3. Our Contributions

In response to the drawbacks identified in the VO methods outlined above, we have made specific improvements. For the issue of overly conservative circular domain structures, we aim to establish an elliptical domain that adequately protects UAVs while reducing the redundancy of the circular domain. Furthermore, to mitigate the adverse effects of sudden velocity shifts on obstacle avoidance decisions, we analyzed the UAV's positional errors under tiny time deflection assumptions. By incorporating this positional uncertainty into the elliptical domain structure, we can continue to generate obstacle avoidance velocities through the absolute velocity obstacle cone (Section 3). Concurrently, due to the computational costs associated with precisely solving the Minkowski sum of the elliptical domain, we propose a discretized elliptical boundary accelerated algorithm using convex polygons (Section 4).

As for addressing velocity oscillations, RVO has already been widely adopted. This article is different from the above research on the fully autonomous obstacle avoidance mode of one or more agents. We provides a solution to the velocity oscillations of UAVs with motion priority. Unlike the previous assumption that both conflicting agents share responsibility for collision avoidance [32], this model designates only one UAV in the conflict as being responsible for adjusting its velocity to avoid potential obstacles, while the other UAV maximizes the usage of its original flight path, maintaining a steady course (Section 5). This approach could be particularly relevant in future urban logistics scenarios where, for example, delivery UAVs tasked with off-site deliveries may be given higher priority over UAVs returning for landing, reflecting typical head-on conflict characteristics. Additionally, the integration of UAVs actively avoiding manned aircrafts in fusion airspace is also a potential application scenario.

3. Custom Elliptic Domain Construction

In this part, we refine the original uniform circle protection domain structure with a fixed radius used in the VO method, replacing it with a custom elliptical domain structure that adapts to the UAV's flight state. It is shown that a circular domain with a given radius may be redundant, and the actual elliptical domain based on the UAV's flight conditions is much more accurate and still ensures that the UAV can avoid obstacles while satisfying the flight performance. To avoid the complexities present in initial studies, we continue to focus on the interaction between two UAVs flying in opposite directions, designated as A and B.

3.1. Comparison of VO in Circular and Elliptic Domains

Based on the review of the theoretical basis and fundamental principles of the velocity obstacle method, we can derive the following basic conclusions:

- The mathematical principle of the velocity obstacle space is the Minkowski sum of the boundary curves of two spatial objects. Geometrically, the Minkowski sum of two colliding entities represents the region swept by object A along the boundary of object B as it moves continuously for one revolution, combined with object B.
- When both objects are circles, their Minkowski sum is a circle with a radius equal to the sum of the radii of the two objects. For circles of the same size, their Minkowski sum is a circle with twice the radius.
- Based on the proof that the Minkowski sum of two circles remains a circle, it can be anticipated that the precise calculation of the Minkowski sum and velocity obstacle cone for two elliptical objects will be more difficult. The reason for this is that in Equation (A1), the radius 'r' becomes the non-uniform semi-axis of the ellipse, making it challenging to simplify the computation of the maximum value. The distances from any point on ellipse A to the farthest point from the center of ellipse B obtained through iterative calculations will vary, indicating that the boundary of the Minkowski sum of two ellipses may not possess simple geometric characteristics. Therefore,

further algorithmic solutions are required to address the velocity obstacle for elliptical boundaries.

3.1.1. Description of UAV Collision Stations

By approximating the obstacle boundary with the translation of the ellipse, a qualitative comparison of the velocity obstacle between circular and elliptic domains can be conducted. A and B are a pair of UAVs flying in opposite directions at the same altitude. If no avoidance measures are taken, they will collide at some point in the future. We set A's spatial position at the coordinate origin, with the line connecting A and B along the horizontal axis (X-axis) and the vertical direction to the X-axis representing the Y-axis, thereby establishing a two-dimensional Cartesian coordinate system. Four types of protective domain combinations were constructed as comparative scenarios, denoted as $\alpha_0, \alpha_1, \alpha_2, \alpha_3$. The structural relationships and collision schematic diagrams for these four scenarios are illustrated in Figures 3–6. Among these, the velocity obstacle spaces generated within the elliptical domains in stations α_2 and α_3 are approximate calculations, while the results in stations α_0 and α_1 are completely precise.

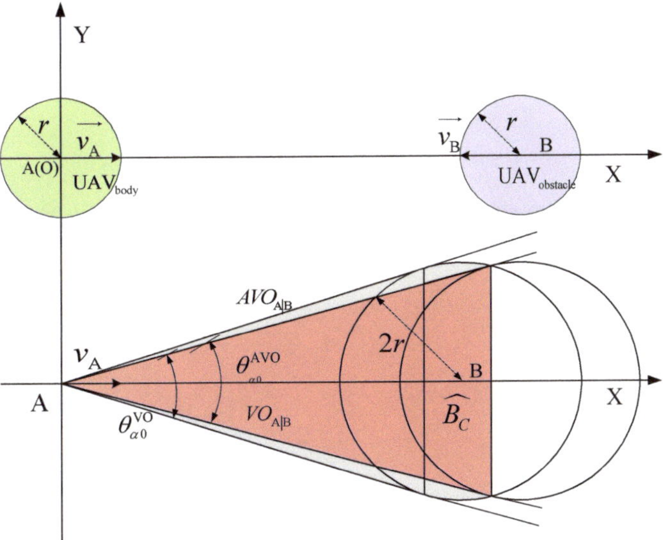

Figure 3. Station α_0: Two of the same circular domain collisions.

Station α_0: Assuming the protective domain of the UAV is a circle of equal size, let A represent the obstacle-avoiding UAV and B represent the obstacle UAV. The velocity obstacle set between A and B forms a large circle with a radius of $2r$. By drawing a tangent line to this large circle from an external point at the origin, we obtain a conical region known as the relative velocity obstacle cone. By translating this relative velocity obstacle cone using the velocity vector, we derive the absolute velocity obstacle cone of A with respect to B. This cone represents the set of velocities that A must avoid in order to evade B.

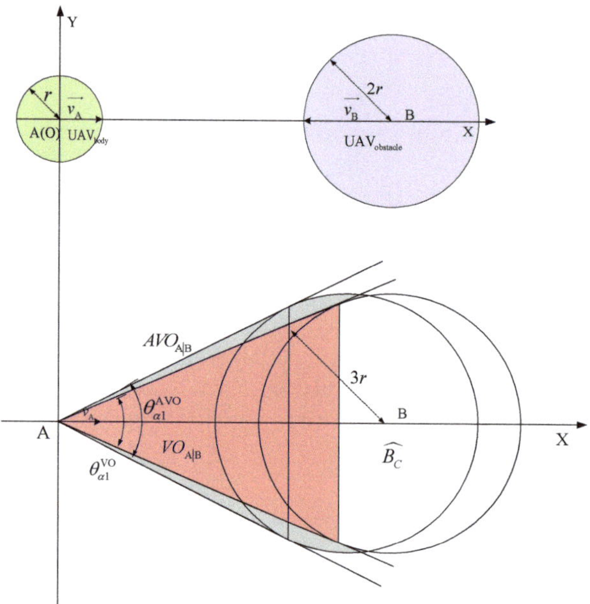

Figure 4. Station α_1 : Two different circular domain collisions.

Station α_1: Keeping the state information of A and B unchanged in Station α_0, we double the radius of B's protective domain to obtain a velocity obstacle set represented by a circular area with a radius of $3r$ after the 'expansion'.

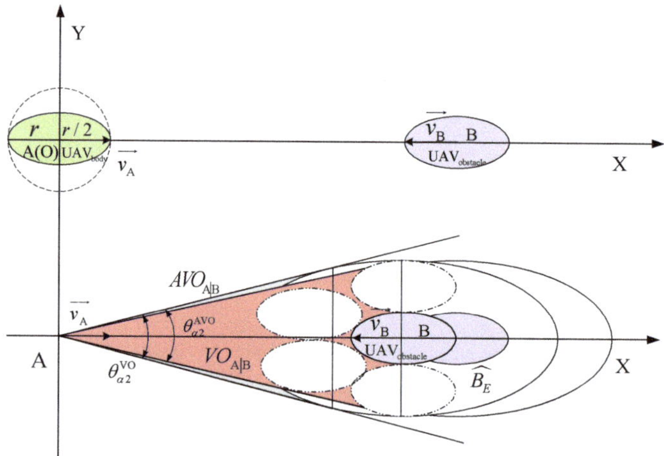

Figure 5. Station α_2: Two of the same internal elliptical collisions.

Station α_2: Assuming the UAV's protective domain is an elliptic area, with A and B being isomorphic, the elliptic domain is contained within the initial circular domain. By geometrically describing the approximate velocity obstacles of both entities, we can simply translate UAV A's elliptic domain several times until it is tangent to the boundary of B. We then envelop several tangential ellipses with a larger elliptic boundary, using this large ellipse to approximate the relative velocity obstacle set of A concerning B. Similarly, by approximating the tangent lines, we obtain the relative velocity obstacle cone, which, after translation, forms the absolute velocity obstacle cone.

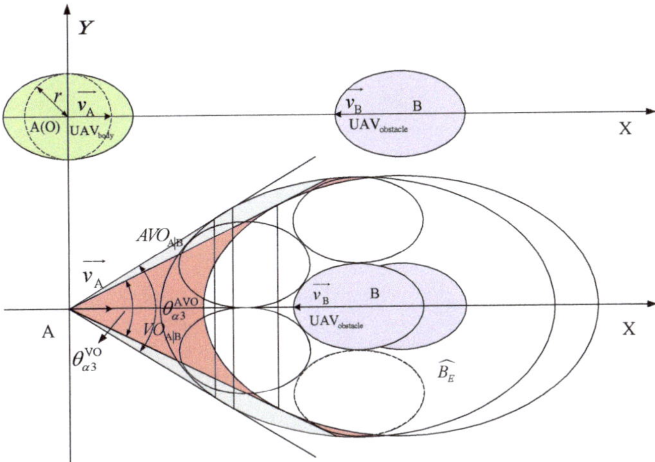

Figure 6. Station α_3: Two of the same external elliptical collisions.

The UAV's protective domain is an elliptical area, with A and B being isomorphic. The elliptical domain encompasses the initial circular domain, with a short semi-axis of r and a long semi-axis of $1.5r$, while all other assumptions remain unchanged.

3.1.2. Comparison of UAV Collision Stations

- Comparison of station α_0 and station α_1: A and B both have circular protective domains. In the same collision scenario, the double protective domain of B in station α_1 will cause a significantly larger velocity obstacle space compared to station α_0. The target avoidance velocity obtained in station α_1 will require a greater angular deviation. Therefore, a critical issue in applying the velocity obstacle principle to ensure collision-free operation for UAVs is determining the appropriate range of these protective domains. It is clear that avoidance decisions and outcomes are sensitive to the initial radius of this area. If the protective domain is too large, it compresses the available free space, leading to increased avoidance costs. Conversely, if the protective domain is too small, the performance requirements for the UAV during avoidance maneuvers increase, along with associated safety risks. Exploring a suitable and safe structure for the protective domain is a primary focus of this research.
- Comparison of station α_0 and station α_2: While maintaining a constant protective distance in the direction of the speed, the protective distance in the normal direction of the speed is reduced. Consequently, the absolute speed obstacle angle is also decreased. This indicates that constructing a collision-free zone in the shape of an elliptical domain with a short axis can not only ensure safe obstacle avoidance but also minimize the utilization of airspace resources.
- Comparison of station α_0 and station α_3: In station α_3, the elliptical domain completely encompasses the circular domain from station α_0, resulting in a velocity obstacle space that also covers the velocity obstacle space obtained from the circular domain. In station α_3, the protective distance in the normal direction of the speed remains at r, while the protective distance in the velocity direction increases by $0.5r$. As a result, the velocity obstacle angle also increases accordingly. This demonstrates that both axes of the ellipse have an impact on the calculated results of the velocity obstacle space.

Based on the comparison above, it is evident that the elliptical structure in station α_2, while maintaining the original protective space in the direction of speed, has corrected the overly conservative circular domain by reducing the size in the normal direction of speed. If there are established safety distance requirements, this dimension should be fixed along the major axis of the ellipse. For the same drone collision event, the avoidance velocity sets calculated from the different protective domain structures in station $\alpha_0, \alpha_1, \alpha_2, \alpha_3$ vary accordingly. Therefore, our objective is to design an elliptical protection domain that ensures the safe flight of the drone while providing a certain degree of redundancy. To the greatest extent possible, we provide a larger protective domain space in the collision direction while minimizing unnecessary dimensions in non-collision directions. Theoretically, this approach can offer the drone a more diverse selection of obstacle avoidance velocity options.

3.2. Elliptic Domain Flight Collision Risk Phase Division

A Cartesian coordinate system was established with the initial position of UAV-A as the origin. The size of the elliptical protection domain is related to the UAV's flight state ensuring that the geometrical structure can still be relieved, even under the least ideal conditions conditions. The model is configured as follows:

In this study, we configure the UAV to update the detection of its surroundings every τ seconds. Upon detecting an obstacle, the UAV will initiate an avoidance maneuver. It is important to note that throughout this research, adjusting the direction of velocity has been consistently used as the primary means of obstacle avoidance. This approach aligns with the ideal collision resolution under the VO principle. An adjustment that alters only the magnitude of the velocity, without changing its direction, may put off the conflict into a subsequent time period, especially when UAV-B is in linear motion. After detecting an obstacle and identifying a new velocity target orientation, the UAV most at risk of collision has approximately τ seconds—excluding the time spent computing the obstacle cone and selecting a new velocity—to reorient towards the target. Consequently, the maximum possible deflection achievable by the UAV within τ seconds sets the minimum threshold necessary for collision avoidance.

Assuming that the maximum deflection angle of the UAV per one second is φ (rad), the minimum threshold on the deflection that can be made within τ is $\varphi_{inf} = \tau\varphi$. In scenarios where UAVs encounter head-on conflicts, if both UAVs simultaneously maneuver to the right with the same constant angular velocity towards a predetermined target direction, the collision can be avoided if their protective domains do not overlap. The entire obstacle avoidance process is divided into two phases: 'Phase 1: Conflict risk phase' and 'Phase 2: Conflict risk elimination phase'. Phase 1 is a deflection flight process, while Phase 2 is a directional process. The active avoidance maneuvers occur during Phase 1. A schematic diagram illustrating this is provided below, where p_{link} is the two-phase transition node. UAV-A and UAV-B have equal elliptic domains.

In the second phase, where the velocities of the two UAVs are parallel but not collinear, overlaps in the protective domains can occur if the steering adjustments from the first phase are inadequate. This overlap might stem from higher longitudinal velocities or smaller intervals between obstacle detections. As illustrated in Figure 7, when both UAVs employ the same avoidance strategy, their directions align at any given moment, and the major and minor axes of their elliptical domains remain parallel, with UAV-B consistently positioned above UAV-A. Therefore, potential overlaps during Phase 1 could manifest as one of three types: frontside, backside, or topside intrusions, as depicted in Figure 8.

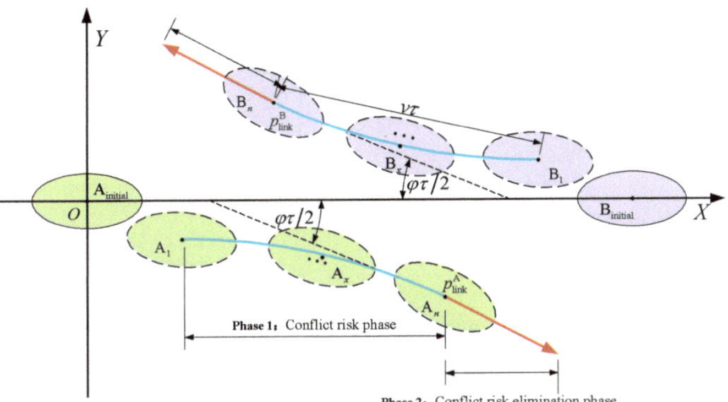

Figure 7. Schematic of UAV conflict risk phase and conflict risk elimination phase flights.

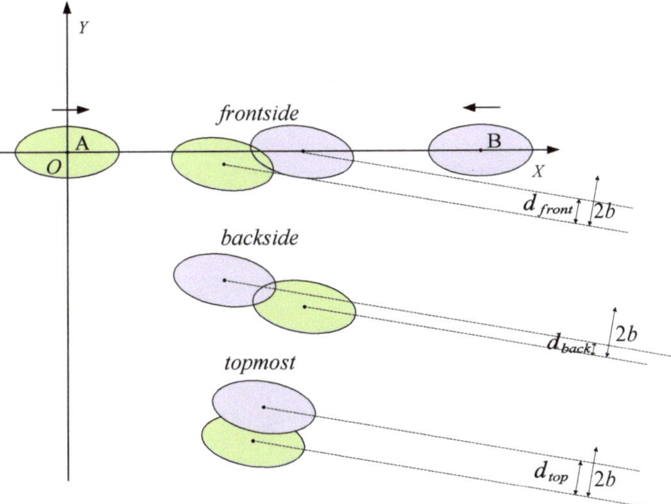

Figure 8. Schematic illustration of possible collisions during the UAVs' conflict risk phase.

When analyzing the three relative positional relationships, collisions can be averted in each scenario as long as the distance between the centers of the two ellipses, when projected along the direction of the minor axis, remains greater than the length of the minor axis ($2b$) throughout Phase 1.

3.3. Collision Risk Phase Uncertainty Analysis

3.3.1. Collision Risk Phase Uncertainty Assumptions

At the critical moment when a UAV anticipates a collision risk using the VO principle, the essence of obstacle avoidance lies in adjusting the velocity direction to fall outside the VO. According to the VO method, the velocity is presumed constant within the interval τ, changing only at the moment a decision is made to avoid an obstacle. This necessitates frequent sensor refreshes [11], which can lead to discrepancies between the actual flight position and the VO theoretical one. This means that at every moment, the actual position of the drone deviates from the estimated position. Such endpoint location uncertainty may continuously accumulate, potentially impacting the design of the elliptical protection domain size. Therefore, we categorize the deflection process into three assumptions and discuss them below.

- *Assumption 1: Segmented Multiple Tiny Deflections*
  Assume that the drone's directional adjustments are linearly varied over each time interval. Segmented, multiple, tiny deflections provide a smoother representation of the actual flight process. This method suggests that the angular velocity during the deflection process remains constant throughout each tiny time interval. The UAV produces consistent angles of deflection over identical, short periods, with the cumulative deflection angle increasing incrementally. Simply put, *Assumption 1* is a gradual deflection process, which corresponds to the blue line in Figure 9.
- *Assumption 2: Deflection Along the Average Deflection Angle*
  Deflection along the average deflection angle means that the UAV is oriented towards the target with the average angle $\bar{\theta}$ of *Assumption 1*. The endpoint of the blue line in Figure 9 represents the flight position obtained from Assumption 1, and the heading angle towards this endpoint from the initial position is denoted as $\bar{\theta}$. The UAV will fly along this direction instantly when the risk of collision is detected, which corresponds to the red dashed line in the middle of Figure 9.
- *Assumption 3: Deflected Along the Target Deflection Angle*
  This assumption is the default assumption of the VO method. It specifies that upon detecting a collision threat, the UAV will immediately navigate in the direction of a velocity selected outside the *AVO*, which is named as the target deflection angle. In simple terms, the drone initially adjusts its heading to fly along the final flight angle defined by *Assumption 1*, which corresponds to the red dashed line at the bottom of Figure 9.

The flight process for a given time interval τ under the three assumptions is demonstrated in Figure 9. It is obvious that the flight endpoints are different across the three different assumptions.

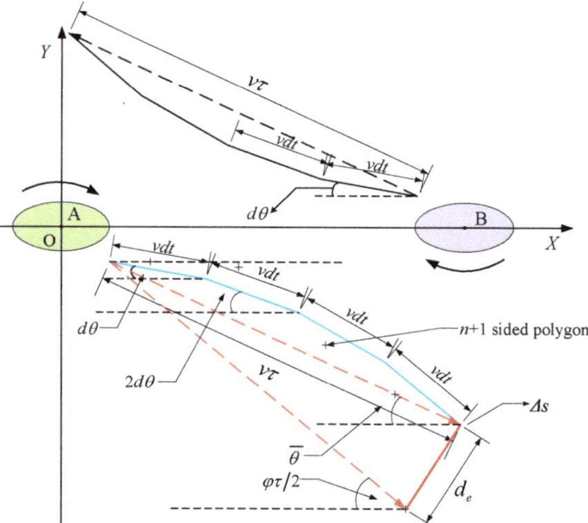

Figure 9. Schematic of flight endpoints under the three assumptions in the conflict risk phase.

3.3.2. Collision Risk Phase Error Expression Derivation

The line velocity of the UAV deflection process is constant in magnitude. Only the direction of the velocity changes, discretizing the UAV deflection process into n linear deflection microelements. The time step within each microelement is τ/n, and the displacement distance is $v\tau/n$. Thus, one UAV will produce a total displacement during the time interval τ:

$$s = \Sigma \frac{v\tau}{n} + \cdots + \frac{v\tau}{n} = v\tau \tag{5}$$

The UAV can be deflected by $\varphi\tau/2$ at a maximum angular velocity during τ. The UAV deflection angles form an arithmetic progression with time microelements. The formula for the general term of the arithmetic series is $a_n(n = 1, 2, \cdots, n)$, the common difference is $d\theta = \varphi\tau/2n$, and the sum of the first n terms is S_n, where n denotes the number of microelements. Then, the construction shown in Figure 9 will form an $n + 1$ sided polygon. The average deflection angle can be expressed as follows:

$$\bar{\theta} = a_1 + \frac{(n+1-2)\pi - (n-1)(\pi - d\theta)}{2} = a_1 + \frac{(n-1)d\theta}{2} \tag{6}$$

Since the deflection from the horizontal is 0 degrees, there is essentially $a_1 = d\theta$. Thus, the above equation can be simplified to:

$$\bar{\theta} = \frac{n+1}{2}d\theta \tag{7}$$

which is exactly the mean of the sum of the n terms of the angular equidistant series:

$$\overline{S_n} = \frac{S_n}{n} = \frac{n[a_1 + a_1 + (n-1)d\theta]}{2n} = a_1 + \frac{(n-1)d\theta}{2} = \bar{\theta} \tag{8}$$

Therefore, $\bar{\theta} = \frac{n+1}{2} \cdot \frac{\varphi\tau}{2n} = \frac{n+1}{4n}\varphi\tau \Rightarrow \lim_{n\to\infty} \bar{\theta} = \frac{\varphi\tau}{4}$

Next, it is proved that the displacement endpoints obtained under *Assumptions 1* and *2* have small errors. Decompose the velocity in each microelement in both the horizontal and vertical directions and accumulate these to obtain the displacement distance in both directions in τ :

$$\begin{cases} \Delta x = v \lim_{n\to\infty} \sum_{i=1}^{n} \cos\left(\frac{\varphi\tau}{2} \cdot \frac{i}{n}\right) \cdot \frac{\tau}{n} = v\tau \int_0^1 \cos(\frac{\varphi\tau}{2} \cdot x)dx = \frac{2v}{\varphi}\sin\frac{\varphi\tau}{2} \\[2mm] \Delta y = v \lim_{n\to\infty} \sum_{i=1}^{n} \sin\left(\frac{\varphi\tau}{2} \cdot \frac{i}{n}\right) \cdot \frac{\tau}{n} = v\tau \int_0^1 \sin(\frac{\varphi\tau}{2} \cdot y)dy = \frac{2v}{\varphi}\left(1 - \cos\frac{\varphi\tau}{2}\right) \end{cases} \tag{9}$$

Then, generate a joint displacement: $s_1 = \sqrt{\Delta x^2 + \Delta y^2} = \frac{2v}{\varphi}\sqrt{2\left(1 - \cos\frac{\varphi\tau}{2}\right)}$

For UAV-A, the coordinates of the end point of the flight by $\bar{\theta}$ are $(\Delta x', \Delta y')$:

$$\begin{cases} \Delta x' = v\tau \cos\frac{\varphi\tau}{4} \\ \Delta y' = v\tau \sin\frac{\varphi\tau}{4} \end{cases} \tag{10}$$

Then, generate a joint displacement: $s_2 = \sqrt{(\Delta x')^2 + (\Delta y')^2} = \tau v$

Align the displacements x and y along the $\bar{\theta}$. The displacement deviation Δs can be expressed as a function of the UAV flight speed v, deflection avoidance time interval τ, and deflection angle per unit time φ. Δs is defined as follows:

$$\Delta s(v, \varphi, \tau) = |s_1 - s_2| = \left|\frac{2}{\varphi}\sqrt{2\left(1 - \cos\frac{\varphi\tau}{2}\right)} - \tau\right| \cdot v \tag{11}$$

The mean value of the displacement deviation is defined as follows:

$$\Delta s' = \frac{1}{n}\sum_{i=1}^{n} v \left|\frac{2}{\varphi_i}\sqrt{2\left(1 - \cos\frac{\varphi_i \cdot dt}{2}\right)} - dt\right| \left(\varphi_i = \frac{\pi}{3}, \frac{\pi}{3} + d\theta, \cdots, 2\pi\right) \tag{12}$$

If the UAV is flying in the direction of the target from the beginning, there is also a distance difference d_e between the endpoints of the displacements in *Assumptions 2* and *3*.

We used the law of cosines in the red isosceles triangle in Figure 9 and obtained the following:

$$d_e(v, \varphi, \tau) = v\tau\sqrt{2\left(1 - \cos\frac{\varphi\tau}{4}\right)} \tag{13}$$

3.3.3. Collision Risk Phase Error Analysis

When the UAV detection time scale is small, the UAV flight endpoint error Δs about *Assumptions 1* and *2* is small compared to the actual displacement scale, while the error d_e about *Assumptions 2* and *3* is much larger. Based on the expressions derived in Section 3.3.2 for Δs and d_e, we plot the changes in their associated variables in Figure 10.

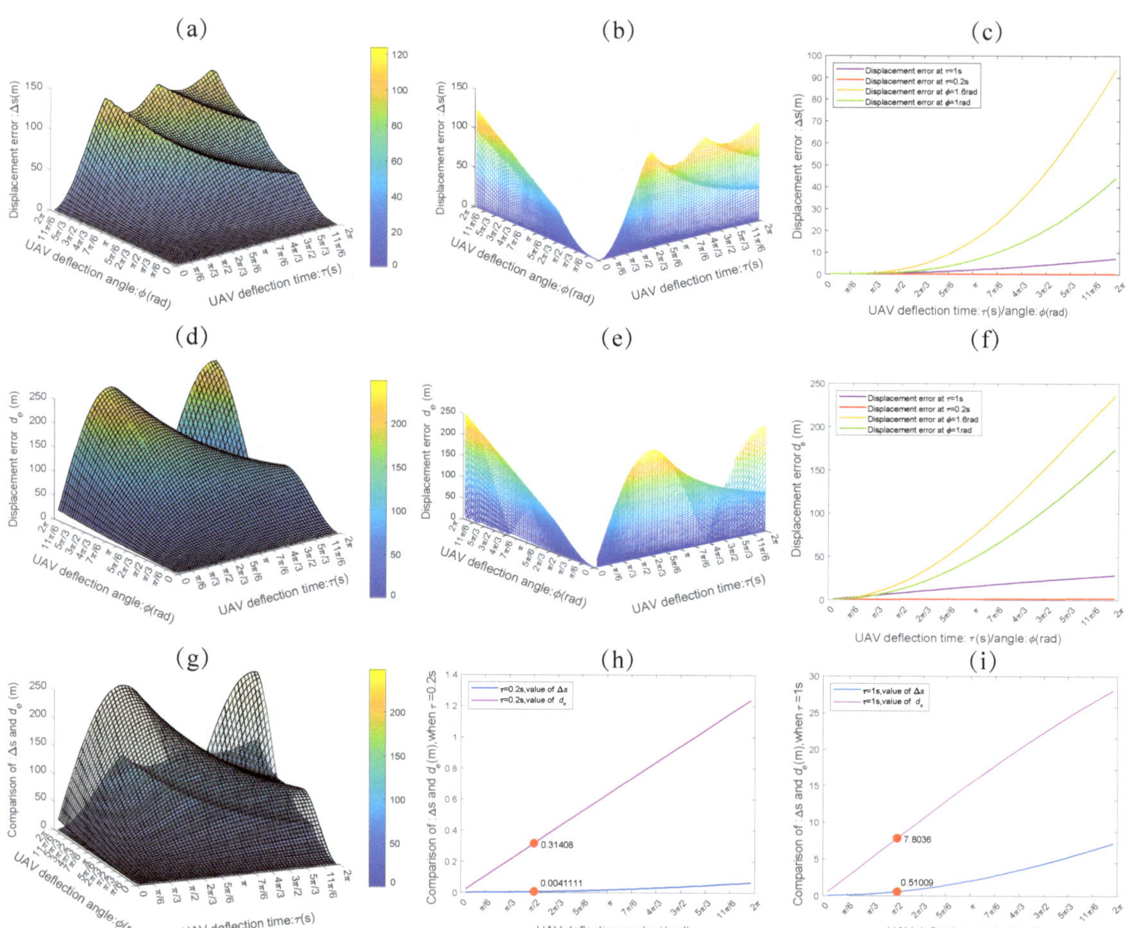

Figure 10. Schematic of displacement endpoint errors Δs and d_e analysis.

Figure 10a shows the variation in the function Δs on $[0, 2\pi]$. Figure 10b shows the projections of the two variables φ, τ on their corresponding axis planes, respectively. It can still be observed qualitatively without exact partial derivatives that the Δs in a certain range around '0' are low (Figure 10a). This indicates that the value of error Δs is not significant when both φ and τ are small. Fixing one of the variables, taking several discrete values for the other variable, and changing their errors, we can see that all four curves are increasing (Figure 10c). This indicates that the partial derivative of the other variable is positive at its corresponding value. The purple and red curves grow slowly (corresponding to τ fixed), while the yellow and green curves grow fast (corresponding to φ fixed), so that time τ has a more pronounced effect on Δs than the deflection angle φ. This reveals that we prioritize the control shortening time. The error distance Δs about *Assumptions 1* and *2* will be lower.

By analyzing the functional relationship for d_e over the same range of independent variables (see Figure 10d–f), we can obtain similar conclusions to Δs, except that most of the same independent variables φ, τ correspond to d_e greater than Δs (see Figure 10g–i). Hence, under *Assumption 3* of the VO method, the position UAV-A reaches after executing the redirection maneuver may not align with its actual position, potentially encroaching into the VO space occupied by UAV-B. Our discussion of this part of the error is similar to the starting point of the NLVO [28–31].

In this paper, we address the fact that our UAV's protection domain is neither pre-determined nor fixed. Consequently, the uncertainty error introduced by this linear assumption can be incorporated into the design of the protection domain. Equivalently, this displacement deviation is added to the original protection domain space so that the we can continue to adjust the UAV under *Assumption 3* [11], simplifying the calculations and skipping the complex flight details of *Assumption 1*.

3.4. Elliptic Domain Size Construction Considering Uncertainty Errors

In Phase 1, as established in this paper, the average displacement deflection angle consistently remains smaller than the target deflection angle. Thus, under *Assumption 3*, the distance between the endpoints of UAVs A and B along the minor axis direction is invariably greater than that observed under *Assumption 2*. The minor axis defined under *Assumption 3* contains the length of the error d_e. Upon analyzing the geometric logic, we conclude the following:

$$\begin{cases} d_b^i = d_{v\tau}^i \cdot \sin \theta_b^i \approx d_\theta^i \cdot \sin \theta_b^i + 2d_e \\ d_{v\tau}^i = \sqrt{\left(y_B^{\tau_i} - y_A^{\tau_i}\right)^2 + \left(x_B^{\tau_i} - x_A^{\tau_i}\right)^2} \\ \theta_b^i = \beta_\tau^i + \frac{\varphi\tau}{2} \cdot i - \frac{\pi}{2} \\ \beta_\tau^i = \arctan\left(\frac{y_B^{\tau_i} - y_A^{\tau_i}}{x_B^{\tau_i} - x_A^{\tau_i}} \right) \end{cases} \tag{14}$$

In the above equation, d_b^i denotes the projected distance between the two ellipses in the direction of the minor axis at the moment τ_i ; $d_{v\tau}^i$, d_θ^i denote the distance between the endpoints of the two UAV displacements under *Assumptions 2* and *3* at each time step τ, respectively; β_τ^i indicates the inclination of the terminal line under *Assumption 3*; β_τ^i indicates the angle between the displacement endpoint line and the direction of the normal velocity under *Assumption 3*. Since it has been proved in the previous section that Δs is obviously smaller than d_e, ignoring Δs, the above equation is approximately equal.

If UAVs A and B are deflected to avoid obstacles at the initial moment under *Assumption 3*, the spatial position at the end of each time τ_i can be expressed as follows:

$$\begin{cases} x_A^{\tau_i} = x_A^{\tau_0} + \sum_{i=1}^{n} v\tau \cdot \cos\left(\frac{\varphi\tau}{2} \cdot i\right) \\ y_A^{\tau_i} = y_A^{\tau_0} - \sum_{i=1}^{n} v\tau \cdot \sin\left(\frac{\varphi\tau}{2} \cdot i\right) \end{cases} \& \begin{cases} x_B^{\tau_i} = x_B^{\tau_0} - \sum_{i=1}^{n} v\tau \cdot \cos\left(\frac{\varphi\tau}{2} \cdot i\right) \\ y_B^{\tau_i} = y_B^{\tau_0} + \sum_{i=1}^{n} v\tau \cdot \sin\left(\frac{\varphi\tau}{2} \cdot i\right) \end{cases} \left(i \in N^+\right) \tag{15}$$

To ensure that the UAV can effectively avoid collisions upon detecting a risk during Phase 1, within the constraints of its physical capabilities, the value of the length of the minor semi-axis of the ellipse ($2b$) must not exceed the minimum separation distance. We define this minimum value as the distance projection along the minor axis direction before the velocities of UAVs A and B are deflected to $\pi/2$. This is recorded as follows:

$$b_{\min}^\tau = \min\{d_b^i/2\} \left(i \in \lfloor \frac{\pi}{2\varphi\tau} \rfloor \& i \in N^+ \right) \tag{16}$$

Based on this, we anticipate that the lateral distance will also be sufficiently large to minimize the possibility of the collision scenarios depicted in Figure 8. Consequently, within each time step τ, we have the following:

$$l^i_{v\tau} = \left(x^{\tau_i}_B - x^{\tau_i}_A\right) \cdot \cos\left(\frac{\varphi\tau}{2} \cdot i\right) > 0 \tag{17}$$

Similarly,

$$a^\tau_{\min} = \min\{l^i_{v\tau}\}/2\left(i \in \lfloor\frac{\pi}{2\varphi\tau}\rfloor \& i \in N^+\right) \tag{18}$$

Since A and B are generally asynchronous in reaching the minimum value, the equation $a^\tau_{\min} > b^\tau_{\min}$ does not necessarily hold. To avoid ambiguity, when $a^\tau_{\min} < b^\tau_{\min}$, we assign $a^\tau_{\min} = 1.5b^\tau_{\min}$. In summary, the custom elliptic domain structure is given by the following constraints:

$$a^\tau_{\min} > b^\tau_{\min} : \begin{cases} a^\tau_{\min} = \dfrac{\min\{l^i_{v\tau}\}}{2} \\ b^\tau_{\min} = \dfrac{\min\{d^i_b\}}{2} \end{cases} \left(i \in N^+\right) \tag{19}$$

$$a^\tau_{\min} \leq b^\tau_{\min} : \begin{cases} a^\tau_{\min} = 1.5b^\tau_{\min} \\ b^\tau_{\min} = \dfrac{\min\{d^i_b\}}{2} \end{cases} \left(i \in N^+\right) \tag{20}$$

For a given velocity–position state, the values calculated above represent the elliptical domain sizes for the pair of reversed conflict UAVs, which are simplified as follows:

$$a = a^\tau_{\min}; b = b^\tau_{\min} \tag{21}$$

The elliptic domain structure we constructed maps the ellipse's major and minor semi-axes to the UAV speed, velocity direction, angular velocity performance, hostile UAV spacing distance, and UAV detection frequency. It has adaptability associated with the real-time status of the UAV. In Phase 1, if equation $d^i_b < 2b$ is satisfied, they will intrude on each other at a certain moment; otherwise, they will never intrude. The projected distance in the minor-axis direction reflects the closest distance between the two in subsequent flights in the non-intrusive state and reflects the extent of mutual invasion between the two in the intrusive state.

4. EVO Algorithm-Based Custom Elliptic Domains

As elliptic domains no longer have the simplicity of calculating the VO space compared to the circle domains, the exact calculation of the Minkowski sum for the elliptic domains in which A and B are located involves either calculating the boundary convolution curves [40–42] or employing the close-form implicit equations [43] with more expensive computational costs. The computation of the tangent line at a point outside the non-circle domain will also become complicated. Lee and Beom H. [20] have accurately derived and computed the VO space and obstacle cones for the minimum area boundary ellipse approximation of obstacles with a high degree of complexity. The conservative linear computational method proposed by Best [10] is more efficient and has been validated under multi-obstacle experimental conditions. In addition to applying the VO method for obstacle avoidance in elliptical agents, Boolean operations on elliptical boundaries have also been utilized to address conflict scenarios in multi-ship encounters [31,44]. In this section, we approximate the boundary of the elliptical domain as a convex polygon, allowing for rapid computation of the VO space. The tangents of this space are efficiently estimated, too.

4.1. EVO Algorithm Preparations

This part discretizes the elliptic boundary in any axial direction to form a convex polygonal structure. According to the obtained custom elliptic domain size, the standard elliptic equation located at the coordinate origin is as follows:

$$\frac{x^2}{a^2} + \frac{y^2}{b^2} = 1 \tag{22}$$

We take the center of the ellipse where UAV-A is located as the coordinate origin and the major and minor semi-axes as the direction of the X- and Y-axes coordinate system. Without a loss of generality, any axial UAV-B ellipse exists in this coordinate system with inclination angle o_B and center spacing distance d_{AB}.

We discretize the boundary of the ellipse on which UAV-A is located by taking the θ_m angle. As a result, there will be $2\pi/\theta_m$ discrete points. Elliptic protected domains are transformed into convex polygons. Each discrete coordinate point can be represented as follows:

$$\begin{cases} x_A^i = a\cos\theta_i = a\cos i * \theta_m (i = 1, \cdots, 2\pi/\theta_m) \\ y_A^i = b\sin\theta_i = b\sin i * \theta_m (i = 1, \cdots, 2\pi/\theta_m) \end{cases} \tag{23}$$

where θ_i is the angle corresponding to every interval θ_m, with 0 degrees in the positive direction of the x-axis and a positive counterclockwise rotation. With UAV-A as a reference, the discretized boundary points of B at any position and axial direction in this coordinate system are equivalent to two transformations of the boundary discretization points sought by A. Rotation transformation first and then translation transformation is equivalent to translation transformation first and then rotation transformation. If we rotate first using the clockwise rotation matrix *Turn*, the mapping is applied accordingly.

$$Turn = \begin{bmatrix} \cos o_B & -\sin o_B \\ \sin o_B & \cos o_B \end{bmatrix} \tag{24}$$

$$\begin{cases} x_{B1}^i = a\cos i * \bar{\theta} \cdot Turn(i = 1, \cdots, 2\pi/\theta_m) \\ y_{B1}^i = b\sin i * \bar{\theta} \cdot Turn(i = 1, \cdots, 2\pi/\theta_m) \end{cases} \tag{25}$$

where o_B is the angle of rotation of B with reference to the direction of the major axis of A. After the rotation is completed, the coordinate points are translated to the location of B to obtain the discrete boundary points of B in the coordinate system:

$$\begin{cases} x_{B2}^i = x_B = x_{B1}^i + d_{AB}\cos o_B \\ y_{B2}^i = y_B = y_{B1}^i + d_{AB}\sin o_B \end{cases} \tag{26}$$

At this point, we have discretized the two elliptical protection domains corresponding to UAVs A and B into convex polygons and established coordinate correspondences for each node. Therefore, we obtain a set of discrete points on the boundaries of the elliptical domains of A and B:

$$\begin{cases} Boundary_A = \left\{ x_A^i, y_A^i \right\} \\ Boundary_B = \left\{ x_{B2}^i, y_{B2}^i \right\} \end{cases} (i = 1, \cdots, 2\pi/\theta_m) \tag{27}$$

4.2. EVO Algorithm Steps

Taking UAV-A as the subject of obstacle avoidance and evaluating the *EAVO* imposed on UAV-B, the following algorithm steps are executed:

- Step 1: Computation of Minkowski sum and convex hull boundary points

We considered the discrete convex polygons of UAVs A and B. The Minkowski sum, imposed by A on B, is calculated. Using Method 1 as described in Appendix A.1 of this paper, we employ the convex polygon convex hull algorithm to derive the convex boundary from the results of the discrete point summation:

$$\Omega_{A \oplus B} = convhull\{Boundary_A + Boundary_B\} \tag{28}$$

We can obtain the convex hull structure by extracting the boundary point index. The 2D convex hull boundary is EVO space ($\widehat{B_E}$). This is the set of velocities we need to satisfy disjointness for obstacle avoidance. The discrete points on the convex hull boundary are marked as $Boundary_{\hat{B}} = \{p_i, i = (1, \cdots, n)\}$, and it may be an elliptical-like structure.

- Step 2: Finding the approximate EVO space tangent line

Since the convex packet boundary is not smooth, the left and right derivatives of each discrete boundary point are not equal to each other. If it is not differentiable, the slope does not exist. Owing to the abundance of discrete boundary nodes and the small intervals between them, we efficiently utilize data from the convex hull algorithm by approximating the slope at any point within the discrete boundary set by using the slope of the line connecting the adjacent front and back nodes:

$$k_i = \frac{\Delta y_i}{\Delta x_i} = \frac{y_{i+1} - y_{i-1}}{x_{i+1} - x_{i-1}} \tag{29}$$

Then, we look for a point (where A is located) outside the convex hull tangent to it. We do this to find any point on the convex hull and the point where the slope of the line connecting A is consistent with k_i. We connect the boundary discrete points with position A, where the linear direction function is expressed in the slope as follows:

$$k_i' = \frac{y_A - y_i}{x_A - x_i} \tag{30}$$

The absolute errors of these slopes can be found for a series of discrete points:

$$\varepsilon_i = |k_i - k_i'| \times 100\% \tag{31}$$

- Step 3: Return EVO space tangent points

Because more than one discrete point is close to the true tangent points within the error allowance, there is a risk of double-counting the tangent points. Therefore, the minimum and maximum slope values within this error range (ε_{agree}) should be considered as the two approximate tangent points of the convex hull at that moment:

$$\begin{cases} p_{con}^1 = \arg\max(\arctan k_i') = (x_{con}^1, y_{con}^1) \\ p_{con}^2 = \arg\min(\arctan k_i') = (x_{con}^2, y_{con}^2) \end{cases} \quad for\ each\ \varepsilon_i < \varepsilon_{agree} \tag{32}$$

where p_{con}^1 and p_{con}^2 are the two tangent points we computed.

The denser the discrete points, the smaller the error. Accordingly, ε_{agree} can be set smaller. The approximation of the tangent line is closer to the true value, but the calculation becomes more time-consuming as a result.

- Step 4: Computation of EAVO

Translate all convex packet boundary points along $\overrightarrow{v_B}$ in τ. The set of convex packet boundary points after translation is $Boundary_{\hat{B}} \oplus \overrightarrow{v_B} \cdot \tau$. The position of the tangent points after translation is $p_{con}^1 = (x_{con}^1, y_{con}^1) \oplus \overrightarrow{v_B} \cdot \tau, p_{con}^2 = (x_{con}^2, y_{con}^2) \oplus \overrightarrow{v_B} \cdot \tau$. We connect the translational tangent points p_{con}^1 and p_{con}^2 to the UAV-A coordinates to construct $EAVO$ at a given moment in time: $EAVO_{A|B}^{\tau} = \left\{v | v \notin EVO_{A|B}^{\tau} \oplus v_B\right\}$.

4.3. EVO Algorithm Obstacle Avoidance Velocity Control

After determining the custom elliptic domain for the UAV, it is essential to establish a control scheme to adjust the velocity for obstacle avoidance. Additionally, we must address potential velocity oscillations that could arise during this process to prevent the UAV from entering a repetitive loop of obstacle avoidance maneuvers and course corrections. To streamline the process and minimize the time and complexity involved, this paper fixes the major and minor semi-axes under the initial τ_0 computation as the ellipse size based on the analysis and discussion in Section 3. This eliminates the need for recalculating the ellipse size with each obstacle encounter.

We utilize the *EAVO* as the basis for obstacle avoidance. UAV-B has a higher navigational priority compared to A. We adjust only the direction of UAV-A's velocity while keeping UAV-B's heading unchanged. The *EAVO* for UAV-A is recalculated at the end of each time step τ_i. If UAV-A's velocity falls within the *EAVO* at point τ_{i-1}, the heading of UAV-A is then adjusted accordingly. It is worth noting that we cannot always regard the boundary velocity of the cone as the adjusted orientation because of the limitations of UAV physical deflection capabilities. Additionally, the accuracy of the tangent line determination is influenced by the number of discrete boundary points, introducing some discrepancies between the actual position and the results of the approximation algorithm. To ensure adequate safety, in scenarios where only a minor deflection is necessary for obstacle avoidance, we deflect A into $[\varphi\tau/2, \varphi\tau]$. Therefore, there are three situations:

$$\tau_{i+1}^{new} = \begin{cases} \begin{cases} o_A^{\tau_i} - \varphi\tau/2, \ o_A^{\tau_i} \in EAVO_{\tau_i} \& 0 < \left|o_{EAVO}^{\tau_i} - o_A^{\tau_i}\right| \le \varphi\tau/2 \Rightarrow \text{Collision} \\ -\left|o_{EAVO}^{\tau_i}\right|, \ o_A^{\tau_i} \in EAVO_{\tau_i} \& \varphi\tau/2 < \left|o_{EAVO}^{\tau_i} - o_A^{\tau_i}\right| \le \varphi\tau \Rightarrow \text{Collision} \end{cases} \\ \left\{o_A^{\tau_i} - \varphi\tau, o_A^{\tau_i} \in EAVO_{\tau_i} \& \left|o_{EAVO}^{\tau_i}\right| > \varphi\tau \Rightarrow \text{Collision}\right. \\ \left\{o_A^{\tau_i}, o_A^{\tau_i} \notin EAVO_{\tau_i} \Rightarrow \text{NoCollision}\right. \end{cases} \tag{33}$$

When $o_A^{\tau_i} \in EAVO_{\tau_i} \& 0 < \left|o_{EAVO}^{\tau_i} - o_A^{\tau_i}\right| \le \varphi\tau/2$, we do not have to deflect at the upper deflection limit to avoid obstacles. When $o_A^{\tau_i} \in EAVO_{\tau_i} \& \left|o_{EAVO}^{\tau_i}\right| > \varphi\tau$, at most, we can only deflect $\varphi\tau$ based on the current velocity. When $o_A^{\tau_i} \notin EAVO_{\tau_i}$, a collision will not occur, and we maintain the UAV at its current velocity. It is worth noting that because of the default clockwise deflection, as long as the obstacle is avoided more than once before τ_i, $o_A^{\tau_i}$ must be the fourth quadrant angle, with one negative value.

4.4. EVO Algorithm Oscillation Elimination in Velocity Reback

In practical applications, after the UAV adjusts its velocity direction to EVO space, it deviates from the established route of the original flight mission. Therefore, the UAV needs to adjust its velocity one or more times to gradually fly towards the target point. Nevertheless, if velocity reback is performed as soon as $o_A^{\tau_i} \notin EAVO_{\tau_i}$ is detected, the UAV will generate velocity oscillations (Section 2.2.3). In the RVO approach, the velocity vectors of UAVs A and B are averaged as a standard translation relative to the apex of the velocity obstacle cone, ensuring the adjusted velocity remains within the feasible velocity intersection of A and B. However, this paper intends not to alter the direction of UAV-B's velocity, allowing it to continue as much as possible on its original course, while UAV-A actively performs avoidance maneuvers. Consequently, the velocity oscillations are transferred primarily to UAV-A, rendering the traditional RVO method inapplicable in this scenario. There are two solutions for this problem we consider:

Solution 1: Find an optimal position at which UAV-A initiates obstacle avoidance to ensure safety. Concurrently, UAVs A and B pass each other, ensuring that subsequent conditions satisfy $o_A^{\tau_i} \notin EAVO_{\tau_i}$. This approach prevents the occurrence of velocity oscillations. However, determining this optimal position introduces another layer of complexity, which this article will not explore in detail at this time.

Solution 2: We impose a constraint on UAV-A to continue moving in the direction of the avoidance velocity until UAVs A and B have passed each other, after completing the avoidance maneuver. This approach simplifies the management of UAV trajectories, ensuring that the avoidance maneuver results in a successful and stable transition. Therefore, we have the constraint $d_{v\tau}^i < d_{v\tau}^{i+1}$ of mutual passing. After completing the initial obstacle avoidance, UAV-A's velocity direction is continuously adjusted towards the endpoint, ensuring a smooth flight without the need for secondary obstacle avoidance maneuvers. The adjustment processes still need to satisfy the limit of deflection:

$$\text{when}\,o_{new}^{\tau_{i+1}} = o_A^{\tau_i} = \begin{cases} o_A^{\tau_i} d_{v\tau}^i > d_{v\tau}^{i+1} \\ o_{target}^i = \arctan \dfrac{y_{goal} - y_A^{\tau_i}}{x_{goal} - x_A^{\tau_i}}\, d_{v\tau}^i < d_{v\tau}^{i+1} \\ o_A^{\tau_i} + \varphi\tau,\ d_{v\tau}^i < d_{v\tau}^{i+1}\,\&o_{target}^i - o_A^{\tau_i} > \varphi\tau \end{cases} \tag{34}$$

These are the rules for adjusting and rebacking the velocity of UAV-A throughout the obstacle avoidance process. We will verify their validity in two scenarios in Section 5.

5. Simulation

5.1. VO and EVO Obstacle Avoidance Evaluation Indicators

In order to distinguish and validate the differences in the resolutions of UAVs in different protection domains for the same conflict scenario, some of the process indicators and overall indicators in obstacle avoidance are selected as evaluation indicators. The details of the indicators and what they represent are shown in Table 1.

Table 1. Evaluation indicators for VO and EVO obstacle avoidance models.

Evaluation Indicators	Implication
Process Indicators	**Indicators of Changes over Time with the Flight Process**
VO space	VO in elliptic domains or circular domains
Flight distance	Distance between UAVs during flight
Occupied area	AVO-occupied area in 2D space
Angle of velocity direction	Change in velocity direction throughout the flight of the UAVs
Detour distance	Detour distance in τ compared to the original flight direction
Overall Indicators	**Indicators for the Entire Flight**
Total detour distance	Detour distance + remaining distance
Single obstacle avoidance time	Average time per calculation of obstacle avoidance direction for UAV-A

5.2. UAV Simulation Parameters

We simulate and analyze a pair of head-on conflict UAVs, A and B. Their flight endpoints are each other's initial locations. We set two collision scenarios for validating the obstacle avoidance performance of the elliptic domains in congested situations, and we refer to the parameter information of the DJI Air 2S UAV to set up an initial state. The parameter information is shown in Table 2.

Table 2. UAV state parameters of simulation experiments.

Parameters of UAV	UAV	v	θ	o	(x_0, y_0)	τ	θ_m	φ
Scenario 1	A	20	0	0	$(0,0)$	0.2	$\pi/360$	$\pi/3$
	B	20	$-\pi$	$-\pi$	$(500,0)$	0.2	$\pi/360$	$\pi/3$
Scenario 2	A	20	0	0	$(0,0)$	0.2	$\pi/360$	$\pi/2$
	B	20	$-\pi$	$-\pi$	$(200,0)$	0.2	$\pi/360$	$\pi/2$

From left to right, the parameters in the table represent the initial state, i.e., the UAV flight velocity (m/s), the direction of the UAV velocity (rad), the direction of the UAV elliptic domain (rad), the horizontal and vertical coordinate positions in the absolute coordinate system of A and B (m), the time slice interval (s), the center angular interval of the discrete boundary point (rad), and the maximum angular velocity (rad/s), respectively.

It is worth noting the small geometric size of the DJI Air/Mavic series UAVs, which is around 1 dm^3. At the same time, to ensure the effectiveness of obstacle avoidance, the distance between the A and B intervals should not be set too far, and φ should not be too large. Otherwise, it will lead to the completion of obstacle avoidance in a fraction of a second. φ is taken as the DJI Air 2S Normal Gear $\pi/2$ and Smooth Gear $\pi/3$ values in Scenarios 1 and 2, respectively.

5.3. Simulations and Conclusions

At the initial moment τ_0, the initial ellipse domain dimensions under the model of Section 3.3 were calculated under the parameters set in Scenarios 1 and 2.

$$\text{Scenarios1} : \begin{cases} a = 39.2 \text{ m} \\ b = 26.2 \text{ m} \end{cases} \text{Scenarios2} : \begin{cases} a = 23.5 \text{ m} \\ b = 15.6 \text{ m} \end{cases} \tag{35}$$

To verify the reliability of the entire constructed AVO/EAVO algorithm described above, the absolute velocity obstacle cone of UAV B at the initial moment in scenario 1 was drawn (Figure 11). It shows that the elliptic domain tangent is essentially accurate, and that the AEVO is enclosed in the AVO. All the calculations and simulations were completed in Matlab R2022a using a 13th Gen Intel(R) Core(TM) i5-13500H 2.60 GHz processor.

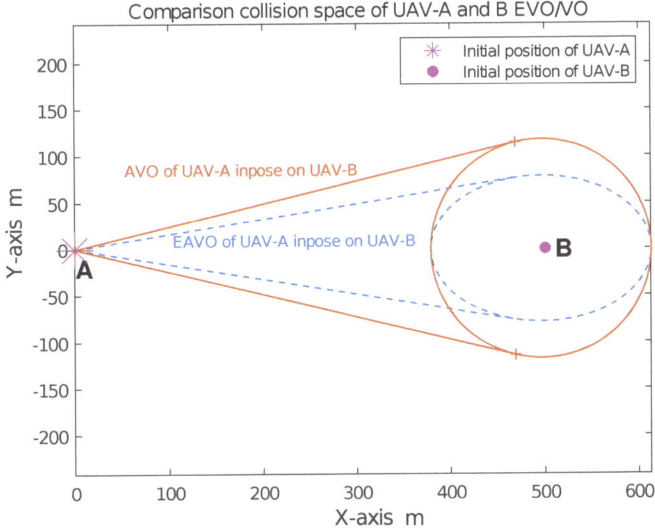

Figure 11. Initial moment *AVO, EAVO* in scenarios 1.

Obstacle avoidance can be ensured by selecting a velocity other than the *AVO/EAVO* at the end of each time slice. Evidently, the elliptic domain offers a range of obstacle avoidance velocities with reduced inclination angles compared to the circle domain.

In the simulation experiments, the existing method of expanding the velocity obstacle space is applied to circular structures to derive the corresponding velocity obstacle cones, while the EVO algorithm proposed in Sections 4.1 and 4.2 of this paper is used for elliptical structures. We simulate the flight process of head-on UAVs A and B according to the obstacle avoidance and velocity reback adjustment strategies in Sections 4.3 and 4.4. The following obstacle avoidance results can be obtained.

5.3.1. Scenarios 1 Simulation Experiment

It can be seen that both protection domain assumptions achieve obstacle avoidance resolution for this pair of UAVs (Figure 12a,b). The flight trajectory of UAV-A under the elliptical assumption is smoother (Figure 12c,d). Based on the distance between A and B throughout the flight (Figure 13a,b), it is clear that the closest distance between the two is much closer under the elliptic domain assumption. Since the length of protection is the same in all circle domain directions, a separation distance of nearly '2*a*' between drones is actually more dangerous. For the elliptic assumption, as the orientation of UAV-A changes, the elliptic protection domain also changes, and the actual distance between A and B is less than '2*a*' but still outside the elliptic domain (Figure 13a green dashed line). Therefore, the result for the circle domain is clearly more conservative.

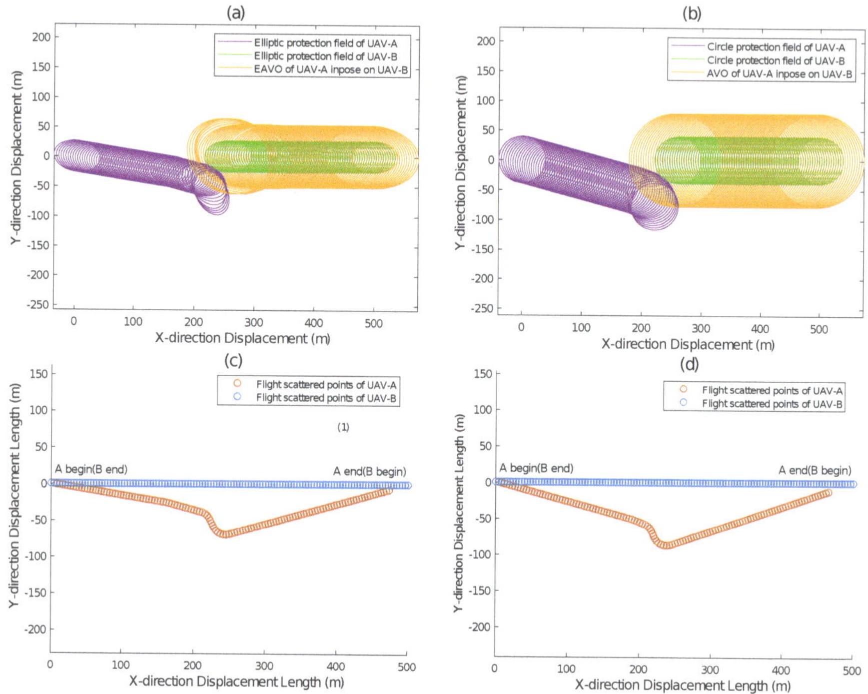

Figure 12. *EAVO/AVO* space and flight position in scenario 1.

It can be seen that the cone angle of the VO of the circle domain assumption is always above the elliptic domain assumption (Figure 14a), indicating that using the circle domain assumption requires a larger angle of deflection to complete the unwinding. The VO of the circle domain assumption occupies a larger area of the space and squeezes more of the free space, while the elliptic domain contributes more of the non-conflicting space (Figure 14c). Since the deflection angle of A under the circle domain assumption is greater (Figure 14b), the resulting detour distance in each simulation interval τ is also larger (Figure 14d). Additionally, it is evident that UAV-A's velocity adjustment does not enter an oscillatory loop. It merely executes a few obstacle avoidance maneuvers before resuming a stable flight. It then continues to monitor for collision risks, making adjustments successively as necessary (Figure 14b).

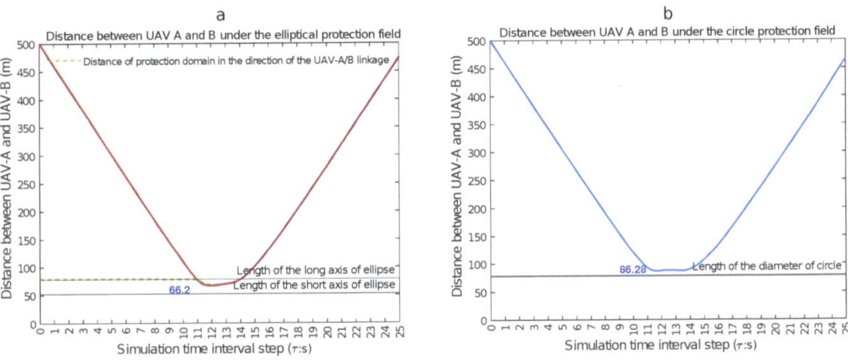

Figure 13. Comparison of flight distances in scenario 1.

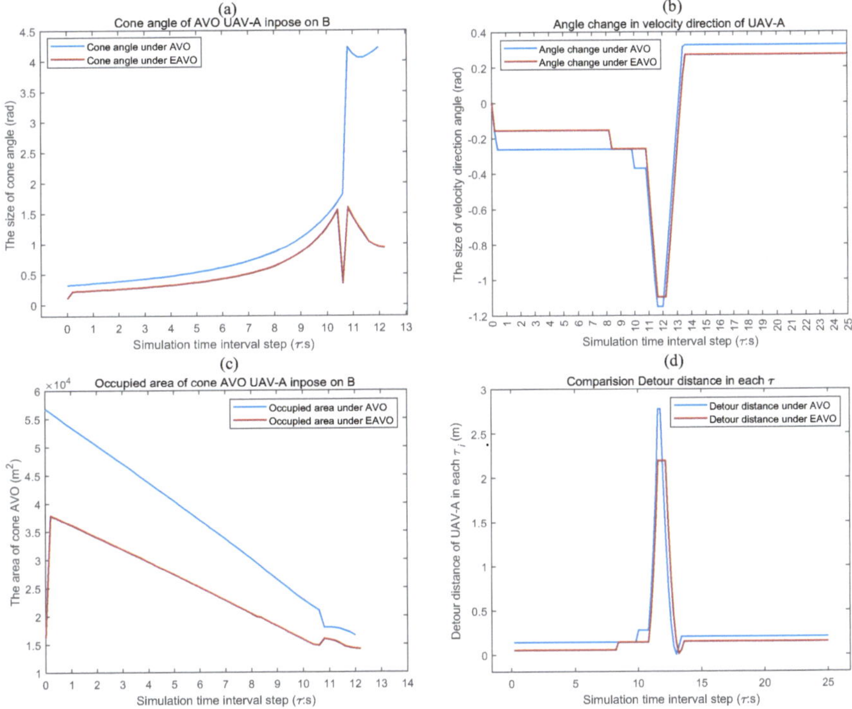

Figure 14. Comparison of process indicators in scenario 1.

The observed results stem from the smaller velocity obstacle space within the elliptical domain, which provides a broader range of optional velocities. This additional velocity selection, which would be unfeasible under the circle domain assumption, will become feasible under the elliptical domain, often requiring only small adjustments. In contrast, under the circle domain hypothesis, a significant portion of the potential obstacle avoidance directions might be prematurely excluded due to the larger space and the greater angle at the top of the obstacle cone.

5.3.2. Scenarios 2 Simulation Experiment

Next, the distance between A and B was adjusted to 200 m. We repeated the procedure according to the experimental conditions in Scenario 2. In Scenario 2, where the spacing is more constrained, the effectiveness of the velocity adjustments by UAV-A within the elliptical domain assumption becomes more apparent. This setup allows UAV-A to avoid obstacles with smaller angular deflections as it approaches UAV-B. However, Figure 15a–d confirms that the purple and green elliptical domains do not overlap at any point, which is facilitated by the lack of inherent arbitrary rotational symmetry to ellipses. Figure 16a,b in the spacing distance curve shows that under the elliptical assumption, UAVs A and B are closer together compared to when operating under the circle domain assumption, falling between the lengths of the major and minor axes. Meanwhile, the circle domain maintains a constant safety interval of 2a at all times, ensuring the distance between A and B remains above this threshold, which necessitates a larger detour.

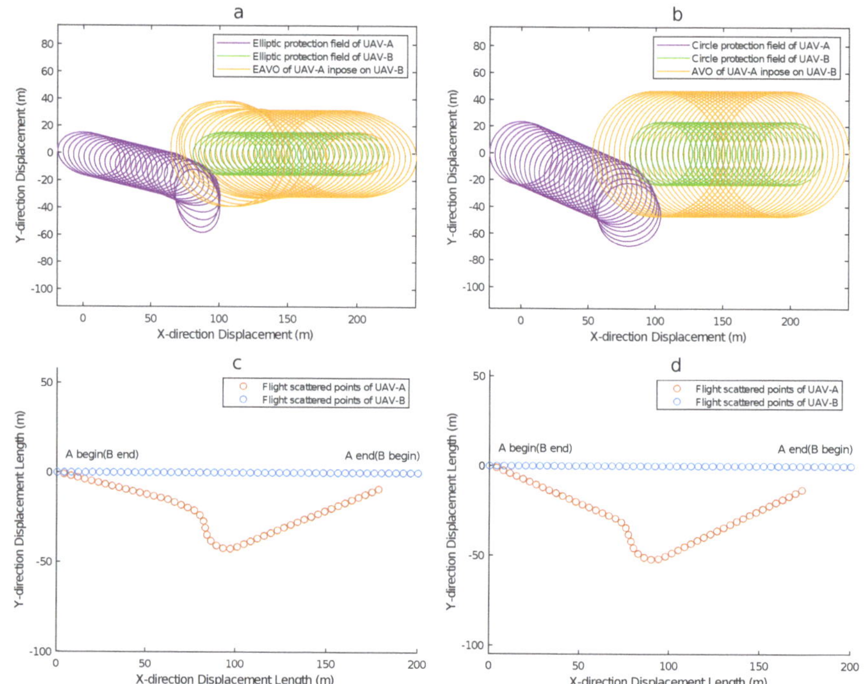

Figure 15. *EAVO/AVO* space and flight position in scenario 2.

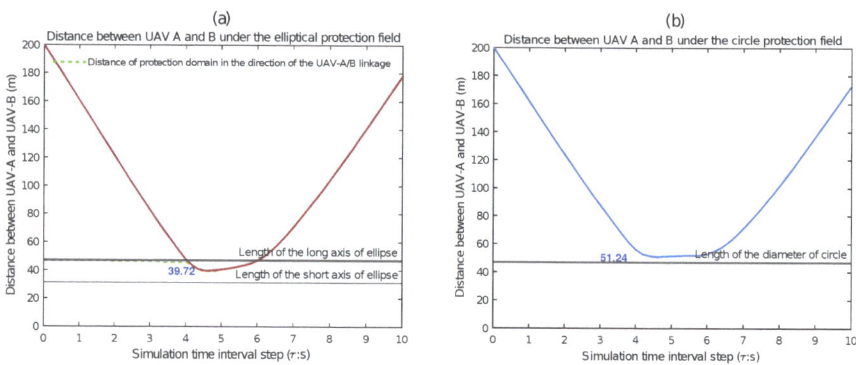

Figure 16. Comparison of flight distances in scenario 2.

The overall indicators for Scenarios 1 and 2 are summarized in Table 3. It can be seen that under the scenarios with different distance scales, the total detour distance of the UAV structured in the elliptic domain is smaller than that in the circle domain throughout the obstacle avoidance process. In terms of the time efficiency of a single resolution, although the VO space computation of the elliptic domain is much more complicated than that of the circle domain, the method of calling the *convhull* function for discrete elliptic boundary nodes proposed in this paper does not significantly lag behind. And, both of them almost reach the critical reback moment when A and B pass by each other at the same time.

Table 3. Overall indicators and velocity reback moment.

Pre-Set Scenario	Domain Hypothesis	Total Detour Distanc (m)	Single Obstacle Avoidance Time (s)	Velocity Reback Moment (s)
Scenario 1	Elliptic domain	54.37	0.0016	$\tau_{reback} = 61\tau = 12.2$
	Circle domain	69.29	0.00056	$\tau_{reback} = 62\tau = 12.4$
Scenario 2	Elliptic domain	44.96	0.0043	$\tau_{reback} = 13\tau = 2.6$
	Circle domain	56.65	0.0014	

Based on Figure 17a–d, we can obtain similar conclusions in Scenario 1. The results of obstacle avoidance under the elliptic domain structure are more superior in Scenario 2, which is reflected in the compression of the total detour distance and the improvement of the computational efficiency in Table 3. Because the initial A and B separation distance is reduced from 500 m to 200 m, the ellipse assumption still saves about 10 m of extra detours.

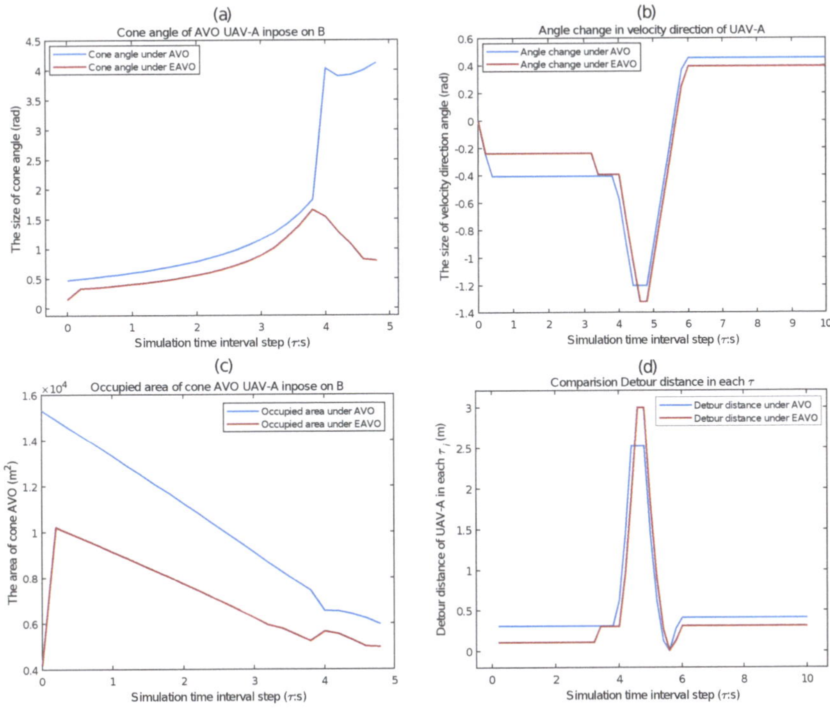

Figure 17. Comparison of process indicators in scenario 2.

We can foresee that when the two UAVs are closer together and the size of the protection domain obtained in Section 3.3 is smaller than the protection domain required by the actual physical properties of the UAVs, a situation may occur where the circle protection domain is unable to avoid obstacles, while the elliptical domain can avoid obstacles. Assuming that this minimum physical protection domain is $a = 20$ m, $b = 10$ m, we test it when the distance between UAVs A and B is 80 or 100 m, and the other parameters are consistent with Scenario 2. The actual experiments demonstrated that when the protection domain fixed the UAV's velocity in the direction of the major axis, it was not effective for obstacle avoidance. The circular domain structure also failed in the 80m scenario. On the contrary, interchanging the values of *a* and *b* yields the desired results (Figure 18a–f). This suggests that timely changes in the axial orientation of the elliptic domain are needed to better accommodate narrow scenarios of obstacle avoidance.

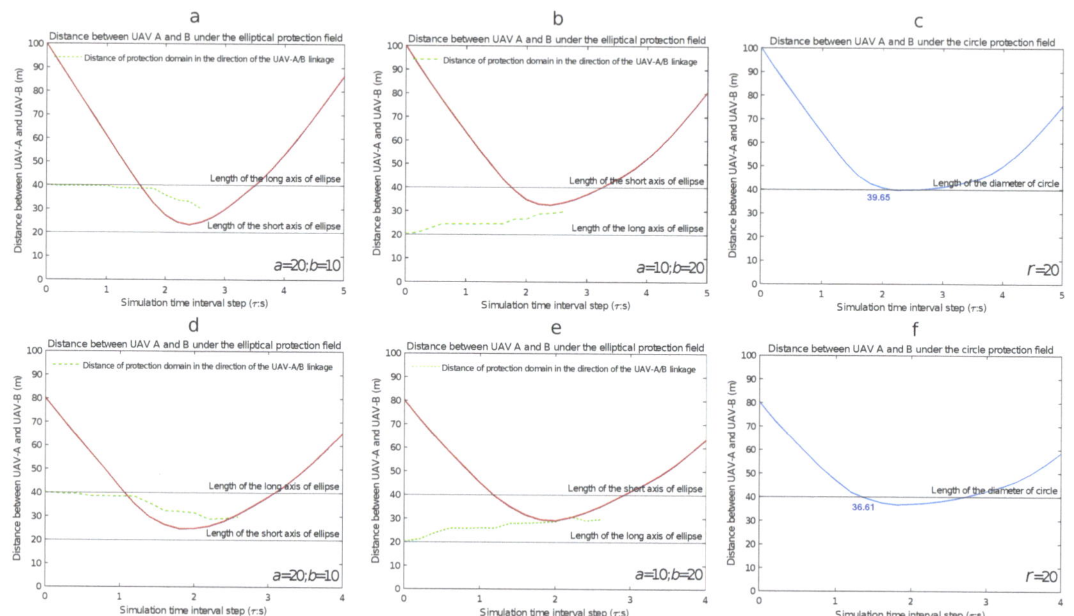

Figure 18. Comparison of protection domain obstacle avoidance results at 100 and 80 m.

5.4. Defects of Custom Elliptic Domains for Proximity UAVs

When the drones are in close proximity, the custom elliptic domain we proposed will also become trapped in a dilemma during obstacle avoidance. During the experimental phase, both elliptic and circle domains produced mutual intrusions at 100 and 80 m. Figure 18 indicates that with the movement of the UAV, our aim to provide a larger protection domain for the velocity direction gradually became inapplicable, and the relative positional relationship between the two should be considered in real time. When swapping the values of *a* and *b* in the elliptic domains, the UAV's obstacle avoidance performance in tight scenarios aligned with our expectations for an elliptic domain (Figure 18). This indicates that the elliptic domain, which depends only on the state of conflict at the initial moment and fixes the direction of the velocity as the major axis of the ellipse, will no longer be applicable as the relative positions of the two UAVs change. Therefore, the ideal elliptic domain should be a dynamic protection domain that constantly changes with the relative positions of the conflicting individuals and should always provide a major-axis protection field for the most significant direction of the conflict.

6. Summary and Outlook

This paper explored the types of protective domains that can scientifically achieve safe obstacle avoidance during UAV flight. Fixed-size circle domains often lead to redundancy, which may be feasible when airspace resources are abundant and conflict scales are small. However, in scenarios where UAVs face limited space for maneuvering or when there are numerous conflicting UAVs, large circle domains with a preset radius may prove ineffective. Consequently, this paper investigated elliptic domains and their algorithms to develop a more generalized and efficient protection framework, enhancing safety in complex or crowded flight environments. For unmanned aerial vehicles (UAVs) in head-on conflict scenarios with a specified separation distance, the size of the preset elliptical protection domain can be determined based on the corresponding UAV performance parameters.

The obstacle avoidance turning angles can be calculated according to the elliptical domain, allowing the UAVs to return to their flight paths and navigate towards the target point after the conflict risk has been eliminated.

Based on the results obtained, in Scenario 1, where the distance between A and B was large, there were some differences between the circle and elliptical domains in terms of the detour distance. In UAV navigation, the circle domain assumption resulted in a larger velocity obstacle area and more conservatively adjusted velocities. In contrast, the elliptic domain occupies less space and lacks the central symmetry of a circle, causing each deflection to alter the shape of the subsequent velocity obstacle. This alteration affects the uniformity of tangential velocity changes, thereby eliminating the conservatism of the circle domain assumption. In Scenario 2, where the proximity between UAVs A and B was less, the elliptical domain still achieved obstacle avoidance with a decreased detour distance. Moreover, due to the less demanding computation times, the time consumed approached that of the circle domain.

To improve the research process in this paper, we plan to explore and solve these technical problems in the future:

- Custom protection domains for arbitrary flight scenarios

In this paper, we simplified the flight process and reduced the complexity of calculating parameters such as ellipse distance and projection length by setting a research scenario where two UAVs collide head-on while simultaneously dodging in the same deflection velocity direction. This scenario restricts the applicability of the derived ellipse protection domain sizes to head-on collisions only. For collisions at varying angles, the complexity of motion and rotation within the elliptic structure may significantly increase. It requires us to develop a design algorithm that is suitable for more general scenarios and fast enough for planning.

- The lower limit of the elliptic protection domain

In this paper, the size of the custom elliptic domain was determined based on interval distance between UAVs A and B, the actual velocity, and the angular velocity. As the interval distance decreases, the model-derived protection domain may become excessively small, suggesting the need for a minimum protection domain size based on the actual physical properties of the UAVs. If the domain size falls below this threshold, the UAVs may disrupt each other's flight stability due to airflow interference. Due to the high costs associated with actual UAV field test flights, this paper had not established a lower limit for the protection domain, which presents certain limitations.

- Exploring elliptical domain applications in more complex experimental scenarios

This study is limited to the examination of head-on collision risks for drones in a two-dimensional scenario. The rationale for this foundational assumption is that collisions between drones flying directly toward each other are deterministic and that their collision points are easily identifiable. This ensures that critical conditions for avoiding collisions can be readily derived, facilitating the analysis of the obstacle avoidance flight process using elliptical domains. It allows for easier consideration of nonlinear velocity adjustments during the deflection process of the velocity obstacle method and the resulting endpoint errors in obstacle avoidance flight positions. In contrast, establishing a quantitative functional relationship between the size of the elliptical protection domain and the obstacle avoidance process for drones in arbitrary flight states, such as non-collinear uniform linear motion or variable speed curve flights, poses significant challenges. Therefore, more complex obstacle environments, such as multiple static obstacles, drones following arbitrary curved flight paths, and multiple drones engaged in obstacle avoidance, have not been addressed in this paper.

Moreover, in real-world flight scenarios for drones, especially those with smaller geometric dimensions, they are highly susceptible to spatial gusts. When considering wind factors, the navigation of the drone may need to satisfy various constraints related to kinematics, dynamics, communication connectivity, and obstacle avoidance. Wind disturbances require the planning process to continuously update and generate new trajectories. Control barrier functions (CBFs) are used to assist in collision avoidance in the face of wind disturbances while alleviating the need to continually recalculate the motion plans [45]. Additionally, Phadke [46] established a disruption model considering obstacles and wind factors across multiple scenarios, and its experimental results provide important insights for assessing whether drones can navigate safely amid environmental disturbances. In this paper, if external factors such as wind are taken into account, assessing the degree of spatial positional uncertainty or increased collision risk level due to wind, quantifying it into the spatial extent of a customized ellipsoidal domain, and superimposing it on an existing ellipsoidal domain is a research direction that can be explored.

In the future, we plan to further improve and refine the elliptic domain setting algorithm to correlate with the real-time relative positions of the conflicting objects with respect to the above two problems. In addition, we plan to explore the best obstacle avoidance position eliminating velocity oscillations and verify the superior conflict resolution performance of the elliptic domains in narrower or crowded multi-UAV situations.

Author Contributions: Conceptualization, Y.W. and Y.L.; methodology, Y.W. and Y.L.; software, Y.W.; validation, Y.W. and Y.L.; formal analysis, S.Z. and D.Z.; investigation, Y.W.; resources, Y.W. and S.Z.; data curation, Y.W. and D.Z.; writing—original draft preparation, Y.W.; writing—review and editing, Y.W.; visualization, Y.W.; supervision, Y.L. and Y.W.; and project administration, Y.L., S.Z. and D.Z. All authors have read and agreed to the published version of the manuscript.

Funding: This research was funded by the Key Research and Development Project of Sichuan Province (No. 2023YFG0163), the China Civil Aviation Safety Capacity Building Fund Project, and the 2023 Fundamental Research Funds for Universities—Doctoral Innovation Capacity Enhancement Program (No. PHD2023-038).

Data Availability Statement: The data are contained within this article. The other data will be made available upon request.

Conflicts of Interest: The authors declare no conflicts of interest.

Abbreviations

The following abbreviations are used in this paper:

UAV	Unmanned aerial vehicle
VO	Velocity obstacles method
EVO	Elliptical velocity obstacles method
RVO	Reciprocal velocity obstacles method
VO	Relative velocity obstacle cone in the circle domain
AVO	Absolute velocity obstacle cone in the circle domian
EVO	Relative velocity obstacle cone in th elliptic domain
EAVO	Absolute velocity obstacle cone in the elliptic domian

Appendix A. Instruction and Analysis

Appendix A.1. Convex Polygons Minkowski Sum methods

Specifically, for a set of points constituted by two convex polygons in a 2D space that satisfies the property of a convex set, the vectors formed by any two points within the set do not exceed the boundary of the convex polygons. A related concept is that of Convex hulls. The intersection S of all convex sets containing the target set X is called the convex hull of X. S is also the smallest convex set containing X. Simply put, a convex hull is the smallest convex shape enclosing a set of points. This gives rise to two methods for computing the Minkowski sum of a convex polygon.

Method 1: For the set of points formed by the boundary points of A and B $Points_A^m$, $Points_B^n$, m and n represent the number of elements in the point set, respectively, and the new point set generated by the corresponding addition contains at most mn elements. The convex hull of the new point set is the Minkowski sum of A and B. Its complexity is $o(mn \log mn)$.

Method 2: For the set of vectors to the boundary of A and B $Boundary_A^m$, $Boundary_B^n$, calculating the Minkowski sum of two convex sets is simply a matter of joining and merging the '$m + n$' edge vectors after sorting them by their polar angles and then connecting and merging them in descending order of polar angle size. It can be guaranteed that the resulting graph is still convex, and the resulting convex hull is the Minkowski sum of A, B. Its complexity is $o((m + n) \log mn)$.

Appendix A.2. VO Space 'Expanding' to Twice under Two Identical Circle Domains

In fact, 'expanding' B to $\widehat{B_C}$ is the result of $\widehat{B_C} = A \oplus B$, and the reason for 'expanding' to twice the original circle can be proved simply by the following procedure:

Assuming that the center of circle B is the origin and the radius is r, the set of boundary points of B can be given by the parametric equation $Point_B = \{r \cos \delta_B, r \sin \delta_B\}$. Ignoring the specific location of A, the Minkowski sum imposed on B remains constant, regardless of where A is located in this coordinate system. In other words, it is only related to the shape of its boundary, whose boundary points can still be expressed by the parametric equation $Point_A = \{r \cos \delta_A, r \sin \delta_A\}$. Then, based on the Minkowski sum calculation **Method 1**, $A \oplus B$ should be equal to the convex packet of the boundary of $Point_A + Point_B$; i.e., formed by the point farthest from the origin as the boundary, The problem is equivalent to finding the maximum value of the following function:

$$L^2 = [r(\cos \delta_A + \cos \delta_B)]^2 + [r(\sin \delta_A + \sin \delta_B)]^2 = 2r^2[1 + \cos(\delta_B - \delta_A)] \tag{A1}$$

It is clear that the above equation obtains its maximum value when $\delta_B = \delta_A$, showing that for any boundary point corresponding to δ_B, the point furthest from the origin is the point at the same location as it, this distance is exactly $2r$. This means that the result of $A \oplus B$ is a great circle after doubling the radius of the expansion of B. It also satisfies the geometrical definition of the concatenation of the region swept by one round of continuous motion of the set A along the margins of B with the set B itself. At the same time, it is clear that elliptic domains are not reducible to this forms. Therefore, the Minkowski sum of two elliptic domains is not simply expandable in the radial direction in general.

Appendix A.3. Descriptions of the Symbols Used in the Text

The following is an explanation of the meanings of some of the symbols, which can help the reader to understand them.

$\widehat{B_C}$	Velocity obstacle space imposed by A on B in the circle domian
$\widehat{B_E}$	Velocity obstacle space imposed by A on B in the elliptic domian
a, b	Length of the major and minor semi-axes of the ellipse
τ	Arbitrary length time slice
τ_i	Time slice sequence
τ_{reback}	Time slice for the UAV to return to the endpoint
φ	Maximum deflection angle of the UAV in one second
Δs	UAV flight endpoint error under *Assumptions 1* and *2*
d_e	UAV flight endpoint error under *Assumptions 2* and *3*
$o_A^{\tau_i}, o_B^{\tau_i}$	Velocities of UAVs A and B at moment τ_i
$o_{EAVO}^{\tau_i}$	Elliptic velocity obstacle tangent direction
d_b^i	Projected distance between the two ellipses in the direction of the minor axis at τ_i

References

1. Fu, Q.; Liang, X.; Zhang, J.; Hou, Y. Cooperative conflict detection and resolution for multiple UAVs using two-layer optimization. *Harbin Gongye Daxue Xuebao/J. Harbin Inst. Technol.* **2020**, *52*, 74–83.
2. Sarim, M.; Radmanesh, M.; Dechering, M.; Kumar, M.; Pragada, R.; Cohen, K. Distributed detect-and-avoid for multiple unmanned aerial vehicles in national air space. *J. Dyn. Syst. Meas. Control* **2019**, *141*, 071014. [CrossRef]

3. Sunberg, Z.N.; Kochenderfer, M.J.; Pavone, M. Optimized and trusted collision avoidance for unmanned aerial vehicles using approximate dynamic programming. In Proceedings of the 2016 IEEE International Conference on Robotics and Automation (ICRA), Stockholm, Sweden, 16–21 May 2016; IEEE: New York, NY, USA, 2016; pp. 1455–1461.

4. Phung, M.D.; Ha, Q.P. Safety-enhanced UAV path planning with spherical vector-based particle swarm optimization. *Appl. Soft Comput.* **2021**, *107*, 107376. [CrossRef]

5. Pehlivanoglu, Y.V.; Pehlivanoglu, P. An enhanced genetic algorithm for path planning of autonomous UAV in target coverage problems. *Appl. Soft Comput.* **2021**, *112*, 107796. [CrossRef]

6. Hao, G.; Lv, Q.; Huang, Z.; Zhao, H.; Chen, W. UAV Path Planning Based on Improved Artificial Potential Field Method. *Aerospace* **2023**, *10*, 562. [CrossRef]

7. Liu, Y.; Zhao, Y. A virtual-waypoint based artificial potential field method for UAV path planning. In Proceedings of the 2016 IEEE Chinese Guidance, Navigation and Control Conference (CGNCC), Nanjing, China, 12–14 August 2016; IEEE: New York, NY, USA, 2016; pp. 949–953.

8. Pan, Z.; Zhang, C.; Xia, Y.; Xiong, H.; Shao, X. An Improved Artificial Potential Field Method for Path Planning and Formation Control of the Multi-UAV Systems. *IEEE Trans. Circuits Syst. II Express Briefs* **2022**, *69*, 1129–1133. [CrossRef]

9. Fiorini, P.; Shiller, Z. Motion planning in dynamic environments using velocity obstacles. *Int. J. Robot. Res.* **1998**, *17*, 760–772 [CrossRef]

10. Best, A.; Narang, S.; Manocha, D. Real-time reciprocal collision avoidance with elliptical agents. In Proceedings of the 2016 IEEE International Conference on Robotics and Automation (ICRA), Stockholm, Sweden, 16–21 May 2016; IEEE: New York, NY, USA, 2016; pp. 298–305.

11. Guo, H.; Guo, X. Local path planning algorithm for UAV based on improved velocity obstacle method. *Hangkong Xuebao/Acta Aeronaut. et Astronaut. Sin.* **2023**, *44*, 327586.

12. Bi, K.; Wu, M.; Zhang, W.; Wen, X.; Du, K. Modeling and analysis of flight conflict network based on velocity obstacle method. *Xi Tong Gong Cheng Yu Dian Zi Ji Shu/Syst. Eng. Electron.* **2021**, *43*, 2163–2173.

13. Zhang, H.; Gan, X.; Li, A.; Gao, Z.; Xu, X. UAV obstacle avoidance and track recovery strategy based on velocity obstacle method. *Xi Tong Gong Cheng Yu Dian Zi Ji Shu/Syst. Eng. Electron.* **2020**, *42*, 1759–1767.

14. Yang, W.; Wen, X.; Wu, M.; Bi, K.; Yue, L. Three-Dimensional Conflict Resolution Strategy Based on Network Cooperative Game. *Symmetry* **2022**, *14*, 1517. [CrossRef]

15. Peng, M.; Meng, W. Cooperative obstacle avoidance for multiple UAVs using spline_VO method. *Sensors* **2022**, *22*, 1947. [CrossRef]

16. Adouane, L.; Benzerrouk, A.; Martinet, P. Mobile robot navigation in cluttered environment using reactive elliptic trajectories. *IFAC Proc. Vol.* **2011**, *44*, 13801–13806. [CrossRef]

17. Braquet, M.; Bakolas, E. Vector field-based collision avoidance for moving obstacles with time-varying elliptical shape. *IFAC-PapersOnLine* **2022**, *55*, 587–592. [CrossRef]

18. Gérin-Lajoie, M.; Richards, C.L.; McFadyen, B.J. The negotiation of stationary and moving obstructions during walking: Anticipatory locomotor adaptations and preservation of personal space. *Mot. Control* **2005**, *9*, 242–269. [CrossRef]

19. Chraibi, M.; Seyfried, A.; Schadschneider, A. Generalized centrifugal-force model for pedestrian dynamics. *Phys. Rev. E* **2010**, *82*, 046111. [CrossRef] [PubMed]

20. Lee, B.H.; Jeon, J.D.; Oh, J.H. Velocity obstacle based local collision avoidance for a holonomic elliptic robot. *Auton. Robot.* **2017**, *41*, 1347–1363. [CrossRef]

21. Wang, G.; Liu, M.; Wang, F.; Chen, Y. A novel and elliptical lattice design of flocking control for multi-agent ground vehicles. *IEEE Control Syst. Lett.* **2022**, *7*, 1159–1164. [CrossRef]

22. Du, Z.; Li, W.; Shi, G. Multi-USV Collaborative Obstacle Avoidance Based on Improved Velocity Obstacle Method. *ASCE-ASME J. Risk Uncertain. Eng. Syst. Part A Civ. Eng.* **2024**, *10*, 04023049. [CrossRef]

23. Mao, S.; Yang, P.; Gao, D.; Bao, C.; Wang, Z. A Motion Planning Method for Unmanned Surface Vehicle Based on Improved RRT Algorithm. *J. Mar. Sci. Eng.* **2023**, *11*, 687. [CrossRef]

24. Singh, Y.; Sharma, S.; Sutton, R.; Hatton, D.; Khan, A. A constrained A* approach towards optimal path planning for an unmanned surface vehicle in a maritime environment containing dynamic obstacles and ocean currents. *Ocean Eng.* **2018**, *169*, 187–201. [CrossRef]

25. Liu, Y.; Bucknall, R. Path planning algorithm for unmanned surface vehicle formations in a practical maritime environment. *Ocean Eng.* **2015**, *97*, 126–144. [CrossRef]

26. Munasinghe, S.R.; Oh, C.; Lee, J.J.; Khatib, O. Obstacle avoidance using velocity dipole field method. In Proceedings of the International Conference on Control, Automation, and Systems, ICCAS, Budapest, Hungary, 26–29 June 2005; pp. 1657–1661.

27. Abdallaoui, S.; Aglzim, E.H.; Kribeche, A.; Ikaouassen, H.; Chaibet, A.; Abid, S.E. Dynamic and Static Obstacles Avoidance Strategies Using Parallel Elliptic Limit-Cycle Approach for Autonomous Robots. In Proceedings of the 2023 11th International Conference on Control, Mechatronics and Automation (ICCMA), Agder, Norway, 1–3 November 2023; IEEE: New York, NY, USA, 2023; pp. 133–138.

28. Shiller, Z.; Large, F.; Sekhavat, S. Motion planning in dynamic environments: Obstacles moving along arbitrary trajectories. In Proceedings of the 2001 ICRA. IEEE International Conference on Robotics and Automation (Cat. No. 01CH37164), Seoul, Republic of Korea, 21–26 May 2001; IEEE: New York, NY, USA, 2001; Volume 4, pp. 3716–3721.

29. Large, F.; Laugier, C.; Shiller, Z. Navigation among moving obstacles using the NLVO: Principles and applications to intelligent vehicles. *Auton. Robot.* **2005**, *19*, 159–171. [CrossRef]

30. Chen, P.; Huang, Y.; Mou, J.; Van Gelder, P. Ship collision candidate detection method: A velocity obstacle approach. *Ocean Eng.* **2018**, *170*, 186–198. [CrossRef]

31. Chen, P.; Huang, Y.; Papadimitriou, E.; Mou, J.; van Gelder, P. An improved time discretized non-linear velocity obstacle method for multi-ship encounter detection. *Ocean Eng.* **2020**, *196*, 106718. [CrossRef]

32. Van den Berg, J.; Lin, M.; Manocha, D. Reciprocal velocity obstacles for real-time multi-agent navigation. In Proceedings of the 2008 IEEE International Conference on Robotics and Automation, Pasadena, CA, USA, 19–23 May 2008; IEEE: New York, NY, USA, 2008; pp. 1928–1935.

33. Van Den Berg, J.; Guy, S.J.; Lin, M.; Manocha, D. Reciprocal n-body collision avoidance. In *Robotics Research: The 14th International Symposium ISRR*; Springer: Berlin/Heidelberg, Germany, 2011; pp. 3–19.

34. Han, R.; Chen, S.; Wang, S.; Zhang, Z.; Gao, R.; Hao, Q.; Pan, J. Reinforcement learned distributed multi-robot navigation with reciprocal velocity obstacle shaped rewards. *IEEE Robot. Autom. Lett.* **2022**, *7*, 5896–5903. [CrossRef]

35. Giese, A.; Latypov, D.; Amato, N.M. Reciprocally-rotating velocity obstacles. In Proceedings of the 2014 IEEE International Conference on Robotics and Automation (ICRA), Hong Kong, China, 31 May–7 June 2014; IEEE: New York, NY, USA, 2014; pp. 3234–3241.

36. JEON, J.D. A Velocity-Based Local Navigation Approach to Collision Avoidance of Elliptic Robots. Ph.D. Thesis, Seoul National University, Seoul, Republic of Korea, 2017.

37. Feurtey, F. *Simulating the Collision Avoidance Behavior of Pedestrians*; The University of Tokyo, School of Engineering, Department of Electronic Engineering: Tokyo, Japan, 2000.

38. Snape, J.; Van Den Berg, J.; Guy, S.J.; Manocha, D. The hybrid reciprocal velocity obstacle. *IEEE Trans. Robot.* **2011**, *27*, 696–706. [CrossRef]

39. Snape, J.; Van Den Berg, J.; Guy, S.J.; Manocha, D. Independent navigation of multiple mobile robots with hybrid reciprocal velocity obstacles. In Proceedings of the 2009 IEEE/RSJ International Conference on Intelligent Robots and Systems, St. Louis, MO, USA, 11–15 October 2009; IEEE: New York, NY, USA, 2009; pp. 5917–5922.

40. Kavraki, L.E. Computation of configuration-space obstacles using the fast Fourier transform. *IEEE Trans. Robot. Autom.* **1995**, *11*, 408–413. [CrossRef]

41. Lee, I.K.; Kim, M.S.; Elber, G. Polynomial/rational approximation of Minkowski sum boundary curves. *Graph. Model. Image Process.* **1998**, *60*, 136–165. [CrossRef]

42. Wein, R. Exact and efficient construction of planar Minkowski sums using the convolution method. In *European Symposium on Algorithms*; Springer: Berlin/Heidelberg, Germany, 2006; pp. 829–840.

43. Yan, Y.; Chirikjian, G.S. Closed-form characterization of the Minkowski sum and difference of two ellipsoids. *Geom. Dedicata* **2015**, *177*, 103–128. [CrossRef]

44. Cheng, Z.; Chen, P.; Mou, J.; Chen, L. Multi-ship Encounter Situation Analysis with the Integration of Elliptical Ship Domains and Velocity Obstacles. *TransNav. Int. J. Mar. Navig. Saf. Od Sea Transp.* **2023**, *17*, 895–902. [CrossRef]

45. Abichandani, P.; Lobo, D.; Muralidharan, M.; Runk, N.; McIntyre, W.; Bucci, D.; Benson, H. Distributed Motion Planning for Multiple Quadrotors in Presence of Wind Gusts. *Drones* **2023**, *7*, 58. [CrossRef]

46. Phadke, A.; Medrano, F.A.; Chu, T.; Sekharan, C.N.; Starek, M.J. Modeling Wind and Obstacle Disturbances for Effective Performance Observations and Analysis of Resilience in UAV Swarms. *Aerospace* **2024**, *11*, 237. [CrossRef]

Disclaimer/Publisher's Note: The statements, opinions and data contained in all publications are solely those of the individual author(s) and contributor(s) and not of MDPI and/or the editor(s). MDPI and/or the editor(s) disclaim responsibility for any injury to people or property resulting from any ideas, methods, instructions or products referred to in the content.

Article

A Multi-Waypoint Motion Planning Framework for Quadrotor Drones in Cluttered Environments

Delong Shi *, Jinrong Shen *, Mingsheng Gao and Xiaodong Yang

College of Information Science and Enginnering, Hohai University, Changzhou 213200, China; gaoms@hhu.edu.cn (M.G.); 221320010019@hhu.edu.cn (X.Y.)
* Correspondence: 221320010014@hhu.edu.cn (D.S.); 19941470@hhu.edu.cn (J.S.)

Abstract: In practical missions, quadrotor drones frequently face the challenge of navigating through multiple predetermined waypoints in cluttered environments where the sequence of the waypoints is not specified. This study presents a comprehensive multi-waypoint motion planning framework for quadrotor drones, comprising multi-waypoint trajectory planning and waypoint sequencing. To generate a trajectory that follows a specified sequence of waypoints, we integrate uniform B-spline curves with a bidirectional A* search to produce a safe, kinodynamically feasible initial trajectory. Subsequently, we model the optimization problem as a quadratically constrained quadratic program (QCQP) to enhance the trackability of the trajectory. Throughout this process, a replanning strategy is designed to ensure the traversal of multiple waypoints. To accurately determine the shortest flight-time waypoint sequence, the fast marching (FM) method is utilized to efficiently establish the cost matrix between waypoints, ensuring consistency with the constraints and objectives of the planning method. Ant colony optimization (ACO) is then employed to solve this variant of the traveling salesman problem (TSP), yielding the sequence with the lowest temporal cost. The framework's performance was validated in various complex simulated environments, demonstrating its efficacy as a robust solution for autonomous quadrotor drone navigation.

Keywords: quadrotor drone; trajectory planning; waypoint sequencing

Citation: Shi, D.; Shen, J.; Gao, M.; Yang, X. A Multi-Waypoint Motion Planning Framework for Quadrotor Drones in Cluttered Environments. *Drones* **2024**, *8*, 414. https://doi.org/10.3390/drones8080414

Academic Editors: Jihong Zhu, Heng Shi, Zheng Chen and Minchi Kuang

Received: 16 July 2024
Revised: 20 August 2024
Accepted: 21 August 2024
Published: 22 August 2024

Copyright: © 2024 by the authors. Licensee MDPI, Basel, Switzerland. This article is an open access article distributed under the terms and conditions of the Creative Commons Attribution (CC BY) license (https://creativecommons.org/licenses/by/4.0/).

1. Introduction

In real-world scenarios such as inspection, surveying, and monitoring tasks, quadrotor drones frequently encounter the need to execute complex missions that involve navigating through multiple waypoints rather than simply targeting a single point. These tasks demand quadrotor drones to navigate safely through cluttered environments while minimizing travel time, where the sequence of waypoints is often not predetermined. Furthermore, multi-waypoint trajectory planning requires quadrotor drones to consider the spatial relationships between these waypoints. Navigation algorithms designed for single-goal navigation may introduce unnecessary path redundancy when applied to multi-waypoint trajectory planning problems, resulting in a waste of time and energy resources. Therefore, efficiently determining the optimal sequence for traversing waypoints and subsequently generating safe trajectories with minimal flight times pose a significant challenge.

Therefore, this paper proposes an advanced quadrotor multi-waypoint motion planning framework, as illustrated in Figure 1. For the multi-waypoint trajectory planning problem, we consider the spatial relationships between waypoints and propose a velocity-adaptive trajectory generation method based on uniform B-spline curves. This method utilizes the bidirectional A* method to obtain a sequence of B-spline control points, generating a kinodynamically feasible initial trajectory. Subsequently, we further optimize the smoothness of the trajectories using quadratically constrained quadratic programming (QCQP), enhancing trajectory trackability and reducing energy consumption. Finally, during the trajectory generation process, we employ a replanning approach to reduce the number of control point variables solved by QCQP, thus improving numerical stability and algorithm real-time performance. To determine the waypoint sequence in this variant of the traveling salesman problem (TSP), we employ a combination of the fast marching (FM) method and ant colony optimization (ACO). First, we compute the velocity map while adhering to the constraints of the proposed planning method. Then, we utilize FM to establish a time cost matrix between every pair of waypoints in the TSP problem. Finally, ACO is utilized to find the waypoint sequence with the lowest time cost.

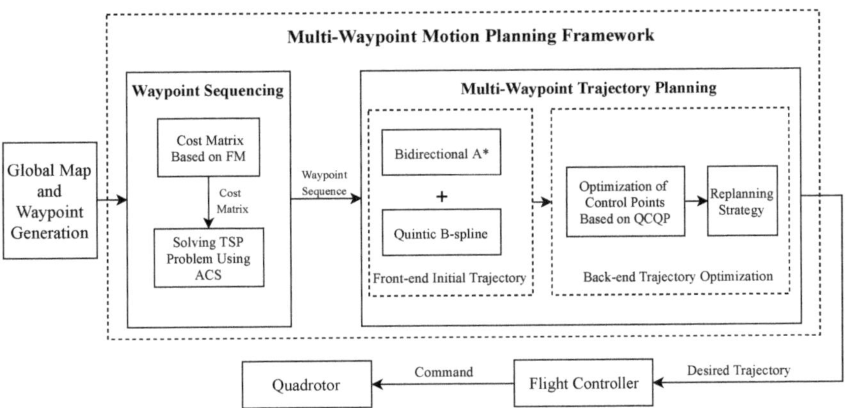

Figure 1. Quadrotor drone multi-waypoint motion planning framework.

We summarize our contributions as follows:

- We design a multi-waypoint trajectory planning method. We refine the bidirectional A* method with kinodynamic constraints, framing trajectory generation as a B-spline control point placement problem to achieve initial kinodynamically feasible trajectories, and QCQP is used to optimize the coordinates of the control points. During this process, the MINVO-based generation of the minimum convex hull of B-spline curves is employed to enlarge the solution space and avoid overly conservative trajectories.
- We design a method for determining waypoint sequences. While maintaining consistency with the objectives and constraints of multi-waypoint trajectory planning, we utilize the FM method to establish a cost matrix. ACO is then applied to solve this variant of TSP, yielding the waypoint sequence with the shortest time.
- We propose a multi-waypoint motion planning framework incorporating the aforementioned two components. We validate the effectiveness of our proposed method and this framework through extensive simulation experiments.

This paper is structured as follows. Section 2 reviews the related work. Section 3 introduces the proposed trajectory planning method, while Section 4 presents the proposed waypoint sequencing method. Section 5 presents the experimental results, followed by the discussion in Section 6 and conclusions in Section 7.

2. Related Works

Due to the differential flatness of quadrotor drones, their entire state can be represented by four flat outputs, with the most widely used being 3D position and yaw angle. The trajectory generation problem for quadrotors is also transformed into a problem of generating time-parameterized curves [1].

In the realm of quadrotor trajectory planning, a hierarchical planning framework is commonly utilized, which involves front-end path searching and back-end trajectory generation. The front-end search comprises search-based and sampling-based methods. Among the search-based approaches, A* [2] and its variations are extensively studied, while popular sampling-based techniques include RRT [3] and PRM [4]. However, these methods typically generate piecewise-linear paths, rendering them unsuitable for direct execution by quadrotor controllers. Consequently, back-end trajectory generation is essential to convert them into time-parameterized curves, such as polynomial curves [5], Bezier curves [6], and B-spline curves [7,8].

However, classical front-end path searching methods focus solely on obstacle avoidance, disregarding the dynamic characteristics of quadrotors. Consequently, the initial paths obtained from front-end path searching may not belong to the same homotopy class as the optimal trajectory in practice. In some cases, they may even fail to produce dynamically feasible trajectories through back-end trajectory generation, resulting in planning failures. Therefore, some scholars integrate the dynamic properties of quadrotors into the front-end path searching process. Reference [9] proposed a kinodynamic search method based on motion primitives, representing the motion dynamic constraints of quadrotors as motion primitives. Then, a graph search method similar to A* was employed to generate a kinodynamically feasible, continuous, and safe trajectory. Reference [7] introduced a search method considering motion dynamics based on the hybrid-state A* algorithm. After generating dynamically feasible initial trajectories, B-spline curves were used to fit and optimize the initial trajectories, further enhancing trajectory smoothness and safety. Reference [8] considered dynamical feasibility using the A* algorithm. By combining B-spline curves, the trajectory generation problem is transformed into a problem of placing B-spline control points. After generating a feasible sequence of control points, QCQP is used to optimize the positions of control points further, thereby improving trajectory smoothness.

Nevertheless, most trajectory planning algorithms are designed for single-goal navigation problems, with fewer algorithms tailored explicitly for multi-waypoint trajectory planning. Compared to single-goal problems, multi-waypoint problems must consider the spatial relationships between different waypoints to generate efficient trajectories. Reference [10] applied B-spline curves to trajectory generation involving multiple waypoints, resulting in shorter and smoother trajectories. However, its flight corridor is predefined and relatively narrow, suitable only for scenarios without obstacles between waypoints. Reference [11] considered the complete quadrotor flight model and proposed an optimization-based multi-waypoint trajectory generation method, aiming to obtain the optimal trajectory in terms of time. However, it did not account for the influence of obstacles. Reference [12] also considered the complete quadrotor model and adopted a hierarchical optimization approach. It first used a variant of PRM to obtain homotopy classes of paths, then guided the trajectory planning of the point-mass model using these homotopy classes, finally using the SST algorithm to plan multi-waypoint trajectories using the complete quadrotor model. Although this method considered obstacles, it suffered from the drawback of excessive computation times, making it unsuitable for real-time planning.

For the problem of finding the optimal waypoint sequence in terms of time, scholars have regarded it as a variant of the traveling salesman problem (TSP) [13–16]. Reference [13] addressed the problem of determining the sequence of multiple target points for autonomous underwater vehicles (AUVs) in complex underwater environments. Based on the A* algorithm, it utilized a path planning method combining coarse-grained modeling and fine-grained modeling to establish a cost matrix and solved the TSP problem using ACO to obtain the shortest path sequence of waypoints. Reference [14] proposed a method combining the adaptive weighted PRM (AWPRM) and the best ant colony system (BACS) to solve the problem of multi-quadrotor waypoint assignment and sequence determination. It used the Dubins method to construct the shortest obstacle-avoiding paths with turn radius constraints but did not consider additional quadrotor dynamic constraints. Both methods aim to minimize distance. Hence, the waypoint sequences obtained may not necessarily be the time-optimal solution after back-end trajectory generation. Reference [15] proposed a solution for the shortest time in multi-region quadrotor surveillance problems, but it required quadrotor drones to fly at maximum speed between multiple regions, which is unrealistic, as it did not consider the dynamic characteristics of quadrotors. The fast marching (FM) method is also employed to calculate the distance costs between multiple task points, which are then optimized using a genetic algorithm (GA) [16]. However, since they do not account for the robot's dynamic characteristics when creating the speed map, it only determines the shortest distance sequence of task points using FM and GA, which does not correspond to the objective of minimizing time in practical tasks. Therefore, to achieve the shortest time waypoint sequence, it is essential to maintain consistency between the constraints of the waypoint sequencing method and the trajectory planning method. This consistency allows for the establishment of a cost matrix that more accurately reflects actual flight times, thereby yielding a more optimal solution.

3. B-Spline-Based Bidirectional A* Trajectory Planning

This paper proposes a hierarchical multi-waypoint trajectory planning method for a given sequence of waypoints. The method comprises a front-end search integrating bidirectional A* and B-spline curves and back-end optimization using quadratically constrained quadratic programming (QCQP). The front-end search transforms the initial trajectory generation problem into a B-spline control point placement problem. By integrating motion dynamic constraints, an improved bidirectional A* algorithm is employed to obtain a sequence of control points that meet the requirements. Consequently, a safe initial trajectory that satisfies the motion dynamic constraints is generated. In the back-end optimization, QCQP is employed to further optimize the positions of the control points, resulting in smoother and more energy-efficient trajectories. Finally, in response to the challenge of optimizing a large number of waypoints, a replanning strategy is utilized to improve the real-time performance of trajectory planning and the numerical stability of QCQP.

3.1. B-Spline Curves

B-spline curves are piecewise polynomial curves. For a B-spline curve of degree k, it is determined by a knot vector $[t_0, t_1 \cdots t_{n+k+1}]$ and control points $[q_0, q_1 \cdots q_n]$ together. The computation method is as follows:

$$x(t) = \sum_{i=0}^{n} q_i B_{i,k}(t) \tag{1}$$

$$B_{i,0}(t) = \begin{cases} 1, & t_i \leq t < t_{i+1} \\ 0, & \text{otherwise} \end{cases} \tag{2}$$

$$B_{i,k}(t) = \frac{t - t_i}{t_{i+k} - t_i} B_{i,k-1}(t) + \frac{t_{i+k+1} - t}{t_{i+k+1} - t_{i+1}} B_{i+1,k-1}(t) \tag{3}$$

where $t \in [t_i, t_{i+1})$ is the interval when the basis function $B_{i,k}(t)$ is nonzero. This indicates that the B-spline curve in $t \in [t_i, t_{i+1})$ is solely determined by the local $k + 1$ control points $[q_i, q_{i+1} \cdots q_{i+k}]$. This property is known as the local control characteristic of B-spline curves, and the B-spline curve in $t \in [t_i, t_{i+1})$ lies entirely within the convex hull formed by $[q_i, q_{i+1} \cdots q_{i+k}]$.

Figure 2 shows where the local quintic B-spline trajectory (red curve) lies within the convex hull formed by six local control points (cyan–blue area). Additionally, B-spline curves also possess derivative properties, meaning the derivative of a B-spline curve remains a B-spline curve.

The unique properties of B-spline curves confer distinct advantages in obstacle avoidance and dynamic constraints. Regarding obstacle avoidance, the convex hull property of B-spline curves ensures that the convex hull formed by control points satisfies the obstacle avoidance constraints when located in obstacle-free regions. Dynamic constraints are typically defined by the maximum values of the quadrotor drone's velocity and acceleration, denoted as $v_{\max}$ and $a_{\max}$, respectively. Leveraging the derivative characteristics, low-order B-spline curves (velocity and acceleration curves) are derived. Subsequently, by utilizing the convex hull property, control points v_i and a_i of the velocity and acceleration curves are constrained not to exceed their maximum values, thereby ensuring dynamic feasibility.

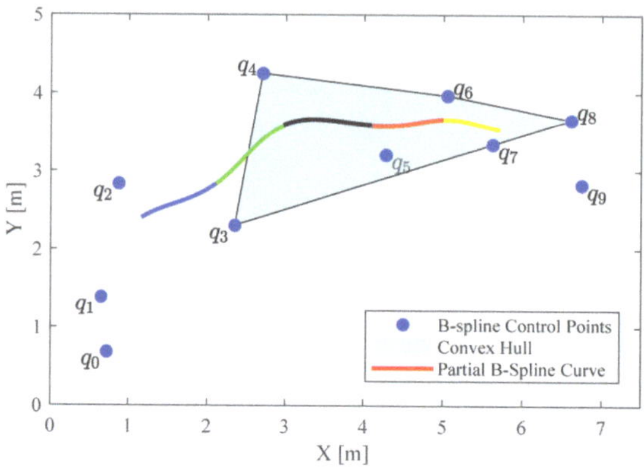

Figure 2. Trajectory represented by a uniform quintic B-spline.

Due to the significant influence of the intermediate two B-spline control points on the local curve, for a quintic B-spline curve, the local control points $\mathbf{Q}_i = [q_i, q_{i+1} \cdots q_{i+5}]^\top$ generate the i-th segment trajectory near the line connecting q_{i+2} and q_{i+3}. It provides favorable conditions for using A* search: ensuring that the line connecting q_{i+2} and q_{i+3} does not intersect with obstacles, it can be approximately assumed that the i-th segment B-spline curve will not collide with obstacles [8].

However, in practical applications, the convex hull formed by the control points of the B-spline curve does not tightly enclose the local segments of the curve. Consequently, requiring the convex hull to be within the feasible region will result in conservative trajectories. Therefore, we use control points generated by the MINVO basis [17] to replace the B-spline control points, minimizing the volume of the convex hull and producing more aggressive trajectories.

As shown in Figure 3, it can be observed that compared to the local B-spline control points $\mathbf{Q}_{bs}$, the convex hull formed by the MINVO control points $\mathbf{Q}_{mv}$ can more tightly envelop the local B-spline curve. Therefore, in the process of verifying dynamic feasibility, we use $\mathbf{Q}_{mv}$ instead of $\mathbf{Q}_{bs}$ for feasibility checks. These two can be converted into each other through matrix operations:

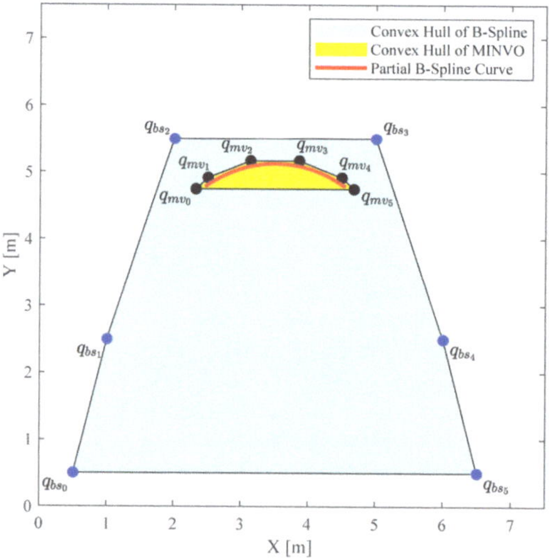

Figure 3. Comparison between MINVO and B-spline convex hull.

$$p(s(t)) = s(t)^\top \mathbf{M}_{bs6} \mathbf{Q}_{bs} = s(t)^\top \mathbf{M}_{mv6} \mathbf{Q}_{mv} \tag{4}$$

$$\mathbf{Q}_{mv} = (\mathbf{M}_{mv6})^{-1} \mathbf{M}_{bs6} \mathbf{Q}_{bs} \tag{5}$$

where $\mathbf{M}_{bs6}$ represents the constant basis matrix for a quintic B-spline curve [18], and $\mathbf{M}_{mv6}$ denotes the corresponding MINVO basis matrix [17].

3.2. B-Spline-Based Bidirectional A* Search

The A* algorithm is a classic heuristic search algorithm that is widely used for finding the shortest path from a starting point to a target point. It combines the benefits of heuristic search and cost functions to efficiently find the shortest path. When using an appropriate heuristic function, A* can guarantee finding the optimal solution from the start point to the end point. Additionally, A* is flexible and can adjust the heuristic function according to different environments and constraints, making it suitable for various application scenarios. However, A* applies to discrete paths and cannot be directly applied to generate continuous trajectories. The core of the A* algorithm lies in the cost function:

$$f(n) = g(n) + h(n) \tag{6}$$

where n represents the current node, $g(n)$ denotes the actual cost from the start point to the current point, and $h(n)$ is the estimated cost from the current point to the target point. In shortest distance problems, distance is typically used to measure cost. Commonly used estimates for $h(n)$ include Euclidean distance, Manhattan distance, and diagonal distance. To ensure that the estimated distance is closer to the actual distance while balancing search efficiency and optimality, $h(n)$ generally employs diagonal distance.

Most current research utilizes discrete path planning algorithms like A* to generate safe initial paths. Subsequently, typical waypoints are selected from these paths, and continuous trajectories are generated using spline interpolation or other smoothing algorithms. However, because the initial paths do not consider dynamic feasibility, and subsequent trajectory smoothing may encounter time allocation issues, these trajectories are often not optimal. Therefore, we employ a dynamic-aware search method. Instead of specific path points, we treat the expanded nodes in the search process as control points of the B-spline curve. During node expansion, we consider safety and dynamic feasibility when choosing the expansion step size, thus avoiding issues with time allocation. By leveraging the local control characteristics of B-spline curves to compute the smoothness cost of local curves, we incorporate the smoothness cost into the heuristic function, thereby generating time-parametrized trajectories that satisfy both avoidance and dynamic constraints, as shown in Figure 4.

Due to the additional computational overhead introduced by dynamic feasibility checks, adaptive step size expansion, and the smoothness cost in the cost function of this search method, the search time is significantly increased. Therefore, we adopt a bidirectional search method [19] to improve the search speed. The bidirectional A* method essentially creates and expands two search trees from the initial and goal states. The starting nodes target the goal nodes, while the goal nodes target the starting nodes. The complete search path is generated when the nodes expanded by the two search trees meet. By simultaneously expanding two search trees, the depth of search for each tree decreases, reducing the number of expanded nodes and decreasing the search time.

We adopt uniform quintic B-spline curves, with waypoints set as $\{w_0, w_1, \cdots, w_{m-1}\}$. Algorithm 1 is the B-spline-based bidirectional A* (BA*-BS) search method used to obtain the kinodynamically feasible initial trajectory between w_j and w_{j+1}. The initial and terminal nodes in the algorithm are determined by **ComputeStartNode()** and **ComputeEndNode()**, respectively. **Expand()** alternately extends the forward and backward search trees, while **Checkdynamic()** is used to determine if the newly expanded nodes satisfy dynamic feasibility. **Cost()** and **HeuristicCost()** are employed to compute the actual cost and heuristic cost of the newly expanded nodes, respectively. Lines 9–18 determine whether the bidirectional search is terminated. Taking the forward search as an example, **Near()** obtains a set I of nodes from the reverse $Close_e$ whose distance from p_{cur} is less than $v_{\max} \cdot \Delta t$. Lines 11–13 find the nodes in I that satisfy the obstacle avoidance requirements, forming the set $Cand.e$ which is then sorted based on distance from p_{cur}. **Retrivepath()** is used to backtrack the control point list, and **Evaluatetraj()** is employed to determine if the returned control point sequence satisfies dynamic feasibility. Finally, based on the returned control point sequence, a trajectory satisfying both obstacle avoidance and dynamic feasibility requirements is generated, combined with the node list of the uniform B-spline.

Algorithm 1: BA*-BS Search Method

Input: state$(w_j), w_{j+1}, w_{j+2}, \Delta t$
Output: CPS $= \{cp_0, cp_1, \cdots, cp_n\}^\top$

1   $p_{start} \leftarrow$ **ComputeStartNode**(**state**(w_j));
2   $p_{end}, p_{end1}, p_{end2} \leftarrow$ **ComputeEndNode**$(w_j, w_{j+1}, w_{j+2}, \Delta t)$;
3   $Open_s \leftarrow Close_s \leftarrow Open_e \leftarrow Close_e \leftarrow \varnothing$;
4   Add$(Open_s, p_{start}, 0)$,Add$(Open_e, p_{start}, 0)$,$i \leftarrow 0$;
5   **while** Size$(Open_s) \cup$ Size$(Open_e)$ **do**
6   **if** $i \bmod 2 = 0$ **then**   // Forward search
7     $p_{cur} \leftarrow$ PopMin$(Open_s)$,Add$(Close_s, p_{cur})$;
8     $I \leftarrow$ **Near**$(p_{cur}, Close_e)$;
9   **if** $I \neq \varnothing$ **then**
10      $Cand.e \leftarrow \varnothing$;
11   **for** p_i in I **do**
12   **if** $(p_i - p_{cur})$.norm$() \leq \min\{d_{esdf}(p_i), d_{esdf}(p_{cur})\}$ **then**
13   Add$(Cand.e, p_i, (p_i - p_{cur})$.norm$())$;
14   **while** Size$(Cand.e) \neq 0$ **do**
15       $p_e \leftarrow$ Pop$(Cand.e)$;
16   **CPS** $\leftarrow$ **Retrivepath**(p_{cur}, p_e);
17   **if** **Evaluatetraj**$(cps) = true$ **then**
18   **return CPS**;
19   **for** p_{nbr} in **Expand**(p_{cur}) **do**
20   **if** $p_{nbr} \notin Close_s$ **then**
21       $LocalCPs \leftarrow$ **Retrive6**(p_{cur});
22   **if** **Checkdynamic**$(LocalCPs) = true$ **then**
23        $cost_{now} \leftarrow$ **Cost**$(p_{cur}) +$ **CostSmmoth**$(LocalCPs) + \Delta t$;
24   **if** $p_{nbr} \in Open_s$ **then**
25   **if** $cost_{now} <$ Cost(p_{nbr}) **then**
26   Update$(p_{nbr}, cost_{now} +$ **HeuristicCost**$(p_{nbr}))$;
27   **else**
28   Add$(Open_s, p_{nbr}, cost_{now} +$ **HeuristicCost**$(p_{nbr}))$;
29   **else**
30     $\cdots$ // Backward search

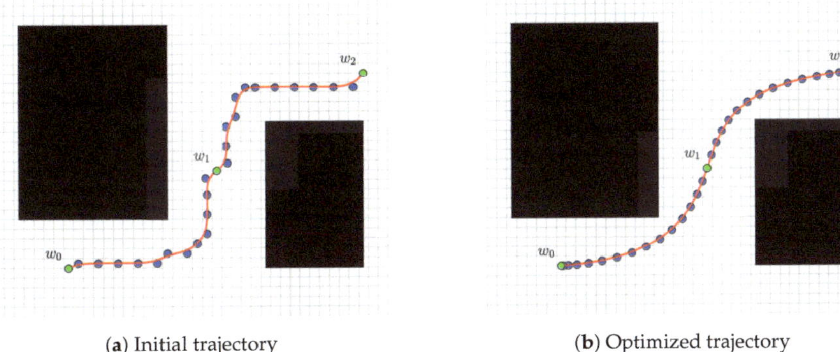

(a) Initial trajectory   (b) Optimized trajectory

Figure 4. Illustration of trajectory generation for three waypoints (green dots). As in (**a**), the sequence of control points (blue dots) is obtained using the BA*-BS search method, and the initial trajectory (red curve) is generated. In (**b**), the position of the control points is optimized using QCQP to generate the final trajectory.

3.2.1. Initial Node and Terminal Node

To ensure continuous state tracking of the quadrotor drone during flight, we determine the initial node p_{start} based on the quadrotor's initial flight state (position and its higher-order derivatives) at waypoint w_j, represented as $\textbf{state} = [p, p', p'', p''', p'''']$. The calculation of the initial node is implemented by the **ComputeStartNode()**:

$$p(s(t)) = \mathbf{s}(t)^\top \mathbf{M}_{bs6} \mathbf{Q}_j$$
$$\mathbf{s}(t) = \begin{bmatrix} 1 & s(t) & s^2(t) & \cdots & s^5(t) \end{bmatrix}^\top \tag{7}$$
$$\mathbf{Q}_j = \begin{bmatrix} q_{j-5} & q_{j-4} & q_{j-3} & \cdots & q_j \end{bmatrix}^\top$$

where p represents the position of the B-spline curve, with $t \in [t_j, t_{j+1}]$, $s(t) = (t - t_j)/\Delta t$, where Δt is the uniform B-spline time interval, $\mathbf{Q}$ denotes the B-spline control points, and $\mathbf{M}_{bs6}$ is the quintic B-spline basis matrix. At the initial moment $s(t) = 0$, we use $\textbf{state} = [p, p', p'', p''', p'''']$ as the five constraints to solve (7), resulting in $[q_0, q_1, q_2, q_3, q_4]^\top$. The coordinate of p_{start} is set to q_4.

Since our method considers the trajectory planning of multiple waypoints, the terminal node selection needs to consider the positional relationships between multiple waypoints. Similar to the initial node, which needs to be calculated based on the flight state, the terminal node must also be determined based on the predetermined terminal flight state. Given that our planning objectives include flight time and smoothness, the flight time would significantly increase if the quadrotor were to hover (with zero higher-order derivatives such as speed and acceleration) at every waypoint. Therefore, we need to design the flight state of the quadrotor at waypoint w_{j+1}. We consider the positional relationships of the initial waypoint w_j, the terminal waypoint w_{j+1}, and the next waypoint w_{j+2}, and we set the velocity of the quadrotor at w_{j+1} as v_{j+1}:

$$v_{j+1} = \left\{ \frac{u_j d_{j+1} + u_{j+1} d_j}{d_j + d_{j+1}} \right\} \cdot v_{\max} \cdot scale \tag{8}$$

$$scale = \tanh\left(\frac{\min(d_j, d_{j+1})}{v_{\max} \cdot \alpha} \right) \tag{9}$$

where u_j is the normalized vector of $\overrightarrow{w_j w_j}$, and d_j is the Euclidean distance between w_j and w_{j+1}. α is the scaling factor used to adjust the influence of waypoint spacing on v. $\{x\}$ denotes normalization of x. The principles for calculating v_{j+1} are as follows:

1. Waypoints closer to the w_{j+1} have a greater impact on v_{j+1}.
2. Make sure that v_{j+1} does not exceed $v_{\max}$. Its magnitude is determined by the minimum distance between w_{j+1} and the adjacent waypoints.

Although the preceding steps set the heuristic velocity v_{j+1} at waypoint w_{j+1}, during the actual execution of the algorithm, substituting v_{j+1} into (7) to solve for the remaining control points may result in significant deviation of the control points from the current trajectory. Therefore, to enhance the robustness of the algorithm and maintain high-order continuity of the trajectory, we employ a safer approach **ComputeEndNode()** to set the terminal node p_{end} and subsequent nodes:

$$\begin{cases} p_{end} = w_{j+1} - v_{j+1} \cdot \Delta t \\ p_{end1} = w_{j+1} \\ p_{end2} = w_{j+1} + v_{j+1} \cdot \Delta t \end{cases} \tag{10}$$

if p_{end} or p_{end2} fall within obstacles, we employ a gradient ascent method based on the Euclidean signed distance field (ESDF) to ensure that p_{end}, p_{end1}, and p_{end2} are all located in obstacle-free regions. Here, p_{end} serves as the terminal node during the search and is also the starting node for the backward search in the bidirectional search. When the bidirectional search is completed, p_{end1} and p_{end2} are sequentially added as control points to the end of the control point sequence, and termination is determined.

As the position of the local B-spline trajectory formed by $[q_i, q_{i+1}, q_{i+2}, q_{i+3}, q_{i+4}, q_{i+5}]^\top$ lies near the line connecting q_{i+2} and q_{i+3}, the endpoint of the trajectory obtained by the search is around p_{end}, while the position of the waypoint is p_{end1}. Therefore, we need an additional control point to ensure that the endpoint of this segment of the trajectory is at p_{end1}. In fact, due to the adoption of the replanning method (Section 3.4), we set the coordinates of the last control point p_{end3} of this trajectory segment based on the first control point returned by the next search.

3.2.2. Adaptive Expansion

Similar to the traditional A* algorithm, the improved A* algorithm also searches for the next expansion node based on 26 directions from the current node (in 3D space). However, the traditional A* algorithm may encounter situations where it comes too close to obstacles due to its fixed expansion step size. Therefore, for environments that are densely populated with obstacles, we adopt an adaptive expansion step size **Expand()**, where the step size is smaller when closer to obstacles and larger when farther away, as shown in Figure 5.

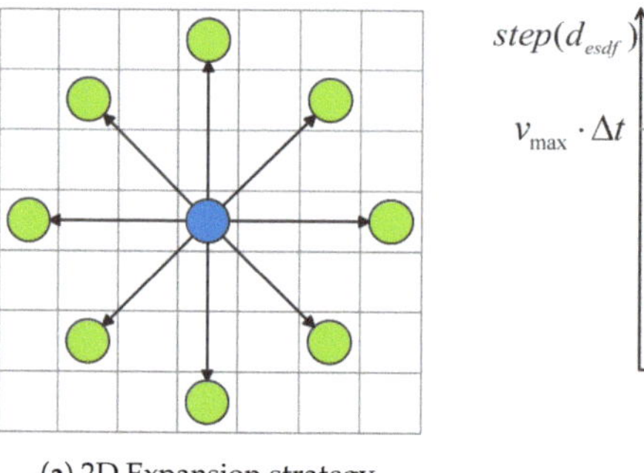
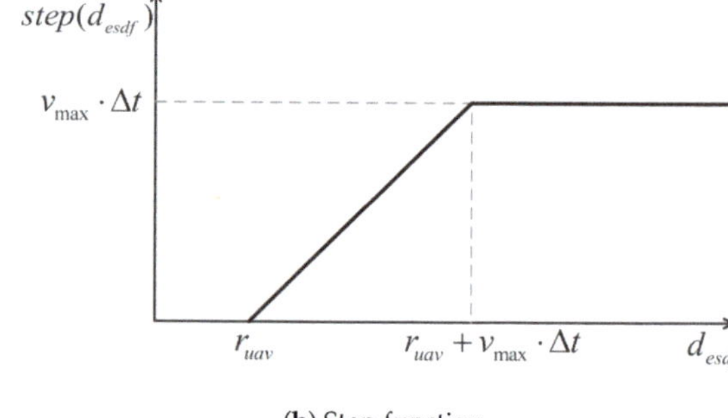

(**a**) 2D Expansion strategy (**b**) Step function

Figure 5. Illustration of non-uniform expansion. In (**a**), the blue dot represents the current node, while the green dots represent the neighbor nodes. The step function is shown in (**b**).

Based on the fixed time interval of the uniform B-spline curve, the quadrotor drone flies at a higher speed when farther away from the obstacles and gradually slows down as the distance from the obstacles decreases, thereby enhancing the safety of the trajectory. The step function is as follows:

$$step(d_{esdf}) = \begin{cases} 0, & d_{esdf} \leqslant r_{uav} \\ d_{esdf} - r_{uav}, & r_{uav} < d_{esdf} \leqslant r_{uav} + v_{\max} \cdot \Delta t \\ v_{\max} \cdot \Delta t, & d_{esdf} \geqslant r_{uav} + v_{\max} \cdot \Delta t \end{cases} \tag{11}$$

where $step(d_{esdf})$ represents the expansion step size, r_{uav} denotes the radius of the quadrotor drone, and d_{esdf} stands for the distance between the current node and the nearest obstacle, obtained from the ESDF.

3.2.3. Cost Function

In contrast to the traditional A* algorithm (6), the cost function is defined as follows:

$$g(p_{cur}) = g(p_{cur.parent}) + \Delta t + \beta_{smooth} \cdot \text{cost}_{smmoth}(p_{cur}) \tag{12}$$

$$h(p_{cur}) = d_{diag}(p_{cur}, p_{end})/(v_{\max}/\eta) \tag{13}$$

where $g(p_{cur})$ and $h(p_{cur})$ denote the actual cost and estimated cost of the current node p_{cur}, respectively, corresponding to the **Cost()** and **HeuristicCost()**. $g(p_{cur})$ consists of two parts: the time cost $n \cdot \Delta t$, and the smoothness cost $\beta_{smooth} \cdot \text{cost}_{smmoth}$. The time cost represents the actual time required from p_{start} to p_{cur}. As the smoothness and energy cost of the trajectory can be represented by the square integral of the trajectory derivatives, the smoothness cost is the cumulative $\int jerk^2 \, dt$ from p_{start} to p_{cur}, where $jerk$ is third-order derivative of the trajectory $\frac{d^3 p(t)}{dt^3}$. $h(p_{cur})$ estimates the time from p_{cur} to p_{end}. $d_{diag}(p_{cur}, p_{end})$ represents the diagonal distance from p_{cur} to p_{end}. Since the quadrotor cannot continuously fly at $v_{\max}$, we use a scaling factor η to avoid underestimating the flight time, typically set to 1–2.

3.2.4. Dynamic Feasibility Checking

The traditional A* algorithm only determines that a node is infeasible when the expansion reaches inside obstacles, whereas our modified A* algorithm considers dynamic constraints. When expanding a node p_{nbr}, if adding this node to the control point list would cause the velocity or acceleration to exceed the maximum values $v_{\max}$, $a_{\max}$, then the node is considered infeasible.

The judging method is as follows: first, **Retrieve6()** backtracks 5 control points to obtain the local control point list $[q_{p_{nbr}-5}, q_{p_{nbr}-4}, \ldots, q_{p_{nbr}}]^\top$. Then, the control points of its velocity and acceleration curves are calculated:

$$v_i = \frac{q_{i+1} - q_i}{\Delta t}, \quad a_i = \frac{v_{i+1} - v_i}{\Delta t} \tag{14}$$

Due to the excessive volume of the convex hull formed by the B-spline control points, we utilize a method similar to (5) to transform them into control points under the MINVO basis, denoted as v_{mv_i} and a_{mv_i}. If the local v_{mv_i} and a_{mv_i} do not exceed $v_{\max}$ and $a_{\max}$, respectively, the new node p_{nbr} is deemed to satisfy the dynamic constraints.

Through the aforementioned steps, the improved A* algorithm considering dynamics can generate a list of B-spline control points that form an efficient time-parameterized trajectory satisfying dynamic constraints and avoiding collisions with obstacles.

3.3. QCQP Optimization

Although we have generated a safe trajectory satisfying dynamic constraints using the BA*-BS algorithm, there is still significant room for improvement in its smoothness. We employ quadratic constraint quadratic programming (QCQP) to further optimize the coordinates of the B-spline control points, thereby enhancing trajectory trackability.

First, we optimize a segment of the trajectory between two waypoints. For the trajectory between w_j and w_{j+1}, the initial control points obtained through the search are denoted as $\mathbf{CPS} = [cp_0, cp_1, \cdots, cp_n]^\top$.

(1) Objective function:

$$f = \sum_{i=0}^{n-5} \int_{t_i}^{t_{i+1}} \left(\frac{d^3(p(t))}{dt^3} \right)^2 dt = \sum_{i=0}^{n-5} \text{Trace}\left(\mathbf{Q}_i \mathbf{M}_{bs6}^\top \mathbf{H} \mathbf{M}_{bs6} \mathbf{Q}_i \right) \tag{15}$$

The objective is to minimize $\int jerk^2 dt$, where $p(t)$ represents the B-spline trajectory, and $\mathbf{Q}_i = [q_i, q_{i+1}, \ldots, q_{i+5}]^\top$. $\mathbf{H}$ is a constant matrix related to Δt.

(2) Dynamic feasibility constraints:

$$abs\left(\mathbf{M}_{mv5}^{-1} \mathbf{M}_{bs5} \mathbf{D}_1 \mathbf{Q}_i \right) \leqslant \mathbf{V}_{\max}, \quad (i = 1, 2, \ldots, n-5)$$

$$abs\left(\mathbf{M}_{mv4}^{-1} \mathbf{M}_{bs4} \mathbf{D}_2 \mathbf{Q}_i \right) \leqslant \mathbf{A}_{\max}, \quad (i = 1, 2, \ldots, n-5) \tag{16}$$

where $\mathbf{D}_1$ and $\mathbf{D}_2$ are the first-order and second-order derivative matrices, respectively. The transformation from B-spline control points to MINVO control points, as achieved through (16), constrains velocity and acceleration within feasible ranges.

(3) Boundary constraints:

The initial and final states need to be constrained to ensure the continuity of the trajectory, where the initial state consists of the positions of the first five control points and the final state is constrained by position and velocity:

$$q_i = cp_i, \quad (i = 0, 1, \cdots, 4) \tag{17}$$

$$\frac{1}{120} \begin{pmatrix} 1 & 26 & 66 & 26 & 1 \\ -5 & -50 & 0 & 50 & 5 \end{pmatrix} \begin{pmatrix} q_{n-4} \\ q_{n-3} \\ q_{n-2} \\ q_{n-1} \\ q_n \end{pmatrix} = \begin{pmatrix} w_{j+1} \\ v_{j+1} \end{pmatrix} \tag{18}$$

Equation (17) imposes constraints on the initial state, while (18) imposes the terminal position and velocity constraints. $[q_{k-4}, q_{k-3}, \cdots, q_k]$ denotes the corresponding control points, ensuring that the position and velocity trajectories pass through w_j and v_j, respectively, while maintaining the continuity of motion.

(4) Safety constraints:

For a quintic uniform B-spline, the i-th segment is entirely determined by the six control points: $[q_i, q_{i+1}, q_{i+2}, q_{i+3}, q_{i+4}, q_{i+5}]$. Due to the significant influence of q_{i+2} and q_{i+3} on the trajectory, the line segment connecting q_{i+2} and q_{i+3} can be approximated as the i-th segment of the B-spline trajectory. Therefore, we can constrain the i-th segment of the B-spline trajectory within a sphere centered at $(q_{i+2} + q_{i+3})/2$ with a radius of $d_{esdf}((q_{i+2} + q_{i+3})/2)$, as shown in Figure 6. According to the conclusion of Section 3.1, we transform the local B-spline control points into local MINVO control points $\left[q_{mv_{i_0}}, q_{mv_{i_1}}, \ldots, q_{mv_{i_5}} \right]$ using (5).

$$\left\| q_{mv_{i_j}} - ct_i \right\|_2 \leqslant r_i, \quad (i = 2, 3, \ldots, n-3; j = 0, 1, \ldots, 5) \tag{19}$$

The center of the sphere $ct_i = (cp_{i+2} + cp_{i+3})/2$ is obtained from the initial control points, and $r_i = d_{esdf}(ct_i)$ represents the distance from ct_i to the nearest obstacle. Since the first five control points are determined by the initial state, i starts from 2.

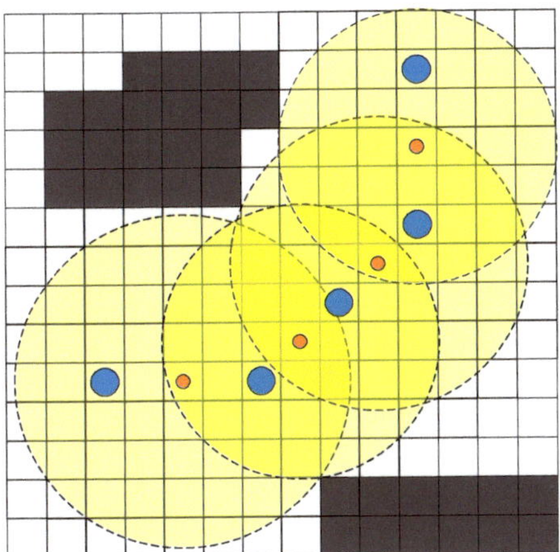

Figure 6. Illustration of collision-free balls (yellow circle) and control points (blue dots). The red circles represent the centers, and the black regions indicate obstacles.

After obtaining all the constraints for trajectory optimization, we express waypoint constraints as linear equality constraints ($\mathbf{A}_{eq}q = b_{eq}$), dynamic feasibility constraints as linear inequality constraints ($\mathbf{A}_{ie}q \leq b_{ie}$), and avoidance constraints as quadratic constraints ($q^\top \mathbf{Q}_{qc}q + \mathbf{A}_{qc}q \leq b_{qc}$). Thus, the trajectory optimization can be reformulated as follows:

$$
\begin{aligned}
\min \quad & q^T \mathbf{Q}_o q \\
\text{s.t.} \quad & \mathbf{A}_{eq}q = b_{eq}, \\
& \mathbf{A}_{ie}q \leq b_{ie}, \\
& q^\top \mathbf{Q}_{qc}q + \mathbf{A}_{qc}q \leq b_{qc}, \\
& q_i \in \Omega_i, \quad i = 0, 1, 2, \ldots, n.
\end{aligned}
\tag{20}
$$

where Ω_i denotes the feasible domain of the variable q_i to be optimized. Solving this problem yields the optimized B-spline control points.

3.4. Replanning Strategy

Based on receding horizon control (RHC) [20], we designed a replanning method to improve the real-time performance while avoiding degradation of QCQP numerical stability due to excessive control point variables. For the waypoint sequence $\{w_0, w_1, \ldots, w_{m-1}\}$, we search and optimize the trajectories between three consecutive waypoints w_j, w_{j+1}, w_{j+2}, $j \in 0, 1, \ldots, m - 3$, at each iteration. To ensure that the trajectory passes through w_{j+1}, we add intermediate waypoint constraints to the QCQP and modify (18) as follows:

$$
\frac{1}{120}
\begin{pmatrix} 1 & 26 & 66 & 26 & 1 \end{pmatrix}
\begin{pmatrix} q_{k-4} \\ q_{k-3} \\ q_{k-2} \\ q_{k-1} \\ q_k \end{pmatrix}
= \begin{pmatrix} w_{j+1} \end{pmatrix}
\tag{21}
$$

$$
\frac{1}{120}
\begin{pmatrix} 1 & 26 & 66 & 26 & 1 \\ -5 & -50 & 0 & 50 & 5 \end{pmatrix}
\begin{pmatrix} q_{n-4} \\ q_{n-3} \\ q_{n-2} \\ q_{n-1} \\ q_n \end{pmatrix}
= \begin{pmatrix} w_{j+2} \\ v_{j+2} \end{pmatrix}
\tag{22}
$$

The control points $[q_{k-4}, q_{k-3}, \ldots, q_k]$ correspond to the local b-spline curve that terminates at w_{j+1}. After optimizing the trajectories between these three consecutive waypoints, we retain only the segment of the trajectory from w_j to w_{j+1}, and we use the last five control points of this segment $[q_{k-4}, q_{k-3}, \ldots, q_k]$ as the initial state for planning the trajectories under the subsequent three waypoints $w_{j+1}, w_{j+2}, w_{j+3}$. Compared to initializing and optimizing the trajectories for all waypoints at once, this local replanning strategy can enhance the real-time performance and robustness of the proposed method.

4. FM-ACO Waypoint Sequencing

We consider the problem of optimizing the traversal sequence of multiple waypoints with the requirement of the shortest flight time as a variant of the traveling salesman problem (TSP). The TSP is a classical combinatorial optimization problem, aiming to find a path where a traveler can visit each city exactly once, then return to the starting city while minimizing the total travel cost. The solution to the TSP problem mainly involves two parts: constructing a cost matrix between every pair of cities and selecting a solution method. The difference in solution methods lies their ability to find the minimum-cost solution in a given cost matrix in a shorter time. Additionally, efficiently constructing a cost matrix that better reflects the actual problem is a more critical part of determining the optimal sequence of waypoints in this variant TSP problem.

For environments containing obstacles or threat zones, scholars mainly adopt two approaches. One still utilizes Euclidean distance to represent travel cost, but it incurs additional travel costs when the line connecting two points intersects with obstacles. This method is heavily influenced by parameters and cannot guarantee the existence of a feasible path between two points. The other approach involves using methods such as A* to generate the shortest discrete path between each pair of waypoints [13], which is then considered as the travel cost. This method also does not consider the dynamic constraints of the quadrotor, leading to potentially infeasible shortest discrete paths and the issue that the shortest path may not be equivalent to the shortest flight time. Moreover, the computational cost required for m waypoints, denoted as $C_m^2 = m(m-1)/2$, indicates that the computational cost will exhibit a quadratic increase with the increase in the number of waypoints.

To address these issues, we model the travel cost between two points using the fast marching (FM) method, transforming the shortest path problem into a variant of the TSP aiming at minimizing flight time. We then utilize ant colony optimization (ACO) to solve the problem, thereby obtaining the optimal sequence of waypoints.

4.1. FM-Constructed Cost Matrix

The fast marching (FM) method [21] is a special case of level set methods, which are numerical techniques used to simulate wavefront propagation. The FM method finds wide applications in fields such as image processing, curve evolution, fluid simulation, and path planning. It simulates the propagation of wavefronts by assuming that the wavefront propagates with a speed f along its normal direction. It computes the time at which the wavefront first reaches a target point, which represents the shortest time for the wavefront to travel from the starting point to the target point. Assuming a propagation speed $f > 0$, indicating that the wavefront only expands outward, and that f is constant over time and only depends on the spatial position, the evolution of the wavefront is described by the following equation:

$$|\nabla T(x)| = \frac{1}{f(x)} \tag{23}$$

where T is a function representing the arrival time, x denotes the position, and $f(x)$ describes the speed at x.

In order to apply the FM method to trajectory planning problems, we need to predefine a velocity map, assigning a speed value $f(x)$ to each grid position in the map. Then, we utilize (23) to simulate the wave expanding from the starting point, thus obtaining the arrival time $T(x)$ for each point on the map. By backtracking along the gradient descent direction of the arrival time from the target point to the starting point, we obtain a path with the minimum arrival time. Unlike other potential field-based methods, the FM method avoids local minima and possesses completeness and consistency [22].

In theory, it has the same time complexity as A*, which is $O(N \log(N))$. However, since our goal is to obtain the flight cost between each pair of waypoints, there is no need for path backtracking, and the number of computations required is also fewer. For example, for the m waypoints $\{w_0, w_1, \cdots, w_{m-1}\}$, we use FM with w_0 as the starting point. By expanding the wavefront once, we obtain the flight time from w_0 to $w_1, w_2, \cdots, w_{m-1}$. It effectively reduces redundant computations, thereby significantly lowering the time cost.

In order to ensure consistency with the planning method, FM uses the same velocity constraints as our trajectory planning method. When defining the velocity map, we maintain consistency with the adaptive expansion step (11) of the trajectory planning method:

$$f(d_{esdf}) = \begin{cases} 0, & d_{esdf} \leq r_{uav} \\ \frac{d_{esdf} - r_{uav}}{\Delta t}, & r_{uav} < d_{esdf} \leq r_{uav} + v_{\max} \cdot \Delta t \\ v_{\max}, & d_{esdf} \geq r_{uav} + v_{\max} \cdot \Delta t \end{cases} \tag{24}$$

where d_{esdf} represents the distance from the current position to the nearest obstacle, Δt is the time interval set for the quintic uniform B-spline curve in trajectory planning, and $f(d_{esdf})$ denotes the velocity at the current position. It is consistent with the calculation method in (11), where $step(d)$ represents the step size for one expansion within Δt time.

In addition to the flight time cost between two waypoints, there is also a turning time cost between two segments of the trajectory. For the three waypoints w_{j-1}, w_j, w_{j+1}, the quadrotor drone often requires some additional time to reduce speed or extend the trajectory in order to adjust its heading while flying from w_{j-1} to w_j. Considering the extreme case where the direction of $\overrightarrow{w_{j-1}w_j}$ is opposite to $\overrightarrow{w_jw_{j+1}}$, the additional time consumed by the quadrotor drone for decelerating to zero velocity at maximum acceleration and then accelerating back to maximum velocity is $v_{\max}/a_{\max}$. Hence, we define the turning time cost T_θ as follows:

$$T_\theta(j) = \frac{1 - \arccos(\theta_j)}{2} \cdot \frac{v_{\max}}{a_{\max}}, \quad \theta_j \in [0, \pi] \tag{25}$$

where θ_j is the angle between $\overrightarrow{w_{j-1}w_j}$ and $\overrightarrow{w_jw_{j+1}}$. Therefore, for $\{w_0, w_1, \cdots, w_{m-1}\}$, the total flight time cost is:

$$T_{sum} = \sum_{j=1}^{m-1} T_D(j) + \sum_{j=1}^{m-2} T_\theta(j) \tag{26}$$

where $T_D(j)$ is the flight time cost from w_{j-1} to w_j calculated by FM.

Compared to methods such as A*, which obtain path length costs, our method aims to maintain consistency with the trajectory planning method regarding flight time cost, thereby establishing a more realistic cost matrix.

4.2. ACO for Sequence Determination

Ant colony optimization (ACO) [23] is a nature-inspired algorithm that is used to solve combinatorial optimization problems. Due to the limited onboard energy of quadrotor drones, trajectory planning for large-scale routes is generally not performed in a single pass. ACO, known for its robustness and adaptability, is widely used in medium-scale TSP problems. Nevertheless, ACO suffers from slow convergence and sensitivity to parameters. To balance computational speed and solution quality, we opt for a variant of ACO known as the ant colony system (ACS) [24] to address our variant TSP problem. ACS significantly improves the convergence speed and solution quality compared to ACO, primarily through three improvements: optimizing path selection rules, the global pheromone update strategy, and the local pheromone update strategy.

(1) Path selection:

ACS provides a more direct strategy for balancing exploration and exploitation:

$$j = \begin{cases} \arg\max_{s \in J_k(i)} \left\{ [\tau(i,s)]^{\alpha} \cdot [\eta(i,s)]^{\beta} \right\}, & \text{if } q \le q_0 \text{ (exploitation)} \\ 0, & \text{otherwise (biased exploration)} \end{cases} \tag{27}$$

where $\tau(i,s)$ represents the pheromone concentration between waypoint i and waypoint s, $\eta(i,s)$ denotes the heuristic function, and $J_k(i)$ is the set of remaining selectable waypoints, α and β are the weighting coefficients for the pheromone concentration and heuristic function, respectively. q is a uniformly distributed random number between 0 and 1, and q_0 is the corresponding threshold ($0 \le q_0 \le 1$). If $q \le q_0$, the best waypoint is selected directly; otherwise, the next waypoint is chosen randomly according to the following:

$$p_k(i,j) = \begin{cases} \dfrac{[\tau(i,j)]^{\alpha} \cdot [\eta(i,j)]^{\beta}}{\sum\limits_{s \in J_k(i)} [\tau(i,s)]^{\alpha} \cdot [\eta(i,s)]^{\beta}}, & \text{if } j \in J_k(i) \\ 0, & \text{otherwise} \end{cases} \tag{28}$$

where $p_k(i,j)$ represents the probability that the k-th ant selects waypoint j given that it is at waypoint i.

For the classical TSP problem, $\eta(i,j)$ is typically set as the reciprocal of the distance between i and j. However, for our variant problem:

$$\eta(i,j) = \frac{1}{T_D(i,j) + T_\theta(r,i,j)} \tag{29}$$

where $T_D(i,j)$ is the flight time cost from waypoint i to waypoint j, and $T_\theta(r,i,j)$ is the turning time cost at waypoint i after selecting waypoint j, where r is the previous waypoint of i.

(2) Global pheromone update strategy:

In ACS, only the ant with the best history can update the pheromone:

$$\begin{cases} \tau(i,j) \leftarrow (1-\varepsilon) \cdot \tau(i,j) + \varepsilon \cdot \Delta\tau(i,j) \\ \Delta\tau_k(i,j) = \begin{cases} \dfrac{1}{T_{sum}(best)}, & \text{if } (i,j) \in \text{global best} \\ 0, & \text{otherwise} \end{cases} \end{cases} \tag{30}$$

where ε is the global pheromone evaporation coefficient, $\Delta\tau_k(i,j)$ represents the amount of pheromone left by the k-th ant on edge (i,j) in this iteration, and $T_{sum}(best)$ represents the flight time cost of the globally best waypoint sequence.

(3) Local pheromone update strategy:

To accelerate the convergence speed, ACS increases the rate of pheromone evaporation, and the pheromone concentration corresponding to each ant's path is updated after the ant traverses a path.

$$\begin{cases} \tau(i,j) \leftarrow (1-\rho) \cdot \tau(i,j) + \rho \cdot \Delta\tau(i,j) \\ \Delta\tau(i,j) = \dfrac{1}{m \cdot T_{greedy}} \end{cases} \tag{31}$$

where ρ is the local pheromone evaporation coefficient, m is the number of waypoints, and T_{greedy} is the total time cost of the waypoint sequence obtained using the greedy algorithm.

5. Experiment and Results

5.1. Experiment Settings

We verify the effectiveness of the multi-waypoint trajectory planning method proposed in Section 3 and the waypoint sequence determination method proposed in Section 4 through simulation experiments. The simulation experiment scene size is 25 m × 25 m × 6 m, with a map resolution of 0.2 m. We randomly generated 15 waypoints (including the specified ending point) and 50 to 100 cylindrical obstacles in the scene. The quadrotor drone starts from a specified starting point, traverses all waypoints, and flies to the specified ending point, with both the initial and final states being hover states. The maximum speed and maximum acceleration of the trajectory are limited to $2\,\mathrm{m/s}$ and $2\,\mathrm{m/s}^2$, respectively. All simulation experiments are conducted using a laptop computer equipped with Windows 11, an i7-12700H processor, and 16 GB of memory. For the trajectory planning part, we use MATLAB R2019a for the simulation and the Mosek solver to solve the QCQP problem. For the waypoint sequence determination part, we use PyCharm Community Edition 2021 for the simulation and utilize the scikit-fmm package to establish the FM cost matrix.

5.2. Trajectory Planning

In this section, we validate the effectiveness of our proposed trajectory planning algorithm using simulations. In the experiments, we predefine the sequence of waypoints. We compare our algorithm with the method proposed by Richter et al. in [5]. We use the A* algorithm to generate the initial path and conduct comparative experiments in various environments.

Figure 7 shows the trajectories generated by three different trajectory planning methods on the same random map. Among them, *polynomial* refers to the method in [5], BA*-BS refers to our proposed trajectory planning method, and A*-BS refers to the trajectory planning method without using a bidirectional search. Figure 8 illustrates the acceleration curves of trajectories generated using three different methods, which are used to assess trajectory smoothness. The figure shows that the acceleration peaks and the rate of change in acceleration (jerk) for the A*-BS and BA*-BS methods are slightly higher compared to the polynomial method, resulting in more aggressive trajectories.

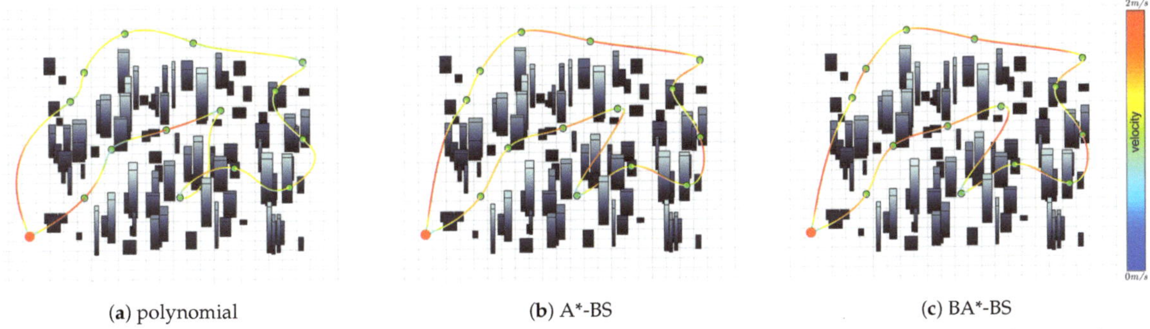

(**a**) polynomial   (**b**) A*-BS   (**c**) BA*-BS

Figure 7. Comparison of different trajectory planning methods on the same random map (green dots represent waypoints, the red dot represents the starting point and black regions indicate cylindrical obstacles).

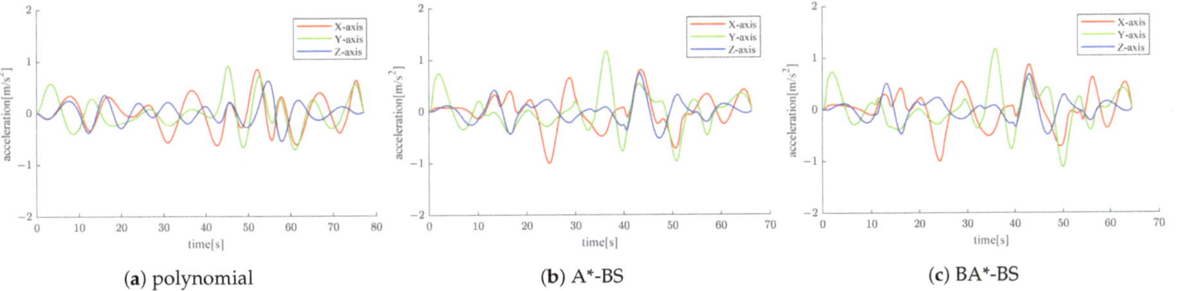

(**a**) polynomial   (**b**) A*-BS   (**c**) BA*-BS

Figure 8. Comparison of trajectory acceleration curves.

Table 1 presents the average data from 50 randomly generated maps, including trajectory time, planning time, energy cost ($\int \mathrm{jerk}^2\,dt$), and trajectory length. The simulation results indicate that all three methods can generate smooth, safe, and dynamically feasible trajectories. However, by comparing the data in Table 1, we found that both A*-BS and

BA*-BS showed significant improvements across several key metrics of trajectory plan ning compared to the polynomial method. BA*-BS performed better than the polynomial method in terms of the average trajectory time and planning time, significantly reducing the time required for trajectory planning and generating faster trajectories, thereby enhancing overall planning efficiency. Although BA*-BS has slightly higher energy consumption due to its bidirectional search strategy, the significant improvements in planning speed and trajectory optimization more than make up for this. Overall, BA*-BS demonstrates superior performance relative to A*-BS and the polynomial method, particularly by en hancing planning speed and generating faster trajectories. We analyze the underlying reasons below.

Table 1. Comparison of trajectory planning methods in 50 random environments.

Method	Avg.Traj.Time (s)	Avg.Planning Time (s)	Avg.Energy (m^2/s^5)	Avg.Length (m)
polynomial	75.521	25.820	20.284	91.927
A*-BS	66.960	7.298	20.672	90.068
1*BA*-BS	66.555	3.865	22.493	89.863

The polynomial method is essentially an extension of the method proposed in ref. [1] It first generates an initial discrete path and extracts waypoints, then it uses the method in ref. [1] to generate time–polynomial trajectories. It iteratively seeks the optimal time allocation, scaling the maximum velocity or acceleration to the boundaries of the dynamic constraints to minimize the trajectory time. However, if the trajectory collides with obstacles during this process, additional waypoints must be added to avoid collisions, resulting in significant time overhead. Moreover, its final result largely depends on the selection of initial discrete paths and waypoints.

Unlike the classic A* method, which does not consider the quadrotor's characteristics, both the A*-BS and BA*-BS methods simultaneously consider dynamic constraints and time optimality during the search process. They generate initial trajectories with dynamic feasibility and solve the problem of time allocation with non-uniform expansion steps Comparing A*-BS and BA*-BS, their average trajectory times are very close. However, since BA*-BS uses a bidirectional A* search, it can complete the search faster in complex environments. The average planning time of BA*-BS is reduced by about 47% in different environments. Although its energy is slightly higher than that of A*-BS, this is acceptable.

Furthermore, an analysis of the simulation results reveals that our proposed BA*-BS method is more likely to identify superior feasible trajectories within narrow passages, as illustrated in Figure 9. This is because traditional polynomial methods treat the turning points of nodes identified by the A* search as mandatory waypoints, necessitating the prior expansion of obstacles to prevent the generation of dynamically infeasible trajectories in overly narrow corridors. As the expansion coefficient increases, the search process tends to favor regions with sparse obstacles, which can result in trajectories that are unlikely to belong to the same homotopy class as the optimal trajectory, thereby precluding the possibility of deriving the optimal trajectory through time optimization. In contrast, our proposed method considers the dynamic feasibility of the trajectory during the search process, thoroughly evaluating whether the current narrow passage adheres to the drone's dynamic constraints. Additionally, our method accounts for the spatial relationships between waypoints, aiming to generate trajectories within the same homotopy class as the optimal trajectory, thereby reducing flight time and enhancing trajectory smoothness.

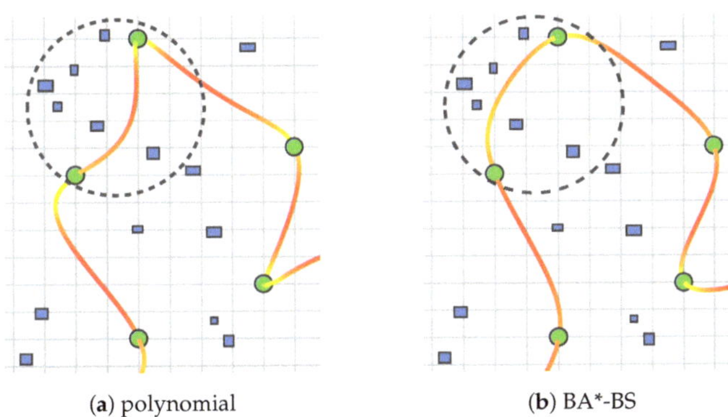

(a) polynomial   (b) BA*-BS

Figure 9. Comparison of trajectories when facing narrow passages using different methods.(green dots represent waypoints, blue regions indicate obstacles).

5.3. Waypoint Sequencing

In this section, we validate the effectiveness of the proposed waypoint sequencing method through simulation experiments. We compare our FM-ACO algorithm, which combines FM and ACO, with the baseline method A*-ACO that we set. First, we separately use FM and classical A* to build the time cost matrix and distance cost matrix. Then, we apply the ACO algorithm to determine the waypoint sequence with the minimum cost, using the same parameter settings recommended in [24]. Finally, we use the same BA*-BS to generate complete multi-waypoint trajectories and compare them. The parameters for the simulation experiments in this section are identical to those in Section 5.1. Similarly, we create 50 random environments for comparative experiments. Figure 10 shows the trajectories generated by the two algorithms for determining the waypoint sequences in one of the random environments. Table 2 presents the average data for these 50 environments. In addition to the trajectory planning data, we also recorded the cost matrix time and ACO time, representing the time taken to build the cost matrix and the time used for ACO iterations, respectively.

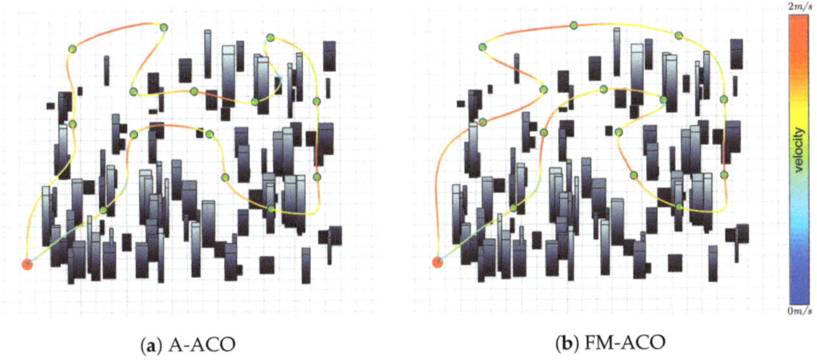

(a) A-ACO   (b) FM-ACO

Figure 10. Comparison of final trajectories using different waypoint sequencing methods.

Based on the data in Table 2, FM-ACO demonstrates significant advantages over the A*-ACO method. For the final trajectories generated by BA*-BS, the waypoint sequences determined by FM-ACO outperform those from A*-ACO in terms of average trajectory time. This indicates that the FM method provides a cost matrix that more accurately reflects actual flight costs compared to the A* method. Additionally, FM-ACO shows improvements in the planning time and energy consumption. This enhancement is due to FM's method of generating a velocity map that maintains constraint consistency with BA*-BS. By assigning lower velocities to grid points closer to obstacles, FM-ACO ensures that larger free spaces have lower cost values, which enhances the smoothness of the final trajectory and reduces planning time.

Table 2. Comparison of waypoint sequencing methods in 50 random environments.

Method	Avg.Traj.Time (s)	Avg.Planning Time (s)	Avg.Energy (m^2/s^5)	Avg.Length (m)	Avg.Cost Matrix Time (s)	Avg.ACO Time (s)
A*-ACO	69.003	8.174	29.758	90.135	9.190	0.520
FM-ACO	67.329	3.898	24.496	90.515	4.127	0.523

Furthermore, FM has a significant advantage in terms of the time it takes to build the cost matrix compared to the A* algorithm. The A* algorithm needs to search for paths between each pair of waypoints separately to obtain distance costs, leading to longer computation times. In contrast, FM can expand the entire map at once using a waypoint as the origin of wavefront expansion, computing the time cost for all other waypoints relative to the origin in a single pass. This approach of expanding the map in a single pass significantly improves the search efficiency. According to the experimental data, FM reduces the time required to build the cost matrix by 55% compared to the A* algorithm.

5.4. Robustness Analysis

In this section, we further analyze the robustness of the proposed method by investigating the effects of varying waypoint numbers and obstacle densities on the proposed method. Since the waypoint sequencing method uses the FM algorithm to expand nodes across the entire map, it can always construct a flight cost matrix between waypoints and solve it using the ACO method, provided that feasible paths exist between waypoints. Therefore, this section will focus primarily on a detailed analysis of the final trajectories generated using the trajectory planning method.

For each environmental configuration, we first generate the optimal waypoint sequence using FM-ACO and then apply the proposed BA*-BS method to produce multi-waypoint trajectories. It is important to note that for a given number of waypoints, the coordinates of the waypoints remain consistent. In all tested scenarios, our method efficiently generates dynamically feasible and safe trajectories within seconds, as demonstrated in Figure 11.

Table 3 presents the planning time (including only the BA*-BS planning time) and trajectory time for trajectories generated using the BA*-BS method under varying numbers of waypoints and obstacle densities. The results indicate that, with the number of waypoints and their coordinates held constant, an increase in obstacle density extends the search time of the BA*-BS method, leading to a longer planning time. Additionally, as the number of obstacles increases, the likelihood of the optimal path being obstructed also increases, resulting in a longer trajectory time. It is important to note that these results are influenced by the distribution of obstacles. For example, even if the number of obstacles increases, they may be concentrated in areas unrelated to the trajectory, which can affect the planning results.

Table 3. Comparison of trajectory times and planning times across different numbers of waypoints (m) and cylindrical obstacle ($N.CO$) configurations.

m	$N.CO = 50$		$N.CO = 100$		$N.CO = 150$	
	Traj.T(s)	Plan.T(s)	Traj.T(s)	Plan.T(s)	Traj.T(s)	Plan.T(s)
5	40.95	0.845	45	2.81	44.55	2.925
10	51.75	1.24	56.7	4.465	61.65	5.673
15	68.85	1.451	70.2	4.055	75.15	5.441

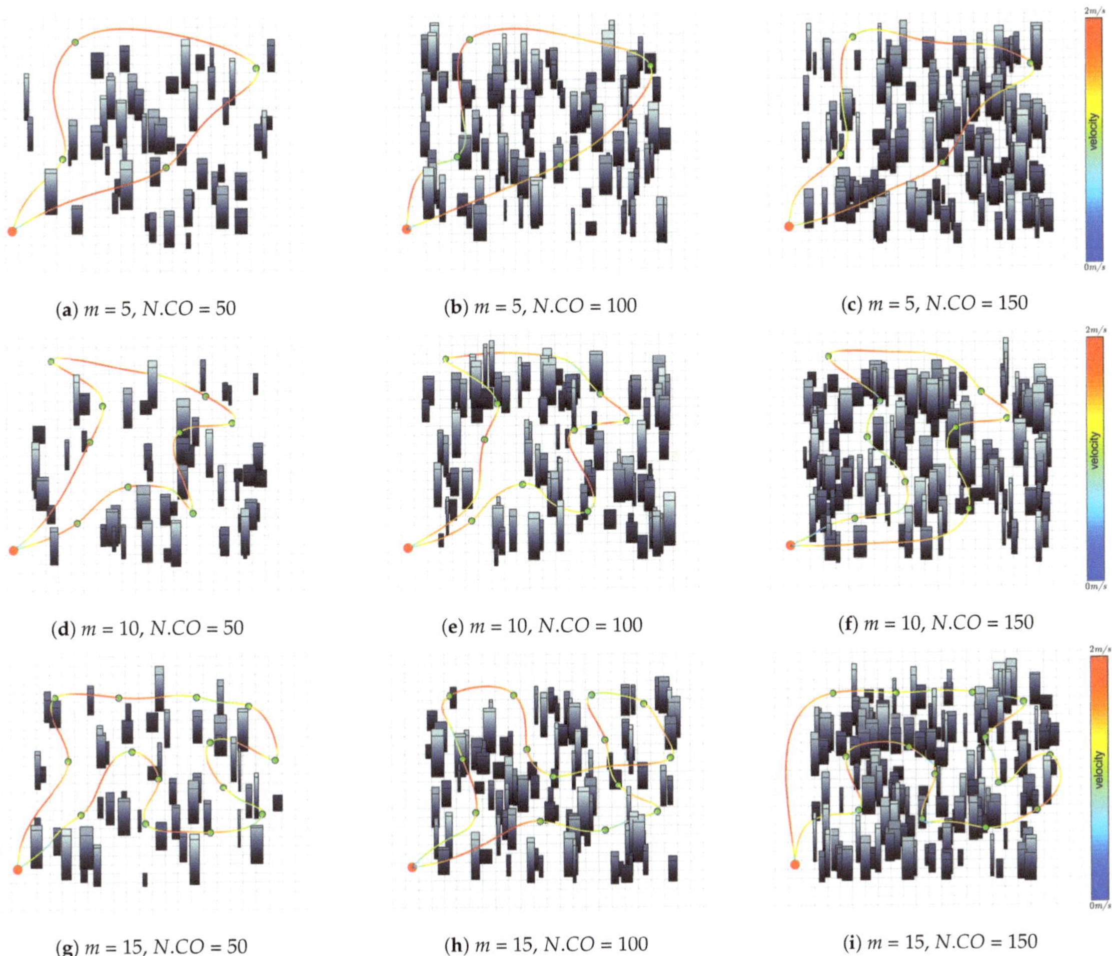

(a) $m = 5$, $N.CO = 50$

(b) $m = 5$, $N.CO = 100$

(c) $m = 5$, $N.CO = 150$

(d) $m = 10$, $N.CO = 50$

(e) $m = 10$, $N.CO = 100$

(f) $m = 10$, $N.CO = 150$

(g) $m = 15$, $N.CO = 50$

(h) $m = 15$, $N.CO = 100$

(i) $m = 15$, $N.CO = 150$

Figure 11. Comparison of final trajectories across different numbers of waypoints (m) and cylindrical obstacle ($N.CO$) configurations.

Furthermore, the number of waypoints also impacts the experimental results. For a map of the same size, there is a positive correlation between the number of waypoints and the trajectory time. Regarding planning time, although the BA*-BS method requires more searches with an increasing number of waypoints, the distance between waypoints decreases, which reduces the number of nodes expanded per search. As a result, the overall planning time does not necessarily increase with the number of waypoints.

5.5. Complexity Analysis

The complexity of an algorithm significantly affects its scalability. In this section, we analyze the time complexity of the algorithm.

According to our theoretical analysis, the primary time consumption of our proposed trajectory planning method, BA*-BS, is attributed to the search process. The time complexity of BA*-BS is $O(m \cdot n \log n)$, where m is the number of waypoints and n is the number of nodes expanded in a single BA*-BS search. Since the search method of BA*-BS is an improvement upon A*, the value of n is influenced by factors such as the distance between waypoints, obstacle density, obstacle distribution, map size, and resolution.

For the waypoint sequencing method, the time complexity of computing the cost matrix using FM is $O(m \cdot (n \log n))$, where m represents the number of waypoints and n is the total number of cells in the three-dimensional grid map. The complexity is relatively insensitive to the density and distribution of obstacles. The time complexity of the ACO iterative solution is $O(m^2)$ (with a fixed number of ants and iterations). However, due to constraints in onboard energy, the number of waypoints traversed in a single mission is typically limited.

Figure 12 illustrates the variation in the FM-ACO computation time with the number of waypoints for a $25\,\text{m} \times 25\,\text{m} \times 6\,\text{m}$ map with a resolution of 0.2 m. Figure 12a shows that the computation time for the cost matrix scales approximately linearly with the number of waypoints. Figure 12b displays the curve for ACO iterations set to 100, which follows an m^2 growth trend.

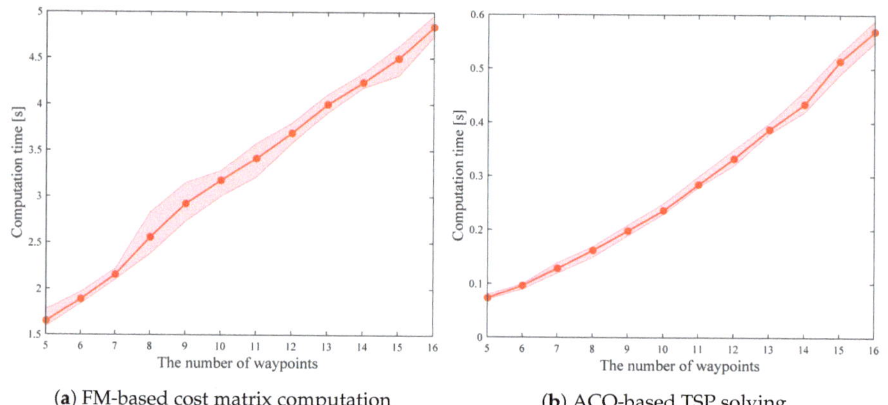

(**a**) FM-based cost matrix computation   (**b**) ACO-based TSP solving

Figure 12. Computation time of FM-ACO for varying numbers of waypoints.

Additionally, the time complexity of ACO is also influenced by its convergence speed. Figure 13 shows the convergence curves of ACO for different waypoint numbers, indicating that ACO converges within 100 iterations when the number of waypoints $m < 20$.

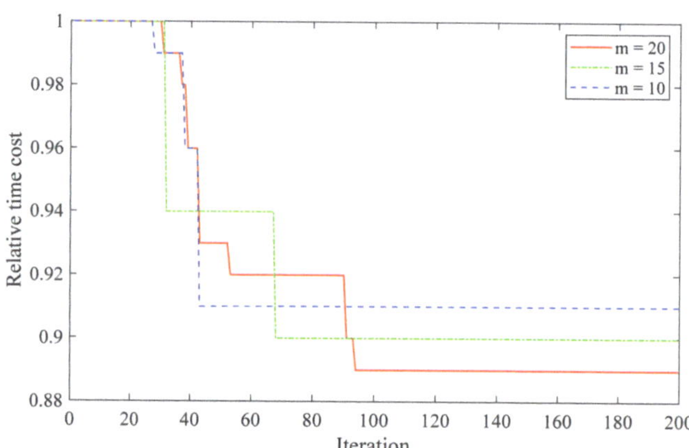

Figure 13. ACO convergence curves.

6. Discussion

This paper proposes a new multi-waypoint motion planning framework to determine the waypoint sequence that minimizes flight time and generates safe, dynamically feasible multi-waypoint trajectories in complex environments with multiple waypoints. However, our research primarily focuses on static environments and fixed waypoints, which may have certain limitations in practical applications. To better meet real-world demands, future work needs to deepen and expand in several directions.

1. Different shapes of obstacles may introduce varying computational complexities. In our experiments, we only used cylindrical obstacles, which, to some extent, facilitate the efficiency of the search process of A*. However, in the real world, obstacles come in various forms, such as maze-like obstacles, which can significantly increase the search time of A*.

2. In reality, waypoints and environments can be dynamic. In our experiments, we only consider static environments and obstacles. In actual missions, dynamic environments and waypoints may be encountered. Future work will focus on improving waypoint sequencing and trajectory planning methods to address these aspects, further enhancing real-time performance and expanding the applicability of our framework.

3. The dynamic model of quadrotor drones can be made more realistic. Our current dynamic model is relatively simple, considering only three-dimensional velocity and acceleration limits. However, in reality, more practical factors need to be considered, including maximum motor speed and thrust, aerodynamic effects, and battery power. These factors significantly impact the quadrotor drone's motion state and trajectory planning.

4. Multi-quadrotor coordination is also an important research direction. In practical applications, multiple quadrotors often need to work together to complete complex tasks. For example, in search and rescue missions, multiple quadrotors need to coordinate searches and task allocation, posing higher demands on waypoint sequencing and trajectory planning. Future work can explore multi-waypoint motion planning problems considering multi-quadrotor coordination, studying efficient coordination strategies and distributed planning algorithms.

Additionally, the performance of the BA*-BS method is influenced by parameters including the time interval Δt of the uniform B-spline and the map resolution. If the map resolution is low, or if the waypoints are too close to obstacles, the gridded control point coordinates might fall within obstacle grids or result in all neighboring nodes being infeasible during the initial node expansion, often leading to planning failure. Increasing the map resolution is the simplest way to reduce such failures, although it comes with a higher computational cost. Furthermore, if the distance between waypoints is too long, it extends the search time and generates longer B-spline control point sequences, which can impact the numerical stability of the QCQP-optimized trajectories. In such cases, increasing the time interval Δt of the uniform B-spline curve can be a feasible solution to improve search efficiency and numerical stability, even though it may slightly compromise optimality.

In conclusion, although the multi-waypoint motion planning framework proposed in this paper addresses the problem of finding the waypoint sequence that minimizes flight time and generates safe multi-waypoint trajectories to some extent, further research and optimization are needed for practical applications. Future work will aim to solve the mentioned challenges, apply the framework to actual quadrotor drone flight missions, and develop a more efficient, practical, and intelligent multi-waypoint motion planning system.

7. Conclusions

This paper presents a novel framework for multi-waypoint motion planning for quadrotor drones. Faced with cluttered environments containing multiple waypoints that must be traversed, we decouple the problem into waypoint sequencing and multi-waypoint trajectory planning. For multi-waypoint trajectory planning, considering the spatial relationships between waypoints, we propose a bidirectional A* search method based on B-splines (BA*-BS) to generate continuous initial trajectories that satisfy dynamic feasibility. Subsequently, the trajectories are further optimized for smoothness using QCQP to enhance the trackability of the trajectories and reduce energy consumption. Moreover, we design a replanning strategy to improve the numerical stability of QCQP and the real-time performance of trajectory planning. For the waypoint sequencing problem, we introduce an FM-ACO method. Initially, a velocity map consistent with the constraints of the backend trajectory planning is established. Then, the fast marching method is employed to construct a time cost matrix. Finally, ACO is used to find the waypoint sequence with the shortest trajectory time. The proposed methods are compared with existing approaches through extensive simulation experiments, demonstrating their effectiveness.

Author Contributions: Conceptualization, D.S., J.S. and M.G.; methodology, D.S., J.S. and M.G.; software, D.S. and X.Y.; validation, D.S. and X.Y.; formal analysis, D.S., J.S. and M.G.; investigation, D.S. and X.Y.; resources, J.S. and M.G.; data curation, D.S. and X.Y.; writing—original draft preparation, D.S.; writing—review and editing, D.S., J.S. and M.G.; visualization, D.S. and X.Y.; supervision, J.S. and M.G.; project administration, J.S. and M.G.; funding acquisition, J.S. All authors have read and agreed to the published version of the manuscript.

Funding: This research was funded by the Jiangsu Provincial Key Research and Development Program (NO.BE2022100).

Data Availability Statement: The original contributions presented in this study are included in the article. Any further inquiries can be directed to the corresponding author.

Conflicts of Interest: The authors declare no conflicts of interest.

References

1. Mellinger, D.; Kumar, V. Minimum snap trajectory generation and control for quadrotors. In Proceedings of the 2011 IEEE International Conference on Robotics and Automation, Shanghai, China, 9–13 May 2011; pp. 2520–2525.
2. Hart, P.E.; Nilsson, N.J.; Raphael, B. A formal basis for the heuristic determination of minimum cost paths. *IEEE Trans. Syst. Sci. Cybern.* **1968**, *4*, 100–107. [CrossRef]
3. Karaman, S.; Frazzoli, E. Sampling-based algorithms for optimal motion planning. *Int. J. Robot. Res.* **2011**, *30*, 846–894. [CrossRef]
4. Kavraki, L.E.; Svestka, P.; Latombe, J.C.; Overmars, M.H. Probabilistic roadmaps for path planning in high-dimensional configuration spaces. *IEEE Trans. Robot. Autom.* **1996**, *12*, 566–580. [CrossRef]
5. Richter, C.; Bry, A.; Roy, N. Polynomial trajectory planning for aggressive quadrotor flight in dense indoor environments. In *Robotics Research: The 16th International Symposium ISRR*; Springer: Berlin/Heidelberg, Germany, 2016; pp. 649–666.
6. Gao, F.; Wu, W.; Lin, Y.; Shen, S. Online safe trajectory generation for quadrotors using fast marching method and bernstein basis polynomial. In Proceedings of the 2018 IEEE International Conference on Robotics and Automation (ICRA), Brisbane, Australia, 21–25 May 2018; pp. 344–351.
7. Zhou, B.; Gao, F.; Wang, L.; Liu, C.; Shen, S. Robust and efficient quadrotor trajectory generation for fast autonomous flight. *IEEE Robot. Autom. Lett.* **2019**, *4*, 3529–3536. [CrossRef]
8. Tang, L.; Wang, H.; Liu, Z.; Wang, Y. A real-time quadrotor trajectory planning framework based on B-spline and nonuniform kinodynamic search. *J. Field Robot.* **2021**, *38*, 452–475. [CrossRef]
9. Liu, S.; Atanasov, N.; Mohta, K.; Kumar, V. Search-based motion planning for quadrotors using linear quadratic minimum time control. In Proceedings of the 2017 IEEE/RSJ international Conference on Intelligent Robots and Systems (IROS), Vancouver, BC, Canada, 24–28 September 2017; pp. 2872–2879.
10. Rousseau, G.; Maniu, C.S.; Tebbani, S.; Babel, M.; Martin, N. Minimum-time B-spline trajectories with corridor constraints. Application to cinematographic quadrotor flight plans. *Control Eng. Pract.* **2019**, *89*, 190–203. [CrossRef]
11. Foehn, P.; Romero, A.; Scaramuzza, D. Time-optimal planning for quadrotor waypoint flight. *Sci. Robot.* **2021**, *6*, eabh1221. [CrossRef] [PubMed]
12. Penicka, R.; Scaramuzza, D. Minimum-time quadrotor waypoint flight in cluttered environments. *IEEE Robot. Autom. Lett.* **2022**, *7*, 5719–5726. [CrossRef]
13. Yu, X.; Chen, W.N.; Gu, T.; Yuan, H.; Zhang, H.; Zhang, J. ACO-A*: Ant colony optimization plus A* for 3-D traveling in environments with dense obstacles. *IEEE Trans. Evol. Comput.* **2018**, *23*, 617–631. [CrossRef]
14. Fu, J.; Sun, G.; Liu, J.; Yao, W.; Wu, L. On hierarchical multi-UAV dubins traveling salesman problem paths in a complex obstacle environment. *IEEE Trans. Cybern.* **2023**, *54*, 123–135. [CrossRef] [PubMed]
15. Tsai, H.C.; Hong, Y.W.P.; Sheu, J.P. Completion time minimization for UAV-enabled surveillance over multiple restricted regions. *IEEE Trans. Mob. Comput.* **2022**, *22*, 6907–6920. [CrossRef]
16. Kong, S.; Wu, Z.; Qiu, C.; Tian, M.; Yu, J. An FM*-Based Comprehensive Path Planning System for Robotic Floating Garbage Cleaning. *IEEE Trans. Intell. Transp. Syst.* **2022**, *23*, 23821–23830. [CrossRef]
17. Tordesillas, J.; How, J.P. MINVO basis: Finding simplexes with minimum volume enclosing polynomial curves. *Comput.-Aided Des.* **2022**, *151*, 103341. [CrossRef]
18. Qin, K. General matrix representations for B-splines. In Proceedings of the Pacific Graphics' 98, Sixth Pacific Conference on Computer Graphics and Applications (Cat. No. 98EX208), Singapore, 26–29 October 1998; pp. 37–43.

19. Pohl, I. *Bi-Directional Search*; IBM TJ Watson Research Center: Yorktown Heights, NY, USA, 1970.
20. Bellingham, J.; Richards, A.; How, J.P. Receding horizon control of autonomous aerial vehicles. In Proceedings of the of the 2002 American control conference (IEEE Cat. No. CH37301), Anchorage, AK, USA, 8–10 May 2002; Volume 5, pp. 3741–3746.
21. Valero-Gomez, A.; Gomez, J.V.; Garrido, S.; Moreno, L. The path to efficiency: Fast marching method for safer, more efficient mobile robot trajectories. *IEEE Robot. Autom. Mag.* **2013**, *20*, 111–120. [CrossRef]
22. LaValle, S.M. *Planning Algorithms*; Cambridge University Press: Cambridge, UK, 2006.
23. Dorigo, M.; Blum, C. Ant colony optimization theory: A survey. *Theor. Comput. Sci.* **2005**, *344*, 243–278. [CrossRef]
24. Dorigo, M.; Gambardella, L.M. Ant colony system: A cooperative learning approach to the traveling salesman problem. *IEEE Trans. Evol. Comput.* **1997**, *1*, 53–66. [CrossRef]

Disclaimer/Publisher's Note: The statements, opinions and data contained in all publications are solely those of the individual author(s) and contributor(s) and not of MDPI and/or the editor(s). MDPI and/or the editor(s) disclaim responsibility for any injury to people or property resulting from any ideas, methods, instructions or products referred to in the content.

Article

Cooperative Path Planning for Multi-UAVs with Time-Varying Communication and Energy Consumption Constraints

Jia Guo *, Minggang Gan and Kang Hu

State Key Laboratory of Intelligent Control and Decision of Complex Systems, School of Automation,
Beijing Institute of Technology, Beijing 100081, China; aganbit@126.com (M.G.); 3120215456@bit.edu.cn (K.H.)
* Correspondence: 3220205114@bit.edu.cn

Abstract: In the field of Unmanned Aerial Vehicle (UAV) path planning, designing efficient, safe and feasible trajectories in complex, dynamic environments poses substantial challenges. Traditional optimization methods often struggle to address the multidimensional nature of these problems, particularly when considering constraints like obstacle avoidance, energy efficiency, and real-time responsiveness. In this paper, we propose a novel algorithm, Dimensional Learning Strategy and Spherical Motion-based Particle Swarm Optimization (DLS-SMPSO), specifically designed to handle the unique constraints and requirements of cooperative path planning for Multiple UAVs (Multi-UAVs). By encoding particle positions as motion paths in spherical coordinates, the algorithm offers a natural and effective approach to navigating multidimensional search spaces. The incorporation of a Dimensional Learning Strategy (DLS) enhances performance by minimizing particle oscillations and allowing each particle to learn valuable information from the global best solution on a dimension-by-dimension basis. Extensive simulations validate the effectiveness of the DLS-SMPSO algorithm, demonstrating its capability to consistently generate optimal paths. The proposed algorithm outperforms other metaheuristic optimization algorithms, achieving a feasibility ratio as high as 97%. The proposed solution is scalable, adaptable, and suitable for real-time implementation, making it an excellent choice for a broad range of cooperative multi-UAV applications.

Keywords: multi-UAVs; cooperative path planning; time-varying communication constraint; DLS-SMPSO; energy consumption constraint

Citation: Guo, J.; Gan, M.; Hu, K. Cooperative Path Planning for Multi-UAVs with Time-Varying Communication and Energy Consumption Constraints. *Drones* **2024**, *8*, 654. https://doi.org/10.3390/drones8110654

Academic Editors: Jihong Zhu, Heng Shi, Zheng Chen and Minchi Kuang

Received: 30 September 2024
Revised: 2 November 2024
Accepted: 4 November 2024
Published: 7 November 2024

Copyright: © 2024 by the authors. Licensee MDPI, Basel, Switzerland. This article is an open access article distributed under the terms and conditions of the Creative Commons Attribution (CC BY) license (https://creativecommons.org/licenses/by/4.0/).

1. Introduction

Unmanned Aerial Vehicles (UAVs) have become increasingly prevalent in a variety of applications, ranging from environmental monitoring [1] and agricultural surveillance [2] to search and rescue operations [3] and military missions [4]. The ability to deploy multi-UAVs in a coordinated manner significantly enhances the efficiency and effectiveness of these missions. However, the complexity of ensuring cooperation among multi-UAVs introduces several challenges, particularly in terms of path planning [5–7] and maintaining reliable communication [8].

In dynamic and unpredictable environments, the communication links between UAVs can fluctuate significantly due to factors such as obstacles, interference, and the mobility of the UAVs themselves. These time-varying communication [9] constraints present a major challenge for effective path planning, as UAVs must continuously adapt their trajectories to maintain connectivity while still accomplishing their mission objectives. Research on communication constraints in multi-UAV systems has primarily focused on ensuring reliable data exchange [10] and maintaining network connectivity [11]. Techniques such as relay placement [12] and network topology optimization [13] have been explored to enhance communication reliability. However, these methods often assume static or predictable environments, which is not always the case in real-world applications. Several studies have proposed integrating communication models with path planning algorithms [14–16] to address time-varying communication constraints. These approaches generally involve real-time evaluation of communication link quality and dynamic adjustments to UAV paths to ensure continuous connectivity. However, despite these advancements, there is still a need for more robust and adaptive methods capable of addressing the complexities of real-time, dynamic environments.

The field of cooperative path planning for multi-UAV systems has been widely explored, with numerous approaches developed to tackle various facets of the problem. Traditional methods like A* and Dijkstra's algorithms have been commonly applied to path planning in static environments [17], but they struggle to perform effectively in dynamic settings where time-varying constraints play a significant role. More recent methods have employed heuristic and metaheuristic approaches, including Genetic Algorithms (GA) [18,19], Ant Colony Optimization (ACO) [20–22], Grey Wolf Optimizer (GWO) [23–25], and Artificial Bee Colony (ABC) [26–28] to enhance path planning capabilities under dynamic conditions. Particle Swarm Optimization (PSO) [29–32] has emerged as a powerful tool for optimization problems, including path planning for UAVs. Its capability to discover near-optimal solutions in complex search spaces makes it well-suited for multi-UAV path planning [33]. However, the standard PSO algorithm lacks the intrinsic ability to manage communication constraints, which are crucial for the coordination and effective operation of UAVs in collaborative missions.

Traditional path-planning algorithms often struggle to address the complexities of time-varying communication constraints, highlighting the need for advanced methods that can handle dynamic network conditions [34] while ensuring robust communication among UAVs. PSO [35], a nature-inspired optimization technique, has demonstrated considerable potential in path planning applications due to its simplicity and effectiveness in navigating complex search spaces. However, standard PSO algorithms require modifications to adequately handle the challenges posed by time-varying communication constraints in multi-UAV systems, ensuring both connectivity and optimal path planning.

To tackle these challenges, various enhanced PSO variants have been developed. These improvements generally fall into several key areas: adaptive parameter control, hybridization with other optimization techniques, the introduction of multi-swarm or cooperative strategies, and the incorporation of novel operators. Ref. [36] presents an Adaptive Quantum-behaved PSO (AQPSO) algorithm, which is applied to UAV path planning tasks, demonstrating notable improvements in both convergence speed and solution quality. Ref. [37] introduces a Cooperative Multiple Swarm PSO (CMSPSO) method, integrating traditional PSO with a cooperative strategy among multiple swarms, resulting in enhanced convergence efficiency and superior solution quality for UAV path planning applications. Ref. [38] proposes an Adaptive Mutation PSO (AMPSO) algorithm, which integrates mutation operators to prevent premature convergence and improve solution diversity, particularly in challenging UAV path planning scenarios. Ref. [39] introduces a Chaotic PSO (CPSO), incorporating chaos theory into the standard PSO algorithm, thereby enhancing global search capabilities and avoiding local minima in UAV path optimization. Ref. [40] proposes a Multi-Objective PSO (MOPSO) approach specifically tailored for UAV path planning, efficiently optimizing multiple conflicting objectives. Meanwhile, Ref. [41] presents a Quantum-behaved PSO (QPSO) algorithm designed for dynamic environments. By incorporating quantum mechanics principles, QPSO significantly enhances the swarm's exploration capabilities, leading to more robust and adaptive path planning under uncertain and fluctuating conditions. Ref. [42] investigates the application of Angle-encoded PSO (APSO) for optimizing UAV deployment in search and rescue missions. This approach encodes the search area using angular parameters, allowing UAVs to swiftly adjust their search patterns in response to changing environmental conditions, which enhances the efficiency and speed of rescue operations. Additionally, Ref. [43] introduces a hybrid PSO-GA method designed to optimize the coverage and connectivity of UAV ad hoc networks. By combining the strengths of PSO and GA, this hybrid algorithm improves network performance, specifically in terms of coverage and communication latency. Refs. [44–46] combine different metaheuristic algorithms to enhance performance in UAV path planning. The above UAV path planning algorithm is summarized as shown in Table 1.

Table 1. The classification of UAV path planning algorithms.

Algorithms	Example
Heuristic	A^*, Dijkstra's, and so on
Metaheuristic	GA, ACO, PSO, ABC, GWO, and so on
Hybrid Metaheuristic	PSO-GA, PSO-GWO, and so on
PSO variants	AQPSO, CMSPSO, AMPSO, CPSO, MOPSO, and so on

To address the challenge of multi-UAV collaborative path planning under communication constraints, Ref. [47] proposed the Comprehensive Learning and Dynamic Multiswarm PSO (CL-DMSPSO) algorithm. This approach facilitates the effective planning of high-quality paths for UAVs, ensuring optimized performance in constrained environments. However, it primarily addresses communication constraints alone. In real-world UAV missions, communication and energy consumption constraints are often interdependent. For instance, maintaining stable communication may require the UAV to follow a path that increases energy consumption, while conserving energy might force the UAV to operate in areas with weaker communication links.

Thus, an effective path-planning algorithm must strike a balance between these constraints, optimizing both communication reliability and energy efficiency. Given the limitations in communication and energy consumption, certain UAVs may be unable to independently perform path planning. Consequently, effective coordination among all UAVs becomes crucial in the path-planning process. In this paper, we introduce the DLS-SMPSO algorithm, specifically designed for cooperative path planning in multi-UAV systems with time-varying communication and energy consumption constraints. The DLS-SMPSO algorithm tackles the complex challenges of UAV path planning by encoding each particle's position as a motion path in spherical coordinates, enabling efficient exploration of the search space. By incorporating a DLS into velocity updates, the algorithm minimizes oscillations, allowing particles to learn from the global best solution dimension by dimension. Simulation results validate the algorithm's feasibility and effectiveness, showcasing its superior performance in handling these constraints.

The key contributions of our work are outlined as follows:

(1) A novel DLS-SMPSO algorithm is proposed to address the challenges of collaborative path planning for multi-UAVs. By simply adjusting the angles in spherical coordinates, the particle's orientation can be modified directly without the need to decompose these changes into cartesian components. As a result, the DLS-SMPSO algorithm can explore the search space more naturally, facilitating smoother transitions and more precise adjustments in particle positions.

(2) In the DLS-SMPSO algorithm, particle positions are encoded as motion paths using spherical coordinates, rather than the conventional cartesian coordinates employed in standard PSO. This spherical encoding is particularly advantageous for UAV path planning, as it allows for more intuitive and direct manipulation of trajectories, resulting in more efficient optimization and improved path generation.

(3) The integration of DLS minimizes particle oscillation during the evolutionary process by enabling each particle to learn from the global best solution in a dimension-by-dimension manner. This strategy helps prevent premature stagnation, leading to a more stable and efficient optimization process. In addition, the algorithm seamlessly incorporates constraint handling mechanisms, such as obstacle avoidance and boundary enforcement, within the optimization process. This guarantees that the generated solutions are not only optimal but also feasible and safe for practical real-world applications.

The rest of this paper is organized as follows: Section 2 offers a comprehensive overview of the multi-UAV path planning problem formulation and reviews related work on objective function design. Section 3 introduces the PSO variant algorithms and details the proposed DLS-SMPSO approach. Section 4 presents the simulation setup and results, demonstrating the effectiveness of our approach. Finally, Section 5 concludes the paper and outlines potential directions for future research.

2. Problem Formulation

In this section, we begin by outlining the cooperative path planning problem for multi-UAVs, incorporating various essential constraints in Section 2.1. Following that, we provide a detailed explanation of the path representation in Section 2.2. Building on this foundation, we formulate the objective function for the specified path planning problem in Section 2.3.

2.1. Problem Description

The UAV path planning problem involves determining optimal paths for n UAVs, each starting from a specific location and aiming to reach Unmanned Ground Vehicle 0 (UGV0). The goal is to minimize various factors such as path length, energy consumption, and collision risks while ensuring that the paths are feasible within the UAV's kinematic and dynamic constraints. Additionally, the paths must avoid both static and dynamic obstacles in the environment. In more detail, the UAV path planning problem entails finding the most efficient trajectories for five UAVs to travel from their respective starting points to UGV0. The cost function to be minimized typically includes critical aspects such as total distance traveled, energy consumption, and the risk of collisions with obstacles or other UAVs. The planned paths must adhere to each UAV's physical limitations, including speed, acceleration, and turning radius, ensuring that the maneuvers are both possible and safe. Obstacle avoidance plays a central role, requiring the UAVs to navigate around static obstacles like buildings and trees, as well as dynamic obstacles such as other moving UAVs or changing environmental factors. The problem also demands consideration of communication constraints [9,48], especially in scenarios where UAVs must maintain connectivity with ground stations or other UAVs. Ultimately, the challenge lies in balancing these multiple objectives achieving paths that are not only optimal in terms of minimizing costs but also robust, feasible, and safe for real-world operations.

The path planning problem for multi-UAV systems can be defined as determining the optimal routes for a fleet of UAVs to travel from their respective starting points to UGV0. These routes must minimize specific cost functions while satisfying constraints such as obstacle avoidance, communication range, and the UAV's dynamic capabilities.

(1) The UAV must avoid collision obstacles in the environment.

(2) The UAV's path must adhere to its turning angle constraints to ensure feasible and safe flight.

(3) The UAV must maintain communication with a UGV0 or other UAVs, which may impose constraints on its path.

(4) The UAV's path must minimize energy consumption to ensure that the mission can be completed within the available battery capacity.

2.2. Path Representation

In the context of multi-UAV path planning, where there are n UAVs and m waypoints, the path representation becomes a more complex but structured task. Each UAV's trajectory is defined by a series of waypoints, where each waypoint represents a specific coordinate in 3D space.

For n UAVs, the paths can be represented as:

$$P_i = [(x_{i1}, y_{i1}, z_{i1}), (x_{i2}, y_{i2}, z_{i2}), \ldots, (x_{im}, y_{im}, z_{im})] \tag{1}$$

where $i = 1, 2, \ldots, n$ denotes the UAV index, and (x_{ij}, y_{ij}, z_{ij}) denotes the j-th waypoint for the i-th UAV. Here, m represents the total number of waypoints that each UAV must navigate through from its starting position to UGV0.

The sequence of waypoints forms a trajectory that the UAV must follow from its start position to UGV0. The challenge lies in ensuring that these waypoints are chosen to minimize a predefined cost function in Equation (1) while satisfying the UAV's operational constraints, avoiding obstacles, and coordinating with other UAVs to prevent collisions. For a UAV's path to be considered feasible, it must meet several essential constraints. These include complying with the UAV's kinematic and dynamic limits, such as maximum speed, acceleration, and turning radius. The path must also guarantee the safe avoidance of obstacles in the environment. Additionally, UAVs must maintain adequate separation to avoid mid-air collisions. The chosen waypoints should facilitate smooth transitions between different path segments, avoiding sharp turns or abrupt maneuvers that could compromise the UAV's stability. Furthermore, the path must accommodate communication requirements, ensuring the UAVs remain within the necessary communication range to maintain control and receive mission updates. By satisfying these conditions, the planned paths will ensure safe, efficient, and successful mission execution.

2.3. Objective Function

The objective function in UAV path planning is a key mathematical tool that defines the mission's goals, such as minimizing total path length, reducing energy consumption, and avoiding collisions with obstacles. It incorporates various criteria affecting the UAV's performance, including kinematic and dynamic constraints, environmental terrain, and static obstacle avoidance. By optimizing this function, the path planning process aims to identify the most efficient and safe routes for all UAVs, ensuring they successfully reach UGV0 while meeting all mission-specific requirements.

In the cooperative path planning problem for multi-UAV systems with time-varying communication constraints, the objective function is crafted to optimize multiple criteria simultaneously. The aim is to determine a set of paths that minimize the overall mission cost while adhering to all constraints. Key factors include path length, safety, energy consumption, communication connectivity, turning angle limitations, and obstacle collision avoidance, ensuring that the mission is both efficient and feasible.

The objective function J can be formulated as a weighted sum of these criteria:

$$J = w_1 J_{\text{length}} + w_2 J_{\text{safety}} + w_3 J_{\text{energy}} + w_4 J_{\text{communication}} + w_5 J_{\text{turning}} + w_6 J_{\text{obstacle}} \qquad (2)$$

where w_1, w_2, w_3, w_4, w_5, and w_6 are weighting factors that balance the relative importance of each criterion. Each component of the objective function is detailed below:

(1) Path length

The path length component aims to minimize the total distance traveled by all UAVs. It is defined as the sum of the lengths of the paths [31] of all UAVs:

$$J_{\text{length}} = \sum_{i=1}^{n} \sum_{j=1}^{m_i-1} \sqrt{\left(x_{i(j+1)} - x_{ij}\right)^2 + \left(y_{i(j+1)} - y_{ij}\right)^2 + \left(z_{i(j+1)} - z_{ij}\right)^2} \qquad (3)$$

where n is the number of UAVs, m_i is the number of waypoints for UAV i, and (x_{ij}, y_{ij}, z_{ij}) are the coordinates of the j-th waypoint for UAV i.

The cost of path length Equation (3) in UAV path planning is a critical element of the overall Objective Function Value (OFV), which serves as a comprehensive measure for evaluating the efficiency and feasibility of a given path. The path length cost reflects the total distance a UAV must travel from its starting point to UGV0, passing through the necessary waypoints. A shorter path length typically results in reduced energy consumption, shorter travel times, and overall more efficient mission execution, which is especially important when UAVs have limited battery life or must operate within strict time constraints. However, focusing solely on path length may not yield the best solution, as other critical mission factors must also be considered. To account for this, the OFV often incorporates a penalty function alongside the path length cost. This penalty function adds terms that impose additional costs for violating specific constraints, such as proximity to obstacles, entering no-fly zones, or straying from predefined safe routes. For example, if a UAV's path passes too close to a hazardous area or another UAV, the penalty function increases the OFV, discouraging such risky paths during the optimization process. By integrating the path length cost in Equation (3) and penalty functions, the objective function becomes more robust, guiding the path planning algorithm not only to minimize the distance traveled but also to adhere to safety, regulatory, and operational constraints. This approach ensures that the resulting paths are not only efficient but also safe and compliant with mission requirements. By balancing various factors, the optimization process delivers paths that are both effective and practical, enhancing the UAV's overall performance in complex and dynamic environments.

(2) Safety

In UAV path planning, safety and feasibility constraints such as Equation (4) are paramount to ensuring secure and effective operations. Safety constraints include avoiding collisions with both static obstacles (e.g., buildings, trees) and dynamic obstacles (e.g., other UAVs and UGV0), adhering to no-fly zones, and maintaining altitude and speed within operational limits. Additionally, UAVs must monitor battery life to ensure mission completion and a safe return while remaining within the communication range of control systems. Feasibility constraints involve navigating diverse terrains, managing payload weight and size, adhering to airspace regulations and local laws, and ensuring that the path planning algorithm is computationally efficient and scalable. These constraints collectively guarantee that UAVs operate safely, efficiently, and in compliance with relevant regulations. This can be represented as [42]:

$$J_{\text{safety}} = \sum_{i=1}^{n-1} (\mathbb{I}[v(m_i, m_i + 1) < v_{\text{min}}] + \mathbb{I}[v(m_i, m_i + 1) > v_{\text{max}}]$$

$$+ \mathbb{I}[W_i > W_{\text{max}}]) \tag{4}$$

where $v(m_i, m_i + 1)$ is the speed between waypoints m_i and $m_i + 1$. v_{min} and v_{max} are the minimum and maximum allowable speeds, respectively. W_i is the weight of the UAV's payload. W_{max} is the maximum allowable payload weight.

(3) Energy consumption

Energy consumption is a critical factor in UAV path planning, particularly for missions requiring long endurance or operations in environments where recharging opportunities are limited. Several factors influence a UAV's energy consumption, including path length, speed, altitude, and maneuvering requirements. Therefore, incorporating and minimizing energy consumption is essential in the objective function, as seen in Equation (5) of the UAV path planning problem. The energy consumption component aims to minimize the energy used by UAVs and can be approximated by the total distance traveled, changes in altitude, and the number of turns along the path. This ensures that the UAVs can complete their missions efficiently while conserving energy. It is defined as:

$$J_{\text{energy}} = \sum_{i=1}^{n} \left(\sum_{j=1}^{m_i-1} \sqrt{\left(x_{i(j+1)} - x_{ij}\right)^2 + \left(y_{i(j+1)} - y_{ij}\right)^2 + \left(z_{i(j+1)} - z_{ij}\right)^2} \right.$$

$$\left. + \alpha \sum_{j=1}^{m_i-1} \left| z_{i(j+1)} - z_{ij} \right| + \beta \sum_{j=1}^{m_i-2} \left| \theta_{i(j+2)} - \theta_{i(j+1)} \right| \right) \tag{5}$$

where α and β are weighting factors for altitude changes and turns, respectively, and θ_{ij} represents the heading angle at waypoint j for UAV i.

(4) Time-varying communication

Time-varying communication constraints in UAV path planning refer to the dynamic nature of communication between UAVs and between UAVs and UGV0. These constraints can fluctuate due to several factors, including environmental changes (such as obstacles or weather conditions), UAV positions, and varying network conditions. As UAVs navigate their paths, the quality of communication links may change [48], leading to periods of reliable communication interspersed with periods of limited or no communication. To ensure smooth operations, the path planning algorithm must account for these time-varying constraints by optimizing routes to maximize communication reliability, ensuring critical data are transmitted during periods of strong communication, as outlined in Equation (6). This often involves adjusting the UAV's positions and flight paths to maintain connectivity while meeting the mission objectives and environmental challenges. The communication component of the path planning model penalizes paths where communication constraints are violated, ensuring that UAVs maintain stable links throughout their mission. It is defined as [9,49]:

$$J_{\text{communication}} = \sum_{k=1}^{T} \sum_{i=1}^{n} \sum_{k=1, j \neq i}^{n} \left(\frac{1}{d_{ij}(k) - d_{\max}} \cdot \mathbb{I}\big(d_{ij}(k) > d_{\max}\big) \right) \tag{6}$$

where T is the total number of time steps, $d_{ij}(k)$ is the distance between UAV i and UAV j at time k, $d_{\max}$ is the maximum allowable communication range, and $\mathbb{I}$ is the indicator function that is 1 if the condition is true and 0 otherwise.

(5) Turning angle limitation

The turning angle limitation is a critical constraint in UAV path planning, restricting the maximum angle at which a UAV can change its direction between consecutive waypoints, as expressed in Equation (7). This constraint ensures that UAV trajectories remain smooth and feasible, avoiding abrupt maneuvers that could compromise stability or lead to increased energy consumption. By integrating this limitation into the path planning algorithm, the resulting flight paths are more realistic and aligned with the UAV's physical capabilities, enhancing both safety and overall mission performance. The turning angle limitation component penalizes large deviations in the heading angle between consecutive waypoints, ensuring UAVs avoid sharp turns that could be challenging to execute and potentially destabilizing. This can be represented as [33]:

$$J_{\text{turning}} = \sum_{i=1}^{n} \sum_{j=1}^{m_i-2} \left(\theta_{i(j+1)} - \theta_{ij} \right)^2 \tag{7}$$

where θ_{ij} is the heading angle at waypoint j for UAV i. The objective is to minimize the sum of squared differences between consecutive heading angles.

(6) Obstacle collision avoidance

Each obstacle is represented as a predefined area or volume within the environment, and the UAV's path must be designed to avoid entering these restricted zones. This constraint is typically enforced by calculating the minimum distance between the UAV and each obstacle at every point along its path, ensuring that this distance stays above a predetermined safety margin throughout the flight. This is achieved by defining each obstacle as a fixed region and imposing the condition that the UAV's coordinates never fall within these regions, thereby maintaining a collision-free trajectory. Additionally, paths that come too close to obstacles are penalized, as described in Equation (8), ensuring safe navigation throughout the mission. This can be represented as [49]:

$$J_{\text{obstacle}} = \sum_{i=1}^{n} \sum_{j=1}^{m_i} \left(\sum_{k=1, k \neq i}^{n} \sum_{l=1}^{m_k} \frac{1}{\left\| (x_{ij}, y_{ij}, z_{ij}) - (x_{kl}, y_{kl}, z_{kl}) \right\|^2 + \epsilon} \right.$$

$$\left. + \sum_{o=1}^{O} \frac{1}{\left\| (x_{ij}, y_{ij}, z_{ij}) - (x_o, y_o, z_o) \right\|^2 + \epsilon} \right) \tag{8}$$

where O is the number of obstacles, (x_0, y_0, z_0) are the coordinates of the obstacles, and ϵ is a small positive constant to avoid division by zero.

Equation (2) highlights several important aspects of the UAV path planning problem. It demonstrates that the problem is a multi-objective optimization task, where the overall objective function comprises multiple weighted cost terms, such as path length, energy consumption, collision risk, and turning angle limitations. Furthermore, the inclusion of penalty functions underscores that constraints like obstacle avoidance and UAV kinematic limitations are integrated into the optimization process. These penalties ensure that the generated paths are not only optimal concerning the objective function but also feasible and safe in real-world applications. Moreover, Equation (2) incorporates constraints such as obstacle avoidance and turning angle limitations via penalty functions, ensuring that the planned paths maintain both optimality and feasibility. The use of weighted sums enables flexibility in prioritizing different mission objectives, making the approach adaptable to real-world scenarios where multiple objectives and constraints must be addressed simultaneously.

3. UAV Path Planning Method

In this part, we first review the standard PSO and its variants, including ASPSO and QPSO, in Sections 3.1–3.3. Then, we introduce a new DLS-SMPSO algorithm in Section 3.4. Finally, the detailed implementation of the path planning method using DLS-SMPSO is proposed in Section 3.5.

3.1. PSO Algorithm

The PSO Algorithm [50,51] is an optimization technique inspired by the social behavior of animals like birds flocking or fish schooling. In this algorithm, a swarm of particles, each representing a potential solution to an optimization problem, moves through the search space to find the best solution. Each particle through Equations (9) and (10) has a position vector $x_i(k)$ and a velocity vector $v_i(k)$, where i is the particle index and k is the iteration number. The movement of each particle is influenced by its own experience (personal best position $p_i(k)$) and the experience of the entire swarm (global best position $g(k)$).

The velocity of each particle is updated using the formula:

$$v_i(k+1) = \omega v_i(k) + c_1 r_1 \cdot (p_i(k) - x_i(k)) + c_2 r_2 \cdot (g(k) - x_i(k)) \tag{9}$$

where ω is the inertia weight that controls the influence of the previous velocity, balancing exploration and exploitation. c_1 and c_2 are cognitive and social coefficients that weigh the particle's personal best position and the global best position, respectively. r_1 and r_2 are random numbers between 0 and 1, introducing stochastic variability.

The new position of each particle is then calculated by updating its current position with the new velocity:

$$x_i(k+1) = x_i(k) + v_i(k+1). \tag{10}$$

At each iteration, the fitness of the new position is evaluated. The personal best position $p_i(k+1)$ is updated if the current position $x_i(k+1)$ offers a better fitness value. Similarly, the global best position $g(k+1)$ is updated if any particle achieves a better fitness than the current global best.

3.2. APSO Algorithm

In APSO [42], each particle's position and velocity are represented by an angle or a set of angles in Equations (11) and (12), rather than by cartesian coordinates or other conventional representations. The position of a particle is encoded as an angle θ_i within a certain range, typically between 0 and 2π. This angle can represent directions, phases, or any other cyclic variables. The velocity and position updates are conducted in angular space, which requires modifications to the standard PSO in Equations (9) and (10).

The velocity $v_i(k)$ in APSO is an angular velocity that determines how quickly and in which direction the particle's angle $\theta_i(k)$ will change:

$$v_i(k+1) = \omega v_i(k) + c_1 r_1 \cdot \left(\theta_{pbest,i} - \theta_i(k) \right) + c_2 r_2 \cdot \left(\theta_{gbest} - \theta_i(k) \right) \tag{11}$$

$$\theta_i(k+1) = \theta_i(k) + v_i(k+1) \tag{12}$$

where $\theta_{pbest,i}$ is the angle corresponding to the personal best position, θ_{gbest} is the angle of the global best position, and ω, c_1, c_2 are the inertia and acceleration coefficients as in standard PSO.

3.3. QPSO Algorithm

In QPSO [41], particles do not have fixed trajectories determined by velocity. Instead of being directly influenced by velocity and position updates, as in traditional PSO, each particle's position in QPSO is governed by a probability distribution derived from quantum mechanics. This allows particles to have a probabilistic range of positions, enabling a broader and more diverse exploration of the search space. In QPSO, each particle is attracted to an "attractor" point, which is a combination of its personal best position and the global best position. This attractor guides the particle's probabilistic position updates, allowing for more flexible and efficient searching within the solution space.

The attractor $P_i(k)$ for a particle i is typically defined as:

$$P_i(k) = [p_i(k) + g_i(k)]/2 \tag{13}$$

where $p_i(k)$ is the personal best position of particle i at time k. $g_i(k)$ is the global best position at time k.

Instead of updating the velocity and position directly, QPSO updates the position using a random number generated from the particle's probability distribution. The position of a particle $x_i(k+1)$ is updated according to:

$$x_i(k+1) = P_i(k) \pm \beta \cdot |m(k) - x_i(k)| \cdot \ln\left(\frac{1}{u}\right) \tag{14}$$

where $m(k)$ is the mean best position of all particles at time k. β is a parameter controlling the convergence speed. u is a uniformly distributed random number in the interval $(0, 1)$. The $\pm$ sign indicates that the particle can move towards or away from the attractor, introducing exploration.

The mean best position $m(k)$ is calculated as:

$$m(k) = \frac{1}{N} \sum_{i=1}^{N} p_i(k) \tag{15}$$

where N is the number of particles in the swarm. This position helps determine the overall direction of the swarm's movement. The particles evolve according to Equations (13)–(15) to converge to the optimal path.

3.4. The Proposed DLS-SMPSO Algorithm

The DLS-SMPSO algorithm extends the traditional PSO by encoding the position of each particle as a series of motion paths, with each path represented by a set of directional vectors. Specifically, if a path consists of waypoints $w_1, w_2, \ldots, w_n$, then the motion vector V_k for the k-th segment of the path is defined as:

$$V_k = w_{k+1} - w_k \tag{16}$$

where $k = 1, 2, \ldots, n - 1$. Each waypoint in the path is defined by spherical coordinates: $w_k = (r_k, \theta_k, \phi_k)$. Radial distance from the origin $r_k \in (0, path_{length})$, azimuthal angle $\theta_k \in (-\pi, \pi)$ in the xy-plane from the x-axis and polar angle $\phi_k \in (-\pi/2, \pi)$ from the z-axis.

From Equation (16), one gets that

$$V_k = (\Delta r_k, \Delta \theta_k, \Delta \phi_k). \tag{17}$$

According to Equation (17), we can deduce velocity update Equations (18)–(20):

$$v^i_{r_k}(k+1) = w \cdot v^i_{r_k}(k) + c_1 \cdot r_1 \cdot (p^i_{r_k} - w^i_{r_k}(k)) + c_2 \cdot r_2 \cdot (g_{r_k} - w^i_{r_k}(k)) + d^i_{r_k}(k) \qquad (18)$$

$$v^i_{\theta_k}(k+1) = w \cdot v^i_{\theta_k}(k) + c_1 \cdot r_1 \cdot (p^i_{\theta_k} - w^i_{\theta_k}(k)) + c_2 \cdot r_2 \cdot (g_{\theta_k} - w^i_{\theta_k}(k)) + d^i_{\theta_k}(k) \qquad (19)$$

$$v^i_{\phi_k}(k+1) = w \cdot v^i_{\phi_k}(k) + c_1 \cdot r_1 \cdot (p^i_{\phi_k} - w^i_{\phi_k}(k)) + c_2 \cdot r_2 \cdot (g_{\phi_k} - w^i_{\phi_k}(k)) + d^i_{\phi_k}(k). \qquad (20)$$

The DLS adjustment is:

$$d^i_{d_k}(k) = \lambda \cdot (g_{d_k} - w^i_{d_k}(k)) \qquad (21)$$

where d_k represents each spherical component (r_k, θ_k, ϕ_k). w is the inertia weight, controlling exploration and exploitation. c_1 and c_2 are cognitive and social coefficients. r_1 and r_2 are random values between 0 and 1. p^i is the personal best position of particle i. g is the global best position among all particles. λ is the DLS learning rate influencing the degree of adjustment.

Updated velocities are used to compute new positions:

$$w^i_{r_k}(k+1) = w^i_{r_k}(k) + v^i_{r_k}(k+1) \qquad (22)$$

$$w^i_{\theta_k}(k+1) = w^i_{\theta_k}(k) + v^i_{\theta_k}(k+1) \qquad (23)$$

$$w^i_{\phi_k}(k+1) = w^i_{\phi_k}(k) + v^i_{\phi_k}(k+1). \qquad (24)$$

To evaluate the associated costs, we will convert Equations (22)–(24) to cartesian coordinates:

$$x_k = w^i_{r_k} \cdot \sin(w^i_{\phi_k}) \cdot \cos(w^i_{\theta_k}) \qquad (25)$$

$$y_k = w^i_{r_k} \cdot \sin(w^i_{\phi_k}) \cdot \sin(w^i_{\theta_k}) \qquad (26)$$

$$z_k = w^i_{r_k} \cdot \cos(w^i_{\phi_k}). \qquad (27)$$

The DLS-SMPSO algorithm is an advanced optimization technique that enhances traditional PSO by encoding particle positions as motion paths, making it particularly effective for trajectory optimization problems like UAV path planning. By incorporating spherical vector-based representation and the DLS, the algorithm significantly improves exploration capabilities, convergence speed, and overall solution quality. Algorithm 1 provides a summary of the implementation details for the DLS-SMPSO algorithm.

3.5. Implementation of the UAV Path Planning Method Using DLS-SMPSO

The encoding process in the DLS-SMPSO algorithm is crucial for representing and manipulating the paths or trajectories in the solution space. Here, we present a detailed description of how the search trajectory is encoded as a series of motion paths and how each path is further encoded as a set of vectors. In DLS-SMPSO, each solution is interpreted as a path or trajectory that, like a UAV, follows through the environment. This path is encoded as a series of motion vectors, each of which represents a movement from one waypoint to another. Additionally, the DLS is incorporated to address the oscillation of particles during the evolution process. By using DLS, each particle assimilates advantageous information from the global optimal solution on a dimension-by-dimension basis. This approach helps to minimize the degradation of particles throughout the evolution and ensures a more robust optimization process.

Algorithm 1 Pseudo code of DLS-SMPSO for UAV path planning.

Initialize: Dimension (f); inertia weight (w); acceleration coefficients ($c1, c2$); max iterations ($iter_{max}$); swarm size (nPop); nonupdating number (c_i); position (x_i); velocity (v_i);

Iterate:

 1: **for** each particle in swarm **do**
 2:      Create random motion-encoded paths w_k;
 3:      Get a set of directional vectors V_k;
 4:      The fitness value of x_i is calculated by Equation (2);
 5:      Calculate p^i based on current fitness values;
 6: **end for**
 7: Set g to the best fit particle;
 8: **for** $i = 1 : iter_{max}$ **do**
 9:      **for** each particle in swarm **do**
10:         Compute velocity v_i using Equations (18)–(20);
11:         Compute new position x_i using Equations (25)–(27);
12:         Calculate the fitness value J by Equation (2);
13:         Update personal best position p^i;
14:      **end for**
15:      **if** $J(x_i) > p^i$ **then**
16:         $c_i = c_i + 1$;
17:         **if** $c_i > d$ **then**
18:           $c_i = 0$;
19:         **else**
20:           **for** $j = 1 : f$ **do**
21:             Substitute the the j-th dimension of x_i with the corresponding j-th
22:             dimension of g, referred to as S;
23:             **if** $J(S) < J(x_i)$ **then**
24:               $S = x_i$;
25:             **end if**
26:           **end for**
27:         **end if**
28:      **end if**
29:      Update global best position g;
30: **end for**

The encoding process in the DLS-SMPSO algorithm plays a critical role in representing and optimizing paths or trajectories in the solution space. Specifically, the algorithm encodes each search trajectory as a series of motion paths, where each path is represented by a set of vectors. These vectors correspond to movements from one waypoint to another. To further enhance the optimization process, the DLS is incorporated, addressing the common issue of particle oscillation during evolution. DLS allows each particle to assimilate beneficial information from the global best solution on a dimension-by-dimension basis. This method reduces the risk of particle degradation during evolution and ensures a more robust and effective optimization process.

Step 1: Randomly generate initial positions and velocities for all particles in spherical coordinates. Evaluate initial fitness and set personal and global bests.

Step 2: Compute new velocities using the above Equations (18)–(20), integrating DLS adjustments (Equation (21)). Update particle positions based on new velocities. Adjust positions and velocities to satisfy all constraints in Equation (2), compute fitness for updated positions in Equations (25)–(27). Update personal and global bests based on new fitness evaluations.

Step 3: Repeat iteration until convergence criteria are met (e.g., maximum iterations, acceptable fitness level).

Step 4: The global best position represents the optimal or near-optimal solution to the optimization problem.

4. Experiments and Analysis

In this section, we begin by introducing the experimental setups for UAV path planning in Section 4.1. Following this, Section 4.2 presents the performance evaluation criteria. Finally, Section 4.3 provides a comparative analysis based on the planned paths within the experimental setups. This is followed by further discussion in Section 4.4.

4.1. Experimental Setups

In the designed scenario for UAV path planning, several key parameters are established to ensure the simulation closely mirrors real-world conditions. Each UAV is set to operate within a speed range of 5 m/s to 25 m/s, balancing agility with stability. The UAVs are constrained by a minimal turning radius of 50 m, which corresponds to a maximal turning angle θ of 45 degrees, ensuring that maneuvers are smooth and within safe limits. The environment is modeled using a Digital Elevation Model (DEM), accurately representing terrain features. All five UAVs depart simultaneously from point (200, 100, 150) and proceed toward the UGV0 at point (800, 800).

Additionally, communication limitations are considered, as noted in Ref. [9]. The path planning algorithm is configured with specific values for the DLS-SMPSO method, including a population size of 50 particles, a maximum of 200 iterations, and parameter settings such as an inertia weight w of 1 and both cognitive c_1 and social c_2 coefficients set to 1.5. For consistency in comparison, all algorithms are implemented using the same set of parameters. The simulation environment is divided into two parts, Case 1 and Case 2, and the parameters are shown in Table 2 and Table 3, respectively.

Table 2. Environment parameter setting: simple environment.

NO.	x/m	y/m	z/m	r/m
1	400	500	100	80
2	600	200	150	70
3	500	350	150	80
4	350	200	150	70
5	700	550	150	70
6	650	750	150	80

Table 3. Environment parameter setting: complex environment.

NO.	x/m	y/m	z/m	r/m
1	450	550	100	80
2	500	350	150	80
3	300	200	150	50
4	650	650	150	70
5	200	300	150	70
6	300	450	150	70
7	450	150	150	70
8	650	250	150	55
9	750	400	150	75
10	850	650	150	70

4.2. Performance Evaluation Criteria

When assessing the performance of UAV path planning algorithms, several essential criteria are typically taken into account to evaluate the effectiveness and efficiency of the proposed solutions. These criteria include the following:

Feasibility Ratio (FR): FR measures the proportion of solutions produced by a UAV path planning algorithm that satisfies all specified constraints, such as obstacle avoidance, compliance with turning angle restrictions, and adherence to energy consumption limits. A higher FR indicates that the algorithm is more effective in generating feasible paths that meet mission-critical requirements.

Best Cost: The best cost metric is a critical evaluation criterion in optimization algorithms, especially for UAV path planning. It reflects the most optimal solution found during the search process, representing the path with the lowest possible cost. This metric is often presented alongside the "Optimal", "Worst", and "Mean" values for a more holistic assessment of the algorithm's performance. The "Worst" cost highlights the least favorable solution encountered, offering valuable insights into the algorithm's robustness and its ability to consistently generate high-quality solutions.

4.3. Comparison Analysis

(1) Case 1: In this section, we present the experimental results obtained from testing the proposed UAV path planning algorithm in a simple and complex flight environment. The aim is to validate the feasibility and effectiveness of the path planning method under controlled conditions with limited obstacles and forward terrain.

Figures 1 and 2 present the top and three-dimensional views, respectively, of the flight path for Case 1. From these images, it is evident that the DLS-SMPSO algorithm effectively designed a path that minimizes the risk of collision and ensures the safety of each UAV by avoiding potential threats. In contrast, the paths generated by the PSO, APSO, QPSO, GWO, and ABC algorithms exhibit shortcomings in terms of safety and communication interference, highlighting the superior performance of DLS-SMPSO in these aspects.

In addition, Figure 3 illustrates the convergence curves of the six algorithms in Case 1. The convergence curve reveals that the DLS-SMPSO algorithm quickly identifies an optimal solution in the early stages. As the search progresses, the DLS-SMPSO algorithm adapts to stricter constraints and gradually converges toward the feasible region, demonstrating excellent convergence performance and the ability to find superior paths. Although it may not appear advantageous compared to the PSO, APSO, QPSO, GWO, and ABC algorithms at first glance, a closer examination of Figure 1a–f in Figure 3 reveals a key distinction: most of the optimizations in Figure 1a–e involve the continuous refinement of a single UAV's path. In contrast, the DLS-SMPSO algorithm, depicted in Figure 1f, excels in collaborative path planning. Notably, in Figure 1f, it can be observed that UAV2 is able to maintain a stable flight path with the assistance of other UAVs, even when communication is interrupted. This underscores the significant advantages of the DLS-SMPSO algorithm in fostering cooperation among UAVs.

Table 4 presents the statistical results of all algorithms under Case 1, with boldface indicating the results of our algorithm. As observed in Table 4, the DLS-SMPSO algorithm outperformed other algorithms in terms of average, worst, and best values; it achieved the highest FR. This suggests that the DLS-SMPSO algorithm is more effective in finding feasible solutions. However, it is important to note that the path selected by UAVs using the DLS-SMPSO algorithm may not always be optimal regarding cooperative cost. This observation leads us to infer that cooperative path planning for multi-UAVs prioritizes relatively optimal flight decisions that align with coordination constraints rather than focusing on achieving an individually optimal path for each UAV.

The experimental results show that the proposed path-planning algorithm successfully guided all UAVs to their targets in a simple environment. The total path length was optimized for each UAV, with the algorithm achieving a high FR across all test cases. The execution time remained within acceptable limits, even as the number of UAVs increased, demonstrating the scalability of the method.

(2) Case 2: In this section, we present the experimental results obtained from testing the proposed UAV path planning algorithm in a complex flight environment. The objective is to evaluate the algorithm's performance under more challenging conditions, which include a higher density of obstacles and elements that simulate real-world scenarios.

Figure 4 provides a detailed 3D view of the best flight paths achieved by the various algorithms in operation, while Figure 5 offers a top view of these corresponding paths. As observed in Figures 4 and 5, when the flight environment becomes more complex, the DLS-SMPSO algorithm enables the UAV to successfully avoid obstacles and complete the path planning task. The convergence curve depicted in Figure 6 indicates that, although the convergence speed of the DLS-SMPSO algorithm may be slower at times, its performance remains consistent with the results obtained in simpler environments. Notably, our algorithm demonstrates a significant advantage in collaborative path planning. Furthermore, the DLS-SMPSO algorithm leverages DLS to learn beneficial information dimension by dimension from the global optimal solution, thereby improving the feasible ratio of the paths. As shown in Table 5, the FR of the DLS-SMPSO algorithm is the best, which verifies the feasibility and superiority of the algorithm.

The experimental results revealed that the proposed path planning algorithm was able to effectively navigate the UAVs through challenging scenarios. The algorithm maintained a high FR, with most UAVs successfully reaching their targets while avoiding obstacles. The results suggest that the proposed UAV path planning algorithm is robust and adaptable to real-world scenarios.

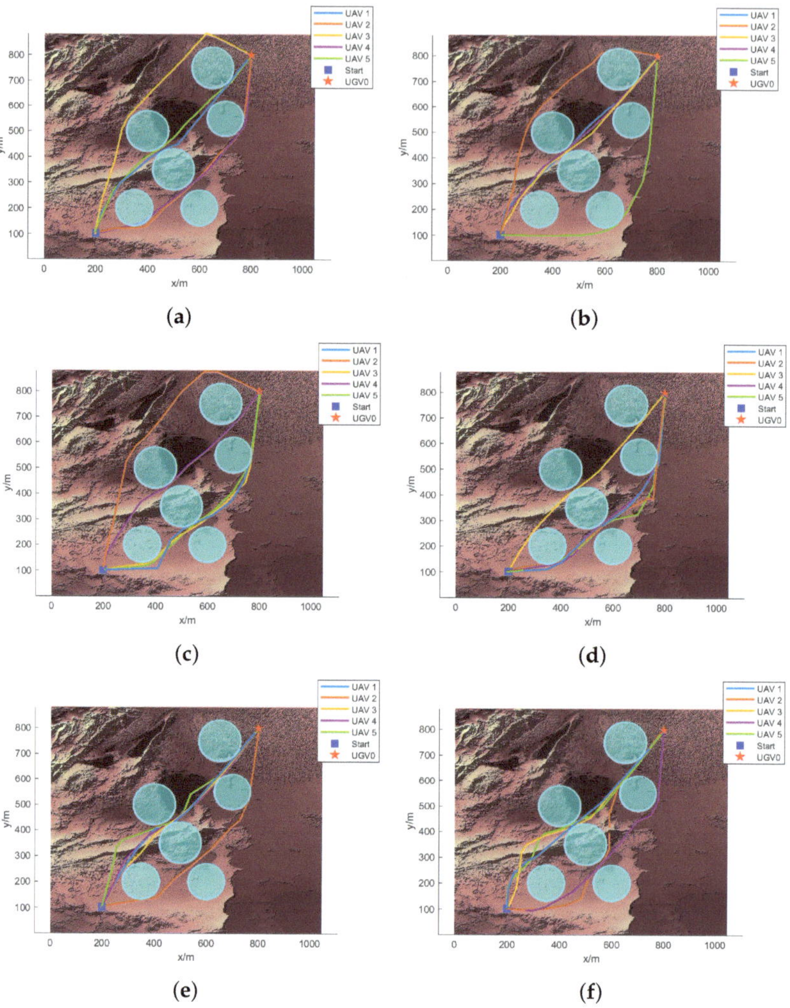

Figure 1. Top view of flight paths for five UAVs in Case 1: (**a**) QPSO; (**b**) GWO; (**c**) ABC; (**d**) PSO; (**e**) APSO; (**f**) DLS-SMPSO.

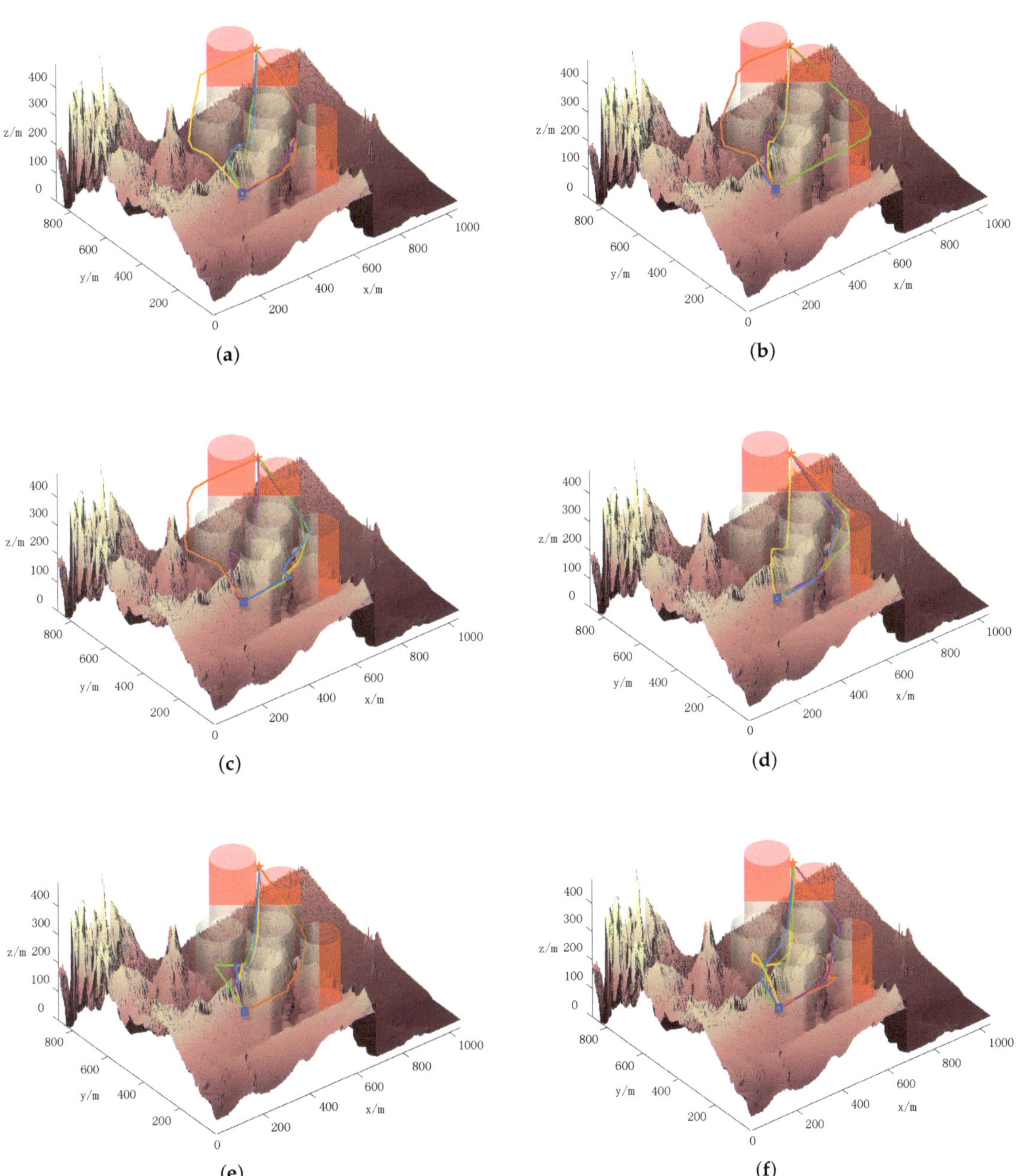

Figure 2. A 3D view of flight paths for five UAVs in Case 1: (**a**) QPSO; (**b**) GWO; (**c**) ABC; (**d**) PSO; (**e**) APSO; (**f**) DLS-SMPSO.

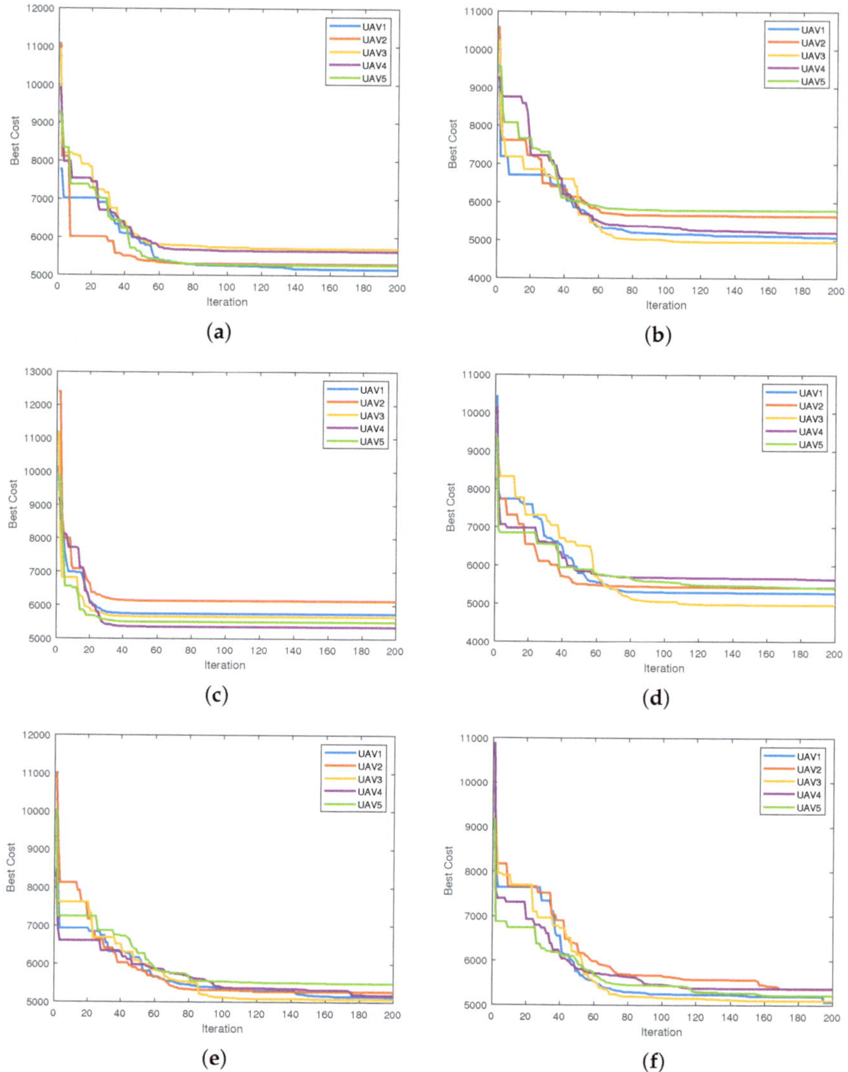

Figure 3. Convergence curves of five UAVs in Case 1: (**a**) QPSO; (**b**) GWO; (**c**) ABC; (**d**) PSO; (**e**) APSO; (**f**) DLS-SMPSO.

Table 4. Result comparison after 200 repetitions in a simple environment.

NO.	1	2	3	4
Indicators	Worst	Optimal	Mean	FR(%)
QPSO	5.4316×10^3	5.2311×10^3	5.3968×10^3	93
GWO	6.1404×10^3	5.3821×10^3	5.8103×10^3	88
ABC	6.0464×10^3	5.4693×10^3	5.7223×10^3	89
PSO	5.8903×10^3	5.1473×10^3	5.8103×10^3	90
APSO	5.8136×10^3	5.2198×10^3	5.8103×10^3	92
DLS-SMPSO	$\mathbf{5.3804 \times 10^3}$	$\mathbf{5.1142 \times 10^3}$	$\mathbf{5.2286 \times 10^3}$	**97**

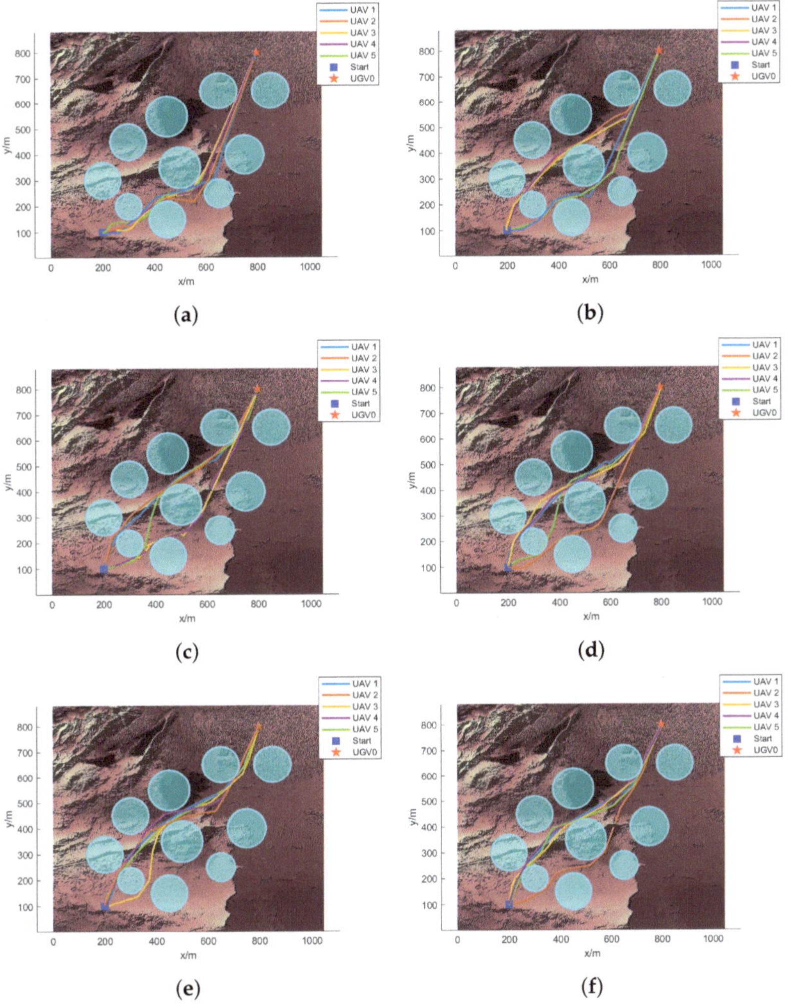

Figure 4. Top view of flight paths for five UAVs in Case 2: (**a**) GWO; (**b**) QPSO; (**c**) APSO; (**d**) PSO; (**e**) ABC; (**f**) DLS-SMPSO.

Table 5. Result comparison after 200 repetitions in a simple environment.

NO.	1	2	3	4
Indicators	Worst	Optimal	Mean	FR(%)
PSO	5.9557×10^3	5.4611×10^3	5.7140×10^3	88
QPSO	5.5405×10^3	5.3727×10^3	5.4199×10^3	91
APSO	5.8203×10^3	5.6814×10^3	5.7140×10^3	90
GWO	7.4486×10^3	6.0255×10^3	6.8329×10^3	85
ABC	6.1732×10^3	5.5040×10^3	5.8287×10^3	87
DLS-SMPSO	$\mathbf{5.3878 \times 10^3}$	$\mathbf{5.1790 \times 10^3}$	$\mathbf{5.2391 \times 10^3}$	**96**

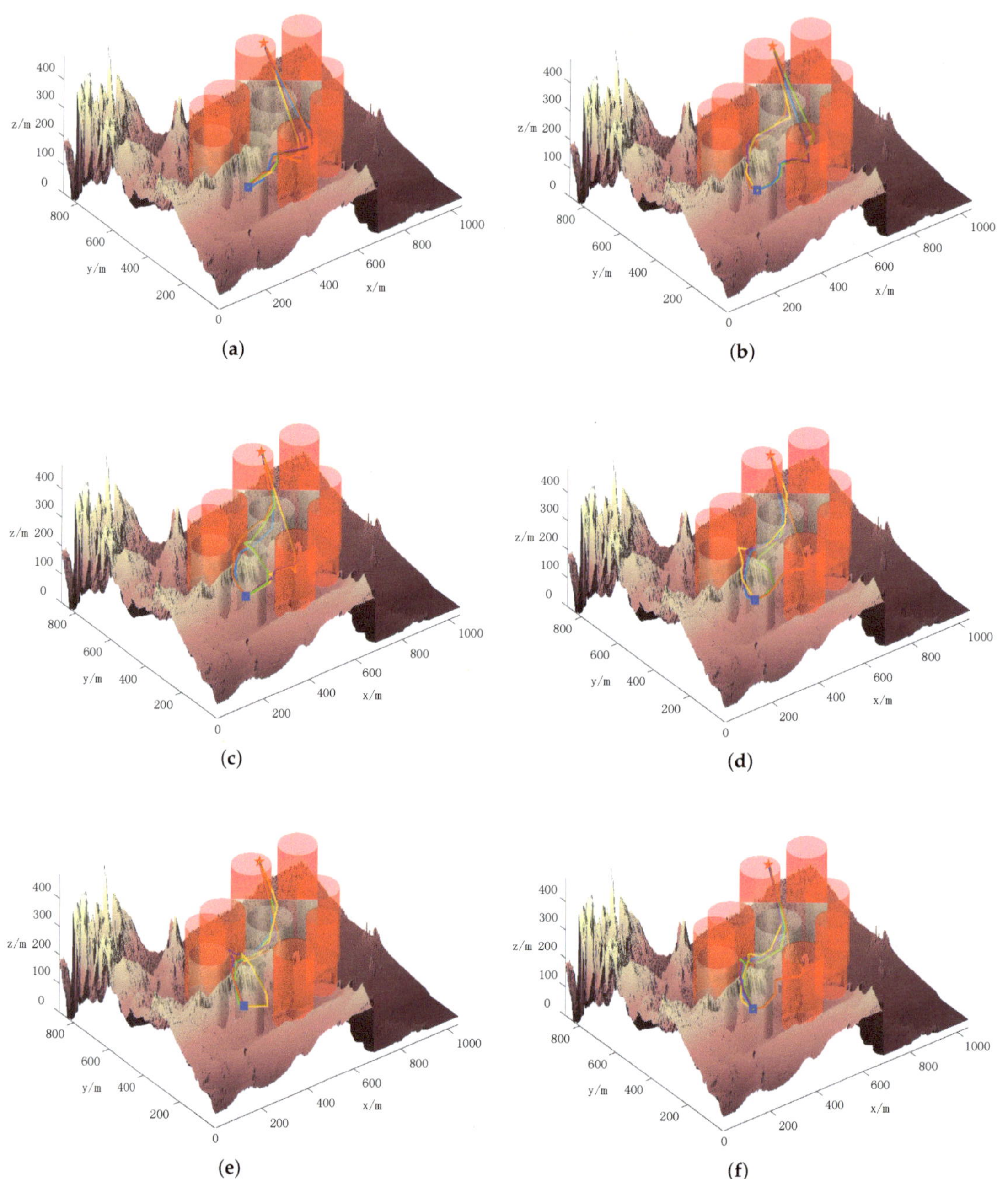

Figure 5. A 3D view of flight paths for five UAVs in Case 2: (**a**) GWO; (**b**) QPSO; (**c**) APSO; (**d**) PSO; (**e**) ABC; (**f**) DLS-SMPSO.

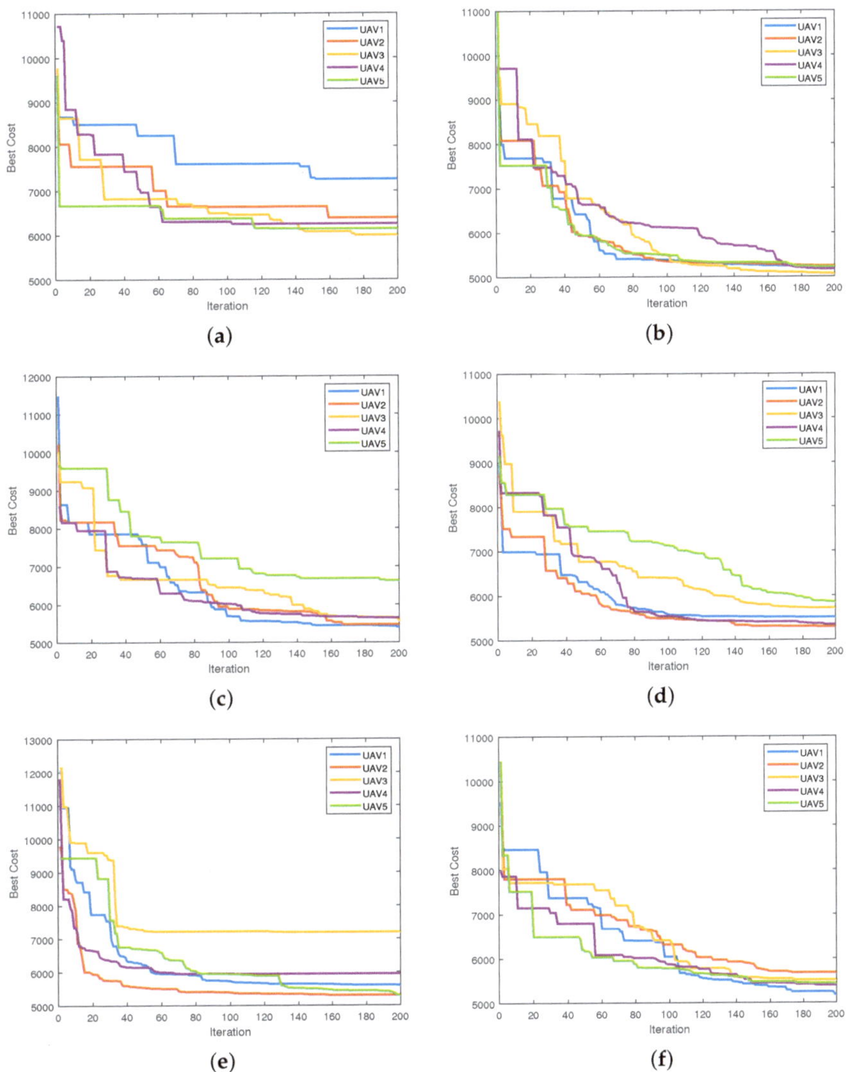

Figure 6. Convergence curves of five UAVs in Case 2: (**a**) GWO; (**b**) QPSO; (**c**) APSO; (**d**) PSO; (**e**) ABC; (**f**) DLS-SMPSO.

4.4. Discussion

Extensive simulations and comparative analyses demonstrate that DLS-SMPSO excels in generating safe, feasible, and optimal paths for multi-UAV operations. The proposed algorithm performs exceptionally well in complex environments with numerous obstacles, as evidenced by its consistently low fitness values. This effectiveness is driven by the transformation of the search space from cartesian to spherical coordinates, enabling more intuitive and flexible exploration. Furthermore, constraints on UAV dynamics, such as turning and climbing angles, are seamlessly integrated into the DLS-SMPSO variables, refining the search space and yielding high-quality solutions.

However, this paper encounters challenges with complex wind patterns and intricate obstacle geometries, which may delay UAV progress or cause them to miss waypoints. Refs. [52,53] explore the modeling of wind and obstacle disturbances to assess UAV swarm resilience, emphasizing performance and adaptability under challenging environmental conditions. These studies offer valuable insights into how dynamic environmental factors impact UAV swarm behavior, which may aid in addressing the issues posed by complex wind patterns and obstacle geometries.

5. Conclusions

In this paper, we tackled the complex issue of cooperative path planning for multi-UAV systems under time-varying communication constraints using the DLS-SMPSO algorithm. The DLS-SMPSO method provides a robust solution by introducing an innovative approach that encodes particle positions as motion paths in spherical coordinates. This, coupled with the DLS, significantly enhances the algorithm's ability to plan cooperative paths, minimizes oscillations during the optimization process, and improves overall stability. The simulation results confirm the feasibility and effectiveness of the proposed approach, demonstrating its potential for real-world applications.

While the proposed approach demonstrates effective path planning capabilities, several limitations highlight opportunities for future enhancement. First, this study evaluates the algorithm with a limited number of agents (five UAVs), constraining the assessment of scalability and collaborative effectiveness in larger, multi-agent scenarios. Future work will focus on extending the algorithm to accommodate a greater number of UAVs, enabling a comprehensive evaluation of its robustness and scalability in more complex swarm configurations.

Furthermore, this study does not account for environmental factors such as complex wind patterns and diverse obstacle geometries, both of which could significantly impact path feasibility and necessitate adaptive re-planning to maintain safe and efficient trajectories. For instance, intricate wind dynamics may alter UAV paths and demand continuous adjustments, while complex obstacle geometries could trap UAVs in confined spaces, testing the limits of the pathfinding strategy. Future research will incorporate these dynamic environmental variables to enhance the algorithm's resilience. Additionally, embedding comprehensive UAV dynamics and considering aspects like inertia and aerodynamics directly into the optimization framework will allow for a more realistic representation of physical constraints, enabling adaptable and accurate path planning. Addressing these factors will help validate and refine the algorithm for a wider range of applications, thereby improving its efficacy in dynamic and unpredictable environments.

Author Contributions: J.G. was responsible for conceptualization, methodology, writing the original draft, and supervising the research. M.G. provided valuable guidance and assisted in refining the methodology. K.H. contributed to editing the manuscript and made critical revisions. All authors have read and agreed to the published version of the manuscript.

Funding: This work was supported by the National Key Research and Development Program of China under Grant 2020YFB1708500.

Institutional Review Board Statement: Not applicable.

Informed Consent Statement: Not applicable.

Data Availability Statement: Data are contained within the article.

Acknowledgments: The authors would like to thank the editor and the anonymous reviewers for their careful reading and valuable suggestions that helped to improve the quality of this article.

Conflicts of Interest: The authors declare no conflicts of interest.

References

1. Asadzadeh, S.; de Oliveira, W.J.; de Souza Filho, C.R. UAV-based remote sensing for the petroleum industry and environmental monitoring: State-of-the-art and perspectives. *J. Pet. Sci. Eng.* **2022**, *208*, 109633. [CrossRef]
2. Fu, R.; Ren, X.; Li, Y.; Wu, Y.; Sun, H.; Al-Absi, M.A. Machine-learning-based uav-assisted agricultural information security architecture and intrusion detection. *IEEE Internet Things J.* **2023**, *10*, 18589–18598. [CrossRef]

3. Martinez-Alpiste, I.; Golcarenarenji, G.; Wang, Q.; Alcaraz-Calero, J.M. Search and rescue operation using UAVs: A case study. *Expert Syst. Appl.* **2021**, *178*, 114937. [CrossRef]

4. Di Giovanni, D.; Fumian, F.; Chierici, A.; Bianchelli, M.; Martellucci, L.; Carminati, G.; Malizia, A.; D'Errico, F.; Gaudio, P. Design of miniaturized sensors for a mission-oriented UAV application: A new pathway for early warning. *Int. J. Saf. Secur. Eng.* **2021**, *11*, 435–444. [CrossRef]

5. Maboudi, M.; Homaei, M.; Song, S.; Malihi, S.; Saadatseresht, M.; Gerke, M. A Review on Viewpoints and Path Planning for UAV-Based 3-D Reconstruction. *IEEE J. Sel. Top. Appl. Earth Obs. Remote. Sens.* **2023**, *16*, 5026–5048. [CrossRef]

6. Bassolillo, S.R.; Raspaolo, G.; Blasi, L.; D'Amato, E.; Notaro, I. Path Planning for Fixed-Wing Unmanned Aerial Vehicles: An Integrated Approach with Theta* and Clothoids. *Drones* **2024**, *8*, 62. [CrossRef]

7. Adam, M.S.; Nordin, R.; Abdullah, N.F.; Abu-Samah, A.; Amodu, O.A.; Alsharif, M.H. Optimizing Disaster Response through Efficient Path Planning of Mobile Aerial Base Station with Genetic Algorithm. *Drones* **2024**, *8*, 272. [CrossRef]

8. Gu, S.; Wang, Y.; Wang, N.; Wu, W. Intelligent optimization of availability and communication cost in satellite-UAV mobile edge caching system with fault-tolerant codes. *IEEE Trans. Cogn. Commun. Netw.* **2020**, *6*, 1230–1241. [CrossRef]

9. Guo, J.; Gan, M.; Hu, K. Relative Localization and Circumnavigation of a UGV0 Based on Mixed Measurements of Multi-UAVs by Employing Intelligent Sensors. *Sensors* **2024**, *24*, 2347. [CrossRef]

10. Gai, K.; Wu, Y.; Zhu, L.; Choo, K.K.R.; Xiao, B. Blockchain-enabled trustworthy group communications in UAV networks. *IEEE Trans. Intell. Transp. Syst.* **2020**, *22*, 4118–4130. [CrossRef]

11. Yanmaz, E. Positioning aerial relays to maintain connectivity during drone team missions. *Ad Hoc Netw.* **2022**, *128*, 102800. [CrossRef]

12. Jiao, L.; Zhang, R.; Liu, M.; Hua, Q.; Zhao, N.; Nallanathan, A.; Wang, X. Placement optimization of UAV relaying for covert communication. *IEEE Trans. Veh. Technol.* **2022**, *71*, 12327–12332. [CrossRef]

13. Zhao, T.; Cao, D.; Yao, J.; Zhang, S. Topology optimization algorithm for UAV formation based on wireless ultraviolet communication. *Photonic Netw. Commun.* **2023**, *45*, 25–36. [CrossRef]

14. Padilla, G.E.G.; Kim, K.J.; Park, S.H.; Yu, K.H. Flight path planning of solar-powered UAV for sustainable communication relay. *IEEE Robot. Autom. Lett.* **2020**, *5*, 6772–6779. [CrossRef]

15. Woosley, B.; Dasgupta, P.; Rogers III, J.G.; Twigg, J. Multi-robot information driven path planning under communication constraints. *Auton. Robot.* **2020**, *44*, 721–737. [CrossRef]

16. Ramaswamy, V.; Moon, S.; Frew, E.W.; Ahmed, N. Mutual information based communication aware path planning: A game theoretic perspective. In Proceedings of the 2016 IEEE/RSJ International Conference on Intelligent Robots and Systems (IROS), Daejeon, Republic of Korea, 9–14 October 2016; IEEE: Piscataway, NJ, USA, 2016; pp. 1823–1828.

17. Marashian, A.; Razminia, A. Mobile robot's path-planning and path-tracking in static and dynamic environments: Dynamic programming approach. *Robot. Auton. Syst.* **2024**, *172*, 104592. [CrossRef]

18. Pehlivanoglu, Y.V.; Pehlivanoglu, P. An enhanced genetic algorithm for path planning of autonomous UAV in target coverage problems. *Appl. Soft Comput.* **2021**, *112*, 107796. [CrossRef]

19. Ab Wahab, M.N.; Nazir, A.; Khalil, A.; Ho, W.J.; Akbar, M.F.; Noor, M.H.M.; Mohamed, A.S.A. Improved genetic algorithm for mobile robot path planning in static environments. *Expert Syst. Appl.* **2024**, *249*, 123762. [CrossRef]

20. Miao, C.; Chen, G.; Yan, C.; Wu, Y. Path planning optimization of indoor mobile robot based on adaptive ant colony algorithm. *Comput. Ind. Eng.* **2021**, *156*, 107230. [CrossRef]

21. Liu, C.; Wu, L.; Xiao, W.; Li, G.; Xu, D.; Guo, J.; Li, W. An improved heuristic mechanism ant colony optimization algorithm for solving path planning. *Knowl. Based Syst.* **2023**, *271*, 110540. [CrossRef]

22. Ma, Y.N.; Gong, Y.J.; Xiao, C.F.; Gao, Y.; Zhang, J. Path planning for autonomous underwater vehicles: An ant colony algorithm incorporating alarm pheromone. *IEEE Trans. Veh. Technol.* **2018**, *68*, 141–154. [CrossRef]

23. Dewangan, R.K.; Shukla, A.; Godfrey, W.W. Three dimensional path planning using Grey wolf optimizer for UAVs. *Appl. Intell.* **2019**, *49*, 2201–2217. [CrossRef]

24. Yu, X.; Jiang, N.; Wang, X.; Li, M. A hybrid algorithm based on grey wolf optimizer and differential evolution for UAV path planning. *Expert Syst. Appl.* **2023**, *215*, 119327. [CrossRef]

25. Liu, X.; Li, G.; Yang, H.; Zhang, N.; Wang, L.; Shao, P. Agricultural UAV trajectory planning by incorporating multi-mechanism improved grey wolf optimization algorithm. *Expert Syst. Appl.* **2023**, *233*, 120946. [CrossRef]

26. Vijitha Ananthi, J.; Subha Hency Jose, P. Optimal design of artificial bee colony based UAV routing (ABCUR) algorithm for healthcare applications. *Int. J. Intell. Unmanned Syst.* **2023**, *11*, 285–295. [CrossRef]

27. Lv, M.; Liu, H.; Li, Y.; Li, L.; Gao, Y. The improved artificial bee colony method and its application on UAV disaster rescue. In *Man-Machine-Environment System Engineering: Proceedings of the 21st International Conference on MMESE: Commemorative Conference for the 110th Anniversary of Xuesen Qian's Birth and the 40th Anniversary of Founding of Man-Machine-Environment System Engineering*, Beijing, China, 23–25 October 2021; Springer: Singapore, 2022; pp. 375–381.

28. Han, Z.; Chen, M.; Zhu, H.; Wu, Q. Ground threat prediction-based path planning of unmanned autonomous helicopter using hybrid enhanced artificial bee colony algorithm. *Def. Technol.* **2024**, *32*, 1–22. [CrossRef]

29. Song, B.; Wang, Z.; Zou, L. An improved PSO algorithm for smooth path planning of mobile robots using continuous high-degree Bezier curve. *Appl. Soft Comput.* **2021**, *100*, 106960. [CrossRef]

30. Abhishek, B.; Ranjit, S.; Shankar, T.; Eappen, G.; Sivasankar, P.; Rajesh, A. Hybrid PSO-HSA and PSO-GA algorithm for 3D path planning in autonomous UAVs. *SN Appl. Sci.* **2020**, *2*, 1–16. [CrossRef]

31. Yu, Z.; Si, Z.; Li, X.; Wang, D.; Song, H. A novel hybrid particle swarm optimization algorithm for path planning of UAVs. *IEEE Internet Things J.* **2022**, *9*, 22547–22558. [CrossRef]

32. Lin, S.; Liu, A.; Wang, J.; Kong, X. An improved fault-tolerant cultural-PSO with probability for multi-AGV path planning. *Expert Syst. Appl.* **2024**, *237*, 121510. [CrossRef]

33. Li, K.; Yan, X.; Han, Y.; Ge, F.; Jiang, Y. Many-objective optimization based path planning of multiple UAVs in oilfield inspection. *Appl. Intell.* **2022**, *52*, 12668–12683. [CrossRef]

34. Li, Z.; Chen, G. Global synchronization and asymptotic stability of complex dynamical networks. *IEEE Trans. Circuits Syst. II Express Briefs* **2006**, *53*, 28–33.

35. Kennedy, J.; Eberhart, R. Particle swarm optimization. In Proceedings of the ICNN'95-International Conference on Neural Networks, Perth, WA, USA, 27 November–1 December 1995; IEEE: Piscataway, NJ, USA, 1995; Volume 4, pp. 1942–1948.

36. Xu, B.; Li, S.; Razzaqi, A.A.; Wang, L.; Jiao, M. A novel ANFIS-AQPSO-GA-Based online correction measurement method for cooperative localization. *IEEE Trans. Instrum. Meas.* **2022**, *71*, 1–17. [CrossRef]

37. Shichao, M.; Xianglun, Z.; Qiang, T.; Zhiyu, L.; Yukun, Y. Research on Cooperative Path Planning and Formation Control for Multiple UAVs. In Proceedings of the Chinese Conference on Swarm Intelligence and Cooperative Control, Nanjing, China, 17–19 November 2023; Springer: Berlin/Heidelberg, Germany, 2023; pp. 52–60.

38. Zhang, J.; Ning, X.; Ma, S. An improved particle swarm optimization based on age factor for multi-AUV cooperative planning. *Ocean Eng.* **2023**, *287*, 115753. [CrossRef]

39. Zhang, X.; Xia, S.; Zhang, T.; Li, X. Hybrid FWPS cooperation algorithm based unmanned aerial vehicle constrained path planning. *Aerosp. Sci. Technol.* **2021**, *118*, 107004. [CrossRef]

40. Chen, Z.; Wu, H.; Chen, Y.; Cheng, L.; Zhang, B. Patrol robot path planning in nuclear power plant using an interval multi-objective particle swarm optimization algorithm. *Appl. Soft Comput.* **2022**, *116*, 108192. [CrossRef]

41. Qian, Q.; Wu, J.; Wang, Z. Optimal path planning for two-wheeled self-balancing vehicle pendulum robot based on quantum-behaved particle swarm optimization algorithm. *Pers. Ubiquitous Comput.* **2019**, *23*, 393–403. [CrossRef]

42. Zhao, R.; Wang, Y.; Xiao, G.; Liu, C.; Hu, P.; Li, H. A method of path planning for unmanned aerial vehicle based on the hybrid of selfish herd optimizer and particle swarm optimizer. *Appl. Intell.* **2022**, *52*, 16775–16798. [CrossRef]

43. Lin, C.; Zhang, X. Application of UAV path planning based on parameter optimization GA-PSO fusion algorithm. *J. Physics Conf. Ser.* **2022**, *2258*, 012018. [CrossRef]

44. Gul, F.; Rahiman, W.; Alhady, S.; Ali, A.; Mir, I.; Jalil, A. Meta-heuristic approach for solving multi-objective path planning for autonomous guided robot using PSO–GWO optimization algorithm with evolutionary programming. *J. Ambient. Intell. Humaniz. Comput.* **2021**, *12*, 7873–7890. [CrossRef]

45. Zhang, H.; Gan, X.; Li, S.; Chen, Z. UAV safe route planning based on PSO-BAS algorithm. *J. Syst. Eng. Electron.* **2022**, *33*, 1151–1160. [CrossRef]

46. Chen, J.; Ye, F.; Li, Y. Travelling salesman problem for UAV path planning with two parallel optimization algorithms. In Proceedings of the 2017 Progress in Electromagnetics Research Symposium-Fall (PIERS-FALL), Singapore, 19–22 November 2017; IEEE: Piscataway, NJ, USA, 2017; pp. 832–837.

47. Xu, L.; Cao, X.; Du, W.; Li, Y. Cooperative path planning optimization for multiple UAVs with communication constraints. *Knowl.-Based Syst.* **2023**, *260*, 110164. [CrossRef]

48. Zhang, C.; Zhang, L.; Zhu, L.; Zhang, T.; Xiao, Z.; Xia, X.G. 3D deployment of multiple UAV-mounted base stations for UAV communications. *IEEE Trans. Commun.* **2021**, *69*, 2473–2488. [CrossRef]

49. Thuy, N.D.T.; Bui, D.N.; Phung, M.D.; Duy, H.P. Deployment of UAVs for optimal multihop ad-hoc networks using particle swarm optimization and behavior-based control. In Proceedings of the 2022 11th International Conference on Control, Automation and Information Sciences (ICCAIS), Hanoi, Vietnam, 21–24 November 2022; IEEE: Piscataway, NJ, USA, 2022; pp. 304–309.

50. Pervaiz, S.; Bangyal, W.H.; Ashraf, A.; Nisar, K.; Haque, M.R.; Ibrahim, A.; Ag, A.; Chowdhry, B.; Rasheed, W.; Rodrigues, J.; et al. Comparative research directions of population initialization techniques using PSO algorithm. *Intell. Autom. Soft Comput.* **2022**, *32*, 1427–1444. [CrossRef]

51. Wang, D.; Tan, D.; Liu, L. Particle swarm optimization algorithm: An overview. *Soft Comput.* **2018**, *22*, 387–408. [CrossRef]

52. Chodnicki, M.; Siemiatkowska, B.; Stecz, W.; Stępień, S. Energy efficient UAV flight control method in an environment with obstacles and gusts of wind. *Energies* **2022**, *15*, 3730. [CrossRef]

53. Phadke, A.; Medrano, F.A.; Chu, T.; Sekharan, C.N.; Starek, M.J. Modeling Wind and Obstacle Disturbances for Effective Performance Observations and Analysis of Resilience in UAV Swarms. *Aerospace* **2024**, *11*, 237. [CrossRef]

Disclaimer/Publisher's Note: The statements, opinions and data contained in all publications are solely those of the individual author(s) and contributor(s) and not of MDPI and/or the editor(s). MDPI and/or the editor(s) disclaim responsibility for any injury to people or property resulting from any ideas, methods, instructions or products referred to in the content.

MDPI AG
Grosspeteranlage 5
4052 Basel
Switzerland
Tel.: +41 61 683 77 34

Drones Editorial Office
E-mail: drones@mdpi.com
www.mdpi.com/journal/drones

Disclaimer/Publisher's Note: The title and front matter of this reprint are at the discretion of the Guest Editors. The publisher is not responsible for their content or any associated concerns. The statements, opinions and data contained in all individual articles are solely those of the individual Editors and contributors and not of MDPI. MDPI disclaims responsibility for any injury to people or property resulting from any ideas, methods, instructions or products referred to in the content.